COURSE OF THEORETICAL PHYSICS

Volume 5

STATISTICAL PHYSICS

I

STATISTICAL PHYSICS

by

L. D. LANDAU AND E. M. LIFSHITZ

INSTITUTE OF PHYSICAL PROBLEMS,
U.S.S.R. ACADEMY OF SCIENCES

Volume 5 of *Course of Theoretical Physics*

Translated from the Russian by
J. B. SYKES AND M. J. KEARSLEY

SECOND REVISED AND ENLARGED EDITION

ADDISON-WESLEY PUBLISHING COMPANY

READING, MASSACHUSETTS — MENLO PARK, CALIFORNIA —
LONDON — DON MILLS, ONTARIO

Pergamon Press Ltd., Headington Hill Hall, Oxford
Pergamon Press Inc., Maxwell House, Fairview Park, Elmsford, New York 10523
Pergamon of Canada Ltd., 207 Queen's Quay West, Toronto 1
Pergamon Press (Aust.) Pty. Ltd., 19a Boundary Street, Rushcutters Bay, N.S.W. 2011, Australia
Vieweg & Sohn GmbH, Burgplatz 1, Braunschweig

First published in English 1958
2nd Impression 1959
3rd Impression 1963
2nd Revised and Enlarged Edition 1969
2nd Impression 1970

U.S.A. edition distributed by Addison-Wesley Publishing Company
Reading, Massachusetts, – Menlo Park, California – London – Don Mills, Ontario

Library of Congress Catalog Card No. 68–18526

Printed in Hungary and reprinted lithographically by A. Wheaton & Co., Exeter

CONTENTS

Preface to the second English edition ix

From the preface to the first English edition xi

Notation xii

I. THE FUNDAMENTAL PRINCIPLES OF STATISTICAL PHYSICS

§1. Statistical distributions 1
§2. Statistical independence 6
§3. Liouville's theorem 9
§4. The significance of energy 11
§5. The statistical matrix 14
§6. Statistical distributions in quantum statistics 20
§7. Entropy 22
§8. The law of increase of entropy 28

II. THERMODYNAMIC QUANTITIES

§ 9. Temperature 33
§10. Macroscopic motion 35
§11. Adiabatic processes 37
§12. Pressure 40
§13. Work and quantity of heat 43
§14. The heat function 45
§15. The free energy and the thermodynamic potential 46
§16. Relations between the derivatives of thermodynamic quantities 49
§17. The thermodynamic scale of temperature 53
§18. The Joule–Thomson process 53
§19. Maximum work 55
§20. Maximum work done by a body in an external medium 57
§21. Thermodynamic inequalities 61
§22. Le Chatelier's principle 63
§23. Nernst's theorem 66
§24. The dependence of the thermodynamic quantities on the number of particles 67
§25. Equilibrium of a body in an external field 70
§26. Rotating bodies 71
§27. Thermodynamic relations in the relativistic region 73

III. THE GIBBS DISTRIBUTION

§28. The Gibbs distribution 76
§29. The Maxwellian distribution 79
§30. The probability distribution for an oscillator 83
§31. The free energy in the Gibbs distribution 86
§32. Thermodynamic perturbation theory 90
§33. Expansion in powers of $\hbar$ 93

§34. The Gibbs distribution for rotating bodies 99
§35. The Gibbs distribution for a variable number of particles 101
§36. The derivation of the thermodynamic relations from the Gibbs distribution 104

IV. IDEAL GASES

§37. The Boltzmann distribution 106
§38. The Boltzmann distribution in classical statistics 108
§39. Molecular collisions 110
§40. Ideal gases not in equilibrium 112
§41. The free energy of an ideal Boltzmann gas 114
§42. The equation of state of an ideal gas 116
§43. Ideal gases with constant specific heat 119
§44. The law of equipartition 122
§45. Monatomic ideal gases 125
§46. Monatomic gases. The effect of the electronic angular momentum 127
§47. Diatomic gases with molecules of unlike atoms. Rotation of molecules 129
§48. Diatomic gases with molecules of like atoms. Rotation of molecules 133
§49. Diatomic gases. Vibrations of atoms 135
§50. Diatomic gases. The effect of the electronic angular momentum 138
§51. Polyatomic gases 140

V. THE FERMI AND BOSE DISTRIBUTIONS

§52. The Fermi distribution 144
§53. The Bose distribution 145
§54. Fermi and Bose gases not in equilibrium 146
§55. Fermi and Bose gases of elementary particles 148
§56. A degenerate electron gas 152
§57. The specific heat of a degenerate electron gas 154
§58. A relativistic degenerate electron gas 157
§59. A degenerate Bose gas 159
§60. Black-body radiation 162

VI. THE CONDENSED STATE

§61. Solids at low temperatures 170
§62. Solids at high temperatures 174
§63. Debye's interpolation formula 177
§64. Thermal expansion of solids 180
§65. Phonons 181
§66. Quantum liquids with Bose-type spectrum 187
§67. Superfluidity 191
§68. Quantum liquids with Fermi-type spectrum 196
§69. The electronic spectra of metals 203
§70. The electronic spectra of solid dielectrics 210
§71. Negative temperatures 211

VII. NON-IDEAL GASES

§72. Deviations of gases from the ideal state 215
§73. Expansion in powers of the density 220
§74. Van der Waals' formula 222

§75. Completely ionised gases 225
§76. The method of correlation functions 228
§77. Calculation of the virial coefficient in quantum mechanics 230
§78. A degenerate "almost ideal" Bose gas 234
§79. A degenerate "almost ideal" Fermi gas with repulsion between the particles 241
§80. A degenerate "almost ideal" Fermi gas with attraction between the particles 246

VIII. PHASE EQUILIBRIUM

§81. Conditions of phase equilibrium 257
§82. The Clapeyron–Clausius formula 261
§83. The critical point 263
§84. Properties of matter near the critical point 266
§85. The law of corresponding states 270

IX. SOLUTIONS

§86. Systems containing different particles 273
§87. The phase rule 274
§88. Weak solutions 275
§89. Osmotic pressure 277
§90. Solvent phases in contact 278
§91. Equilibrium with respect to the solute 281
§92. Evolution of heat and change of volume on dissolution 283
§93. The mutual interaction of solutes 285
§94. Solutions of strong electrolytes 287
§95. Mixtures of ideal gases 289
§96. Mixtures of isotopes 291
§97. Vapour pressure over concentrated solutions 294
§98. Thermodynamic inequalities for solutions 296
§99. Equilibrium curves 300
§100. Examples of phase diagrams 306
§101. Intersection of singular curves on the equilibrium surface 310
§102. Gases and liquids 312

X. CHEMICAL REACTIONS

§103. The condition for chemical equilibrium 316
§104. The law of mass action 317
§105. Heat of reaction 320
§106. Ionisation equilibrium 323
§107. Equilibrium with respect to pair production 324

XI. PROPERTIES OF MATTER AT VERY HIGH DENSITY

§108. The equation of state of matter at high density 327
§109. Equilibrium of bodies of large mass 330
§110. The energy of a gravitating body 336
§111. Equilibrium of a "neutron" sphere 338

XII. FLUCTUATIONS

§112. The Gaussian distribution 343
§113. The Gaussian distribution for more than one variable 345

§114. Fluctuations of the fundamental thermodynamic quantities 348
§115. Fluctuations in an ideal gas 354
§116. Poisson's formula 356
§117. Fluctuations in solutions 358
§118. Correlations of fluctuations 359
§119. Fluctuations at the critical point 362
§120. Correlations of fluctuations in an ideal gas 365
§121. Correlations of fluctuations in time 370
§122. The symmetry of the kinetic coefficients 374
§123. The dissipative function 378
§124. Time correlations of the fluctuations of more than one variable 381
§125. The generalised susceptibility 384
§126. Non-thermodynamic fluctuations of a single variable 391
§127. Non-thermodynamic fluctuations of more than one variable 395

XIII. THE SYMMETRY OF CRYSTALS

§128. Symmetry of particle configuration in a body 401
§129. Symmetry with respect to orientation of molecules 404
§130. Symmetry elements of a crystal lattice 405
§131. The Bravais lattice 406
§132. Crystal systems 408
§133. Crystal classes 412
§134. Space groups 414
§135. The reciprocal lattice 416
§136. Irreducible representations of space groups 418

XIV. PHASE TRANSITIONS OF THE SECOND KIND

§137. Phase transitions of the second kind 424
§138. The discontinuity of specific heat 429
§139. Change in symmetry in a phase transition of the second kind 433
§140. Isolated and critical points of continuous transition 445
§141. Phase transitions of the second kind in a two-dimensional lattice 447

XV. SURFACES

§142. Surface tension 455
§143. Surface tension of crystals 458
§144. Surface pressure 460
§145. Surface tension of solutions 462
§146. Surface tension of solutions of strong electrolytes 463
§147. Adsorption 465
§148. Wetting 467
§149. The angle of contact 470
§150. Nucleation in phase transitions 471
§151. Fluctuations in the curvature of long molecules 475
§152. The impossibility of the existence of phases in one-dimensional systems 478

Index 481

PREFACE TO THE SECOND ENGLISH EDITION

FOR this edition the book has been enlarged and the treatment in some places revised. The revision has, however, been incomplete to the extent that the book does not include topics related to the successful application in recent years of the methods of quantum field theory to statistical physics.

The reason for this is that we have always attempted to construct the *Course of Theoretical Physics* as describing a single science, interlinking the discussion of its various branches in the different volumes. According to the general plan of the Course, this volume should have followed the one on quantum field theory, and the discussion of the above-mentioned methods in this book should have been based on the development of them in the previous volume. Since the latter has not yet been completed, it was not possible to include these methods in the present edition.

To my profound regret, L. D. Landau, my teacher and friend, has been prevented by injuries received in a road accident from personally contributing to the preparation of this new edition.

April 1966 E. M. LIFSHITZ

The publishers learnt with deep regret of the death of Professor L. D. Landau in April 1968 while this volume was in press.

FROM THE PREFACE TO THE FIRST ENGLISH EDITION

THE present volume of the *Theoretical Physics* series is devoted to an exposition of statistical physics and thermodynamics. These two subjects are firmly interconnected, and in our opinion it is rational to present them together as one whole.

As in the other volumes, we have endeavoured, on the one hand, to state the general principles as clearly as possible, and, on the other hand, to present their many specific applications as fully as possible. However, the present book does not contain the theory of electric and magnetic properties of matter, which are treated in another volume which is dealing with the electrodynamics of material media. Similarly, problems of non-equilibrium phenomena are not treated; we propose to consider these in a separate volume.

We have not included in this book the various theories of ordinary liquids and of strong solutions, which to us appear neither convincing nor useful.

We do not share the view, which one encounters sometimes, that statistical physics is the least well-founded branch of theoretical physics (as regards its basic principles). We believe that the difficulties are created artificially, because the problems are often not stated sufficiently rationally. If one talks from the very beginning about the statistical distribution for small parts of a system (subsystems) and not for a closed system as a whole, then one avoids the whole question of the ergodic or similar hypotheses, which are not really essential for physical statistics.

Moscow

L. D. LANDAU
E. M. LIFSHITZ

NOTATION

OPERATORS are denoted by a circumflex

Phase space

p, q generalised momenta and co-ordinates
$dp\,dq = dp_1\,dp_2 \ldots dp_s\,dq_1\,dq_2 \ldots dq_s$ volume element in phase space (with s degrees of freedom)
$d\Gamma = dp\,dq/(2\pi\hbar)^s$
$\int' \ldots d\Gamma$ integral over all physically different states

Thermodynamic quantities

T temperature
V volume
P pressure
E energy
S entropy
$W = E+PV$ heat function
$F = E-TS$ free energy
$\Phi = E-TS+PV$ thermodynamic potential
$\Omega = -PV$ thermodynamic potential
C_p, C_v specific heats
c_p, c_v molecular specific heats
N number of particles
μ chemical potential
α surface-tension coefficient
$\mathfrak{s}$ area of interface

In all formulae the temperature is expressed in energy units; the method of converting to degrees is described in footnotes to §§9 and 42.

CHAPTER I

THE FUNDAMENTAL PRINCIPLES OF STATISTICAL PHYSICS

§1. Statistical distributions

Statistical physics, often called for brevity simply *statistics*, consists in the study of the special laws which govern the behaviour and properties of macroscopic bodies (that is, bodies formed of a very large number of individual particles, such as atoms and molecules). To a considerable extent the general character of these laws does not depend on the mechanics (classical or quantum) which describes the motion of the individual particles in a body, but their substantiation demands a different argument in the two cases. For convenience of exposition we shall begin by assuming that classical mechanics is everywhere valid.

In principle, we can obtain complete information concerning the motion of a mechanical system by constructing and integrating the equations of motion of the system, which are equal in number to its degrees of freedom. But if we are concerned with a system which, though it obeys the laws of classical mechanics, has a very large number of degrees of freedom, the actual application of the methods of mechanics involves the necessity of setting up and solving the same number of differential equations, which in general is impracticable. It should be emphasised that, even if we could integrate these equations in a general form, it would be completely impossible to substitute in the general solution the initial conditions for the velocities and co-ordinates of the particles, if only because of the amount of time and paper that would be needed.

At first sight we might conclude from this that, as the number of particles increases, so also must the complexity and intricacy of the properties of the mechanical system, and that no trace of regularity can be found in the behaviour of a macroscopic body. This is not so, however, and we shall see below that, when the number of particles is very large, new types of regularity appear.

These *statistical laws* resulting from the very presence of a large number of particles forming the body cannot in any way be reduced to purely mechanical laws. One of their distinctive features is that they cease to have meaning when applied to mechanical systems with a small number of degrees of freedom. Thus, although the motion of systems with a very large number of degrees of freedom obeys the same laws of mechanics as that of systems consisting of a small number of particles, the existence of many degrees of freedom results in laws of a different kind.

The importance of statistical physics in many other branches of theoretical physics is due to the fact that in Nature we continually encounter macroscopic bodies whose behaviour can not be fully described by the methods of mechanics alone, for the reasons mentioned above, and which obey statistical laws.

In proceeding to formulate the fundamental problem of classical statistics, we must first of all define the concept of *phase space*, which will be constantly used hereafter.

Let a given macroscopic mechanical system have s degrees of freedom: that is, let the position of points of the system in space be described by s co-ordinates, which we denote by q_i, the suffix i taking the values 1, 2, ..., s. Then the state of the system at a given instant will be defined by the values at that instant of the s co-ordinates q_i and the s corresponding velocities $\dot{q}_i$. In statistics it is customary to describe a system by its co-ordinates and momenta p_i, not velocities, since this affords a number of very important advantages. The various states of the system can be represented mathematically by points in *phase space* (which is, of course, a purely mathematical concept); the co-ordinates in phase space are the co-ordinates and momenta of the system considered. Every system has its own phase space, with a number of dimensions equal to twice the number of degrees of freedom. Any point in phase space, corresponding to particular values of the co-ordinates q_i and momenta p_i of the system, represents a particular state of the system. The state of the system changes with time, and consequently the point in phase space representing this state (which we shall call simply the *phase point* of the system) moves along a curve called the *phase trajectory*.

Let us now consider a macroscopic body or system of bodies, and assume that the system is closed, i.e. does not interact with any other bodies. A part of the system, which is very small compared with the whole system but still macroscopic, may be imagined to be separated from the rest; clearly, when the number of particles in the whole system is sufficiently large, the number in a small part of it may still be very large. Such relatively small but still macroscopic parts will be called *subsystems*. A subsystem is again a mechanical system, but not a closed one; on the contrary, it interacts in various ways with the other parts of the system. Because of the very large number of degrees of freedom of the other parts, these interactions will be very complex and intricate. Thus the state of the subsystem considered will vary with time in a very complex and intricate manner.

An exact solution for the behaviour of the subsystem can be obtained only by solving the mechanical problem for the entire closed system, i.e. by setting up and solving all the differential equations of motion with given initial conditions, which, as already mentioned, is an impracticable task. Fortunately, it is just this very complicated manner of variation of the state of subsystems which, though rendering the methods of mechanics inapplicable, allows a different approach to the solution of the problem.

A fundamental feature of this approach is the fact that, because of the extreme complexity of the external interactions with the other parts of the system, during a sufficiently long time the subsystem considered will be many times in every possible state. This may be more precisely formulated as follows. Let $\Delta p\, \Delta q$ denote some small "volume" of the phase space of the subsystem, corresponding to co-ordinates q_i and momenta p_i lying in short intervals Δq_i and Δp_i. We can say that, in a sufficiently long time T, the extremely intricate phase trajectory passes many times through each such volume of phase space. Let Δt be the part of the total time T during which the subsystem was in the given volume of phase space $\Delta p\, \Delta q$.† When the total time T increases indefinitely, the ratio $\Delta t/T$ tends to some limit

$$w = \lim_{T \to \infty} \Delta t/T. \tag{1.1}$$

This quantity may clearly be regarded as the probability that, if the subsystem is observed at an arbitrary instant, it will be found in the given volume of phase space $\Delta p\, \Delta q$.

On taking the limit of an infinitesimal phase volume‡

$$dq\, dp = dq_1\, dq_2 \ldots dq_s\, dp_1\, dp_2 \ldots dp_s, \tag{1.2}$$

we can define the probability dw of states represented by points in this volume element, i.e. the probability that the co-ordinates q_i and momenta p_i have values in given infinitesimal intervals between q_i, p_i and $q_i + dq_i$, $p_i + dp_i$. This probability dw may be written

$$dw = \varrho(p_1, \ldots, p_s, q_1, \ldots, q_s)\, dp\, dq, \tag{1.3}$$

where $\varrho = (p_1, \ldots, p_s, q_1, \ldots, q_s)$ is a function of all the co-ordinates and momenta; we shall usually write for brevity $\varrho(p, q)$ or even ϱ simply. The function ϱ, which represents the "density" of the probability distribution in phase space, is called the *statistical distribution function*, or simply the *distribution function*, for the body concerned. This function must obviously satisfy the *normalisation condition*

$$\int \varrho\, dp\, dq = 1 \tag{1.4}$$

† For brevity, we shall usually say, as is customary, that the system "is in the volume $\Delta p\, \Delta q$ of phase space", meaning that the system is in states represented by phase points in that volume.

‡ In what follows we shall always use the conventional notation dp and dq to denote the products of the differentials of all the momenta and all the co-ordinates of the system respectively.

(the integral being taken over all phase space), which simply expresses the fact that the sum of the probabilities of all possible states must be unity.

The following circumstance is extremely important in statistical physics. The statistical distribution of a given subsystem does not depend on the initial state of any other small part of the same system, since over a sufficiently long time the effect of this initial state will be entirely outweighed by the effect of the much larger remaining parts of the system. It is also independent of the initial state of the particular small part considered, since in time this part passes through all possible states, any of which can be taken as the initial state. Without having to solve the mechanical problem for a system (taking account of initial conditions), we can therefore find the statistical distribution for small parts of the system.

The determination of the statistical distribution for any subsystem is in fact the fundamental problem of statistical physics. In speaking of "small parts" of a closed system, we must bear in mind that the macroscopic bodies with which we have to deal are usually themselves such "small parts" of a large closed system consisting of these bodies together with the external medium which surrounds them.

If this problem is solved and the statistical distribution for a given subsystem is known, we can calculate the probabilities of various values of any physical quantities which depend on the states of the subsystem (i.e. on the values of its co-ordinates q and momenta p). We can also calculate the mean value of any such quantity $f(p, q)$, which is obtained by multiplying each of its possible values by the corresponding probability and integrating over all states. Denoting the averaging by a bar, we can write

$$\bar{f} = \int f(p, q)\varrho(p, q)\, \mathrm{d}p\, \mathrm{d}q, \tag{1.5}$$

from which the mean values of various quantities can be calculated by using the statistical distribution function.

The averaging with respect to the distribution function (called *statistical averaging*) frees us from the necessity of following the variation with time of the actual value of the physical quantity $f(p, q)$ in order to determine its mean value. It is also obvious that, by the definition (1.1) of the probability, the statistical averaging is exactly equivalent to a time averaging. The latter would involve following the variation of the quantity with time, establishing the function $f = f(t)$, and determining the required mean value as

$$\bar{f} = \lim_{T\to\infty} \frac{1}{T} \int_0^T f(t)\, \mathrm{d}t.$$

The foregoing discussion shows that the deductions and predictions concerning the behaviour of macroscopic bodies which are made possible by statistical physics are probabilistic. In this respect statistical physics differs from (classical) mechanics, the deductions of which are entirely deterministic. It should be emphasised, however, that the probabilistic nature of the results of classical statistics is not an inherent property of the objects considered, but simply arises from the fact that these results are derived from much less information than would be necessary for a complete mechanical description (the initial values of the co-ordinates and momenta are not needed).

In practice, however, when statistical physics is applied to macroscopic bodies, its probabilistic nature is not usually apparent. The reason is that, if any macroscopic body (in external conditions independent of time) is observed over a sufficiently long period of time, it is found that all physical quantities† describing the body are practically constant (and equal to their mean values) and undergo appreciable changes relatively very rarely.‡ This result, which is fundamental to statistical physics, follows from very general considerations (to be discussed in §2) and becomes more and more nearly valid as the body considered becomes more complex and larger. In terms of the statistical distribution, we can say that, if by means of the function $\varrho(q, p)$ we construct the probability distribution function for various values of the quantity $f(p, q)$, this function will have an extremely sharp maximum for $f = \bar{f}$, and will be appreciably different from zero only in the immediate vicinity of this point.

Thus, by enabling us to calculate the mean values of quantities describing macroscopic bodies, statistical physics enables us to make predictions which are valid to very high accuracy for by far the greater part of any time interval which is long enough for the effect of the initial state of the body to be entirely eliminated. In this sense the predictions of statistical physics become practically determinate and not probabilistic. (For this reason, we shall henceforward almost always omit the bar when using mean values of macroscopic quantities.)

If a closed macroscopic system is in a state such that in any macroscopic subsystem the "macroscopic" physical quantities are to a high degree of accuracy equal to their mean values, the system is said to be in a state of *statistical equilibrium* (or *thermodynamic* or *thermal equilibrium*). It is seen from the foregoing that, if a closed macroscopic system is observed for a

† We mean, of course, macroscopic quantities describing the body as a whole or macro scopic parts of it, but not individual particles.

‡ We may give an example to illustrate the very high degree of accuracy with which this is true. If we consider a region in a gas which contains, say, only 1/100 gram-molecule, we find that the mean relative variation of the energy of this quantity of matter from its mean value is only $\sim 10^{-11}$. The probability of finding (in a single observation) a relative deviation of the order of 10^{-6}, say, is given by a fantastically small number, $\sim 10^{-3\times 10^{15}}$

sufficiently long period of time, it will be in a state of statistical equilibrium for much the greater part of this period. If, at any initial instant, a closed macroscopic system was not in a state of statistical equilibrium (if, for example, it was artificially disturbed from such a state by means of an external interaction and then left to itself, becoming again a closed system), it will necessarily enter an equilibrium state. The time within which it will reach statistical equilibrium is called the *relaxation time.* In using the term "sufficiently long" intervals of time, we have meant essentially times long compared with the relaxation time.

The theory of processes relating to the attainment of an equilibrium state is called *kinetics.* It is not part of statistical physics proper, which deals only with systems in statistical equilibrium, and which is the subject of this book.†

§2. Statistical independence

The subsystems discussed in §1 are not themselves closed systems; on the contrary, they are subject to the continuous interaction of the remaining parts of the system. But since these parts, which are small in comparison with the whole of the large system, are themselves macroscopic bodies also, we can still suppose that over not too long intervals of time they behave approximately as closed systems. For the particles which mainly take part in the interaction of a subsystem with the surrounding parts are those near the surface of the subsystem; the relative number of such particles, compared with the total number of particles in the subsystem, decreases rapidly when the size of the subsystem increases, and when the latter is sufficiently large the energy of its interaction with the surrounding parts will be small in comparison with its internal energy. Thus we may say that the subsystems are *quasi-closed.* It should be emphasised once more that this property holds only over not too long intervals of time. Over a sufficiently long interval of time, the effect of interaction of subsystems, however weak, will ultimately appear. Moreover, it is just this relatively weak interaction which leads finally to the establishment of statistical equilibrium.

The fact that different subsystems may be regarded as weakly interacting has the result that they may also be regarded as statistically independent. By *statistical independence* we mean that the state of one subsystem does not affect the probabilities of various states of the other subsystems.

Let us consider any two subsystems, and let $dp^{(1)}dq^{(1)}$ and $dp^{(2)}dq^{(2)}$ be volume elements in their phase spaces. If we regard the two subsystems together as one composite subsystem, then the statistical independence of the

† Problems of kinetics appear only in §§122 and 123.

subsystems signifies mathematically that the probability of the composite subsystem's being in its phase volume element $dp^{(12)}dq^{(12)} = dp^{(1)}dq^{(1)} \cdot dp^{(2)}dq^{(2)}$ can be written as the product of the probabilities for the two subsystems to be respectively in $dp^{(1)}dq^{(1)}$ and $dp^{(2)}dq^{(2)}$, each of these probabilities depending only on the co-ordinates and momenta of the subsystem concerned. Thus we can write

$$\varrho_{12}\, dp^{(12)}dq^{(12)} = \varrho_1\, dp^{(1)}dq^{(1)} \cdot \varrho_2\, dp^{(2)}dq^{(2)},$$

or

$$\varrho_{12} = \varrho_1 \varrho_2, \tag{2.1}$$

where ϱ_{12} is the statistical distribution of the composite subsystem, and ϱ_1, ϱ_2 the distribution functions of the separate subsystems. A similar relation is valid for a group of several subsystems.†

The converse statement is clearly also true: if the probability distribution for a compound system is a product of factors, each of which depends only on quantities describing one part of the system, then the parts concerned are statistically independent, and each factor is proportional to the probability of the state of the corresponding part.

If f_1 and f_2 are two physical quantities relating to two different subsystems, then from (2.1) and the definition (1.5) of mean values it follows immediately that the mean value of the product $f_1 f_2$ is equal to the product of the mean values of the quantities f_1 and f_2 separately:

$$\overline{f_1 f_2} = \bar{f}_1 \cdot \bar{f}_2. \tag{2.2}$$

Let us consider a quantity f relating to a macroscopic body or to a part of it. In the course of time this quantity varies, fluctuating about its mean value. We may define a quantity which represents the average range of this fluctuation. The mean value of the difference $\Delta f = f - \bar{f}$ is not suitable for this purpose, since the quantity f varies from its mean value in both directions, and the difference $f - \bar{f}$, which is alternately positive and negative, has mean value zero regardless of how often f undergoes considerable deviations from its mean value. The required characteristic may conveniently be defined as the mean square of this difference. Since $(\Delta f)^2$ is always positive, its mean value tends to zero only if $(\Delta f)^2$ itself tends to zero; that is, the mean value is small only when the probability of considerable deviations of f from $\bar{f}$ is small. The quantity $\sqrt{[\overline{(\Delta f)^2}]} = \sqrt{[\overline{(f-\bar{f})^2}]}$ is called the *root-mean-square (r.m.s.) fluctuation* of the quantity f. It may be noted that

$$\begin{aligned}\overline{(\Delta f)^2} &= \overline{f^2 - 2f\bar{f} + \bar{f}^2}\\ &= \overline{f^2} - 2\bar{f}\bar{f} + \bar{f}^2,\end{aligned}$$

† Provided, of course, that these subsystems together still form only a small part of the whole closed system.

whence

$$\overline{(\Delta f)^2} = \overline{f^2} - (\bar{f})^2, \tag{2.3}$$

i.e. the r.m.s. fluctuation is determined by the difference between the mean square of the quantity and the square of its mean.

The ratio $\sqrt{[\overline{(\Delta f)^2}]}/\bar{f}$ is called the *relative fluctuation* of the quantity f. The smaller this ratio is, the more negligible is the proportion of time during which the body is in states where the deviation of f from its mean value is a considerable fraction of the mean value.

We shall show that the relative fluctuations of physical quantities decrease rapidly when the size of the bodies (that is, the number of particles) to which they relate increases. To prove this, we first note that the majority of quantities of physical interest are additive. This property is a consequence of the fact that the various parts of a body are quasi-closed systems, and signifies that the value of such a quantity for the whole body is the sum of its values for the various (macroscopic) parts of the body. For example, since the internal energies of these parts are, as shown above, large compared with their interaction energies, it is sufficiently accurate to assume that the energy of the whole body is equal to the sum of the energies of its parts.

Let f be such an additive quantity. We imagine the body concerned to be divided into a large number N of approximately equal small parts. Then $f = \sum_{i=1}^{N} f_i$, where the quantities f_i relate to the individual parts, and likewise for the mean value we have

$$\bar{f} = \sum_{i=1}^{N} \bar{f}_i.$$

It is clear that, as the number of parts increases, $\bar{f}$ increases approximately in proportion to N. Let us also determine the r.m.s. fluctuation of f. We have

$$\overline{(\Delta f)^2} = \overline{(\sum_i \Delta f_i)^2}.$$

Because of the statistical independence of the different parts of the body, the mean values of the products $\Delta f_i \Delta f_k$ are

$$\overline{\Delta f_i \Delta f_k} = \overline{\Delta f_i} \cdot \overline{\Delta f_k} = 0 \qquad (i \neq k),$$

since each $\overline{\Delta f_i} = 0$. Hence

$$\overline{(\Delta f)^2} = \sum_{i=1}^{N} \overline{(\Delta f_i)^2}. \tag{2.4}$$

It follows that, as N increases, the mean square $\overline{(\Delta f)^2}$ also increases in proportion to N. The relative fluctuation is therefore inversely proportional to $\sqrt{N}$:

$$\frac{\sqrt{[\overline{(\Delta f)^2}]}}{\bar{f}} \sim \frac{1}{\sqrt{N}}. \tag{2.5}$$

On the other hand, if we consider a homogeneous body to be divided into parts of a given small size, it is clear that the number of parts will be proportional to the total number of particles (molecules) in the body. Hence the result can also be stated by saying that the relative fluctuation of any additive quantity f decreases inversely as the square root of the number of particles in a macroscopic body, and so, when the number of these is sufficiently large, the quantity f itself may be regarded as practically constant in time and equal to its mean value. This conclusion has already been used in §1.

§3. Liouville's theorem

Let us now return to a further study of the properties of the statistical distribution function, and suppose that a subsystem is observed over a very long interval of time, which we divide into a very large (in the limit, infinite) number of equal short intervals between instants t_1, t_2, ... At each of these instants the subsystem considered is represented in its phase space by a point A_1, A_2, ... The set of points thus obtained is distributed in phase space with a density which in the limit is everywhere proportional to the distribution function $\varrho(p, q)$. This follows from the significance of the latter function as determining the probabilities of various states of the subsystem.

Instead of considering points representing states of one subsystem at different instants t_1, t_2, ..., we may consider simultaneously, in a purely formal manner, a very large (in the limit, infinite) number of exactly identical subsystems†, which at some instant, say $t = 0$, are in states represented by the points A_1, A_2, ...

We now follow the subsequent movement of the phase points which represent the states of these subsystems over a not too long interval of time, such that a quasi-closed subsystem may with sufficient accuracy be regarded as closed. The movement of the phase points will then obey the equations of motion, which involve the co-ordinates and momenta only of the particles in the subsystem.

It is clear that at any instant t these points will be distributed in phase space according to the same distribution function $\varrho(p, q)$, in just the same way as at $t = 0$. In other words, as the phase points move about in the course of time, they remain distributed with a density which is constant at any given point and is proportional to the corresponding value of ϱ.

† Such an imaginary set of identical systems is usually called a *statistical ensemble*.

This movement of phase points may be formally regarded as a steady flow of a "gas" in phase space of $2s$ dimensions, and the familiar equation of continuity may be applied, which expresses the constancy of the total number of "particles" (in this case, phase points) in the gas. The ordinary equation of continuity is

$$\partial \varrho/\partial t + \operatorname{div}(\varrho \mathbf{v}) = 0,$$

where ϱ is the density and $\mathbf{v}$ the velocity of the gas. For steady flow, where $\partial \varrho/\partial t = 0$, we have

$$\operatorname{div}(\varrho \mathbf{v}) = 0.$$

For a space of many dimensions, this will become

$$\sum_{i=1}^{2s} \frac{\partial}{\partial x_i}(\varrho v_i) = 0.$$

In the present case the "co-ordinates" x_i are the co-ordinates q and momenta p, and the "velocities" $v_i = \dot{x}_i$ are the time derivatives $\dot{q}$ and $\dot{p}$ given by the equations of motion. Thus we have

$$\sum_{i=1}^{s} \left[\frac{\partial}{\partial q_i}(\varrho \dot{q}_i) + \frac{\partial}{\partial p_i}(\varrho \dot{p}_i) \right] = 0.$$

Expanding the derivatives gives

$$\sum_{i=1}^{s} \left[\dot{q}_i \frac{\partial \varrho}{\partial q_i} + \dot{p}_i \frac{\partial \varrho}{\partial p_i} \right] + \varrho \sum_{i=1}^{s} \left[\frac{\partial \dot{q}_i}{\partial q_i} + \frac{\partial \dot{p}_i}{\partial p_i} \right] = 0. \tag{3.1}$$

With the equations of motion in HAMILTON's form:

$$\dot{q}_i = \partial H/\partial p_i, \qquad \dot{p}_i = -\partial H/\partial q_i,$$

where $H = H(p, q)$ is the Hamiltonian for the subsystem considered, we see that

$$\partial \dot{q}_i/\partial q_i = \partial^2 H/\partial q_i \partial p_i = -\partial \dot{p}_i/\partial p_i.$$

The second term in (3.1) is therefore identically zero. The first term is just the total time derivative of the distribution function. Thus

$$\frac{d\varrho}{dt} = \sum_{i=1}^{s} \left(\frac{\partial \varrho}{\partial q_i} \dot{q}_i + \frac{\partial \varrho}{\partial p_i} \dot{p}_i \right) = 0. \tag{3.2}$$

We therefore reach the important conclusion that the distribution function is constant along the phase trajectories of the subsystem. This is *Liouville's theorem*. Since quasi-closed subsystems are under discussion, the result is valid only for not too long intervals of time, during which the subsystem behaves as if closed, to a sufficient approximation.

§4. The significance of energy

It follows at once from LIOUVILLE's theorem that the distribution function must be expressible entirely in terms of combinations of the variables p and q which remain constant when the subsystem moves as a closed subsystem. These combinations are the *mechanical invariants* or *integrals of the motion*, which are the first integrals of the equations of motion. We may therefore say that the distribution function, being a function of the mechanical invariants, is itself an integral of the motion.

It proves possible to restrict very considerably the number of integrals of the motion on which the distribution function can depend. To do this, we must take into account the fact that the distribution ϱ_{12} for a combination of two subsystems is equal to the product of the distribution functions ϱ_1 and ϱ_2 of the two subsystems separately: $\varrho_{12} = \varrho_1\varrho_2$ (see (2.1)). Hence

$$\log \varrho_{12} = \log \varrho_1 + \log \varrho_2, \tag{4.1}$$

so that the logarithm of the distribution function is an additive quantity We therefore reach the conclusion that the logarithm of the distribution function must be not merely an integral of the motion, but an additive integral of the motion.

As we know from mechanics, there exist only seven independent additive integrals of the motion: the energy, the three components of the momentum vector and the three components of the angular momentum vector. We shall denote these quantities for the ath subsystem (as functions of the co-ordinates and momenta of the particles in it) by $E_a(p, q)$, $\mathbf{P}_a(p, q)$, $\mathbf{M}_a(p, q)$ respectively. The only additive combination of these quantities is a linear combination of the form

$$\log \varrho_a = \alpha_a + \beta E_a(p, q) + \boldsymbol{\gamma}\cdot\mathbf{P}_a(p, q) + \boldsymbol{\delta}\cdot\mathbf{M}_a(p, q) \tag{4.2}$$

with constant coefficients α_a, β, $\boldsymbol{\gamma}$, $\boldsymbol{\delta}$, of which β, $\boldsymbol{\gamma}$, $\boldsymbol{\delta}$ must be the same for all subsystems in a given closed system.

We shall return in Chapter III to a detailed study of the distribution (4.2); here we need note only the following points. The coefficient α_a is just the normalisation constant, given by the condition $\int \varrho_a \, dp^{(a)} \, dq^{(a)} = 1$. The constants β, $\boldsymbol{\gamma}$, $\boldsymbol{\delta}$, involving seven independent quantities altogether, may be determined from the seven constant values of the additive integrals of the motion for the whole closed system. Thus we reach a conclusion very important in statistical physics. The values of the additive integrals of the motion (energy, momentum and angular momentum) completely define the statistical properties of a closed system, i.e. the statistical distribution of any of its subsystems, and therefore the mean values of any physical quantities relating to them. These seven additive integrals of the motion replace the unimaginable multiplicity of data (initial conditions) which would be required in the approach from mechanics.

The above arguments enable us at once to set up a simple distribution function suitable for describing the statistical properties of a closed system. Since, as we have seen, the values of non-additive integrals of the motion do not affect these properties, the latter can be described by any function ϱ which depends only on the values of the additive integrals of the motion for the system and which satisfies LIOUVILLE's theorem. The simplest such function is ϱ = constant for all points in phase space which correspond to given constant values of the energy (E_0), momentum ($\mathbf{P}_0$) and angular momentum ($\mathbf{M}_0$) of the system (regardless of the values of the non-additive integrals) and $\varrho = 0$ at all other points. It is clear that a function so defined will certainly remain constant along a phase trajectory of the system, i.e. will satisfy LIOUVILLE's theorem.

This formulation, however, is not quite exact. The reason is that the points defined by the equations

$$E(p, q) = E_0, \quad \mathbf{P}(p, q) = \mathbf{P}_0, \quad \mathbf{M}(p, q) = \mathbf{M}_0 \tag{4.3}$$

form a manifold of only $2s-7$ dimensions, not $2s$ like the phase volume. Consequently, if the integral $\int \varrho \, \mathrm{d}p \, \mathrm{d}q$ is to be different from zero, the function $\varrho(p, q)$ must become infinite at these points. The correct way of writing the distribution function for a closed system is

$$\varrho = \text{constant} \times \delta(E-E_0)\delta(\mathbf{P}-\mathbf{P}_0)\delta(\mathbf{M}-\mathbf{M}_0). \tag{4.4}$$

The presence of the delta functions† ensures that ϱ is zero at all points in phase space where one or more of the quantities E, $\mathbf{P}$, $\mathbf{M}$ is not equal to the given value E_0, $\mathbf{P}_0$ or $\mathbf{M}_0$. The integral of ϱ over the whole of a phase volume which includes all or part of the above-mentioned manifold of points is finite. The distribution (4.4) is called *microcanonical*.‡

The momentum and angular momentum of a closed system depend on its motion as a whole (uniform translation and uniform rotation). We can therefore say that the statistical state of a system executing a given motion depends only on its energy. In consequence, energy is of exceptional importance in statistical physics.

In order to exclude the momentum and angular momentum from the subsequent discussion we may use the following device. We imagine the system

† The definition and properties of the delta function are given, for example, in *Quantum Mechanics*, §5.

‡ It should be emphasised once more that this distribution is not the true statistical distribution for a closed system. Regarding it as the true distribution is equivalent to asserting that, in the course of a sufficiently long time, the phase trajectory of a closed system passes arbitrarily close to every point of the manifold defined by equations (4.3). But this assertion (called the *ergodic hypothesis*) is certainly not true in general.

to be enclosed in a rigid "box" and take co-ordinates such that the "box" is at rest. Under these conditions the momentum and angular momentum are not integrals of the motion, and the only remaining additive integral of the motion is the energy. The presence of the "box", on the other hand, clearly does not affect the statistical properties of small parts of the system (subsystems). Thus for the logarithms of the distribution functions of the subsystems, instead of (4.2), we have the still simpler expressions

$$\log \varrho_a = \alpha_a + \beta E_a(p, q). \tag{4.5}$$

The microcanonical distribution for the whole system is

$$\varrho = \text{constant} \times \delta(E - E_0). \tag{4.6}$$

So far we have assumed that the closed system as a whole is in statistical equilibrium; that is, we have considered it over times long compared with its relaxation time. In practice, however, it is usually necessary to discuss a system over times comparable with or even short relative to the relaxation time. For large systems this can be done, owing to the existence of what are called *partial* (or *incomplete*) *equilibria* as well as the complete statistical equilibrium of the entire closed system. Such equilibria occur because the relaxation time increases with the size of the system, and so the separate small parts of the system attain the equilibrium state considerably more quickly than equilibrium is established between these small parts. This means that each small part of the system is described by a distribution function of the form (4.2), with the parameters β, $\boldsymbol{\gamma}$, $\boldsymbol{\delta}$ of the distribution having different values for different parts. In such a case the system is said to be in *partial equilibrium*. In the course of time, the partial equilibrium gradually becomes complete, and the parameters β, $\boldsymbol{\gamma}$, $\boldsymbol{\delta}$ for each small part slowly vary and finally become equal throughout the closed system.

Another kind of partial equilibrium is also of frequent occurrence, namely that resulting from a difference in the rates of the various processes occurring in the system, not from a considerable difference in relaxation time between the system and its small parts. One obvious example is the partial equilibrium in a mixture of several substances involved in a chemical reaction. Owing to the comparative slowness of chemical reactions, equilibrium as regards the motion of the molecules will be reached, in general, considerably more rapidly than equilibrium as regards reactions of molecules, i.e. as regards the composition of the mixture. This enables us to regard the partial equilibria of the mixture as equilibria at a given (actually non-equilibrium) chemical composition.

The existence of equilibria leads to the concept of *macroscopic states* of a system. Whereas a mechanical microscopic description of the system specifies the co-ordinates and momenta of every particle in the system, a macroscopic description is one which specifies the mean values of the physical quantities determining a particular partial equilibrium, for instance the mean values of quantities describing separate sufficiently small but macroscopic parts of the system, each of which may be regarded as being in a separate equilibrium.

§5. The statistical matrix

Turning now to the distinctive features of quantum statistics, we may note first of all that the purely mechanical approach to the problem of determining the behaviour of a macroscopic body in quantum mechanics is of course just as hopeless as in classical mechanics. Such an approach would require the solution of SCHRÖDINGER's equation for a system consisting of all the particles in the body, a problem still more hopeless, one might even say, than the integration of the classical equations of motion. But even if it were possible in some particular case to find a general solution of SCHRÖDINGER's equation, it would be utterly impossible to select and write down the particular solution satisfying the precise conditions of the problem and specified by particular values of an enormous number of different quantum numbers. Moreover, we shall see below that for a macroscopic body the concept of stationary states itself becomes to some extent arbitrary, a fact of fundamental significance.

Let us first elucidate some purely quantum-mechanical features of macroscopic bodies as compared with systems consisting of a relatively small number of particles.

These features amount to an extremely high density of levels in the energy eigenvalue spectrum of a macroscopic body. The reason for this is easily seen if we note that, because of the very large number of particles in the body, a given quantity of energy can, roughly speaking, be "distributed" in innumerable ways among the various particles. The relation between this fact and the high density of levels becomes particularly clear if we take as an example a macroscopic body consisting of a "gas" of N particles which do not interact at all, enclosed in some volume. The energy levels of such a system are just the sums of the energies of the individual particles, and the energy of each particle can range over an infinite series of discrete values.† It is clear that, on choosing in all possible ways the values of the N terms in this sum, we shall obtain a very large number of possible values of the energy of the system in any appreciable finite part of the spectrum, and these values will therefore lie very close together.

† The separations between successive energy levels of a single particle are inversely proportional to the square of the linear dimensions L of the volume enclosing it ($\sim \hbar^2/mL^2$, where m is the mass of the particle and $\hbar$ the quantum constant).

It may be shown (see (7.18)) that the number of levels in a given finite range of the energy spectrum of a macroscopic body increases exponentially with the number of particles in the body, and the separations between levels are given by numbers of the form 10^{-N} (where N is a number of the order of the number of particles in the body), whatever the units, since a change in the unit of energy has no effect on such a fantastically small number.†

In consequence of the extremely high density of levels, a macroscopic body in practice can never be in a strictly stationary state. First of all, it is clear that the value of the energy of the system will always be "broadened" by an amount of the order of the energy of interaction between the system and the surrounding bodies. The latter is very large in comparison with the separations between levels, not only for quasi-closed subsystems but also for systems which from any other aspect could be regarded as strictly closed. In Nature, of course, there are no completely closed systems, whose interaction with any other body is exactly zero; and whatever interaction does exist, even if it is so small that it does not affect other properties of the system, will still be very large in comparison with the infinitesimal intervals in the energy spectrum.

In addition to this, there is another fundamental reason why a macroscopic body in practice cannot be in a stationary state. It is known from quantum mechanics that the state of a quantum-mechanical system described by a wave function is the result of some process of interaction of the system with another system which obeys classical mechanics to a sufficient approximation. In this respect the occurrence of a stationary state implies particular properties of the system. Here we must distinguish between the energy E of the system before the interaction and the energy E' of the state which results from the interaction. The uncertainties ΔE and $\Delta E'$ in the quantities E and E' are related to the duration Δt of the interaction process by the formula‡

$$|\Delta E' - \Delta E| \sim \hbar/\Delta t.$$

The two errors ΔE and $\Delta E'$ are in general of the same order of magnitude, and analysis shows that we cannot make $\Delta E' \ll \Delta E$. We can therefore say that $\Delta E' \sim \hbar/\Delta t$. In order that the state may be regarded as stationary, the uncertainty $\Delta E'$ must certainly be small in comparison with the separations between adjoining levels. Since the latter are extremely small, we see that, in order to bring the macroscopic body into a particular stationary state, an

† It should be mentioned that this discussion is inapplicable to the initial part of the energy spectrum; the separations between the first few energy levels of a macroscopic body may even be independent of the size of the body, as for instance in the electron spectrum in a dielectric (see §70). This point, however, does not affect the subsequent conclusions: when referred to a single particle, the separations between the first few levels for a macroscopic body are negligibly small, and the high density of levels mentioned in the text is reached for very small values of the energy relative to a single particle.

‡ See *Quantum Mechanics*, §44.

extremely long time $\Delta t \sim \hbar/\Delta E'$ would be necessary. In other words, we again conclude that strictly stationary states of a macroscopic body cannot exist.

To describe the state of a macroscopic body by a wave function at all is impracticable, since the available data concerning the state of such a body are far short of the complete set of data necessary to establish its wave function. Here the position is somewhat similar to that which occurs in classical statistics, where the impossibility of taking account of the initial conditions for every particle in a body makes impossible an exact mechanical description of its behaviour; the analogy is imperfect, however, since the impossibility of a complete quantum-mechanical description and the lack of a wave function describing a macroscopic body may, as we have seen, possess a much more profound significance.

The quantum-mechanical description based on an incomplete set of data concerning the system is effected by means of what is called a *density matrix*.† A knowledge of this matrix enables us to calculate the mean value of any quantity describing the system, and also the probabilities of various values of such quantities. The incompleteness of the description lies in the fact that the results of various kinds of measurement which can be predicted with a certain probability from a knowledge of the density matrix might be predictable with greater or even complete certainty from a complete set of data for the system, from which its wave function could be derived.

We shall not pause to write out here the formulae of quantum mechanics relating to the density matrix in the co-ordinate representation, since this representation is seldom used in statistical physics, but we shall show how the density matrix may be obtained directly in the energy representation, which is necessary for statistical applications.

Let us consider some subsystem, and define its "stationary states" as the states obtained when all interactions of the subsystem with the surrounding parts of a closed system are entirely neglected. Let $\psi_n(q)$ be the normalised wave functions of these states (without the time factor), q conventionally denoting the set of all co-ordinates of the subsystem, and the suffix n the set of all quantum numbers which distinguish the various stationary states; the energies of these states will be denoted by E_n.

Let us assume that at some instant the subsystem is in a completely described state with wave function Ψ. The latter may be expanded in terms of the functions $\psi_n(q)$, which form a complete set; we write the expansion as

$$\Psi = \sum_n c_n \psi_n. \tag{5.1}$$

The mean value of any quantity f in this state can be calculated from the

† See *Quantum Mechanics*, §14.

coefficients c_n by means of the formula

$$\bar{f} = \sum_n \sum_m c_n{}^* c_m f_{nm}, \tag{5.2}$$

where

$$f_{nm} = \int \psi_n{}^* \hat{f} \psi_m \,dq \tag{5.3}$$

are the matrix elements of the quantity f ($\hat{f}$ being the corresponding operator).

The change from the complete to the incomplete quantum-mechanical description of the subsystem may be regarded as a kind of averaging over its various Ψ states. In this averaging, the products $c_n^* c_m$ give a double set (two suffixes) of quantities, which we denote by w_{mn} and which cannot be expressed as products of any quantities forming a single set. The mean value of f is now given by

$$\bar{f} = \sum_m \sum_n w_{mn} f_{nm}. \tag{5.4}$$

The set of quantities w_{mn} (which in general are functions of time) is the density matrix in the energy representation; in statistical physics it is called the *statistical matrix*.†

If we regard the w_{mn} as the matrix elements of some *statistical operator* w, then the sum $\sum_n w_{mn} f_{nm}$ will be a diagonal matrix element of the operator product $\hat{w}\hat{f}$, and mean value $\bar{f}$ becomes the trace (sum of diagonal elements) of this operator:

$$\bar{f} = \sum_n (\hat{w}\hat{f})_{nn} = \text{tr}\,(\hat{w}\hat{f}). \tag{5.5}$$

This formula has the advantage of enabling us to calculate with any complete set of orthonormal wave functions: the trace of an operator is independent of the particular set of functions with respect to which the matrix elements are defined.‡

The other expressions of quantum mechanics which involve the quantities c_n are similarly modified, the products $c_n^* c_m$ being everywhere replaced by the "averaged values" w_{mn}:

$$c_n{}^* c_m \to w_{mn}.$$

For example, the probability that the subsystem is in the nth state is equal to the corresponding diagonal element w_{nn} of the density matrix (instead of the squared modulus $c_n^* c_n$). It is evident that these elements, which we shall denote by w_n, are always positive:

$$w_n = w_{nn} > 0, \tag{5.6}$$

† The energy representation is mentioned here, as being the one generally used in statistical physics. We have not so far, however, made direct use of the fact that the ψ_n are wave functions of stationary states. It is therefore clear that the same method could be used to define the density matrix with respect to any complete set of wave functions.

‡ See *Quantum Mechanics*, §12.

and satisfy the normalisation condition

$$\operatorname{tr} \hat{w} = \sum_n w_n = 1 \tag{5.7}$$

(corresponding to the condition $\sum_n |c_n|^2 = 1$).

It must be emphasised that the averaging over various Ψ states, which we have used in order to illustrate the transition from a complete to an incomplete quantum-mechanical description, has only a very formal significance. In particular, it would be quite incorrect to suppose that the description by means of the density matrix signifies that the subsystem can be in various Ψ states with various probabilities and that the averaging is over these probabilities. Such a treatment would be in conflict with the basic principles of quantum mechanics.

The states of a quantum-mechanical system that are described by wave functions are sometimes called *pure states*, as distinct from *mixed states*, which are described by a density matrix. Care should, however, be taken not to misunderstand the latter term in the way indicated above.

The averaging by means of the statistical matrix according to (5.4) has a twofold nature. It comprises both the averaging due to the probabilistic nature of the quantum description (even when as complete as possible) and the statistical averaging necessitated by the incompleteness of our information concerning the object considered. For a pure state only the first averaging remains, but in statistical cases both types of averaging are always present. It must be borne in mind, however, that these constituents cannot be separated; the whole averaging procedure is carried out as a single operation, and cannot be represented as the result of successive averagings, one purely quantum-mechanical and the other purely statistical.

The statistical matrix in quantum statistics takes the place of the distribution function in classical statistics. The whole of the discussion in the previous sections concerning classical statistics and the, in practice, deterministic nature of its predictions applies entirely to quantum statistics also. The proof given in §2 that the relative fluctuations of additive physical quantities tend to zero as the number of particles increases made no use of any specific properties of classical mechanics, and so remains entirely valid in the quantum case. We can therefore again assert that macroscopic quantities remain practically equal to their mean values.

In classical statistics the distribution function $\varrho(p, q)$ gives directly the probability distribution of the various values of the co-ordinates and momenta of the particles of the body. In quantum statistics this is no longer true; the quantities w_n give only the probabilities of finding the body in a particular quantum state, with no direct indication of the values of the co-ordinates and momenta of the particles.

From the very nature of quantum mechanics, the statistics based on it can deal only with the determination of the probability distribution for the co-ordinates and momenta separately, not together, since the co-ordinates and momenta of a particle cannot simultaneously have definite values. The required probability distributions must reflect both the statistical uncertainty and the uncertainty inherent in the quantum-mechanical description. To find these distributions, we repeat the arguments given above. We first assume that the body is in a pure quantum state with the wave function (5.1). The probability distribution for the co-ordinates is given by the squared modulus

$$|\Psi|^2 = \sum_n \sum_m c_n{}^* c_m \psi_n{}^* \psi_m,$$

so that the probability that the co-ordinates have values in a given interval $dq = dq_1\, dq_2 \ldots dq_s$ is $dw_q = |\Psi|^2\, dq$. For a mixed state, the products $c_n{}^* c_m$ are replaced by the elements w_{mn} of the statistical matrix, and $|\Psi|^2$ thus becomes

$$\sum_n \sum_m w_{mn} \psi_n{}^* \psi_m.$$

By the definition of the matrix elements,

$$\sum_m w_{mn} \psi_m = \hat{w} \psi_n,$$

and so

$$\sum_n \sum_m w_{mn} \psi_n{}^* \psi_m = \sum_n \psi_n{}^* \hat{w} \psi_n.$$

Thus we have the following formula for the co-ordinate probability distribution:

$$dw_q = \sum_n \psi_n{}^* \hat{w} \psi_n \cdot dq. \tag{5.8}$$

In this expression the functions ψ_n may be any complete set of normalised wave functions.

Let us next determine the momentum probability distribution. The quantum states in which all the momenta have definite values correspond to free motion of all the particles. We denote the wave functions of these states by $\psi_p(q)$, the suffix p conventionally representing the set of values of all the momenta. As we know, the diagonal elements of the density matrix are the probabilities that the system is in the corresponding quantum states. Hence, having determined the density matrix with respect to the set of functions ψ_p, we obtain the required momentum probability distribution from the formula†

$$dw_p = w_{pp}\, dp = dp \cdot \int \psi_p{}^* \hat{w} \psi_p\, dq, \tag{5.9}$$

where $dp = dp_1\, dp_2 \ldots dp_s$.

† The functions ψ_p are assumed normalised by the delta function of all the momenta.

It is interesting that both distributions (co-ordinate and momentum) can be obtained by integrating the same function,

$$I = \psi_p^*(q)\hat{w}\psi_p(q). \tag{5.10}$$

Integration of this expression with respect to q gives the momentum distribution, and with respect to p gives the co-ordinate distribution (the expression (5.8) with the functions ψ_p as the complete set of wave functions). It should be emphasised, however, that this does not mean that the function (5.10) may be regarded as a probability distribution for co-ordinates and momenta simultaneously: the expression (5.10) is complex, quite apart from the fact that such a view would conflict with the basic principles of quantum mechanics.

§6. Statistical distributions in quantum statistics

In quantum mechanics a theorem can be proved which is entirely analogous to LIOUVILLE's theorem derived in §3 on the basis of classical mechanics.

To do this, we first derive a general equation of quantum mechanics which gives the time derivative of the statistical matrix of any (closed) system.† Following the method used in §5, we first assume that the system is in a pure state with a wave function represented in the form of a series (5.1). Since the system is closed, its wave function will have the same form at all subsequent instants, but the coefficients c_n will depend on the time, being proportional to factors $e^{-iE_nt/\hbar}$. We therefore have

$$\frac{\partial}{\partial t}(c_n{}^*c_m) = \frac{i}{\hbar}(E_n-E_m)c_n{}^*c_m.$$

The change to the statistical matrix in the general case of mixed states is now effected by replacing the products $c_n{}^*c_m$ by w_{mn}, and this gives the required equation:

$$\dot{w}_{mn} = (i/\hbar)(E_n-E_m)w_{mn}. \tag{6.1}$$

This equation can be written in a general operator form by noticing that

$$(E_n-E_m)w_{mn} = \sum_l (w_{ml}H_{ln}-H_{ml}w_{ln}),$$

where H_{mn} are the matrix elements of the Hamiltonian $\hat{H}$ of the system; this matrix is diagonal in the energy representation, which we are using. Hence

$$\hat{\dot{w}} = (i/\hbar)(\hat{w}\hat{H}-\hat{H}\hat{w}). \tag{6.2}$$

It should be pointed out that this expression differs in sign from the usual quantum-mechanical expression for the operator of the time derivative of a quantity.

† In §5 the density matrix of a subsystem was discussed, having regard to its fundamental applications in statistical physics, but a density matrix can of course also be used to describe a closed system in a mixed state.

We see that, if the time derivative of the statistical matrix is zero, the operator $\hat{w}$ must commute with the Hamiltonian of the system. This result is the quantum analogue of LIOUVILLE's theorem: in classical mechanics the requirement of a stationary distribution function has the result that w is an integral of the motion, while the commutability of the operator of a quantity with the Hamiltonian is just the condition, in quantum mechanics, that that quantity is conserved.

In the energy representation used here, the condition is particularly simple: (6.1) shows that the matrix w_{mn} must be diagonal, again in accordance with the usual matrix condition that a quantity is conserved in quantum mechanics, namely that the matrix of such a quantity can be brought to diagonal form simultaneously with the Hamiltonian.

As in §3, we can now apply the results obtained to quasi-closed subsystems, for intervals of time during which they behave to a sufficient approximation as closed systems. Since the statistical distributions (or in this case the statistical matrices) of subsystems must be stationary, by the definition of statistical equilibrium, we first of all conclude that the matrices w_{mn} are diagonal for all subsystems.† The problem of determining the statistical distribution therefore amounts to a calculation of the probabilities $w_n = w_{nn}$, which represent the "distribution function" in quantum statistics. Formula (5.4) for the mean value of any quantity f becomes simply

$$\bar{f} = \sum_n w_n f_{nn}, \tag{6.3}$$

and contains only the diagonal matrix elements f_{nn}.

Next, using the facts that w must be a quantum-mechanical "integral of the motion" and that the subsystems are quasi-independent, we find in an entirely similar way to the derivation of (4.5) that the logarithm of the distribution function for subsystems must be of the form

$$\log w_n^{(a)} = \alpha^{(a)} + \beta E_n^{(a)}, \tag{6.4}$$

where the index a corresponds to the various subsystems. Thus the probabilities w_n can be expressed as a function of the energy level alone: $w_n = w(E_n)$.

Finally, the discussion in §4 concerning the significance of additive integrals of the motion, and in particular the energy, as determining all the statistical properties of a closed system, remains entirely valid. This again enables us to set up for a closed system a simple distribution function suitable for describing its statistical properties though (as in the classical case) certainly not the true distribution function.

† Since this statement involves neglecting the interactions between subsystems, it is more precise to say that the non-diagonal elements w_{mn} tend to zero as the relative importance of these interactions decreases, and therefore as the number of particles in the subsystems increases.

To formulate mathematically this "quantum microcanonical distribution" we must use the following device. The energy spectra of macroscopic bodies being "almost continuous", we make use of the concept of the number of quantum states of a closed system which "belong" to a particular infinitesimal range of values of its energy.† We denote this number by $d\Gamma$; it plays a part analogous to that of the phase volume element $dp\, dq$ in the classical case.

If we regard a closed system as consisting of subsystems, and neglect the interaction of the latter, every state of the whole system can be described by specifying the states of the individual subsystems, and the number $d\Gamma$ is a product

$$d\Gamma = \prod_a d\Gamma_a \tag{6.5}$$

of the numbers $d\Gamma_a$ of the quantum states of the subsystems (such that the sum of the energies of the subsystems lies in the specified interval of energy of the whole system).

We can now formulate the microcanonical distribution analogously to the classical expression (4.6), writing

$$dw = \text{constant} \times \delta(E - E_0) \prod_a d\Gamma_a \tag{6.6}$$

for the probability dw of finding the system in any of the $d\Gamma$ states.

§7. Entropy

Let us consider a closed system for a period of time long compared with its relaxation time; this implies that the system is in complete statistical equilibrium.

The following discussion will be given first of all for quantum statistics. Let us divide the system into a large number of macroscopic parts (subsystems) and consider any one of them. Let w_n be the distribution function for this subsystem; to simplify the formulae we shall at present omit from w_n (and other quantities) the suffix indicating the subsystem. By means of the function w_n we can, in particular, calculate the probability distribution of the various values of the energy E of the subsystem. We have seen that w_n may be written as a function of the energy alone, $w_n = w(E_n)$ (6.4). In order to obtain the probability $W(E)\, dE$ that the subsystem has an energy between E and $E+dE$, we must multiply $w(E)$ by the number of quantum states with energies in this interval; here we use the same idea of a "broadened" energy spectrum as was mentioned at the end of §6. Let $\Gamma(E)$ denote the number of quantum states with energies less than or equal to E. Then the required

† It will be remembered that in §4 we agreed to ignore entirely the momentum and angular momentum of the system as a whole, for which purpose it is sufficient to consider a system enclosed in a rigid "box" with co-ordinates such that the box is at rest.

number of states with energy between E and $E+\mathrm{d}E$ can be written

$$\frac{\mathrm{d}\Gamma(E)}{\mathrm{d}E}\,\mathrm{d}E,$$

and the energy probability distribution is

$$W(E) = \frac{\mathrm{d}\Gamma(E)}{\mathrm{d}E}\,w(E). \tag{7.1}$$

The normalisation condition

$$\int W(E)\,\mathrm{d}E = 1$$

signifies geometrically that the area under the curve $W = W(E)$ is unity.

In accordance with the general statements in §1, the function $W(E)$ has a very sharp maximum at $E = \bar{E}$, being appreciably different from zero only in the immediate neighbourhood of this point. We may define the "width" ΔE of the curve $W = W(E)$ as the width of a rectangle whose height is equal to the value of the function $W(E)$ at the maximum and whose area is unity:

$$W(\bar{E})\,\Delta E = 1. \tag{7.2}$$

Using the expression (7.1), we can write this definition as

$$w(\bar{E})\,\Delta\Gamma = 1, \tag{7.3}$$

where

$$\Delta\Gamma = \frac{\mathrm{d}\Gamma(\bar{E})}{\mathrm{d}E}\,\Delta E \tag{7.4}$$

is the number of quantum states corresponding to the interval ΔE of energy. The quantity $\Delta\Gamma$ thus defined may be said to represent the "degree of broadening" of the macroscopic state of the subsystem with respect to its microscopic states. The interval ΔE is equal in order of magnitude to the mean fluctuation of energy of the subsystem.

These definitions can be immediately applied to classical statistics, except that the function $w(E)$ must be replaced by the classical distribution function ϱ, and $\Delta\Gamma$ by the volume of the part of phase space defined by the form a

$$\varrho(\bar{E})\,\Delta p\,\Delta q = 1. \tag{7.5}$$

The phase volume $\Delta p\,\Delta q$, like $\Delta\Gamma$, represents the size of the region of phase space in which the subsystem will almost always be found.

It is not difficult to establish the relation between $\Delta\Gamma$ in quantum theory and $\Delta p\,\Delta q$ in the limit of classical theory. In the quasi-classical case, which is close to classical mechanics, a correspondence can be set up between the volume of a region of phase space and the "corresponding" number of quantum states†: we can say that a "cell" of volume $(2\pi\hbar)^s$ (where s is the number of

† See *Quantum Mechanics*, §48.

degrees of freedom of the system) "corresponds" in phase space to each quantum state. It is therefore clear that in the quasi-classical case the number of states $\Delta\Gamma$ may be written

$$\Delta\Gamma = \Delta p\,\Delta q/(2\pi\hbar)^s, \tag{7.6}$$

where s is the number of degrees of freedom of the subsystem considered. This formula gives the required relation between $\Delta\Gamma$ and $\Delta p\Delta q$.

The quantity $\Delta\Gamma$ is called the *statistical weight* of the macroscopic state of the subsystem, and its logarithm

$$S = \log \Delta\Gamma \tag{7.7}$$

is called the *entropy* of the subsystem. In the case of classical statistics the corresponding expression is

$$S = \log \frac{\Delta p \Delta q}{(2\pi\hbar)^s}. \tag{7.8}$$

The entropy thus defined is dimensionless, like the statistical weight itself. Since the number of states $\Delta\Gamma$ is not less than unity, the entropy cannot be negative. The concept of entropy is one of the most important in statistical physics.

It is apposite to mention that, if we adhere strictly to the standpoint of classical statistics, the concept of the "number of microscopic states" cannot be defined at all, and we should have to define the statistical weight simply as $\Delta p\Delta q$. But this quantity, like any volume in phase space, has the dimensions of the product of s momenta and s co-ordinates, i.e. the sth power of action ((erg·sec)^s). The entropy, defined as $\log \Delta p\Delta q$, would then have the peculiar dimensions of the logarithm of action. This means that the entropy would change by an additive constant when the unit of action changed: if the unit were changed by a factor a, $\Delta p\Delta q$ would become $a^s\Delta p\Delta q$, and $\log \Delta p\Delta q$ would become $\log \Delta p\Delta q + s \log a$. In purely classical statistics, therefore, the entropy is defined only to within an additive constant which depends on the choice of units, and only differences of entropy, i.e. changes of entropy in a given process, are definite quantities independent of the choice of units.

This accounts for the appearance of the quantum constant $\hbar$ in the definition (7.8) of the entropy for classical statistics. Only the concept of the number of discrete quantum states, which necessarily involves a non-zero quantum constant, enables us to define a dimensionless statistical weight and so to give an unambiguous definition of the entropy.

We may write the definition of the entropy in another form, expressing it directly in terms of the distribution function. According to (6.4), the logarithm of the distribution function of a subsystem has the form

$$\log w(E_n) = \alpha + \beta E_n.$$

Since this expression is linear in E_n, the quantity

$$\log w(\bar{E}) = \alpha + \beta\bar{E}$$

can be written as the mean value $\overline{\log w(E_n)}$. The entropy $S = \log \Delta\Gamma = -\log w(\bar{E})$ (from (7.3)) can therefore be written

$$S = -\overline{\log w(E_n)}, \tag{7.9}$$

i.e. the entropy can be defined as minus the mean logarithm of the distribution function of the subsystem. From the significance of the mean value,

$$S = -\sum_n w_n \log w_n; \tag{7.10}$$

this expression can be written in a general operator form independent of the choice of the set of wave functions with respect to which the statistical matrix elements are defined:†

$$S = -\operatorname{tr}(\hat{w} \log \hat{w}). \tag{7.11}$$

Similarly, in classical statistics, the definition of the entropy can be written

$$\begin{aligned} S &= -\overline{\log [(2\pi\hbar)^s \varrho]} \\ &= -\int \varrho \log [(2\pi\hbar)^s \varrho] \, dp \, dq. \end{aligned} \tag{7.12}$$

Let us now return to the closed system as a whole, and let $\Delta\Gamma_1, \Delta\Gamma_2, \dots$ be the statistical weights of its various subsystems. If each of the subsystems can be in one of $\Delta\Gamma_a$ quantum states, this gives

$$\Delta\Gamma = \prod_a \Delta\Gamma_a \tag{7.13}$$

as the number of different states of the whole system. This is called the statistical weight of the closed system, and its logarithm is the entropy S of the system. Clearly

$$S = \sum_a S_a, \tag{7.14}$$

i.e. the entropy thus defined is additive: the entropy of a composite system is equal to the sum of the entropies of its parts.

For a clear understanding of the way in which entropy is defined, it is important to bear in mind the following point. The entropy of a closed system (whose total energy we denote by E_0) in complete statistical equilibrium can also be defined directly, without dividing the system into subsystems. To do this, we imagine that the system considered is actually only a small part of a

† In accordance with the general rules, the operator $\log \hat{w}$ must be understood as an operator whose eigenvalues are equal to the logarithms of the eigenvalues of the operator $\hat{w}$, and whose eigenfunctions are the same as those of $\hat{w}$.

fictitious very large system (called in this connection a *thermostat* or *heat bath*). The thermostat is assumed to be in complete equilibrium, in such a way that the mean energy of the system considered (which is now a non-closed subsystem of the thermostat) is equal to its actual energy E_0. Then we can formally assign to the system a distribution function of the same form as for any subsystem of it, and by means of this distribution determine its statistical weight $\Delta\Gamma$, and therefore the entropy, directly from the same formulae (7.3)–(7.12) as were used for subsystems. It is clear that the presence of the thermostat has no effect on the statistical properties of individual small parts (subsystems) of the system considered, which in any case are not closed and are in equilibrium with the remaining parts of the system. The presence of the thermostat therefore does not alter the statistical weights $\Delta\Gamma_a$ of these parts, and the statistical weight defined in the way just described will be the same as that previously defined as the product (7.13).

So far we have assumed that the closed system is in complete statistical equilibrium. We must now generalise the above definitions to systems in arbitrary macroscopic states (partial equilibria).

Let us suppose that the system is in some state of partial equilibrium, and consider it over time intervals Δt which are small compared with the relaxation time for complete equilibrium. Then the entropy must be defined as follows. We imagine the system divided into parts so small that their respective relaxation times are small compared with the intervals Δt (remembering that the relaxation times in general decrease with decreasing size of the system). During the time Δt such systems may be regarded as being in their own particular equilibrium states, described by certain distribution functions. We can therefore apply to them the previous definition of the statistical weights $\Delta\Gamma_a$, and so calculate their entropies S_a. The statistical weight $\Delta\Gamma$ of the whole system is then defined as the product (7.13), and the corresponding entropy S as the sum of the entropies S_a.

It should be emphasised, however, that the entropy of a non-equilibrium system, defined in this way as the sum of the entropies of its parts (satisfying the above condition), cannot now be calculated by means of the thermostat concept without dividing the system into parts. At the same time this definition is unambiguous in the sense that further division of the subsystems into even smaller parts does not alter the value of the entropy, since each subsystem is already in "complete" internal equilibrium.

In particular, attention should be drawn to the significance of time in the definition of entropy. The entropy is a quantity which describes the average properties of a body over some non-zero interval of time Δt. If Δt is given, to determine S we must imagine the body divided into parts so small that their relaxation times are small in comparison with Δt. Since these parts must also themselves be macroscopic, it is clear that when the intervals Δt are too short the concept of entropy becomes meaningless; in particular, we cannot speak of its instantaneous value.

Having thus given a complete definition of the entropy, let us now ascertain the most important properties and the fundamental physical significance of this quantity. To do so, we must make use of the microcanonical distribution, according to which a distribution function of the form (6.6) may be used to describe the statistical properties of a closed system:

$$\mathrm{d}w = \text{constant} \times \delta(E-E_0) \cdot \prod_a \mathrm{d}\Gamma_a.$$

Here $\mathrm{d}\Gamma_a$ may be taken as the differential of the function $\Gamma_a(E_a)$, which represents the number of quantum states of a subsystem with energies less than or equal to E_a. We can write $\mathrm{d}w$ as

$$\mathrm{d}w = \text{constant} \times \delta(E-E_0) \cdot \prod_a (\mathrm{d}\Gamma_a/\mathrm{d}E_a)\, \mathrm{d}E_a. \tag{7.15}$$

The statistical weight $\Delta\Gamma_a$, by definition, is a function of the mean energy $\bar{E}_a$ of the subsystem; the same applies to $S_a = S_a(\bar{E}_a)$. Let us now formally regard $\Delta\Gamma_a$ and S_a as functions of the actual energy E_a (the same functions as they really are of $\bar{E}_a$). Then we can replace the derivatives $\mathrm{d}\Gamma_a(E_a)/\mathrm{d}E_a$ in (7.15) by the ratios $\Delta\Gamma_a/\Delta E_a$, where $\Delta\Gamma_a$ is a function of E_a in this sense, and ΔE_a the interval of energy corresponding to $\Delta\Gamma_a$ (also a function of E_a). Finally, replacing $\Delta\Gamma_a$ by $e^{S_a(E_a)}$, we obtain

$$\mathrm{d}w = \text{constant} \times \delta(E-E_0) e^S \prod_a \mathrm{d}E_a/\Delta E_a, \tag{7.16}$$

where

$$S = \sum_a S_a(E_a)$$

is the entropy of the whole closed system, regarded as a function of the exact values of the energies of its parts. The factor e^S, whose exponent is an additive quantity, is a very rapidly varying function of the energies E_a. In comparison with this function, the energy dependence of the quantity $\prod \Delta E_a$ is quite unimportant, and we can therefore replace (7.16) with very high accuracy by

$$\mathrm{d}w = \text{constant} \times \delta(E-E_0) e^S \prod_a \mathrm{d}E_a. \tag{7.17}$$

But $\mathrm{d}w$ expressed in a form proportional to the product of all the differential-$\mathrm{d}E_a$ is just the probability that all the subsystems have energies in given inters vals between E_a and $E_a+\mathrm{d}E_a$. Thus we see that this probability is determined by the entropy of the system as a function of the energies of the subsystems; the factor $\delta(E-E_0)$ ensures that the sum $E = \sum E_a$ has the given value E_0 of the energy of the system. This property of the entropy, as we shall see later, is the basis of its applications in statistical physics.

We know that the most probable values of the energies E_a are their mean values $\bar{E}_a$. This means that the function $S(E_1, E_2, \ldots)$ must have its maximum possible value (for a given value of the sum $\sum E_a = E_0$) when $E_a = \bar{E}_a$. But the $\bar{E}_a$ are just the values of the energies of the subsystems which correspond to complete statistical equilibrium of the system. Thus we reach the important conclusion that the entropy of a closed system in a state of complete statistical equilibrium has its greatest possible value (for a given energy of the system).

Finally, we may mention another interesting interpretation of the function $S = S(E)$, the entropy of any subsystem or closed system; in the latter case it is assumed that the system is in complete equilibrium, so that its entropy may be expressed as a function of its total energy alone. The statistical weight $\Delta\Gamma = e^{S(E)}$, by definition, is the number of energy levels in the interval ΔE which describes in a certain way the width of the energy probability distribution. Dividing ΔE by $\Delta\Gamma$, we obtain the mean separation between adjoining levels in this interval (near the energy E) of the energy spectrum of the system considered. Denoting this distance by $D(E)$, we can write

$$D(E) = \Delta E \cdot e^{-S(E)}. \tag{7.18}$$

Thus the function $S(E)$ determines the density of levels in the energy spectrum of a macroscopic system. Since the entropy is additive, we can say that the mean separations between the levels of a macroscopic body decrease exponentially with increasing size of the body (i.e. with increasing number of particles in it).

§8. The law of increase of entropy

If a closed system is not in a state of statistical equilibrium, its macroscopic state will vary in time, until ultimately the system reaches a state of complete equilibrium. If each macroscopic state of the system is described by the distribution of energy between the various subsystems, we can say that the sequence of states successively traversed by the system corresponds to more and more probable distributions of energy. This increase in probability is in general very considerable, because it is exponential, as shown in §7. We have seen that the probability is given by e^S, the exponent being an additive quantity, the entropy of the system. We can therefore say that the processes occurring in a non-equilibrium closed system do so in such a way that the system continually passes from states of lower to those of higher entropy until finally the entropy reaches the maximum possible value, corresponding to complete statistical equilibrium.

Thus, if a closed system is at some instant in a non-equilibrium macroscopic state, the most probable consequence at later instants is a steady increase in the entropy of the system. This is the *law of increase of entropy* or *second law of thermodynamics*, discovered by CLAUSIUS; its statistical explanation was given by BOLTZMANN.

In speaking of the "most probable" consequence, we must remember that in reality the probability of transition to states of higher entropy is so enormous in comparison with that of any appreciable decrease in entropy that in practice the latter can never be observed in Nature. Ignoring decreases in entropy due to negligible fluctuations, we can therefore formulate the law of increase of entropy as follows: if at some instant the entropy of a closed system does not have its maximum value, then at subsequent instants the entropy will not decrease; it will increase or at least remain constant.

There is no doubt that the foregoing simple formulations accord with reality; they are confirmed by all our everyday observations. But when we consider more closely the problem of the physical nature and origin of these laws of behaviour, substantial difficulties arise, which to some extent have not yet been overcome.

Firstly, if we attempt to apply statistical physics to the entire Universe, regarded as a single closed system, we immediately encounter a glaring contradiction between theory and experiment. According to the results of statistics, the universe ought to be in a state of complete statistical equilibrium. More precisely, any finite region of it, however large, should have a finite relaxation time and should be in equilibrium. Everyday experience shows us, however, that the properties of Nature bear no resemblance to those of an equilibrium system; and astronomical results show that the same is true throughout the vast region of the Universe accessible to our observation.

We might try to overcome this contradiction by supposing that the part of the Universe which we observe is just some huge fluctuation in a system which is in equilibrium as a whole. The fact that we have been able to observe this huge fluctuation might be explained by supposing that the existence of such a fluctuation is a necessary condition for the existence of an observer (a condition for the occurrence of biological evolution). This argument, however, is easily disproved, since a fluctuation within, say, the volume of the solar system only would be very much more probable, and would be sufficient to allow the existence of an observer.

The escape from this contradiction is to be sought in the general theory of relativity. The reason is that, when large regions of the Universe are considered, the gravitational fields present become important. According to the general theory of relativity, these fields are just a change in the space-time metric, described by the metric tensor g_{ik}. When the statistical properties of bodies are discussed, the metric properties of space-time may in a sense be regarded as "external conditions" to which the bodies are subject. The statement that a closed system must, over a sufficiently long time, reach a state of equilibrium, applies of course only to a system in steady external conditions. The metric tensor g_{ik} is in general a function not only of the co-ordinates but also of time, so that the "external conditions" are by no means steady in this case. Here it is important that the gravitational field cannot itself be included in a closed system, since the conservation laws which are, as we have seen, the foundation of statistical physics would then reduce to identities. For this reason, in the general theory of relativity, the Universe as a whole must be regarded not as a closed system but as a system in a variable gravitational field. Consequently the application of the law of increase of entropy does not prove that statistical equilibrium must necessarily exist.

Thus this aspect of the problem of the Universe as a whole indicates the physical basis of the apparent contradictions. There are, however, other difficulties in understanding the physical nature of the law of increase of entropy.

Classical mechanics itself is entirely symmetrical with respect to the two directions of time. The equations of mechanics remain unaltered when the time t is replaced by $-t$; if these equations allow any particular motion, they will therefore allow the reverse motion, in which the mechanical system passes through the same configurations in the reverse order. This symmetry must naturally be preserved in a statistics based on classical mechanics. Hence, if any particular process is possible which is accompanied by an increase in the entropy of a closed macroscopic system, the reverse process must also be possible, in which the entropy of the system decreases. The formulation of the law of increase of entropy given above does not itself contradict this symmetry, since it refers only to the most probable consequence of a macroscopically described state. In other words, if some non-equilibrium macroscopic state is given, the law of increase of entropy asserts only that, out of all the microscopic states which meet the given macroscopic description, the great majority lead to an increase of entropy at subsequent instants.

A contradiction arises, however, if we look at another aspect of the problem. In formulating the law of increase of entropy, we have referred to the most probable consequence of a macroscopic state given at some instant. But this state must itself have resulted from some other states by means of processes occurring in Nature. The symmetry with respect to the two directions of time means that, in any macroscopic state arbitrarily selected at some instant $t = t_0$, we can say not only that much the most probable consequence at $t > t_0$ is an increase in entropy, but also that much the most probable origin of the state was from states of greater entropy; that is, the presence of a minimum of entropy as a function of time at the arbitrarily chosen instant $t = t_0$ is much the most probable.

This assertion, of course, is not at all equivalent to the law of increase of entropy, according to which the entropy never decreases (apart from entirely negligible fluctuations) in any closed systems which actually occur in Nature. And it is precisely this general formulation of the law of increase of entropy which is confirmed by all natural phenomena. It must be emphasised that it is certainly not equivalent to the formulation given at the beginning of this section, as it might appear to be. In order to derive one formulation from the other, it would be necessary to use the concept of an observer who artificially "creates" a closed system at some instant, so that the problem of its previous behaviour does not arise. Such a dependence of the laws of physics on the nature of an observer is quite inadmissible, of course.

At the present time it is not certain whether the law of increase of entropy thus formulated can be derived on the basis of classical mechanics. It may be noted that, because the equations of classical mechanics are invariant under time reversal, we can consider only the deduction that the entropy varies monotonically. In order to derive a law of monotonic increase, we should have to define the future as the direction of time in which the entropy increases, and the problem would then arise of proving that this definition of the future and the past is the same as the definition used in quantum mechanics (see below).

It is more reasonable to suppose that the law of increase of entropy in the above general formulation arises from quantum effects.

The fundamental equation of quantum mechanics, namely SCHRÖDINGER's equation, is itself symmetrical under time reversal, provided that the wave function Ψ is also replaced by Ψ^*. This means that, if at some instant $t = t_1$ the wave function $\Psi = \Psi(t_1)$ is given, and if according to SCHRÖDINGER's equation it should become $\Psi(t_2)$ at some other instant t_2, then the change from $\Psi(t_1)$ to $\Psi(t_2)$ is reversible; in other words, if $\Psi = \Psi^*(t_2)$ at the initial instant t_1, then $\Psi = \Psi^*(t_1)$ at t_2.

However, despite this symmetry, quantum mechanics does in fact involve an important non-equivalence of the two directions of time. This appears in connection with the interaction of a quantum object with a system which with sufficient accuracy obeys the laws of classical mechanics, a process of fundamental significance in quantum mechanics. If two interactions A and B with a given quantum object occur in succession, then the statement that the probability of any particular result of process B is determined by the result of process A can be valid only if process A occurred earlier than process B.†

Thus in quantum mechanics there is a physical non-equivalence of the two directions of time, and the "macroscopic" expression of this may in fact be the law of increase of entropy. Up to the present, however, this relation has not been at all convincingly shown to exist in reality. If this is indeed the origin of the law of increase of entropy, there must exist an inequality involving the quantum constant $\hbar$ which ensures the validity of the law and is satisfied in the real world (and probably satisfied by a very wide margin).

Summarising, we may repeat the general formulation of the law of increase of entropy: in all closed systems which occur in Nature, the entropy never

† See also *Quantum Mechanics*, §7.

decreases; it increases, or at least remains constant. In accordance with these two possibilities, all processes involving macroscopic bodies are customarily divided into *irreversible* and *reversible* processes. The former comprise those which are accompanied by an increase of entropy of the whole closed system; the reverse processes cannot occur, since the entropy would then have to decrease. Reversible processes are those in which the entropy of the closed system remains constant†, and which can therefore take place in the reverse direction. A strictly reversible process is, of course, an ideal limiting case; processes actually occurring in Nature can be reversible only to within a certain degree of approximation.

† It must be emphasised that the entropies of the individual parts of the system need not remain constant also.

CHAPTER II

THERMODYNAMIC QUANTITIES

§9. Temperature

Thermodynamic physical quantities are those which describe macroscopic states of bodies. They include some which have both a thermodynamic and a purely mechanical significance, such as energy and volume. There are also, however, quantities of another kind, which appear as a result of purely statistical laws and have no meaning when applied to non-macroscopic systems, for example entropy.

In what follows we shall define a number of relations between thermodynamic quantities which hold good whatever the particular bodies to which these quantities relate. These are called *thermodynamic relations*.

When thermodynamic quantities are discussed, the negligible fluctuations to which they are subject are usually of no interest. Accordingly, we shall entirely ignore such fluctuations, and regard the thermodynamic quantities as varying only with the macroscopic state of the body.†

Let us consider two bodies in thermal equilibrium with each other, forming a closed system. Then the entropy S of this system has its maximum value (for a given energy E of the system). The energy E is the sum of the energies E_1 and E_2 of the two bodies: $E = E_1+E_2$. The same applies to the entropy S of the system, and the entropy of each body is a function of its energy: $S = S_1(E_1)+S_2(E_2)$. Since $E_2 = E-E_1$, E being a constant, S is really a function of one independent variable, and the necessary condition for a maximum may be written

$$\frac{dS}{dE_1} = \frac{dS_1}{dE_1}+\frac{dS_2}{dE_2}\frac{dE_2}{dE_1}$$

$$= \frac{dS_1}{dE_1}-\frac{dS_2}{dE_2} = 0,$$

whence

$$dS_1/dE_1 = dS_2/dE_2.$$

This conclusion can easily be generalised to any number of bodies in equilibrium with one another.

† Fluctuations of thermodynamic quantities will be discussed in a separate chapter (Chapter XII).

Thus, if a system is in a state of thermodynamic equilibrium, the derivative of the entropy with respect to the energy is the same for every part of it, i.e. is constant throughout the system. A quantity which is the reciprocal of the derivative of the entropy S of a body with respect to its energy E is called the *absolute temperature* T (or simply the *temperature*) of the body:

$$\mathrm{d}S/\mathrm{d}E = 1/T. \tag{9.1}$$

The temperatures of bodies in equilibrium with one another are therefore equal: $T_1 = T_2$.

Like the entropy, the temperature is seen to be a purely statistical quantity, which has meaning only for macroscopic bodies.

Let us next consider two bodies forming a closed system but not in equilibrium with each other. Their temperatures T_1 and T_2 are then different. In the course of time, equilibrium will be established between the bodies, tand their temperatures will gradually become equal. During this process, heir total entropy $S = S_1 + S_2$ must increase, i.e. its time derivative is positive:

$$\begin{aligned} \frac{\mathrm{d}S}{\mathrm{d}t} &= \frac{\mathrm{d}S_1}{\mathrm{d}t} + \frac{\mathrm{d}S_2}{\mathrm{d}t} \\ &= \frac{\mathrm{d}S_1}{\mathrm{d}E_1}\frac{\mathrm{d}E_1}{\mathrm{d}t} + \frac{\mathrm{d}S_2}{\mathrm{d}E_2}\frac{\mathrm{d}E_2}{\mathrm{d}t} > 0. \end{aligned}$$

Since the total energy is conserved, $\mathrm{d}E_1/\mathrm{d}t + \mathrm{d}E_2/\mathrm{d}t = 0$, and so

$$\frac{\mathrm{d}S}{\mathrm{d}t} = \left(\frac{\mathrm{d}S_1}{\mathrm{d}E_1} - \frac{\mathrm{d}S_2}{\mathrm{d}E_2}\right)\frac{\mathrm{d}E_1}{\mathrm{d}t} = \left(\frac{1}{T_1} - \frac{1}{T_2}\right)\frac{\mathrm{d}E_1}{\mathrm{d}t} > 0.$$

Let the temperature of the second body be greater than that of the first ($T_2 > T_1$). Then $\mathrm{d}E_1/\mathrm{d}t > 0$, and $\mathrm{d}E_2/\mathrm{d}t < 0$. In other words, the energy of the second body decreases and that of the first increases. This property of the temperature may be formulated as follows: energy passes from bodies at higher temperature to bodies at lower temperature.

The entropy S is a dimensionless quantity. The definition (9.1) therefore shows that the temperature has the dimensions of energy, and so can be measured in energy units, for example ergs. In ordinary circumstances, however, the erg is too large a quantity, and in practice the temperature is customarily measured in its own units, called *degrees Kelvin* or simply *degrees.* The conversion factor between ergs and degrees, i.e. the number of ergs per degree, is called *Boltzmann's constant* and is usually denoted by k; its value is†

$$k = 1.38 \times 10^{-16}\ \mathrm{erg/deg}.$$

† For reference, we may also give the conversion coefficient between degrees and electron-volts:

$$1\ \mathrm{eV} = 11{,}606\ \mathrm{deg}.$$

In all subsequent formulae the temperature will be assumed measured in energy units. To convert to the temperature measured in degrees, in numerical calculations, we need only replace T by kT. The continual use of the factor k, whose only purpose is to indicate the conventional units of temperature measurement, would merely complicate the formulae.

If the temperature is in degrees, the factor k is usually included in the definition of entropy:

$$S = k \log \Delta\Gamma, \tag{9.2}$$

instead of (7.7), in order to avoid the appearance of k in the general relations of thermodynamics. Then formula (9.1) defining the temperature, and therefore all the general thermodynamic relations derived subsequently in this chapter, are unaffected by the change to degrees.

Thus the rule for conversion to degrees is to substitute in all formulae

$$T \to kT, \quad S \to S/k. \tag{9.3}$$

§10. Macroscopic motion

As distinct from the microscopic motion of molecules, the *macroscopic motion* is that in which the various macroscopic parts of a body participate as a whole. Let us consider the possibility of macroscopic motion in a state of thermodynamic equilibrium.

Let the body be divided into a large number of small (but macroscopic) parts, and let M_a, E_a and $\mathbf{P}_a$ denote the mass, energy and momentum of the ath part. The entropy S_a of each part is a function of its internal energy, i.e. the difference between its total energy E_a and the kinetic energy $P_a^2/2M_a$ of its macroscopic motion.† The total entropy of the body can therefore be written

$$S = \sum_a S_a(E_a - P_a^2/2M_a). \tag{10.1}$$

We shall assume that the body is a closed system. Then its total momentum and angular momentum are conserved, as well as its energy:

$$\sum_a \mathbf{P}_a = \text{constant}, \quad \sum_a \mathbf{r}_a \times \mathbf{P}_a = \text{constant}, \tag{10.2}$$

where $\mathbf{r}_a$ is the radius vector of the ath part. In a state of equilibrium, the total entropy S of the body as a function of the momenta $\mathbf{P}_a$ has a maximum subject to the conditions (10.2). Using the familiar LAGRANGE's method of undetermined multipliers, we find the necessary conditions for a maximum

† The fact that the entropy of a body is a function only of its internal energy follows at once from GALILEO's relativity principle; the number of quantum states, and therefore the statistical weight (whose logarithm is the entropy), must be the same in all inertial frames of reference, and in particular that in which the body is at rest.

by equating to zero the derivatives with respect to $\mathbf{P}_a$ of the sum

$$\sum_a \{S_a + \mathbf{a}\cdot\mathbf{P}_a + \mathbf{b}\cdot\mathbf{r}_a \times \mathbf{P}_a\}, \tag{10.3}$$

where **a** and **b** are constant vectors. Differentiation of S_a with respect to $\mathbf{P}_a$ gives†, by the definition of the temperature,

$$\frac{\partial}{\partial \mathbf{P}_a} S_a\left(E_a - \frac{P_a^2}{2M_a}\right) = -\frac{\mathbf{P}_a}{M_a T} = -\frac{\mathbf{v}_a}{T},$$

where $\mathbf{v}_a = \mathbf{P}_a/M_a$ is the velocity of the ath part of the body. Differentiation of (10.3) therefore gives

$$-\mathbf{v}_a/T + \mathbf{a} + \mathbf{b}\times\mathbf{r}_a = 0,$$

or

$$\mathbf{v}_a = \mathbf{u} + \boldsymbol{\Omega}\times\mathbf{r}_a, \tag{10.4}$$

where $\mathbf{u} = T\mathbf{a}$ and $\boldsymbol{\Omega} = T\mathbf{b}$ are constant vectors.

This result has a simple physical significance. If the velocities of all the parts of a body are given by formula (10.4) with the same **u** and $\boldsymbol{\Omega}$, this means that we have a translational motion of the body as a whole with constant velocity **u** and a rotation of the body as a whole with constant angular velocity $\boldsymbol{\Omega}$. Thus we arrive at the important result that in thermodynamic equilibrium a closed system can execute only a uniform translational and rotational motion as a whole. No internal macroscopic motion is possible in a state of equilibrium.

In what follows we shall usually consider bodies at rest, and the energy E will accordingly be the internal energy of the body.

So far we have made use only of the necessary condition for a maximum of entropy as a function of the momenta, but not of the sufficient condition to be imposed on the second derivatives. It is easy to see that the latter condition leads to the very important result that the temperature must be positive: $T > 0$.‡ To deduce this, it is not in fact necessary to calculate the second derivatives; instead, we can argue as follows.

Let us consider a body forming a closed system, at rest as a whole. If the temperature were negative, the entropy would increase with decreasing argument. Since the entropy tends to increase, the body would spontaneously seek to break up into dispersing parts (with total momentum $\sum \mathbf{P}_a = 0$), so that the argument of each S_a in the sum (10.1) should take its least possible value. In other words, bodies in equilibrium could not exist with $T < 0$.

† The derivative with respect to a vector is to be understood as another vector whose components are equal to the derivatives with respect to the components of the first vector.

‡ The temperature $T = 0$ (absolute zero) corresponds to -273.15 °C.

The following point should be noted, however. Although the temperature of a body or any part of it can never be negative, there may exist incomplete equilibria in which the temperature corresponding to a particular group of degrees of freedom of the body is negative. This is further discussed in §71.

§11. Adiabatic processes

Among the various kinds of external interactions to which a body is subject, those which consist in a change in the external conditions form a special group. By "external conditions" we mean in a wide sense various external fields. In practice the external conditions are most often determined by the fact that the body must have a prescribed volume. In one sense this case may also be regarded as a particular type of external field, since the walls which limit the volume are equivalent in effect to a potential barrier which prevents the molecules in the body from escaping.

If the body is subject to no interactions other than changes in external conditions, it is said to be *thermally isolated.* It must be emphasised that, although a thermally isolated body does not interact directly with any other bodies, it is not in general a closed system, and its energy may vary with time.

In a purely mechanical way, a thermally isolated body differs from a closed system only in that its Hamiltonian (the energy) depends explicitly on the time: $E = E(p, q, t)$, because of the variable external field. If the body also interacted directly with other bodies, it would have no Hamiltonian of its own, since the interaction would depend not only on the co-ordinates of the molecules of the body in question but also on those of the molecules in the other bodies.

This leads to the result that the law of increase of entropy is valid not only for closed systems but also for a thermally isolated body, since here we regard the external field as a completely specified function of co-ordinates and time, and in particular neglect the reaction of the body on the field. That is, the field is a purely mechanical and not a statistical object, whose entropy can in this sense be taken as zero. This proves the foregoing statement.

Let us suppose that a body is thermally isolated, and is subject to external conditions which vary sufficiently slowly. Such a process is said to be *adiabatic.* We shall show that, in an adiabatic process, the entropy of the body remains unchanged, i.e. the process is reversible.

We shall describe the external conditions by certain parameters which are given functions of time. For example, suppose that there is only one such parameter, which we denote by λ. The time derivative $\mathrm{d}S/\mathrm{d}t$ of the entropy will depend in some manner on the rate of variation $\mathrm{d}\lambda/\mathrm{d}t$ of the parameter λ. Since $\mathrm{d}\lambda/\mathrm{d}t$ is small, we can expand $\mathrm{d}S/\mathrm{d}t$ in powers of $\mathrm{d}\lambda/\mathrm{d}t$.

The zero-order term in this expansion, which does not involve $d\lambda/dt$, is zero, since if $d\lambda/dt = 0$ then $dS/dt = 0$ also, because the entropy of a closed system in thermodynamic equilibrium must remain constant under constant external conditions. The first-order term, which is proportional to $d\lambda/dt$, must also be zero, since this term changes sign with $d\lambda/dt$, whereas dS/dt is always positive, according to the law of increase of entropy. Hence it follows that the expansion of dS/dt begins with the second-order term, i.e. for small $d\lambda/dt$ we have

$$dS/dt = A(d\lambda/dt)^2,$$

or

$$dS/d\lambda = A\, d\lambda/dt.$$

Thus, when $d\lambda/dt$ tends to zero, so does $dS/d\lambda$, which proves that the adiabatic process is reversible.

It must be emphasised that, although an adiabatic process is reversible, not every reversible process is adiabatic. The condition for a process to be reversible requires only that the total entropy of the whole of a closed system be constant, while the entropies of its individual parts may either increase or decrease. In an adiabatic process, a stronger condition holds: the entropy of a body which is only a part of a closed system also remains constant.

We have defined an adiabatic process as one which is sufficiently slow. More precisely, we can say that the external conditions must change so slowly that at any instant the body may be regarded as being in a state of equilibrium corresponding to the prevailing external conditions. That is, the process must be slow in comparison with the processes leading to the establishment of equilibrium in the body concerned.†

We may derive a formula to calculate by a purely thermodynamic method various mean values. To do so, we assume that a body undergoes an adiabatic process, and determine the time derivative dE/dt of its energy. By definition, the thermodynamic energy is

$$E = \overline{E(p, q; \lambda)},$$

where $E(p, q; \lambda)$ is the Hamiltonian of the body, depending on λ as a parameter. We know from mechanics that the total time derivative of the

† In practice this may be a very weak condition, so that the "slow" adiabatic process may be quite a "fast" one. For example, in the expansion of a gas, say in a cylinder with a piston moving outwards, the speed of the piston need be small only compared with the velocity of sound in the gas, i.e. it may in practice be very large.

In general textbooks on physics an adiabatic expansion (or compression) is often defined as one which is "sufficiently rapid". This refers to a different aspect of the problem: the process must occur so rapidly that the body cannot exchange heat with the surrounding medium. Thus the condition in question is one which will in practice ensure that the body is thermally isolated, and the condition of slowness compared with processes leading to the establishment of equilibrium is tacitly assumed satisfied.

Hamiltonian is equal to its partial time derivative†:

$$\frac{dE(p, q; \lambda)}{dt} = \frac{\partial E(p, q; \lambda)}{\partial t}.$$

In the present case $E(p, q; \lambda)$ depends explicitly on the time through $\lambda(t)$, and we can therefore write

$$\frac{dE(p, q; \lambda)}{dt} = \frac{\partial E(p, q; \lambda)}{\partial \lambda} \frac{d\lambda}{dt}.$$

Since the operations of averaging over the statistical distribution and differentiating with respect to time can clearly be interchanged, we have

$$\frac{dE}{dt} = \overline{\frac{dE(p, q; \lambda)}{dt}} = \overline{\frac{\partial E(p, q; \lambda)}{\partial \lambda}} \frac{d\lambda}{dt}; \tag{11.1}$$

the derivative $d\lambda/dt$ is a given function of time, and can be taken outside the averaging.

It is very important that, since the process is adiabatic, the mean value of the derivative $\partial E(p, q; \lambda)/\partial \lambda$ in (11.1) can be taken as the mean value over the statistical distribution corresponding to equilibrium for a given value of the parameter λ, i.e. for the external conditions prevailing at a given instant.

The derivative dE/dt can also be written in another form by regarding the thermodynamic quantity E as a function of the entropy S of the body and the external parameters λ. Since, in an adiabatic process, the entropy S remains constant, we have

$$\frac{dE}{dt} = \left(\frac{\partial E}{\partial \lambda}\right)_S \frac{d\lambda}{dt}, \tag{11.2}$$

where the subscript to the parenthesis indicates that the derivative is taken for constant S.

Comparison of (11.1) and (11.2) shows that

$$\overline{\frac{\partial E(p, q; \lambda)}{\partial \lambda}} = \left(\frac{\partial E}{\partial \lambda}\right)_S. \tag{11.3}$$

This is the required formula. It enables us to calculate thermodynamically the mean values (over the equilibrium statistical distribution) of quantities of the form $\partial E(p, q; \lambda)/\partial \lambda$. Such quantities are continually encountered when studying the properties of macroscopic bodies, and in consequence formula (11.3) is of great importance in statistical physics. It appears in the calculation of various forces acting on a body (the parameters λ being the co-ordinates of

† See *Mechanics*, §40.

a particular part of the body; see §12), the calculation of the magnetic or electric moment of bodies (the parameters λ being the magnetic or electric field strengths), and so on.

The arguments given here for classical mechanics are entirely applicable to the quantum theory, except that the energy $E(p, q; \lambda)$ must be everywhere replaced by the Hamiltonian operator $\hat{H}$. Then formula (11.3) becomes

$$\overline{\frac{\partial \hat{H}}{\partial \lambda}} = \left(\frac{\partial E}{\partial \lambda}\right)_S, \tag{11.4}$$

the bar denoting complete statistical averaging (which automatically includes the quantum averaging).

§12. Pressure

The energy E of a body, as a thermodynamic quantity, has the property of being additive: the energy of the body is equal to the sum of the energies of its individual (macroscopic) parts.† Another fundamental thermodynamic quantity, the entropy, also has this property.

The additivity of the energy and the entropy leads to the following very important result. If a body is in thermal equilibrium, we can say that, for a given energy, the entropy depends only on the volume of the body, and not on its shape; the same is true of the energy for a given entropy.‡ For a change in the shape of the body can be regarded as a rearrangement of its individual parts, and so the entropy and energy, being additive, will remain unchanged. Here, of course, it is assumed that the body is not in an external field of force, so that the motion of the parts of the body in space does not involve a change in their energy.

Thus the macroscopic state of a body at rest in equilibrium is entirely determined by only two quantities, for example the volume and the energy. All other thermodynamic quantities can be expressed as functions of these two. Of course, because of this mutual dependence of the various thermodynamic quantities, any other pair could be regarded as the independent variables.

Let us now calculate the force exerted by a body on the surface bounding its volume. According to the formulae of mechanics, the force acting on a

† Insofar as we neglect the energy of interaction of these parts; this is not permissible if we are interested in effects arising from the presence of interfaces between different bodies. Chapter XV deals with this topic.

‡ It should be mentioned that these statements are applicable in practice to liquids and gases but not to solids. A change in shape (deformation) of a solid involves the doing of work, so that the energy of the body is changed. This is because the deformed state of the solid is, strictly speaking, an incomplete thermodynamic equilibrium (but the relaxation time for the establishment of complete equilibrium is so long that in many respects the deformed body behaves as if in equilibrium).

surface element ds is

$$F = -\partial E(p, q; \mathbf{r})/\partial \mathbf{r},$$

where $E(p, q; \mathbf{r})$ is the energy of the body as a function of the co-ordinates and momenta of its particles and of the radius vector of the surface element considered, which here acts as an external parameter. Averaging this equation and using formula (11.3), we obtain

$$\overline{\mathbf{F}} = -\overline{\frac{\partial E(p, q; \mathbf{r})}{\partial \mathbf{r}}} = -\left(\frac{\partial E}{\partial \mathbf{r}}\right)_S = -\left(\frac{\partial E}{\partial V}\right)_S \frac{\partial V}{\partial \mathbf{r}},$$

where V is the volume. Since the change in volume is $\mathbf{ds}\cdot\mathbf{dr}$, we have $\partial V/\partial \mathbf{r} = \mathbf{ds}$, the surface element, and so

$$\overline{\mathbf{F}} = -(\partial E/\partial V)_S \, \mathbf{ds}.$$

Hence we see that the mean force on a surface element is normal to the element and proportional to its area (*Pascal's law*). The magnitude of the force per unit area is

$$P = -(\partial E/\partial V)_S. \qquad (12.1)$$

This quantity is called the *pressure*.

In defining the temperature by formula (9.1) we were essentially considering a body which is not in direct contact with any other bodies, and in particular is not surrounded by any external medium. Under these conditions it was possible to speak of the change in energy and entropy of the body without making more specific the nature of the process. In the general case of a body in an external medium, or surrounded by the walls of a vessel, formula (9.1) must be made more precise. For if during the process the volume of the body changes, this will necessarily affect the state of the bodies in contact with it, and in order to define the temperature we should have to take into consideration at the same time all the bodies in contact (for example, both the body in question and the vessel containing it). If it is desired to define the temperature in terms of thermodynamic quantities for the given body only, its volume must be regarded as constant. In other words, the temperature is defined as the derivative of the energy of the body with respect to its entropy, taken at constant volume:

$$T = (\partial E/\partial S)_V. \qquad (12.2)$$

The equations (12.1), (12.2) can also be written together as a relation between differentials:

$$\mathrm{d}E = T\,\mathrm{d}S - P\,\mathrm{d}V. \qquad (12.3)$$

This is one of the most important relations in thermodynamics.

The pressures of bodies in equilibrium with one another are equal. This follows immediately from the fact that thermal equilibrium necessarily presupposes mechanical equilibrium; in other words, the forces exerted on each other by any two of these bodies at their surface of contact must be equal in magnitude and opposite in direction, and thus balance.

The equality of pressures in equilibrium can also be derived from the condition of maximum entropy, in the same way as the equality of temperatures was shown in §9. To do this, we consider two parts, in contact, of a closed system in equilibrium. One necessary condition for the entropy to be a maximum is that it should be a maximum with respect to a change in the volumes V_1 and V_2 of these two parts when the states of the other parts undergo no change (this means, in particular, that V_1+V_2 remains constant). If the entropies of the two parts are S_1 and S_2, we have

$$\frac{\partial S}{\partial V_1} = \frac{\partial S_1}{\partial V_1}+\frac{\partial S_2}{\partial V_2}\frac{\partial V_2}{\partial V_1} = \frac{\partial S_1}{\partial V_1}-\frac{\partial S_2}{\partial V_2} = 0.$$

From the relation (12.3) in the form

$$\mathrm{d}S = \frac{1}{T}\,\mathrm{d}E+\frac{P}{T}\,\mathrm{d}V$$

it is seen that $\partial S/\partial V = P/T$, and so $P_1/T_1 = P_2/T_2$. Since the temperatures T_1 and T_2 are the same in equilibrium, we therefore find that the pressures are equal, $P_1 = P_2$.

It must be remembered that, when thermal equilibrium is established, the equality of pressures (i.e. mechanical equilibrium) is reached much more rapidly than that of temperatures, and so cases are often met with in which the pressure is constant throughout a body but the temperature is not. The reason is that the non-constancy of pressure is due to the presence of uncompensated forces; these bring about macroscopic motion so as to equalise the pressure much more rapidly than the equalisation of temperature, which does not involve macroscopic motion.

It is easy to see that the pressure must be positive in any equilibrium state: when $P > 0$ we have $(\partial S/\partial V)_E > 0$, and the entropy could increase only by an expansion of the body, which is prevented by the surrounding bodies. If $P < 0$, however, then we should have $(\partial S/\partial V)_E < 0$, and the body would spontaneously contract so as to increase its entropy.

There is, however, an important difference between the requirements of positive temperature and positive pressure. Bodies of negative temperature would be completely unstable and cannot exist in Nature. States (non-equilibrium) of negative pressure can exist in Nature with restricted stability. The reason is that the spontaneous contraction of the body involves "detaching" it from the walls of the vessel or the formation of cavities within it, that is, the formation of a new surface, and this leads to the possibility of the existence of negative pressures in what are called *metastable states*.†

§13. Work and quantity of heat

The external forces applied to a body can do *work* on it, which is determined, according to the general rules of mechanics, by the products of these forces and the displacements which they cause. This work may serve to bring the body into a state of macroscopic motion (or in general to change its kinetic energy), or to move the body in an external field (for instance, to raise it against gravity). We shall, however, be mainly interested in cases where the volume of a body is changed as a result of work done on it (i.e. the external forces compress the body but leave it at rest as a whole).

We shall everywhere regard as positive an amount of work R done on a given body by external forces. Negative work ($R < 0$) will correspondingly mean that the body itself does work (equal to $|R|$) on some external objects (for example, in expanding).

Bearing in mind that the force per unit area of the surface of the body is the pressure, and that the product of the area of a surface element and its displacement is the volume swept out by it, we find that the work done on the body per unit time when its volume changes is

$$\mathrm{d}R/\mathrm{d}t = -P\,\mathrm{d}V/\mathrm{d}t; \tag{13.1}$$

in compression, $\mathrm{d}V/\mathrm{d}t < 0$, so that $\mathrm{d}R/\mathrm{d}t > 0$. This formula is applicable to both reversible and irreversible processes; only one condition need be satisfied, namely that throughout the process the body must be in a state of mechanical equilibrium, i.e. at each instant the pressure must be constant throughout the body.

If the body is thermally isolated, the whole of the change in its energy is due to the work done on it. In the general case of a body not thermally isolated, in addition to the work done, the body gains or loses energy by direct transfer from or to other bodies in contact with it. This part of the change in energy is called the quantity of *heat* Q gained or lost by the body. Thus the change in the energy of the body per unit time may be written

$$\frac{\mathrm{d}E}{\mathrm{d}t} = \frac{\mathrm{d}R}{\mathrm{d}t} + \frac{\mathrm{d}Q}{\mathrm{d}t}. \tag{13.2}$$

Like the work, the heat will be regarded as positive if gained by the body from external sources.

†These are defined in §21. Negative pressures are further discussed in §83.

The energy E in (13.2) must, in general, be understood as the total energy of the body, including the kinetic energy of its macroscopic motion. We shall, however, usually consider the work corresponding to the change in volume of a body at rest, in which case the energy reduces to the internal energy of the body.

Under conditions where the work is defined by formula (13.1), we have for the quantity of heat

$$\frac{\mathrm{d}Q}{\mathrm{d}t} = \frac{\mathrm{d}E}{\mathrm{d}t} + P\frac{\mathrm{d}V}{\mathrm{d}t}. \tag{13.3}$$

Let us assume that at every instant throughout the process the body may be regarded as being in a state of thermal equilibrium corresponding to its energy and volume at that instant; it must be emphasised that this does not mean that the process is necessarily reversible, since the body may not be in equilibrium with surrounding bodies. Then, from the relation (12.3), which gives the differential of the function $E(S, V)$, the energy of the body in the equilibrium state, we can put

$$\frac{\mathrm{d}E}{\mathrm{d}t} = T\frac{\mathrm{d}S}{\mathrm{d}t} - P\frac{\mathrm{d}V}{\mathrm{d}t}.$$

Comparison with (13.3) shows that

$$\mathrm{d}Q/\mathrm{d}t = T\,\mathrm{d}S/\mathrm{d}t. \tag{13.4}$$

The work $\mathrm{d}R$ and the quantity of heat $\mathrm{d}Q$ gained by the body in an infinitesimal change of state are not the total differentials of any quantities.† Only the sum $\mathrm{d}Q+\mathrm{d}R$, i.e. the change in energy $\mathrm{d}E$, is a total differential. We can therefore speak of the energy E in a given state, but not, for example, of the quantity of heat which a body possesses in a given state. In other words, the energy of the body cannot be divided into thermal and mechanical parts; this is possible only when considering the change in energy. The change in energy when a body goes from one state to another can be divided into the quantity of heat gained or lost by the body and the work done on it or by it. This division is not uniquely determined by the initial and final states of the body, but depends also on the nature of the process itself. That is, the work and the quantity of heat are functions of the process undergone by the body and not only of its initial and final states. This is seen particularly when the body undergoes a cyclic process, starting and finishing in the same state. The change in energy is then zero, but the body may gain or lose a quantity of heat or work. Mathematically this corresponds to the fact that the integral of the total differential $\mathrm{d}E$ around a closed circuit is zero, but the integral of $\mathrm{d}Q$ or $\mathrm{d}R$, which are not total differentials, is not zero.

† In this sense the notation $\mathrm{d}R$ and $\mathrm{d}Q$ is not quite precise, and we therefore avoid it as far as possible.

The quantity of heat which must be gained in order to raise the temperature of the body by one unit (for example, one degree) is called its *specific heat*. This clearly depends on the conditions under which the heating takes place. A distinction is usually made between the specific heat at constant volume C_v and that at constant pressure C_p. Clearly

$$C_v = T(\partial S/\partial T)_V, \tag{13.5}$$

$$C_p = T(\partial S/\partial T)_P. \tag{13.6}$$

Let us consider cases where formula (13.4) for the quantity of heat is inapplicable, but at the same time it is possible to establish certain inequalities for this quantity. There exist processes in which the body is not in thermal equilibrium, although the temperature (and pressure) are constant throughout the body; for example, chemical reactions in a homogeneous mixture of reactants. Owing to the irreversible process (the chemical reaction) occurring in the body, its entropy increases independently of the heat gained, and so we can say that the inequality

$$\mathrm{d}Q/\mathrm{d}t < T\,\mathrm{d}S/\mathrm{d}t \tag{13.7}$$

holds.

Another case where a similar inequality can be stated is an irreversible process in which the body goes from one equilibrium state to another neighbouring one but is not in equilibrium during the process.† Then the inequality

$$\delta Q < T\delta S \tag{13.8}$$

holds between the quantity of heat δQ gained by the body in this process and its entropy change δS.

§14. The heat function

If the volume of a body remains constant during a process, then $\mathrm{d}Q = \mathrm{d}E$, i.e. the quantity of heat gained by the body is equal to the change in its energy. If the process occurs at constant pressure, the quantity of heat can be written as the differential

$$\mathrm{d}Q = \mathrm{d}(E+PV) = \mathrm{d}W \tag{14.1}$$

of a quantity

$$W = E+PV, \tag{14.2}$$

called the *heat function* of the body.‡ The change in the heat function in processes occurring at constant pressure is therefore equal to the quantity of heat gained by the body.

† An example is the *Joule–Thomson process* (see §18) with a small change in pressure.
‡ Also called the *enthalpy* or *heat content*.

It is easy to find an expression for the total differential of the heat function. Putting $\mathrm{d}E = T\,\mathrm{d}S - P\,\mathrm{d}V$ and $\mathrm{d}W = \mathrm{d}E + P\,\mathrm{d}V + V\,\mathrm{d}P$, we have

$$\mathrm{d}W = T\,\mathrm{d}S + V\,\mathrm{d}P. \tag{14.3}$$

From this it follows that

$$T = (\partial W/\partial S)_P, \qquad V = (\partial W/\partial P)_S. \tag{14.4}$$

If the body is thermally isolated (which, it will be remembered, does not imply that it is a closed system), $\mathrm{d}Q = 0$, and (14.1) shows that, in processes occurring at constant pressure and involving a thermally isolated body,

$$W = \text{constant}, \tag{14.5}$$

i.e. the heat function is conserved.

The specific heat C_v can be written, using the relation $\mathrm{d}E = T\,\mathrm{d}S - P\,\mathrm{d}V$, as

$$C_v = (\partial E/\partial T)_V. \tag{14.6}$$

Similarly, we have for the specific heat C_p

$$C_p = (\partial W/\partial T)_P.$$

Thus we see that at constant pressure the heat function has properties similar to those of the energy at constant volume.

§15. The free energy and the thermodynamic potential

The work done on a body in an infinitesimal isothermal reversible change of state can be written as a differential:

$$\begin{aligned}\mathrm{d}R = \mathrm{d}E - \mathrm{d}Q &= \mathrm{d}E - T\,\mathrm{d}S \\ &= \mathrm{d}(E - TS)\end{aligned}$$

or

$$\mathrm{d}R = \mathrm{d}F, \tag{15.1}$$

where

$$F = E - TS \tag{15.2}$$

is another function of the state of the body, called the *free energy*. Thus the work done on the body in a reversible isothermal process is equal to the change in its free energy.

Let us find the differential of the free energy. Substituting $\mathrm{d}E = T\,\mathrm{d}S - P\,\mathrm{d}V$ and $\mathrm{d}F = \mathrm{d}E - T\,\mathrm{d}S - S\,\mathrm{d}T$, we have

$$\mathrm{d}F = -S\,\mathrm{d}T - P\,\mathrm{d}V. \tag{15.3}$$

Hence it is evident that

$$S = -(\partial F/\partial T)_V, \qquad P = -(\partial F/\partial V)_T. \tag{15.4}$$

Using the relation $E = F+TS$, we can express the energy in terms of the free energy as

$$\begin{aligned} E &= F-T(\partial F/\partial T)_V \\ &= -T^2\left(\frac{\partial}{\partial T}\frac{F}{T}\right)_V \end{aligned} \tag{15.5}$$

Formulae (12.1), (12.2), (14.4) and (15.4) show that, if we know any of the quantities E, W and F as a function of the corresponding two variables and take its partial derivatives, we can determine all the remaining thermodynamic quantities. For this reason E, W and F are sometimes called *thermodynamic potentials* (by analogy with the mechanical potential) or *characteristic functions*: the energy E with respect to the variables S, V; the heat function W with respect to S, P; the free energy F with respect to V, T.

We still lack a thermodynamic potential with respect to the variables P, T. To derive this we substitute in (15.3) $P\,\mathrm{d}V = \mathrm{d}(PV)-V\,\mathrm{d}P$, take $\mathrm{d}(PV)$ to the left-hand side of the equation, and obtain

$$\mathrm{d}\Phi = -S\,\mathrm{d}T+V\,\mathrm{d}P, \tag{15.6}$$

with a new quantity

$$\begin{aligned} \Phi &= E-TS+PV \\ &= F+PV \\ &= W-TS, \end{aligned} \tag{15.7}$$

called the *thermodynamic potential* (in a restricted sense of the term).†

From (15.6) we clearly have

$$S = -(\partial\Phi/\partial T)_P, \qquad V = (\partial\Phi/\partial P)_T. \tag{15.8}$$

The heat function is expressed in terms of Φ in the same way as E in terms of F:

$$\begin{aligned} W &= \Phi-T(\partial\Phi/\partial T)_P \\ &= -T^2\left(\frac{\partial}{\partial T}\frac{\Phi}{T}\right)_P. \end{aligned} \tag{15.9}$$

If there are other parameters λ_i besides the volume which define the state of the system, the expression for the differential of the energy must be augmented by terms proportional to the differentials $\mathrm{d}\lambda_i$:

$$\mathrm{d}E = T\,\mathrm{d}S-P\,\mathrm{d}V+\sum_i \Lambda_i\,\mathrm{d}\lambda_i, \tag{15.10}$$

where the Λ_i are some functions of the state of the body. Since the transformation to other potentials does not affect the variables λ_i, it is clear that similar

† In Western literature, the functions F and Φ are often called respectively the *Helmholtz free energy* and the *Gibbs free energy*.

terms will be added to the differentials F, Φ, W:

$$dF = -S\,dT - P\,dV + \sum_i \Lambda_i\,d\lambda_i,$$

etc. Hence the quantities Λ_i can be obtained by differentiation with respect to λ_i of any of these potentials (it must be remembered which other variables are treated as constant in the differentiation). Using also formula (11.3), we can write down the analogous relation

$$\overline{\frac{\partial E(p, q; \lambda)}{\partial \lambda}} = \left(\frac{\partial F}{\partial \lambda}\right)_{T, V}, \tag{15.11}$$

which expresses the mean value of the derivative of the Hamiltonian with respect to any parameter as the derivative of the free energy with respect to that parameter (and similar relations involving the derivatives of Φ and W).

The following point may be noted. If the values of the parameters λ_i change slightly, the quantities E, F, W and Φ will also undergo small changes. It is evident that these changes will be equal if each is considered for the appropriate pair of constant quantities:

$$(\delta E)_{S,V} = (\delta F)_{T,V} = (\delta W)_{S,P} = (\delta \Phi)_{T,P}. \tag{15.12}$$

The free energy and the thermodynamic potential have a very important property which determines the direction in which they change in various irreversible processes. From the inequality (13.7), substituting dQ/dt from (13.3), we obtain

$$\frac{dE}{dt} + P\frac{dV}{dt} < T\frac{dS}{dt}. \tag{15.13}$$

Let us assume that the process is isothermal and occurs at constant volume (T = constant, V = constant). Then this inequality may be written

$$\frac{d(E-TS)}{dt} = \frac{dF}{dt} < 0. \tag{15.14}$$

Thus irreversible processes occurring at constant temperature and constant volume are accompanied by a decrease in the free energy of the body.

Similarly, for P = constant and T = constant the inequality (15.13) becomes

$$d\Phi/dt < 0; \tag{15.15}$$

that is, irreversible processes occurring at constant temperature and constant pressure are accompanied by a decrease in the thermodynamic potential.†

† It should be remembered that in both cases the processes in question are those (such as chemical reactions) for which the body is not in equilibrium, so that its state is not uniquely defined by the temperature and the volume (or pressure).

Correspondingly, in a state of thermal equilibrium the free energy and the thermodynamic potential have minimum values, the former with respect to all changes of state with T and V constant, and the latter with respect to changes of state with T and P constant.

PROBLEM

How can the mean kinetic energy of the particles in a body be calculated if the formula for its free energy is known?

SOLUTION. The Hamiltonian function (or, in the quantum case, the Hamiltonian operator) may be written in the form $E(p, q) = U(q)+K(p)$, where $U(q)$ is the potential energy of interaction of the particles in the body, and $K(p)$ their kinetic energy. The latter is a quadratic function of the momenta, inversely proportional to the particle mass m (for a body consisting of identical particles). Regarding m as a parameter, we can therefore write

$$\frac{\partial E(p, q; m)}{\partial m} = -\frac{1}{m} K(p).$$

Then, applying formula (15.11), we obtain the mean kinetic energy $K = \overline{K(p)}$:

$$K = -m(\partial F/\partial m)_{T, V}.$$

§16. Relations between the derivatives of thermodynamic quantities

In practice the most convenient, and the most widely used, pairs of thermodynamic variables are T, V and T, P. It is therefore necessary to transform various derivatives of the thermodynamic quantities with respect to one another to different variables, both dependent and independent.

If V and T are used as independent variables, the results of the transformation can be conveniently expressed in terms of the pressure P and the specific heat C_v (as functions of V and T). The equation which relates the pressure, volume and temperature is called the *equation of state* for a given body. Thus the purpose of the formulae in this case is to make it possible to calculate various derivatives of thermodynamic quantities from the equation of state and the specific heat C_v.

Similarly, when P and T are taken as the basic variables the results of the transformation should be expressed in terms of V and C_p (as functions of P and T).

Here it must be remembered that the dependence of C_v on V or of C_p on P (but not on the temperature) can itself be determined from the equation of state. It is easily seen that the derivative $(\partial C_v/\partial V)_T$ can be transformed so that it is defined in terms of the function $P(V, T)$. Using the fact that $S = -(\partial F/\partial T)_V$, we have

$$\left(\frac{\partial C_v}{\partial V}\right)_T = T\frac{\partial^2 S}{\partial V \partial T} = -T\frac{\partial^3 F}{\partial V \partial T^2}$$

$$= -T\frac{\partial^2}{\partial T^2}\left(\frac{\partial F}{\partial V}\right)_T,$$

and since $(\partial F/\partial V)_T = -P$, we have the required formula

$$(\partial C_v/\partial V)_T = T(\partial^2 P/\partial T^2)_V. \tag{16.1}$$

Similarly we find

$$(\partial C_p/\partial P)_T = -T(\partial^2 V/\partial T^2)_P, \tag{16.2}$$

formulae (15.8) being used in the calculation.

We shall show how some of the thermodynamic derivatives most often encountered may be transformed.

The derivatives of the entropy with respect to volume or pressure can be calculated from the equation of state by means of the following formulae, which are a direct consequence of the expressions for the differentials of the thermodynamic quantities. We have

$$\left(\frac{\partial S}{\partial V}\right)_T = -\frac{\partial}{\partial V}\left(\frac{\partial F}{\partial T}\right)_V = -\frac{\partial}{\partial T}\left(\frac{\partial F}{\partial V}\right)_T$$

or

$$(\partial S/\partial V)_T = (\partial P/\partial T)_V. \tag{16.3}$$

Similarly

$$\left(\frac{\partial S}{\partial P}\right)_T = -\frac{\partial}{\partial P}\left(\frac{\partial \Phi}{\partial T}\right)_P = -\frac{\partial}{\partial T}\left(\frac{\partial \Phi}{\partial P}\right)_T$$

or

$$(\partial S/\partial P)_T = -(\partial V/\partial T)_P. \tag{16.4}$$

The derivative $(\partial E/\partial V)_T$ is calculated from the equation $\mathrm{d}E = T\,\mathrm{d}S - P\,\mathrm{d}V$ as

$$\left(\frac{\partial E}{\partial V}\right)_T = T\left(\frac{\partial S}{\partial V}\right)_T - P$$

or, substituting (16.3),

$$\left(\frac{\partial E}{\partial V}\right)_T = T\left(\frac{\partial P}{\partial T}\right)_V - P. \tag{16.5}$$

Similarly we can derive

$$\left(\frac{\partial E}{\partial P}\right)_T = -T\left(\frac{\partial V}{\partial T}\right)_P - P\left(\frac{\partial V}{\partial P}\right)_T, \tag{16.6}$$

$$\left(\frac{\partial W}{\partial V}\right)_T = T\left(\frac{\partial P}{\partial T}\right)_V + V\left(\frac{\partial P}{\partial V}\right)_T, \quad \left(\frac{\partial W}{\partial P}\right)_T = V - T\left(\frac{\partial V}{\partial T}\right)_P, \tag{16.7}$$

$$\left(\frac{\partial E}{\partial T}\right)_P = C_p - P\left(\frac{\partial V}{\partial T}\right)_P, \quad \left(\frac{\partial W}{\partial T}\right)_V = C_v + V\left(\frac{\partial P}{\partial T}\right)_V. \tag{16.8}$$

Finally, we shall show how the specific heat C_v may be calculated from the specific heat C_p and the equation of state, using T and P as the basic variables. Since $C_v = T(\partial S/\partial T)_V$, we have to transform the derivative

$(\partial S/\partial T)_V$ to different independent variables. A transformation of this type is most simply effected by the use of Jacobians.† We write

$$\begin{aligned} C_v &= T(\partial S/\partial T)_V \\ &= T\partial(S, V)/\partial(T, V) \\ &= T\frac{\partial(S, V)/\partial(T, P)}{\partial(T, V)/\partial(T, P)} \\ &= T\frac{(\partial S/\partial T)_P(\partial V/\partial P)_T-(\partial S/\partial P)_T(\partial V/\partial T)_P}{(\partial V/\partial P)_T} \\ &= C_p-T\frac{(\partial S/\partial P)_T(\partial V/\partial T)_P}{(\partial V/\partial P)_T}. \end{aligned}$$

Substituting (16.4), we obtain the required formula:

$$C_p-C_v = -T[(\partial V/\partial T)_P]^2/(\partial V/\partial P)_T. \tag{16.9}$$

Similarly, transforming $C_p = T(\partial S/\partial T)_P$ to the variables T, V, we can derive the formula

$$C_p-C_v = -T[(\partial P/\partial T)_V]^2/(\partial P/\partial V)_T. \tag{16.10}$$

The derivative $(\partial P/\partial V)_T$ is negative: in an isothermal expansion of a body, its pressure always decreases. This will be rigorously proved in §21. It therefore follows from (16.10) that for all bodies

$$C_p > C_v. \tag{16.11}$$

In adiabatic expansion (or contraction) of a body its entropy remains constant. The relation between the temperature, volume and pressure of the body in an adiabatic process is therefore determined by various derivatives taken at

† The Jacobian $\partial(u, v)/\partial(x, y)$ is defined as the determinant

$$\frac{\partial(u, v)}{\partial(x, y)} = \begin{vmatrix} \partial u/\partial x & \partial u/\partial y \\ \partial v/\partial x & \partial v/\partial y \end{vmatrix}. \tag{I}$$

It clearly has the following properties:

$$\frac{\partial(v, u)}{\partial(x, y)} = -\frac{\partial(u, v)}{\partial(x, y)}, \tag{II}$$

$$\frac{\partial(u, y)}{\partial(x, y)} = \left(\frac{\partial u}{\partial x}\right)_y. \tag{III}$$

The following relations also hold:

$$\frac{\partial(u, v)}{\partial(x, y)} = \frac{\partial(u, v)}{\partial(t, s)}\cdot\frac{\partial(t, s)}{\partial(x, y)}, \tag{IV}$$

$$\frac{\mathrm{d}}{\mathrm{d}t}\frac{\partial(u, v)}{\partial(x, y)} = \frac{\partial(\mathrm{d}u/\mathrm{d}t, v)}{\partial(x, y)}+\frac{\partial(u, \mathrm{d}v/\mathrm{d}t)}{\partial(x, y)}. \tag{V}$$

constant entropy. We shall derive formulae whereby these derivatives may be calculated from the equation of state of the body and its specific heat.

For the derivative of the temperature with respect to volume we have, changing to independent variables V, T,

$$\left(\frac{\partial T}{\partial V}\right)_S = \frac{\partial(T, S)}{\partial(V, S)} = \frac{\partial(T, S)/\partial(V, T)}{\partial(V, S)/\partial(V, T)}$$
$$= -\frac{(\partial S/\partial V)_T}{(\partial S/\partial T)_V}$$
$$= -\frac{T}{C_v}\left(\frac{\partial S}{\partial V}\right)_T,$$

or, substituting (16.3),

$$\left(\frac{\partial T}{\partial V}\right)_S = -\frac{T}{C_v}\left(\frac{\partial P}{\partial T}\right)_V. \tag{16.12}$$

Similarly we find

$$\left(\frac{\partial T}{\partial P}\right)_S = \frac{T}{C_p}\left(\frac{\partial V}{\partial T}\right)_P. \tag{16.13}$$

These formulae show that, according as the thermal expansion coefficient $(\partial V/\partial T)_P$ is positive or negative, the temperature of the body falls or rises in an adiabatic expansion.†

Let us next calculate the adiabatic compressibility $(\partial V/\partial P)_S$ of the body, writing

$$\left(\frac{\partial V}{\partial P}\right)_S = \frac{\partial(V, S)}{\partial(P, S)} = \frac{\partial(V, S)/\partial(V, T)}{\partial(P, S)/\partial(P, T)}\cdot\frac{\partial(V, T)}{\partial(P, T)} = \frac{(\partial S/\partial T)_V}{(\partial S/\partial T)_P}\cdot\left(\frac{\partial V}{\partial P}\right)_T$$

or

$$\left(\frac{\partial V}{\partial P}\right)_S = \frac{C_v}{C_p}\left(\frac{\partial V}{\partial P}\right)_T. \tag{16.14}$$

The inequality $C_p > C_v$ therefore implies that the adiabatic compressibility is always smaller in absolute value than the isothermal compressibility.

Using formulae (16.9) and (16.10), we can derive from (16.14) the relations

$$\left(\frac{\partial V}{\partial P}\right)_S = \left(\frac{\partial V}{\partial P}\right)_T + \frac{T}{C_p}\left[\left(\frac{\partial V}{\partial T}\right)_P\right]^2, \tag{16.15}$$

$$\left(\frac{\partial P}{\partial V}\right)_S = \left(\frac{\partial P}{\partial V}\right)_T - \frac{T}{C_v}\left[\left(\frac{\partial P}{\partial T}\right)_V\right]^2. \tag{16.16}$$

† In §21 it will be shown rigorously that C_v is always positive, and therefore so is C_p

§17. The thermodynamic scale of temperature

We shall show how a thermodynamic scale of temperature may be constructed, at least in principle, using for this purpose an arbitrary body whose equation of state is not assumed known *a priori*. The problem is thus to establish by means of this body the relation $T = T(\tau)$ between the absolute scale of temperature T and some purely arbitrary scale τ defined by an arbitrarily calibrated "thermometer".

To do this, we start from the following relation (in which all quantities refer to the body in question):

$$(\partial Q/\partial P)_T = T(\partial S/\partial P)_T = -T(\partial V/\partial T)_P,$$

where (16.4) has been used. Since τ and T are in one-to-one relation, it does not matter whether the derivative is written for constant T or constant τ. The derivative $(\partial V/\partial T)_P$ may be written as

$$\left(\frac{\partial V}{\partial T}\right)_P = \left(\frac{\partial V}{\partial \tau}\right)_P \frac{d\tau}{dT}.$$

Then

$$\left(\frac{\partial Q}{\partial P}\right)_\tau = -T\left(\frac{\partial V}{\partial \tau}\right)_P \frac{d\tau}{dT},$$

or

$$\frac{d \log T}{d\tau} = -\frac{(\partial V/\partial \tau)_P}{(\partial Q/\partial P)_\tau}. \tag{17.1}$$

The right-hand side involves quantities which can be measured directly as functions of the arbitrary temperature τ: $(\partial Q/\partial P)_\tau$ is the quantity of heat which must be supplied to the body in order to maintain its temperature constant during expansion, and the derivative $(\partial V/\partial \tau)_P$ is determined by the change in volume of the body on heating. Thus formula (17.1) gives the solution of the problem and can be used to determine the required relation $T = T(\tau)$.

Here it must be remembered that the integration of (17.1) determines $\log T$ only to within an additive constant. The temperature T is therefore determined only to within an arbitrary constant factor. This is as it should be, of course: the choice of the units of measurement of the absolute temperature remains arbitrary, which is equivalent to the presence of an arbitrary factor in the function $T = T(\tau)$.

§18. The Joule–Thomson process

Let us consider a process which consists in a gas (or liquid) at pressure P_1 being steadily transferred to a vessel where its pressure is P_2. By "steadily" we mean that the pressures P_1 and P_2 remain constant throughout the process.

Such a process may be diagrammatically represented as a passage of the gas through a porous partition (a in Fig. 1), the constancy of pressure on either side of the partition being maintained by pistons moving inward and outward in an appropriate manner. If the holes in the partition are sufficiently small, the macroscopic flow velocity of the gas may be taken as zero. We shall also assume that the gas is thermally isolated from the external medium.

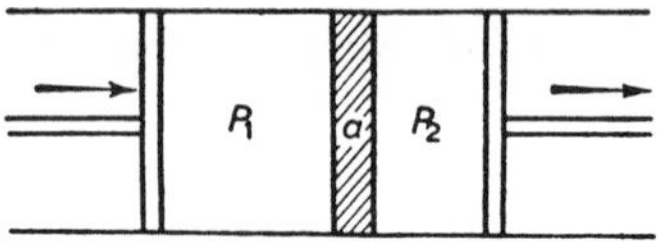

FIG. 1

This process is called a *Joule–Thomson process*. It must be emphasised that it is an irreversible process, as may be seen simply from the presence of the partition with very small holes, which creates a large amount of friction and destroys the velocity of the gas.

Let a quantity of gas, occupying a volume V_1 at pressure P_1, pass (thermally isolated) into the volume V_2, the pressure becoming equal to P_2. The change in energy $E_2 - E_1$ of this gas is equal to the work P_1V_1 done on the gas to move it out of the volume V_1, minus the work P_2V_2 done by the gas in occupying the volume V_2 at pressure P_2. Thus $E_2 - E_1 = P_1V_1 - P_2V_2$, or

$$E_1 + P_1V_1 = E_2 + P_2V_2,$$

that is,

$$W_1 = W_2. \tag{18.1}$$

Thus the heat function of the gas is conserved in a Joule–Thomson process.

The change in temperature caused by a small change of pressure in a Joule–Thomson process is given by the derivative $(\partial T/\partial P)_W$ taken with the heat function constant. We may transform this derivative to independent variables P and T:

$$\left(\frac{\partial T}{\partial P}\right)_W = \frac{\partial(T, W)}{\partial(P, W)} = \frac{\partial(T, W)/\partial(P, T)}{\partial(P, W)/\partial(P, T)} = -\frac{(\partial W/\partial P)_T}{(\partial W/\partial T)_P},$$

whence, by means of formulae (14.7) and (16.7), we obtain

$$\left(\frac{\partial T}{\partial P}\right)_W = \frac{1}{C_p}\left[T\left(\frac{\partial V}{\partial T}\right)_P - V\right]. \tag{18.2}$$

The change in entropy is given by the derivative $(\partial S/\partial P)_W$. From the relation $dW = T\,dS + V\,dP$, written in the form $dS = dW/T - V\,dP/T$, we have

$$(\partial S/\partial P)_W = -V/T. \tag{18.3}$$

This quantity is always negative, as it should be: the change of a gas to a lower pressure by an irreversible Joule–Thomson process results in an increase in entropy.

We may add a few words concerning a process in which a gas originally in one of two communicating vessels expands into the other vessel; this process, of course, is not a steady one, the pressures in the two vessels varying until they become equal. When a gas expands into a vacuum in this way, its energy E is conserved. If, as a result of the expansion, the total volume is changed only slightly, the change in temperature is given by the derivative $(\partial T/\partial V)_E$. On converting this derivative to independent variables V, T, we obtain the formula

$$\left(\frac{\partial T}{\partial V}\right)_E = \frac{1}{C_v}\left[P - T\left(\frac{\partial P}{\partial T}\right)_V\right]. \tag{18.4}$$

The change in entropy is given by

$$(\partial S/\partial V)_E = P/T. \tag{18.5}$$

The entropy increases on expansion, i.e. with increasing V, as it should.

§19. Maximum work

Let us consider a thermally isolated system consisting of several bodies not in thermal equilibrium with one another. While equilibrium is being established, the system may do work on some external objects. The transition to equilibrium may, however, occur in different ways, and the final equilibrium states of the system will of course also be different; in particular, its energy and entropy will be different.

Accordingly, the total work which can be got from a non-equilibrium system will depend on the manner in which equilibrium is established, and we may ask how the equilibrium state must be reached in order that the system should do the maximum possible amount of work. Here we are concerned with the work done because the system is not in equilibrium; that is, we must exclude any work done by a general expansion of the system, since this work could also be done by a system in equilibrium. We shall therefore assume that the total volume of the system is unchanged by the process (although it may vary during the process).

Let the original energy of the system be E_0, and the energy in the equilibrium state, as a function of the entropy of the system in that state, be $E(S)$. Since the system is thermally isolated, the work which it does is just the change in energy:

$$|R| = E_0 - E(S);$$

we write $|R|$, since $R < 0$ in accordance with convention if work is done by the system.

Differentiating $|R|$ with respect to the entropy S of the final state, we have

$$\partial|R|/\partial S = -(\partial E/\partial S)_V = -T,$$

where T is the temperature of the final state; the derivative is taken with the volume of the system in its final state constant (the same as in the initial state). We see that this derivative is negative, i.e. $|R|$ decreases with increasing S. The entropy of a thermally isolated system cannot decrease, and the greatest possible value of $|R|$ therefore occurs if S remains constant throughout the process.

Thus we conclude that the system does maximum work when its entropy remains constant, i.e. when the process of reaching equilibrium is reversible.

Let us determine the maximum work which can be done when a small quantity of energy is transferred between two bodies at different temperatures T_1 and T_2, with $T_2 > T_1$. First of all, it must be emphasised that, if the energy transfer occurred directly between the bodies on contact, no work would be done. The process would be irreversible, the entropy of the two bodies increasing by $\delta E(1/T_1 - 1/T_2)$, where δE is the amount of energy transferred.

Consequently, in order to achieve a reversible transfer of energy and so maximise the work, some further body (the *working medium*) must be brought into the system and caused to execute a reversible cyclic process. This process must be carried out in such a way that the bodies between which direct transfer of energy occurs are at the same temperature. The working medium at temperature T_2 is brought into contact with the body at that temperature and receives a certain amount of energy from it isothermally. It is then adiabatically cooled to T_1, releases energy at this temperature to the body at T_1, and finally is adiabatically returned to its original state. In the expansions involved in this process the working medium does work on external objects. The cyclic process just described is called a *Carnot cycle.*

To calculate the resulting maximum work, we first note that the working medium may be ignored, since it is returned to its initial state at the end of the process. Let the hotter body 2 lose an amount of energy $-\delta E_2 = -T_2\,\delta S_2$, and body 1 gain energy $\delta E_1 = T_1\,\delta S_1$. Since the process is reversible, the sum of the entropies of the two bodies remains constant, i.e. $\delta S_1 = -\delta S_2$. The work done is equal to the decrease in the total energy of the two bodies, i.e.

$$\begin{aligned}|\delta R|_{\max} &= -\delta E_1 - \delta E_2 = -T_1\,\delta S_1 - T_2\,\delta S_2\\ &= -(T_2 - T_1)\,\delta S_2,\end{aligned}$$

or

$$|\delta R|_{\max} = \frac{T_2 - T_1}{T_2}\,|\delta E_2|. \tag{19.1}$$

The ratio of the work done to the amount of energy expended is called the *efficiency* η. The maximum efficiency when energy is transferred from a hotter to a cooler body is, from (19.1),

$$\eta_{\text{max}} = (T_2 - T_1)/T_2. \tag{19.2}$$

A more convenient quantity is the *utilisation coefficient n*, defined as the ratio of the work done to the maximum work which can be obtained in given conditions. Clearly

$$n = \eta/\eta_{\text{max}}. \tag{19.3}$$

§20. Maximum work done by a body in an external medium

Let us now consider a different formulation of the maximum-work problem. Let a body be in an external medium whose temperature T_0 and pressure P_0 differ from the temperature T and pressure P of the body. The body can do work on some object, assumed thermally isolated both from the medium and from the body. The medium, together with the body in it and the object on which work is done, forms a closed system. The volume and energy of the medium are so large that the change in these quantities due to processes involving the body does not lead to any appreciable change in the temperature and pressure of the medium, which may therefore be regarded as constant.

If the medium were absent, the work done by the body on the thermally isolated object, for a given change in state of the body (i.e. for given initial and final states) would be completely defined, and equal to the change in the energy of the body. The presence of the medium which also takes part in the process makes the result indefinite, and the question arises of the maximum work which the body can do for a given change in its state.

If a body does work on an external object in a transition from one state to another, then in the reverse transition from the second state to the first some external source of work must do work on the body. A transition in which the body does the maximum work $|R|_{\text{max}}$ corresponds to a reverse transition which requires the external source to do the minimum work R_{min}. These must obviously be the same, so that the calculation of the one is equivalent to that of the other, and we shall speak below of the work done on the body by a thermally isolated external source of work.

During the process, the body may exchange heat and work with the medium. The work done on the body by the medium must of course be subtracted from the total work done on the body, since we are concerned only with the work done by the external source. Thus the total change ΔE in the energy of the body in some (not necessarily small) change in its state consists of three parts: the work R done on the body by the external source, the work done by the medium, and the heat gained from the medium. As already mentioned, owing to the large size of the medium its temperature and pressure may be taken as constant, and the work done by it on the body is therefore $P_0 \Delta V_0$, while the heat given up by it is $-T_0 \Delta S_0$ (the suffix zero indicates quantities pertaining to the medium, while those for the body have no suffix). Thus

$$\Delta E = R+P_0\Delta V_0-T_0\Delta S_0.$$

Since the total volume of the medium and the body remains constant, $\Delta V_0 = -\Delta V$, and the law of increase of entropy shows that $\Delta S+\Delta S_0 \geqslant 0$; the entropy of the thermally isolated source of work does not vary. Thus $\Delta S_0 \geqslant -\Delta S$. From $R = \Delta E-P_0\Delta V_0+T_0\Delta S_0$ we therefore find

$$R \geqslant \Delta E-T_0\Delta S+P_0\Delta V. \tag{20.1}$$

The equality occurs for a reversible process. Thus we again conclude that the change occurs with minimum expenditure of work, and the reverse change with maximum work, if it occurs reversibly. The value of the minimum work is

$$R_{\text{min}} = \Delta(E-T_0S+P_0V) \tag{20.2}$$

(T_0 and P_0, being constants, can be placed after Δ), i.e. this work is equal to the change in the quantity $E-T_0S+P_0V$. For maximum work the formula must be written with the opposite sign:

$$|R|_{\text{max}} = -\Delta(E-T_0S+P_0V), \tag{20.3}$$

since the initial and final states are interchanged.

If the body is in an equilibrium state at every instant during the process (but not, of course, in equilibrium with the medium), then for an infinitesimal change in its state formula (20.2) may be written differently. Substituting $\mathrm{d}E = T\,\mathrm{d}S-P\,\mathrm{d}V$ in $\mathrm{d}R_{\text{min}} = \mathrm{d}E-T_0\,\mathrm{d}S+P_0\,\mathrm{d}V$, we find

$$\mathrm{d}R_{\text{min}} = (T-T_0)\,\mathrm{d}S-(P-P_0)\,\mathrm{d}V. \tag{20.4}$$

Two important particular cases may be noted. If the volume and temperature of the body remain constant, the latter being equal to the temperature of the medium, (20.2) gives $R_{\text{min}} = \Delta(E-TS)$, or

$$R_{\text{min}} = \Delta F, \tag{20.5}$$

i.e. the minimum work is equal to the change in the free energy of the body. Secondly, if the temperature and pressure of the body are constant and equal to T_0 and P_0, we have

$$R_{\text{min}} = \Delta\Phi, \tag{20.6}$$

i.e. the work done by the external source is equal to the change in the thermodynamic potential of the body.

It should be emphasised that in both these particular cases the body concerned must be one not in equilibrium, so that its state is not defined by T and V (or P) alone; otherwise, the constancy of these quantities would mean that no process could occur at all. We must consider, for example, a chemical reaction in a mixture of reacting substances, a process of dissolution, or the like.

Let us now assume that a body in an external medium is left to itself and no work is done on it. Spontaneous irreversible processes will occur in the body and bring it into equilibrium. In the inequality (20.1) we must now put $R = 0$, and so

$$\Delta(E - T_0 S + P_0 V) \leqslant 0. \tag{20.7}$$

This means that the processes occurring in the body will cause the quantity $E - T_0 S + P_0 V$ to decrease, and it will reach a minimum at equilibrium.

In particular, for spontaneous processes at constant temperature $T = T_0$ and constant pressure $P = P_0$, the thermodynamic potential Φ of the body decreases, and for processes at constant temperature $T = T_0$ and constant volume of the body its free energy F decreases. These results have already been derived by a different approach in §15. It may be noted that the derivation given here does not essentially assume that the temperature and volume (or pressure) of the body remain constant throughout the process: we may say that the thermodynamic potential (or free energy) of a body decreases as a result of any process for which the initial and final temperature and pressure (or volume) are the same (and equal to the temperature and pressure of the medium), even if they vary during the process.

Another thermodynamic significance may also be ascribed to the minimum work. Let S_t be the total entropy of the body and the medium. If the body is in equilibrium with the medium, S_t is a function of their total energy E_t:

$$S_t = S_t(E_t).$$

If the body is not in equilibrium with the medium, their total entropy differs from $S_t(E_t)$ for the same value of the total energy E_t by some amount $\Delta S_t < 0$. In Fig. 2 the continuous line shows the function $S_t(E_t)$ and the vertical

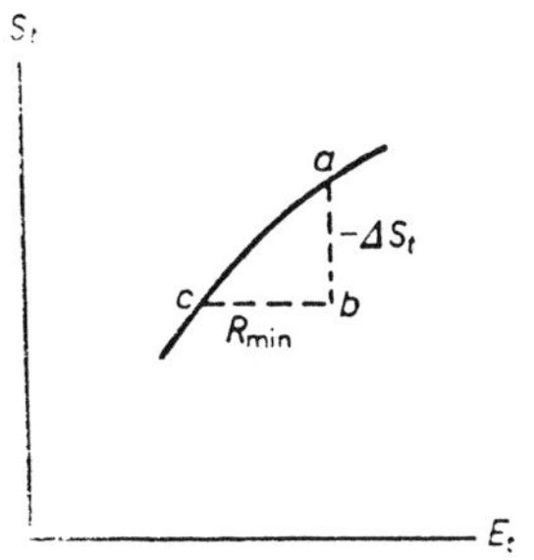

FIG. 2

segment *ab* is $-\Delta S_t$. The horizontal segment *bc* is the change in the total energy when the body goes reversibly from the state of equilibrium with the medium to the state corresponding to the point *b*. In other words, this segment represents the minimum work which must be done by some external source to bring the body from the state of equilibrium with the medium to the state considered; the equilibrium state in question (the point *c* in Fig. 2) is, of course, not the same as that corresponding to the given value of E_t (point *a*).

Since the body is a very small part of the whole system, the processes involving it cause only a negligible relative change in the total energy and entropy. Fig. 2 therefore shows that

$$\Delta S_t = -\frac{\mathrm{d}S_t(E_t)}{\mathrm{d}E_t} R_{\min}.$$

But the derivative $\mathrm{d}E_t/\mathrm{d}S_t$ is the equilibrium temperature of the system, i.e. the temperature T_0 of the medium. Thus

$$\Delta S_t = -\frac{R_{\min}}{T_0} = -\frac{1}{T_0}(\Delta E - T_0 \Delta S + P_0 \Delta V). \tag{20.8}$$

This formula determines the amount by which the entropy of a closed system (body+medium) differs from its greatest possible value if the body is not in equilibrium with the medium; ΔE, ΔS and ΔV are here the differences between the energy, entropy and volume of the body and their values in a state of complete equilibrium.

§21. Thermodynamic inequalities

In deriving the conditions of thermal equilibrium from that of maximum entropy, we have so far considered only the first derivatives. By equating to zero the derivatives with respect to energy and volume, we have deduced in §§9 and 12 the equality of temperature and pressure in all parts of the body as the conditions of equilibrium. But the vanishing of the first derivatives is only a necessary condition for an extremum and does not ensure that the entropy is in fact a maximum. The determination of the sufficient conditions for a maximum involves, of course, an examination of the second derivative of the function.

Such an examination is, however, more conveniently carried out not from the condition of maximum entropy of a closed system but from another equivalent condition.† Let us consider some small but macroscopic part of the

† As regards the dependence of the entropy on the momenta of macroscopic motion, we have already investigated the conditions to be imposed on both the first and the second derivatives (§10), obtaining in this way the conditions that internal macroscopic motions in the body be absent and that the temperature be positive.

body concerned. With respect to this part, the remainder of the body may be regarded as an external medium. Then, as shown in §20, we can state that in equilibrium the quantity

$$E - T_0 S + P_0 V$$

has a minimum, where E, S and V are the energy, entropy and volume of the part considered, and T_0, P_0 the temperature and pressure of the medium, i.e. of the remainder of the body. Clearly T_0 and P_0 are also the temperature and pressure of the part considered when in equilibrium.

Thus in any small deviation from equilibrium the change in the quantity $E - T_0 S + P_0 V$ must be positive, i.e.

$$\delta E - T_0\,\delta S + P_0\,\delta V > 0. \tag{21.1}$$

In other words, the minimum work which must be done to bring this part of the body from equilibrium to any neighbouring state is positive.

In what follows the equilibrium values will be implied for any coefficients appearing in the deviations of thermodynamic quantities from their equilibrium values, and the zero suffixes will therefore be omitted.

Expanding δE as a series (regarding E as a function of S and V), we have as far the second-order terms

$$\delta E = \frac{\partial E}{\partial S}\delta S + \frac{\partial E}{\partial V}\delta V + \tfrac{1}{2}\left[\frac{\partial^2 E}{\partial S^2}(\delta S)^2 + 2\frac{\partial^2 E}{\partial S \partial V}\delta S \delta V + \frac{\partial^2 E}{\partial V^2}(\delta V)^2\right].$$

But $\partial E/\partial S = T$, $\partial E/\partial V = -P$, so that the first-order terms are $T\,\delta S - P\,\delta V$, and cancel when δE is substituted in (21.1). Thus we obtain the condition

$$\frac{\partial^2 E}{\partial S^2}(\delta S)^2 + 2\frac{\partial^2 E}{\partial S \partial V}\delta S\,\delta V + \frac{\partial^2 E}{\partial V^2}(\delta V)^2 > 0. \tag{21.2}$$

If such an inequality holds for arbitrary δS and δV, two conditions must be satisfied:†

$$\partial^2 E/\partial S^2 > 0, \tag{21.3}$$

$$\frac{\partial^2 E}{\partial S^2}\frac{\partial^2 E}{\partial V^2} - \left(\frac{\partial^2 E}{\partial S \partial V}\right)^2 > 0. \tag{21.4}$$

For $\partial^2 E/\partial S^2$ we have

$$\partial^2 E/\partial S^2 = (\partial T/\partial S)_V = T/C_v.$$

The condition (21.3) therefore becomes $T/C_v > 0$, and since $T > 0$,

$$C_v > 0, \tag{21.5}$$

i.e. the specific heat at constant volume is always positive.

† The special case where the equality sign holds in (21.4) will be discussed in §84.

The condition (21.4) may be written in terms of the Jacobian

$$\frac{\partial[(\partial E/\partial S)_V, (\partial E/\partial V)_S]}{\partial(S, V)} > 0,$$

or

$$\partial(T, P)/\partial(S, V) < 0.$$

Changing to the variables T and V, we have

$$\frac{\partial(T, P)}{\partial(S, V)} = \frac{\partial(T, P)/\partial(T, V)}{\partial(S, V)/\partial(T, V)} = \frac{(\partial P/\partial V)_T}{(\partial S/\partial T)_V} = \frac{T}{C_v}\left(\frac{\partial P}{\partial V}\right)_T < 0.$$

Since $C_v > 0$, this is equivalent to the condition

$$(\partial P/\partial V)_T < 0, \tag{21.6}$$

i.e. an increase in volume at constant temperature is always accompanied by a decrease in pressure.

The conditions (21.5) and (21.6) are called *thermodynamic inequalities.* States in which these conditions are not satisfied are unstable and cannot exist in Nature.

It has already been noted in §16 that from the inequality (21.6) and formula (16.10) we always have $C_p > C_v$. From (21.5) we can therefore conclude that

$$C_p > 0 \tag{21.7}$$

always.

The fact that C_v and C_p are positive means that the energy is a monotonically increasing function of temperature at constant volume, and the heat function behaves similarly at constant pressure. The entropy increases monotonically with temperature at either constant volume or constant pressure.

The conditions (21.5), (21.6), which have been derived for an arbitrary small part of a body, are of course valid for the whole body also, since in equilibrium the temperatures and pressures of all parts of the body are the same. Here it is assumed that the body is homogeneous (only such bodies have been considered so far). It must be emphasised that the fulfilment of the conditions (21.5), (21.6) depends on the homogeneity of the body. We can, for example, consider a body whose particles are held together by gravitational forces. Such a body will clearly be inhomogeneous, having a higher density towards the centre, and the specific heat of the body as a whole may be less than zero, so that its temperature rises as its energy decreases. We may note that this does not contradict the result that the specific heat is positive for every small part of the body, since in these conditions the energy of the whole body is not equal to the sum of the energies of its parts; there is also the energy of the gravitational interaction between these parts.

The inequalities derived above are conditions of equilibrium, but their fulfilment is not sufficient for the equilibrium to be completely stable. There can exist states such that the entropy decreases for an infinitesimal deviation from the state and the body then returns to its initial state, whereas for a finite deviation the entropy may be greater than in the original state. After such a finite deviation the body does not return to its original state, but will tend to pass to some other equilibrium state corresponding to a maximum entropy greater than that in the original state. Accordingly, we must distinguish between *metastable* and *stable* equilibrium states. A body in a metastable state may not return to it after a sufficient deviation. Although a metastable state is stable within certain limits, the body will always leave it sooner or later for another state which is stable, corresponding to the greatest of the possible maxima of entropy. A body which is displaced from this state will always eventually return to it.

§22. Le Chatelier's principle

Let us consider a closed system consisting of a body and a medium surrounding it. Let S be the total entropy of the system, and y a quantity pertaining to the body, such that the condition for S to be a maximum relative to y, i.e. $\partial S/\partial y = 0$, signifies that the body itself is in equilibrium, though it is not necessarily in equilibrium with the medium. Also, let x be another thermodynamic quantity pertaining to the same body, such that if both $\partial S/\partial y = 0$ and $\partial S/\partial x = 0$ the body is not only in internal equilibrium but also in equilibrium with the medium.

We shall use the notation

$$X = -\partial S/\partial x, \qquad Y = -\partial S/\partial y. \tag{22.1}$$

In complete thermodynamic equilibrium the entropy S must be a maximum. For this, besides the conditions

$$X = 0, \qquad Y = 0, \tag{22.2}$$

the conditions

$$(\partial X/\partial x)_y > 0, \qquad (\partial Y/\partial y)_x > 0 \tag{22.3}$$

and

$$\left(\frac{\partial X}{\partial x}\right)_y \left(\frac{\partial Y}{\partial y}\right)_x - \left[\left(\frac{\partial X}{\partial y}\right)_x\right]^2 > 0 \tag{22.4}$$

must be satisfied.

Let us now assume that the equilibrium of the body with the medium is destroyed by some small external interaction, the quantity x being somewhat changed and the condition $X = 0$ no longer satisfied; we assume that y is not directly affected by the interaction in question. Let the change in x be Δx.

Then the change in X at the instant of interaction is

$$(\Delta X)_y = (\partial X/\partial x)_y \, \Delta x.$$

The change in x at constant y leads, of course, to a violation of the condition $Y = 0$ also, i.e. of internal equilibrium of the body. When equilibrium is again restored, the quantity $X \equiv \Delta X$ will be

$$(\Delta X)_{Y=0} = (\partial X/\partial x)_{Y=0} \, \Delta x,$$

where the derivative is taken at constant $Y (= 0)$.

To compare the two values of ΔX, using the properties of Jacobians, we have

$$\left(\frac{\partial X}{\partial x}\right)_{Y=0} = \frac{\partial(X, Y)}{\partial(x, Y)} = \frac{\partial(X, Y)/\partial(x, y)}{\partial(x, Y)/\partial(x, y)} = \left(\frac{\partial X}{\partial x}\right)_y - \frac{[(\partial X/\partial y)_x]^2}{(\partial Y/\partial y)_x}.$$

The denominator of the second term in this expression is positive by the condition (22.3); using also (22.4), we find that

$$(\partial X/\partial x)_y > (\partial X/\partial x)_{Y=0} > 0, \tag{22.5}$$

or

$$|(\Delta X)_y| > |(\Delta X)_{Y=0}|. \tag{22.6}$$

The inequality (22.5) or (22.6) forms the content of what is called *Le Chatelier's principle.*

We shall regard the change Δx of the quantity x as a measure of the external interaction acting on the body, and ΔX as a measure of the change in properties of the body resulting from this interaction. The inequality (22.6) shows that, when the internal equilibrium of the body is restored after the external interaction which disturbed it, the value of ΔX is reduced. Thus LE CHATELIER'S principle may be formulated as follows: an external interaction which disturbs the equilibrium brings about processes in the body which tend to reduce the effects of this interaction.

The above may be illustrated by some examples.

First of all, it is convenient to modify somewhat the definition of the quantities X and Y by using formula (20.8), according to which the change in entropy of the system (medium + body) is $-R_{\text{min}}/T_0$, where T_0 is the temperature of the medium and R_{min} the minimum work needed to bring the body from a state of equilibrium with the medium to the state in question. We can therefore write

$$X = \frac{1}{T_0}\frac{\partial R_{\text{min}}}{\partial x}, \qquad Y = \frac{1}{T_0}\frac{\partial R_{\text{min}}}{\partial y}. \tag{22.7}$$

For an infinitesimal change in the state of the body we have (see (20.4))

$$dR_{\text{min}} = (T - T_0)\, dS - (P - P_0)\, dV;$$

here and below all quantities without suffix relate to the body, and those with suffix 0 to the medium.

Let x be the entropy S of the body. Then $X = (T - T_0)/T_0$. The equilibrium condition $X = 0$ gives $T = T_0$, i.e. the temperatures of the body and the medium are equal. The inequalities (22.5) and (22.6) become

$$(\partial T/\partial S)_y > (\partial T/\partial S)_{Y=0} > 0, \tag{22.8}$$

$$|(\Delta T)_y| > |(\Delta T)_{Y=0}|. \tag{22.9}$$

The significance of these inequalities is as follows. The change in x (the entropy of the body) means that a quantity of heat is given to or taken from the body. This destroys the equilibrium of the body itself and, in particular, changes its temperature by $(\Delta T)_y$. The restoration of equilibrium in the body has the result that the absolute value of the change in temperature decreases, becoming $(\Delta T)_{Y=0}$, i.e. it is as if the result of the interaction which brings the body out of equilibrium were reduced. We can say that heating or cooling a body brings about processes in it which tend to lower or raise the temperature respectively.

Now let x be the volume V of a body. Then $X = -(P - P_0)/T$. In equilibrium $X = 0$, i.e. $P = P_0$. The inequalities (22.5) and (22.6) give

$$(\partial P/\partial V)_y < (\partial P/\partial V)_{Y=0} < 0, \tag{22.10}$$

$$|(\Delta P)_y| > |(\Delta P)_{Y=0}|. \tag{22.11}$$

If the body is disturbed from equilibrium by a change in its volume at constant temperature, then, in particular, its pressure is changed; the restoration of equilibrium in the body leads to a decrease in the absolute value of the change in pressure. Since a decrease in the volume of the body causes an increase in its pressure, and *vice versa*, we can say the decreasing or increasing the volume of a body brings about processes in it which tend to lower or raise the pressure respectively.

Later we shall meet with numerous applications of these results (to solutions, chemical reactions and so on).

It may also be noted that, if y in the inequalities (22.8) is taken to be the volume of the body, we have

$$(\partial T/\partial S)_y = (\partial T/\partial S)_V = T/C_v,$$
$$(\partial T/\partial S)_{y=0} = (\partial T/\partial S)_P = T/C_p,$$

since the condition $Y = 0$ then denotes $P = P_0$, i.e. constant pressure. Thus we again obtain the already familiar inequalities $C_p > C_v > 0$. Similarly, if in (22.10) y is taken as the entropy of a body, the condition $Y = 0$ implies that the temperature is constant, $T = T_0$, and we find

$$(\partial P/\partial V)_S < (\partial P/\partial V)_T < 0,$$

another result already known.

§23. Nernst's theorem

The fact that the specific heat C_v is positive means that the energy is a monotonically increasing function of the temperature. Conversely, when the temperature falls the energy decreases monotonically, and therefore, when the temperature has its least possible value, i.e. at absolute zero, a body must be in the state of least possible energy. If we regard the energy of a body as the sum of the energies of the parts into which it may be imagined to be divided, we can say that each of these parts will also be in the state of least energy; it is clear that the minimum value of the sum must correspond to the minimum value of each term.

Thus at absolute zero any part of the body must be in a particular quantum state, the ground state. In other words, the statistical weights of these parts are equal to unity, and therefore so is their product, i.e. the statistical weight of the macroscopic state of the body as whole. The entropy of the body, being the logarithm of its statistical weight, is therefore zero.

We consequently reach the important result that the entropy of any body vanishes at the absolute zero of temperature. This is called *Nernst's theorem*†.

It should be emphasised that this theorem is a deduction from quantum statistics, in which the concept of discrete quantum states is of essential importance. The theorem can not be proved in purely classical statistics, where the entropy is determined only to within an arbitrary additive constant (see §7).

NERNST's theorem enables us to draw conclusions also concerning the behaviour of certain other thermodynamic quantities as $T \to 0$.

For instance, it is easy to see that for $T = 0$ the specific heats C_p and C_v both vanish:

$$C_p = C_v = 0 \quad \text{for} \quad T = 0. \tag{23.1}$$

This follows immediately from the definition of the specific heat in the form

$$\begin{aligned} C &= T\partial S/\partial T \\ &= \partial S/\partial \log T. \end{aligned}$$

When $T \to 0$, $\log T \to \infty$, and since S tends to a finite limit, namely zero, it is clear that the derivative tends to zero.

The thermal expansion coefficient also tends to zero:

$$(\partial V/\partial T)_P = 0 \quad \text{for} \quad T = 0. \tag{23.2}$$

For this derivative is equal to the derivative $-(\partial S/\partial P)_T$ (see (16.4)), which vanishes for $T = 0$, since $S = 0$ for $T = 0$ and any pressure.

† To avoid misunderstandings we should emphasise that this refers to the temperature tending to zero with other conditions remaining unchanged—say at constant volume, or at constant pressure. If, on the other hand, the temperature of a gas tends to zero while its density decreases without limit, for example, the entropy need not tend to zero.

Similarly, we can see that

$$(\partial P/\partial T)_V = 0 \quad \text{for} \quad T = 0. \tag{23.3}$$

The entropy usually vanishes, for $T \to 0$, according to a power law, i.e. as $S = aT^n$, where a is a function of pressure or volume. In this case, clearly, the specific heats and $(\partial V/\partial T)_P$, $(\partial P/\partial T)_V$ will tend to zero in the same way (with the same value of n).

Finally, it may be seen that the difference $C_p - C_v$ tends to zero more rapidly than the specific heats themselves, i.e.

$$(C_p - C_v)/C_p = 0 \quad \text{for} \quad T = 0. \tag{23.4}$$

For let the entropy tend to zero as $S \sim T^n$ for $T \to 0$. From formula (16.9) we then see that $C_p - C_v \sim T^{2n+1}$, so that $(C_p - C_v)/C_p \sim T^{n+1}$; it should be borne in mind that the compressibility $(\partial V/\partial P)_T$ is in general finite and not zero when $T = 0$.

If the specific heat of a body is known for all temperatures, the entropy can be calculated by integration, and NERNST's theorem gives the value of the constant of integration. For example, the dependence of the entropy on temperature for a given pressure is determined by

$$S = \int_0^T (C_p/T)\,\mathrm{d}T. \tag{23.5}$$

The corresponding formula for the heat function is

$$W = W_0 + \int_0^T C_p\,\mathrm{d}T, \tag{23.6}$$

where W_0 is the value of the heat function for $T = 0$. Similarly, for the thermodynamic potential $\Phi = W - TS$ we have

$$\Phi = W_0 + \int_0^T C_p\,\mathrm{d}T - T\int_0^T \frac{C_p}{T}\,\mathrm{d}T. \tag{23.7}$$

§24. The dependence of the thermodynamic quantities on the number of particles

As well as the energy and entropy, such thermodynamic quantities as F, Φ and W also have the property of additivity, as follows directly from their definitions if we bear in mind that the pressure and temperature are constant throughout a body in equilibrium. From this property we can draw certain conclusions concerning the manner in which each of these quantities depends on the number of particles in the body. Here we shall consider bodies consisting of identical particles (molecules); all the results can be immediately generalised to mixtures of different particles (see §86).

The additivity of a quantity signifies that, when the amount of matter (and therefore the number N of particles) is changed by a given factor, the quantity is changed by the same factor. In other words, we can say that an additive thermodynamic quantity must be a homogeneous function of the first order with respect to the additive variables.

Let us express the energy of the body as a function of the entropy, volume, and number of particles. Since S and V are themselves additive, this function must be of the form

$$E = Nf(S/N,\ V/N), \tag{24.1}$$

the most general homogeneous function of the first order in N, S and V.

The free energy F is a function of N, T and V. Since the temperature is constant throughout the body, and the volume is additive, a similar argument gives

$$F = Nf(V/N,\ T). \tag{24.2}$$

In exactly the same way we have for the heat function W, expressed as a function of N, S and the pressure P,

$$W = Nf(S/N,\ P). \tag{24.3}$$

Finally, the thermodynamic potential as a function of N, P and T is

$$\Phi = Nf(P,\ T). \tag{24.4}$$

In the foregoing discussion we have essentially regarded the number of particles as a parameter which has a given constant value for each body. We shall now formally consider N as a further independent variable. Then the expressions for the differentials of the thermodynamic potentials must include terms proportional to $\mathrm{d}N$. For example, the total differential of the energy will be written

$$\mathrm{d}E = T\,\mathrm{d}S - P\,\mathrm{d}V + \mu\,\mathrm{d}N, \tag{24.5}$$

where μ denotes the partial derivative

$$\mu = (\partial E/\partial N)_{S,\ V}. \tag{24.6}$$

The quantity μ is called the *chemical potential* of the body. Similarly we have

$$\mathrm{d}W = T\,\mathrm{d}S + V\,\mathrm{d}P + \mu\,\mathrm{d}N, \tag{24.7}$$

$$\mathrm{d}F = -S\,\mathrm{d}T - P\,\mathrm{d}V + \mu\,\mathrm{d}N, \tag{24.8}$$

$$\mathrm{d}\Phi = -S\,\mathrm{d}T + V\,\mathrm{d}P + \mu\,\mathrm{d}N, \tag{24.9}$$

with the same μ. These formulae show that

$$\mu = (\partial W/\partial N)_{S,\,P} = (\partial F/\partial N)_{T,\,V} = (\partial \Phi/\partial N)_{P,\,T}, \tag{24.10}$$

i.e. the chemical potential can be obtained by differentiating any of the quantities E, W, F and Φ with respect to the number of particles, but the result is expressed in terms of different variables in each case.

Differentiating Φ in the form (24.4), we find that $\mu = \partial\Phi/\partial N = f(P, T)$, i.e.

$$\Phi = N\mu. \tag{24.11}$$

Thus the chemical potential of a body (consisting of identical particles) is just its thermodynamic potential per molecule. When expressed as a function of P and T, the chemical potential is independent of N. Thus we can immediately write down for the differential of the chemical potential

$$\mathrm{d}\mu = -s\,\mathrm{d}T + v\,\mathrm{d}P, \tag{24.12}$$

where s and v are the entropy and volume per molecule.

If we consider (as we have usually done hitherto) a definite amount of matter, the number of particles in it is a given constant, while the volume is variable. Let us now take a certain volume within the body, and consider the matter enclosed therein; the number of particles N will now be variable, and the volume V constant. Then, for example, equation (24.8) reduces to

$$\mathrm{d}F = -S\,\mathrm{d}T + \mu\,\mathrm{d}N.$$

Here the independent variables are T and N. We may define a thermodynamic potential such that the second independent variable is μ, not N. To do so, we substitute $\mu\,\mathrm{d}N = \mathrm{d}(\mu N) - N\,\mathrm{d}\mu$, obtaining

$$\mathrm{d}(F - \mu N) = -S\,\mathrm{d}T - N\,\mathrm{d}\mu.$$

But $\mu N = \Phi$, and $F - \Phi = -PV$. Thus the new thermodynamic potential (denoted by Ω) is just

$$\Omega = -PV, \tag{24.13}$$

and

$$\mathrm{d}\Omega = -S\,\mathrm{d}T - N\,\mathrm{d}\mu. \tag{24.14}$$

The number of particles is obtained by differentiating Ω with respect to the chemical potential at constant temperature and volume:

$$N = -(\partial\Omega/\partial\mu)_{T,\,V} = V(\partial P/\partial\mu)_{T,\,V}. \tag{24.15}$$

In the same way as we proved the equality of small changes in E, W, F and Φ (with the appropriate pairs of quantities constant (see (15.12)), we can easily show that the change $(\delta\Omega)_{T,\mu,V}$ at constant T, μ, V has the same property:

$$(\delta E)_{S,\,V,\,N} = (\delta F)_{T,\,V,\,N} = (\delta\Phi)_{T,\,P,\,N} = (\delta W)_{S,\,P,\,N} = (\delta\Omega)_{T,\,V,\,\mu}. \tag{24.16}$$

Finally, as in §§15 and 20 for the free energy and the thermodynamic potential, we may show that the work in a reversible process occurring at constant T, V and μ is equal to the change in the potential Ω. In a state of thermal equilibrium the potential Ω is a minimum with respect to any change of state at constant T, V, μ.

PROBLEM

Derive an expression for the specific heat C_v in terms of the variables T, μ, V.

SOLUTION. We transform the derivative $C_v = T(\partial S/\partial T)_{V,N}$ to the variables T, V, μ, writing (with V regarded as a constant throughout)

$$\left(\frac{\partial S}{\partial T}\right)_N = \frac{\partial(S, N)}{\partial(T, N)} = \frac{\partial(S, N)/\partial(T, \mu)}{\partial(T, N)/\partial(T, \mu)} = \left(\frac{\partial S}{\partial T}\right)_\mu - \frac{(\partial S/\partial\mu)_T(\partial N/\partial T)_\mu}{(\partial N/\partial\mu)_T}.$$

But $(\partial S/\partial\mu)_T = -\partial^2\Omega/\partial T\partial\mu = (\partial N/\partial T)_\mu$, and therefore

$$C_v = T\left\{\left(\frac{\partial S}{\partial T}\right)_\mu - \frac{[(\partial N/\partial T)_\mu]^2}{(\partial N/\partial\mu)_T}\right\}.$$

§25. Equilibrium of a body in an external field

Let us consider a body in an external field which is constant in time. The different parts of the body are in different conditions, and the body will therefore be inhomogeneous. One of the conditions of equilibrium of such a body is again that the temperature should be constant throughout it, but the pressure will now vary from point to point.

To derive the second condition of equilibrium, let us consider two adjoining volumes in the body and maximise their entropy $S = S_1+S_2$ when the remainder of the body is in a fixed state. One necessary condition for a maximum is that the derivative $\partial S/\partial N_1$ should be zero. Since the total number of particles N_1+N_2 in these two parts of the body is regarded as constant, we have

$$\frac{\partial S}{\partial N_1} = \frac{\partial S_1}{\partial N_1} + \frac{\partial S_2}{\partial N_2}\frac{\partial N_2}{\partial N_1} = \frac{\partial S_1}{\partial N_1} - \frac{\partial S_2}{\partial N_2} = 0.$$

The equation $\mathrm{d}E = T\,\mathrm{d}S + \mu\,\mathrm{d}N$, written in the form

$$\mathrm{d}S = \frac{\mathrm{d}E}{T} - \frac{\mu}{T}\,\mathrm{d}N,$$

shows that the derivative $\partial S/\partial N$ for constant E and T is $-\mu/T$. Thus $\mu_1/T_1 = \mu_2/T_2$. But in equilibrium $T_1 = T_2$, so that $\mu_1 = \mu_2$. We therefore conclude that in equilibrium in an external field, in addition to the constancy of temperature, we must have

$$\mu = \text{constant}, \tag{25.1}$$

i.e. the chemical potential of every part of the body must be the same. The chemical potential of each part is a function of its temperature and pressure, as well as of the parameters which define the external field. If there is no field, the constancy of μ and T necessarily implies that of the pressure.

In a gravitational field the potential energy u of a molecule is a function only of the co-ordinates x, y, z of its centre of gravity (and not of the arrangement of the atoms within the molecule). In this case the change in the thermodynamic quantities for the body amounts to adding to its energy the potential energy of the molecules in the field. In particular, the chemical potential (the thermodynamic potential per molecule) has the form $\mu = \mu_0 + u(x, y, z)$, where $\mu_0(P, T)$ is the chemical potential in the absence of the field. Thus the condition of equilibrium in a gravitational field may be written

$$\mu_0(P, T) + u(x, y, z) = \text{constant}. \tag{25.2}$$

In particular, in a uniform gravitational field $u = mgz$ (where m is the mass of a molecule, g the acceleration due to gravity, and z the vertical co-ordinate). Differentiating equation (25.2) with respect to the co-ordinate z at constant temperature, we have $v\,\mathrm{d}P = -mg\,\mathrm{d}z$, where $v = (\partial\mu_0/\partial P)_T$ is the specific volume. For small changes in pressure, v may be regarded as constant. Substituting the density $\varrho = m/v$ and integrating, we obtain

$$P = \text{constant} - \varrho g z,$$

the customary formula for the hydrostatic pressure in an incompressible fluid.

§26. Rotating bodies

In a state of thermal equilibrium, as we have seen in §10, only a uniform translational motion and a uniform rotation of a body as a whole are possible. The uniform translational motion needs no special treatment, since by GALILEO's relativity principle it has no effect on the mechanical properties of the body, nor therefore on its thermodynamic properties, and the thermodynamic quantities are unchanged except that the energy of the body is increased by its kinetic energy.

Let us consider a body in uniform rotation round a fixed axis with angular velocity $\boldsymbol{\Omega}$. Let $E(p, q)$ be the energy of the body in a fixed co-ordinate system and $E'(p, q)$ the energy in a co-ordinate system rotating with the body. We know from mechanics that these quantities are related by

$$E'(p, q) = E(p, q) - \boldsymbol{\Omega}\cdot\mathbf{M}(p, q), \tag{26.1}$$

where $\mathbf{M}(p, q)$ is the angular momentum of the body.†

† See *Mechanics*, §39. Although the derivation of formula (39.13) is based on classical mechanics, in quantum theory exactly the same relations apply to the operators of the corresponding quantities. Hence all the thermodynamic relations derived below are independent of which mechanics describes the motion of the particles in the body.

Thus the energy $E'(p, q)$ depends on the angular velocity $\boldsymbol{\Omega}$ as a parameter, and

$$\partial E'(p, q)/\partial \boldsymbol{\Omega} = -\mathbf{M}(p, q).$$

Averaging this equation over the statistical distribution and using formula (11.3), we obtain

$$(\partial E'/\partial \boldsymbol{\Omega})_S = -\mathbf{M}, \tag{26.2}$$

where $E' = \overline{E'(p, q)}$, $\mathbf{M} = \overline{\mathbf{M}(p, q)}$ are the mean (thermodynamic) energy and angular momentum of the body.

From this relation we can write down the differential of the energy of a rotating body of given volume:

$$\mathrm{d}E' = T\,\mathrm{d}S - \mathbf{M}\cdot\mathrm{d}\boldsymbol{\Omega}. \tag{26.3}$$

Similarly, for the free energy $F' = E' - TS$ (in the rotating co-ordinate system) we have

$$\mathrm{d}F' = -S\,\mathrm{d}T - \mathbf{M}\cdot\mathrm{d}\boldsymbol{\Omega}. \tag{26.4}$$

Averaging equation (26.1) gives

$$E' = E - \mathbf{M}\cdot\boldsymbol{\Omega}. \tag{26.5}$$

Differentiating this equation and substituting (26.3), we obtain the differential of the energy in the fixed co-ordinate system:

$$\mathrm{d}E = T\,\mathrm{d}S + \boldsymbol{\Omega}\cdot\mathrm{d}\mathbf{M}. \tag{26.6}$$

Correspondingly, for the free energy $F = E - TS$

$$\mathrm{d}F = -S\,\mathrm{d}T + \boldsymbol{\Omega}\cdot\mathrm{d}\mathbf{M}. \tag{26.7}$$

Thus in these relations the independent variable is not the angular velocity but the angular momentum, and

$$\boldsymbol{\Omega} = (\partial E/\partial \mathbf{M})_S = (\partial F/\partial \mathbf{M})_T. \tag{26.8}$$

As we know from mechanics, a uniform rotation is in a certain sense equivalent to the presence of two fields of force, centrifugal and Coriolis. The centrifugal forces are proportional to the size of the body, as they involve the distance from the axis of rotation; the Coriolis forces are independent of the size of the body. For this reason the effect of the Coriolis forces on the thermodynamic properties of a rotating macroscopic body is entirely negligible in comparison with that of the centrifugal forces, and the former can usually be neglected.† The condition of thermal equilibrium of a rotating body is therefore obtained by simply substituting for $u(x, y, z)$ in (25.2) the centrifugal energy of the particles:

$$\mu_0(P, T) - \tfrac{1}{2}m\Omega^2 r^2 = \text{constant}, \tag{26.9}$$

† It may be shown that in classical statistics the Coriolis forces do not affect the statistical properties of the body; see §34.

where μ_0 is the chemical potential of the body at rest, m the mass of a molecule, and r the distance from the axis of rotation.

For the same reason, the total energy E of a rotating body may be written as the sum of its internal energy (here denoted by E_{in}) and its kinetic energy of rotation:

$$E = E_{\text{in}} + M^2/2I, \tag{26.10}$$

where I is the moment of inertia of the body with respect to the axis of rotation. It should be remembered that rotation in general changes the distribution of mass in the body, and so the moment of inertia and internal energy of the body are themselves in general functions of Ω (or of M). They may be regarded as constants independent of Ω only when the rotation is sufficiently slow.

Let us consider an isolated uniformly rotating solid with a given mass distribution. Since the entropy of a body is a function of its internal energy, we have in this case $S = S(E - M^2/2I)$. Because the body is a closed system, its total energy and angular momentum are conserved, and the entropy must have the maximum value possible for the given M and E. We therefore conclude that the equilibrium rotation of the body takes place about the axis with respect to which the moment of inertia has the greatest possible value. This assumes that the axis of rotation is necessarily a principal axis of inertia of the body, but the latter result is evident: if the body rotates about an axis other than a principal axis of inertia, then, as we know from mechanics, the axis of rotation will itself precess in space, and the rotation will be non-uniform, and therefore not an equilibrium rotation.

§27. Thermodynamic relations in the relativistic region

Relativistic mechanics leads to a number of changes in the usual thermodynamic relations. Here we shall discuss the most interesting of these changes.

If the microscopic motion of the particles forming a body becomes relativistic, the general thermodynamic relations are unchanged, but the application of relativity theory to this case leads to an important inequality between the pressure and energy of the body:

$$P < E/3V, \tag{27.1}$$

where E is the energy of the body including the rest energy of the particles in it.†

The changes caused by the general theory of relativity in the conditions of thermal equilibrium, taking account of the gravitational field of the body itself, are of fundamental importance. Let us consider a macroscopic body at

† See *The Classical Theory of Fields*, §35.

rest; its gravitational field is, of course, constant. In a constant gravitational field we must distinguish the conserved energy E_0 of any small part of the body from the energy E measured by an observer situated at a given point. These two quantities are related by†

$$E_0 = E\sqrt{-g_{00}},$$

where g_{00} is the time component of the metric tensor. But, from the sense of the proof given in §9 that the temperature is constant throughout a body in equilibrium, it is clear that the quantity obtained by differentiating the entropy with respect to the conserved energy E_0 must be constant. The temperature T measured by an observer situated at a given point in space is, however, obtained by differentiating the entropy with respect to the energy E, and will therefore be different at different points in the body.

To derive a quantitative relation, we note that the entropy, by definition, depends only on the internal state of the body and so is unchanged by the presence of a gravitational field (provided that this field does not affect the internal properties of the body, a condition which is always satisfied in practice). The derivative with respect to entropy of the conserved energy E_0 is therefore $T\sqrt{-g_{00}}$, and so one of the conditions of thermal equilibrium is that

$$T\sqrt{-g_{00}} = \text{constant} \tag{27.2}$$

throughout the body.

A similar change occurs in the second condition of equilibrium, the constancy of the chemical potential. The latter is defined as the derivative of the energy with respect to the number of particles. Since this number is of course unaffected by a gravitational field, we have for the chemical potential measured at any given point a relation of the same kind as for the temperature:

$$\mu\sqrt{-g_{00}} = \text{constant}. \tag{27.3}$$

We may note that the relations (27.2), (27.3) may be written

$$\begin{aligned} T &= \text{constant} \times dx^0/ds, \\ \mu &= \text{constant} \times dx^0/ds, \end{aligned} \tag{27.4}$$

which enable us to consider the body not only in the frame of reference in which it is at rest but also in those where it is moving (rotating as a whole). The derivative dx^0/ds must be taken along the world line described by the point considered in the body.

In a weak (Newtonian) gravitational field, $g_{00} = -1-2\phi/c^2$, where ϕ is the gravitational potential.‡ Substituting this expression in (27.2) and taking the square root, we have to the same approximation

$$T = \text{constant} \times (1-\phi/c^2). \tag{27.5}$$

† See *The Classical Theory of Fields*, §89 (formula (89.9) with $v = 0$ and $E = mc^2$).

‡ See *The Classical Theory of Fields*, §87.

Since $\phi < 0$, this shows that in equilibrium the temperature is higher at points in the body where $|\phi|$ is greater, i.e. within the body. In the limit of non-relativistic mechanics ($c \to \infty$), (27.5) becomes $T =$ constant, as it should.

We can similarly transform the condition (27.3), bearing in mind that the relativistic chemical potential, in the limit of classical mechanics, does not directly become the ordinary (non-relativistic) expression for the chemical potential in the absence of a field, which we now denote by μ_0, but μ_0+mc^2, where mc^2 is the rest energy of a particle of the body. Thus we have

$$\mu\sqrt{-g_{00}} \cong (\mu_0+mc^2)(1+\phi/c^2)$$
$$\cong \mu_0+mc^2+m\phi,$$

so that the condition (27.3) becomes

$$\mu_0+m\phi = \text{constant};$$

this agrees with (25.2), as it should.

Finally, we may mention a useful relation which follows immediately from the conditions (27.2) and (27.3). Dividing one by the other, we find that $\mu/T =$ constant, and hence

$$\mathrm{d}\mu/\mu = \mathrm{d}T/T.$$

From (24.12), at constant volume (equal to unity) we have

$$\mathrm{d}P = S\,\mathrm{d}T+N\,\mathrm{d}\mu,$$

where S and N are the entropy and number of particles in unit volume of the body. Substituting $\mathrm{d}T = (T/\mu)\,\mathrm{d}\mu$ and noting that $\mu N+ST = \Phi+ST = \varepsilon+P$, where ε is the energy per unit volume, gives†

$$\mathrm{d}\mu/\mu = \mathrm{d}P/(\varepsilon+P). \tag{27.6}$$

† In the non-relativistic case this relation becomes a trivial identity. Putting $\mu \cong mc^2$ $\varepsilon \cong \varrho c^2 \gg P$ (where ϱ is the density), we get $\mathrm{d}\mu = v\,\mathrm{d}P$ with $v = m/\varrho$ the volume per particle; this is as it should be for $T =$ constant.

CHAPTER III

THE GIBBS DISTRIBUTION

§28. The Gibbs distribution

LET us now turn to the problem stated in Chapter I of finding the distribution function for a subsystem, i.e. any macroscopic body which is a small part of some large closed system. The most convenient and general method of approaching the solution of this problem is based on the application of the microcanonical distribution to the whole system.

Distinguishing the body in question from the rest of the closed system, we may consider the system as consisting of these two parts. The rest of the system will be called the "medium" in relation to the body.

The microcanonical distribution (6.6) can be written in the form

$$\mathrm{d}w = \text{constant} \times \delta(E+E'-E^{(0)})\,\mathrm{d}\Gamma\,\mathrm{d}\Gamma', \tag{28.1}$$

where E, $\mathrm{d}\Gamma$ and E', $\mathrm{d}\Gamma'$ relate to the body and the medium respectively, and $E^{(0)}$ is the given value of the energy of the closed system, which must be equal to the sum $E+E'$ of the energies of the body and the medium.

Our object is to find the probability w_n of a state of the whole system such that the body concerned is in some definite quantum state (with energy E_n), i.e. a microscopically defined state. The microscopic state of the medium is of no interest, so that we shall suppose this to be in a macroscopically defined state. Let $\Delta\Gamma'$ be the statistical weight of the macroscopic state of the medium, and let $\Delta E'$ be the range of values of the energy of the medium corresponding to the range $\Delta\Gamma'$ of quantum states in the sense discussed in §7.

The required probability w_n can be found by taking $\mathrm{d}\Gamma = 1$ in (28.1), putting $E = E_n$ and integrating with respect to Γ':

$$w_n = \text{constant} \times \int \delta(E_n+E'-E^{(0)})\,\mathrm{d}\Gamma'.$$

Let $\Gamma'(E')$ be the total number of quantum states of the medium with energy not exceeding E'. Since the integrand depends only on E', we can change to integration with respect to E', putting $\mathrm{d}\Gamma' = \big(\mathrm{d}\Gamma'(E')/\mathrm{d}E'\big)\,\mathrm{d}E'$. The derivative $\mathrm{d}\Gamma'/\mathrm{d}E'$ is replaced (cf. §7) by

$$\mathrm{d}\Gamma'/\mathrm{d}E' = e^{S'(E')}/\Delta E',$$

where $S'(E')$ is the entropy of the medium as a function of its energy; $\Delta E'$ is, of course, also a function of E'. Thus we have

$$w_n = \text{constant} \times \int \frac{e^{S'}}{\Delta E'} \delta(E' + E_n - E^{(0)})\, dE'.$$

Owing to the presence of the delta function, the result of the integration is simply to replace E' by $E^{(0)} - E_n$:

$$w_n = \text{constant} \times \left(\frac{e^{S'}}{\Delta E'}\right)_{E' = E^{(0)} - E_n} \tag{28.2}$$

We now use the fact that, since the body is small, its energy E_n is small in comparison with $E^{(0)}$. The quantity $\Delta E'$ undergoes only a very small relative change when E' varies slightly, and so in $\Delta E'$ we can simply put $E' = E^{(0)}$; it then becomes a constant independent of E_n. In the exponential factor $e^{S'}$, we must expand $S'(E^{(0)} - E_n)$ in powers of E_n as far as the linear term:

$$S'(E^{(0)} - E_n) = S'(E^{(0)}) - E_n\, dS'(E^{(0)})/dE^{(0)}.$$

The derivative of the entropy S' with respect to energy is just $1/T$, where T is the temperature of the system; the temperatures of the body and the medium are the same, since the system is assumed to be in equilibrium.

Thus we have finally for w_n the expression

$$w_n = Ae^{-E_n/T}, \tag{28.3}$$

where A is a normalisation constant independent of E_n. This is one of the most important formulae in statistical physics. It gives the statistical distribution of any macroscopic body which is a comparatively small part of a large closed system. The distribution (28.3) is called the *Gibbs distribution* or *canonical distribution;* it was discovered by GIBBS for classical statistics in 1901.

The normalisation constant A is given by the condition $\sum w_n = 1$, whence

$$\frac{1}{A} = \sum_n e^{-E_n/T}. \tag{28.4}$$

The mean value of any physical quantity f pertaining to the body can be calculated by means of the Gibbs distribution, using the formula

$$\begin{aligned} \bar{f} &= \sum_n w_n f_{nn} \\ &= \sum_n f_{nn} e^{-E_n/T} \Big/ \sum_n e^{-E_n/T}. \end{aligned} \tag{28.5}$$

In classical statistics an expression exactly corresponding to (28.3) is obtained for the distribution function in phase space:

$$\varrho(p, q) = Ae^{-E(p,q)/T}, \tag{28.6}$$

where $E(p, q)$ is the energy of the body as a function of its co-ordinates and momenta.† The normalisation constant A is given by the condition

$$\int \varrho \, dp \, dq = A \int e^{-E(p,q)/T} \, dp \, dq = 1. \tag{28.7}$$

In practice, cases are frequently encountered where it is not the entire microscopic motion of the particles which is quasi-classical, but only the motion corresponding to some of the degrees of freedom, whereas the motion with respect to the remaining degrees of freedom is quantised (for example, the translational motion of the molecules may be quasi-classical while the motion of the atoms within the molecules is quantised). Then the energy levels of the body may be written as functions of the "quasi-classical" co-ordinates and momenta: $E_n = E_n(p, q)$, where n denotes the set of quantum numbers defining the "quantised part" of the motion, for which p and q are parameters. The Gibbs distribution formula then becomes

$$dw_n(p, q) = Ae^{-E_n(p,q)/T} \, dp_{\mathrm{cl}} \, dq_{\mathrm{cl}}, \tag{28.8}$$

where $dp_{\mathrm{cl}} \, dq_{\mathrm{cl}}$ is the product of differentials of the "quasi-classical" co-ordinates and momenta.

Finally, the following comment is necessary concerning the group of problems which may be solved by means of the Gibbs distribution. We have spoken of the latter throughout as the statistical distribution for a subsystem, as in fact it is. It is very important to note, however, that this same distribution can quite successfully be used also to determine the fundamental statistical properties of bodies forming closed systems, since such properties of a body as the values of the thermodynamic quantities or the probability distributions for the co-ordinates and velocities of its individual particles are clearly independent of whether we regard the body as a closed system or as being placed in an imaginary thermostat (§7). But in the latter case the body becomes a "subsystem" and the Gibbs distribution is immediately applicable to it. The difference between bodies forming closed and non-closed systems when the Gibbs distribution is used appears essentially only in the treatment of the fairly unimportant problem of fluctuations in the total energy of the body. The Gibbs distribution gives for the mean fluctuation of this quantity a non-zero value, which is meaningful for a body in a medium but is entirely spurious for a closed system, since the energy of such a body is by definition constant and does not fluctuate.

The possibility of applying the Gibbs distribution (in the manner described) to closed systems is also seen from the fact that this distribution hardly differs

† To avoid misunderstanding, let us mention once more that the w_n (or ϱ) are monotonic functions of energy and need not have maxima for $E = \bar{E}$. It is the distribution function with respect to energy, obtained by multiplying w_n by $d\Gamma(E)/dE$, which has a sharp maximum at $E = \bar{E}$.

from the microcanonical distribution, while being very much more convenient for practical calculations. For the microcanonical distribution is, roughly speaking, equivalent to regarding as equally probable all microstates of the body which correspond to a given value of its energy. The canonical distribution is "spread" over a certain range of energy values, but the width of this range (of the order of the mean fluctuation of energy) is negligible for a macroscopic body.

§29. The Maxwellian distribution

The energy $E(p, q)$ in the Gibbs distribution formula of classical statistics can always be written as the sum of two parts: the kinetic energy and the potential energy. The first of these is a quadratic function of the momenta of the atoms†, and the second is a function of their co-ordinates, the form of which depends on the law of interaction between the particles within the body (and on the external field, if any). If the kinetic and potential energies are denoted by $K(p)$ and $U(q)$ respectively, then $E(p, q) = K(p)+U(q)$, and the probability $\mathrm{d}w = \varrho(p, q)\,\mathrm{d}p\,\mathrm{d}q$ becomes

$$\mathrm{d}w = Ae^{-U(q)/T}e^{-K(p)/T}\,\mathrm{d}p\,\mathrm{d}q,$$

i.e. is the product of two factors, one of which depends only on the co-ordinates and the other only on the momenta. This means that the probabilities for momenta (or velocities) and co-ordinates are independent, in the sense that any particular values of the momenta do not influence the probabilities of the various values of the co-ordinates, and *vice versa*. Thus the probability of the various values of the momenta can be written

$$\mathrm{d}w_p = ae^{-K(p)/T}\,\mathrm{d}p, \tag{29.1}$$

and the probability distribution for the co-ordinates is

$$\mathrm{d}w_q = be^{-U(q)/T}\,\mathrm{d}q. \tag{29.2}$$

Since the sum of the probabilities of all possible values of the momenta must be unity (and the same applies to the co-ordinates), each of the probabilities $\mathrm{d}w_p$ and $\mathrm{d}w_q$ must be normalised, i.e. their integrals over all possible values of the momenta and co-ordinates respectively for the body concerned must be equal to unity. From these conditions we can determine the constants a and b in (29.1) and (29.2).

Let us consider the probability distribution for the momenta, and once again emphasise the very important fact that in classical statistics this distribution does not depend on the nature of the interaction of particles within

† It is assumed that Cartesian co-ordinates are used.

the system or on the nature of the external field, and so can be expressed in a form applicable to all bodies.†

The kinetic energy of the whole body is equal to the sum of the kinetic energies of each of the atoms composing it, and the probability again falls into a product of factors, each depending on the momenta of only one atom. This again shows that the momentum probabilities of different atoms are independent, i.e. the momentum of one does not affect the probabilities of various momenta of any other. We can therefore write the probability distribution for the momenta of each atom separately.

For an atom of mass m the kinetic energy is $(p_x^2+p_y^2+p_z^2)/2m$, where p_x, p_y, p_z are the Cartesian components of its momentum, and the probability distribution is

$$dw_{\mathbf{p}} = ae^{-(p_x^2+p_y^2+p_z^2)/2mT}\, dp_x\, dp_y\, dp_z.$$

The constant a is given by the normalisation condition. By means of the formula

$$\int_{-\infty}^{\infty} e^{-\alpha x^2}\, dx = \sqrt{(\pi/\alpha)}$$

we find

$$a\int_{-\infty}^{\infty}\int_{-\infty}^{\infty}\int_{-\infty}^{\infty} e^{-(p_x^2+p_y^2+p_z^2)/2mT}\, dp_x\, dp_y\, dp_z$$

$$= a\left(\int_{-\infty}^{\infty} e^{-p^2/2mT}\, dp\right)^3$$

$$= a(2\pi mT)^{3/2} = 1,$$

whence $a = (2\pi mT)^{-3/2}$, and the momentum probability distribution takes the final form

$$dw_{\mathbf{p}} = \frac{1}{(2\pi mT)^{3/2}} e^{-(p_x^2+p_y^2+p_z^2)/2mT}\, dp_x\, dp_y\, dp_z. \tag{29.3}$$

Changing from momenta to velocities ($\mathbf{p} = m\mathbf{v}$), we can write the corresponding velocity distribution as

$$dw_{\mathbf{v}} = \left(\frac{m}{2\pi T}\right)^{3/2} e^{-m(v_x^2+v_y^2+v_z^2)/2T}\, dv_x\, dv_y\, dv_z. \tag{29.4}$$

This is the *Maxwellian distribution* (MAXWELL 1860). It again consists of a product of three independent factors

$$dw_{v_x} = \sqrt{\frac{m}{2\pi T}} e^{-mv_x^2/2T} dv_x, \ldots, \tag{29.5}$$

each of which gives the probability distribution for a single velocity component.

† In quantum statistics this statement is not quite true in general.

If the body consists of molecules (e.g. a polyatomic gas), then together with the Maxwellian distribution for the individual atoms there is a similar distribution for the translational motion of each molecule as a whole: from the kinetic energy of the molecule we can separate a term which gives the energy of the translational motion, and so the required distribution separates in the form (29.4), where m must now be taken as the total mass of the molecule, and v_x, v_y, v_z as the velocity components of its centre of mass. It should be emphasised that the Maxwellian distribution for the translational motion of molecules can be valid quite independently of the nature of the motion of the atoms within the molecule (and the rotation of the molecule), and in particular when a quantised description of the latter is necessary.†

The expression (29.4) is written in terms of Cartesian co-ordinates in "velocity space". If we change from Cartesian to spherical polar co-ordinates, the result is

$$dw_{\mathbf{v}} = \left(\frac{m}{2\pi T}\right)^{3/2} e^{-mv^2/2T} v^2 \sin\theta \, d\theta \, d\phi \, dv, \tag{29.6}$$

where v is the absolute magnitude of the velocity, and θ and ϕ the polar angle and azimuthal angle which determine the direction of the velocity. Integration with respect to angle gives the probability distribution for the absolute magnitude of the velocity:

$$dw_v = 4\pi \left(\frac{m}{2\pi T}\right)^{3/2} e^{-mv^2/2T} v^2 \, dv. \tag{29.7}$$

It is sometimes convenient to use cylindrical co-ordinates in velocity space. Then

$$dw_{\mathbf{v}} = \left(\frac{m}{2\pi T}\right)^{3/2} e^{-m(v_z^2+v_r^2)/2T} \, v_r \, dv_r \, dv_z \, d\phi, \tag{29.8}$$

where v_z is the velocity component along the z-axis, v_r the component perpendicular to that axis, and ϕ the angle which gives the direction of this component.

Let us calculate the mean kinetic energy of an atom. According to the definition of the mean, and using (29.5), we find for any Cartesian velocity component‡

$$\overline{v_x^2} = \sqrt{\frac{m}{2\pi T}} \int_{-\infty}^{\infty} v_x^2 e^{-mv_x^2/2T} \, dv_x$$
$$= T/m. \tag{29.9}$$

† The Maxwellian distribution clearly applies also to the Brownian motion of particles suspended in a liquid.

‡ For reference we shall give the values of the integrals of the form

$$I_n = \int_0^{\infty} e^{-\alpha x^2} x^n \, dx,$$

The mean value of the kinetic energy of the atom is therefore $3T/2$, or $3kT/2$ when the temperature is measured in degrees. We can thus say that the mean kinetic energy of all the particles in the body in classical statistics is always $3NT/2$, where N is the total number of atoms.

PROBLEMS

PROBLEM 1. Find the mean value of the nth power of the absolute magnitude of the velocity.

SOLUTION. Using (29.7), we find

$$\overline{v^n} = 4\pi \left(\frac{m}{2\pi T}\right)^{3/2} \int_0^\infty e^{-mv^2/2T} v^{n+2}\, dv = \frac{2}{\sqrt{\pi}}\left(\frac{2T}{m}\right)^{n/2} \Gamma\left(\frac{n+3}{2}\right).$$

In particular, if n is even ($= 2r$), then

$$\overline{v^{2r}} = (T/m)^r (2r+1)!!;$$

if $n = 2r+1$, then

$$\overline{v^{2r+1}} = \frac{2}{\sqrt{\pi}}\left(\frac{2T}{m}\right)^{(2r+1)/2} (r+1)!.$$

PROBLEM 2. Find the mean square fluctuation of the velocity.

SOLUTION. We have

$$\overline{(\Delta v)^2} \equiv \overline{(v-\bar{v})^2} = \overline{v^2} - \bar{v}^2.$$

The result of Problem 1 with $n = 1$ and $n = 2$ gives

$$\overline{(\Delta v)^2} = (T/m)(3 - 8/\pi).$$

PROBLEM 3. Find the mean energy, the mean square energy, and the mean square fluctuation of the kinetic energy of an atom.

---

which often occur in applications of the Maxwellian distribution. The substitution $\alpha x^2 = y$ gives

$$I_n = \tfrac{1}{2}\alpha^{-(n+1)/2} \int_0^\infty e^{-y} y^{(n-1)/2}\, dy = \tfrac{1}{2}\alpha^{-(n+1)/2}\Gamma(\tfrac{1}{2}n + \tfrac{1}{2}),$$

where $\Gamma(x)$ is the gamma function. In particular, if $n = 2r$ with $r > 0$, then

$$I_{2r} = \frac{(2r-1)!!}{2^{r+1}} \sqrt{\frac{\pi}{\alpha^{2r+1}}},$$

where $(2r-1)!! = 1\cdot 3\cdot 5\cdot \ldots (2r-1)$. If $r = 0$, then

$$I_0 = \tfrac{1}{2}\sqrt{(\pi/\alpha)}.$$

If $n = 2r+1$, then

$$I_{2r+1} = r!/2\alpha^{r+1}.$$

The same integral from $-\infty$ to $+\infty$ is zero if $n = 2r+1$ and twice the integral from 0 to ∞ if $n = 2r$.

SOLUTION. From the results of Problem 1 we find

$$\bar{\varepsilon} = \tfrac{1}{2} m\overline{v^2} = 3T/2,$$
$$\overline{\varepsilon^2} = \tfrac{1}{4} m^2\overline{v^4} = 15T^2/4,$$
$$\overline{(\Delta\varepsilon)^2} = \overline{\varepsilon^2} - \bar{\varepsilon}^2 = 3T^2/2.$$

PROBLEM 4. Find the probability distribution for the kinetic energy of an atom.

SOLUTION.

$$\mathrm{d}w_\varepsilon = \frac{2}{\sqrt{(\pi T^3)}} e^{-\varepsilon/T}\sqrt{\varepsilon}\,\mathrm{d}\varepsilon.$$

PROBLEM 5. Find the probability distribution for the angular velocities of rotation of molecules.

SOLUTION. Just as for translational motion, we can write the probability distribution for the rotation of each molecule separately (in classical statistics). The kinetic energy of rotation of a molecule regarded as a rigid body (which is permissible, owing to the smallness of the atomic vibrations within the molecule) is

$$\varepsilon_{\text{rot}} = \tfrac{1}{2}(I_1\Omega_1^2 + I_2\Omega_2^2 + I_3\Omega_3^2) = \tfrac{1}{2}\left(\frac{M_1^2}{I_1} + \frac{M_2^2}{I_2} + \frac{M_3^2}{I_3}\right),$$

where I_1, I_2, I_3 are the principal moments of inertia, Ω_1, Ω_2, Ω_3 are the components of the angular velocity along the principal axes of inertia, and $M_1 = I_1\Omega_1$, $M_2 = I_2\Omega_2$, $M_3 = I_3\Omega_3$ are the components of the angular momentum, which act as generalised momenta with respect to the velocities Ω_1, Ω_2, Ω_3. The normalised probability distribution for the angular-momentum components is

$$\mathrm{d}w_{\mathbf{M}} = (2\pi T)^{-3/2}(I_1I_2I_3)^{-1/2}\exp\left[-\frac{1}{2T}\left(\frac{M_1^2}{I_1} + \frac{M_2^2}{I_2} + \frac{M_3^2}{I_3}\right)\right]\mathrm{d}M_1\,\mathrm{d}M_2\,\mathrm{d}M_3$$

and for the angular velocity

$$\mathrm{d}w_{\mathbf{\Omega}} = (2\pi T)^{-3/2}(I_1I_2I_3)^{1/2}\exp\left[-\frac{1}{2T}(I_1\Omega_1^2 + I_2\Omega_2^2 + I_3\Omega_3^2)\right]\mathrm{d}\Omega_1\,\mathrm{d}\Omega_2\,\mathrm{d}\Omega_3.$$

PROBLEM 6. Find the mean squares of the absolute magnitudes of the angular velocity and angular momentum of a molecule.

SOLUTION. The above distributions give

$$\overline{\Omega^2} = T\left(\frac{1}{I_1} + \frac{1}{I_2} + \frac{1}{I_3}\right),$$
$$\overline{M^2} = T(I_1 + I_2 + I_3).$$

§30. The probability distribution for an oscillator

Let us consider a body whose atoms are executing small oscillations about some equilibrium positions. They may be atoms in a crystal or in a gas molecule; in the latter case the motion of the molecule as a whole does not affect the oscillations of the atoms within it and so does not influence the results.

As we know from mechanics, the Hamiltonian (the energy) of a system consisting of an arbitrary number of particles executing small oscillations can be written as a sum:

$$E(p, q) = \tfrac{1}{2}\sum_\alpha (p_\alpha^2 + \omega_\alpha^2 q_\alpha^2),$$

where q_α are what are called the *normal co-ordinates* of the oscillations (equal to zero at points of equilibrium), $p_\alpha = \dot{q}_\alpha$ are the corresponding generalised momenta, and ω_α are the oscillation frequencies. In other words, $E(p, q)$ is a sum of independent terms, each corresponding to a separate normal oscillation (or, as we say, to an *oscillator*). In quantum mechanics the same is true of the Hamiltonian operator of the system, so that each oscillator is independently quantised and the energy levels of the system are given by the sums

$$\sum_\alpha \hbar\omega_\alpha(n_\alpha + \tfrac{1}{2}),$$

the n_α being integers.

As a result of these facts the Gibbs distribution for the whole system is a product of independent factors, each giving the statistical distribution for a separate oscillator. In consequence we shall consider a single oscillator in what follows.

Let us determine the probability distribution for the co-ordinate q of an oscillator†; the suffix α which gives the number of the oscillator will be omitted henceforward. In classical statistics the solution to this problem would be very simple: since the potential energy of the oscillator is $\frac{1}{2}\omega^2 q^2$, the probability distribution is

$$dw_q = Ae^{-\omega^2 q^2/2T}\, dq,$$

or, determining A from the normalisation condition,

$$dw_q = \frac{\omega}{\sqrt{(2\pi T)}} e^{-\omega^2 q^2/2T}\, dq; \tag{30.1}$$

the integration with respect to q may be taken from $-\infty$ to $+\infty$, since the integral is rapidly convergent.

Let us now consider the solution of this problem in the quantum case. Let $\psi_n(q)$ be the wave functions of the stationary states of the oscillator, corresponding to the energy levels

$$\varepsilon_n = \hbar\omega(n + \tfrac{1}{2}).$$

If the oscillator is in the nth state, the quantum probability distribution for its co-ordinate is given by ψ_n^2 (in the present case the functions ψ_n are real, and so we write simply ψ_n^2 instead of the squared modulus $|\psi_n|^2$). The required statistical probability distribution is obtained by multiplying ψ_n^2 by the probability w_n of finding the oscillator in the nth state, and then summing over all possible states.

According to the Gibbs distribution,

$$w_n = ae^{-\varepsilon_n/T},$$

† The normal co-ordinate has the dimensions $\text{cm}\cdot\text{g}^{1/2}$.

where a is a constant. Thus we have the formula

$$dw_q = a\,dq \sum_{n=0}^{\infty} e^{-\varepsilon_n/T} \psi_n^2, \tag{30.2}$$

which is, of course, entirely in agreement with the general formula (5.8).

To calculate the sum, we can proceed as follows. With the notation $dw_q = \varrho_q\,dq$, we form the derivative

$$\frac{d\varrho_q}{dq} = 2a \sum_{n=0}^{\infty} e^{-\varepsilon_n/T} \psi_n \frac{d\psi_n}{dq}.$$

Using the momentum operator $\hat{p} = -i\hbar\,d/dq$ and the fact that the oscillator momentum has non-zero matrix elements† only for transitions with $n \to n \pm 1$, we can write

$$\begin{aligned}\frac{d\psi_n}{dq} &= \frac{i}{\hbar}\hat{p}\psi_n \\ &= \frac{i}{\hbar}(p_{n-1,\,n}\psi_{n-1} + p_{n+1,\,n}\psi_{n+1}) \\ &= \frac{\omega}{\hbar}(q_{n-1,\,n}\psi_{n-1} - q_{n+1,\,n}\psi_{n+1}).\end{aligned}$$

Here we have used the relations

$$p_{n-1,\,n} = -i\omega q_{n-1,\,n}, \qquad p_{n+1,\,n} = i\omega q_{n+1,\,n}$$

between the momentum and co-ordinate matrix elements. Thus

$$\frac{d\varrho_q}{dq} = \frac{2a\omega}{\hbar}\left\{\sum_{n=0}^{\infty} q_{n-1,\,n}\psi_n\psi_{n-1}e^{-\varepsilon_n/T} - \sum_{n=0}^{\infty} q_{n+1,\,n}\psi_n\psi_{n+1}e^{-\varepsilon_n/T}\right\}.$$

In the first sum we change the summation suffix from n to $n+1$ and use the relations

$$\varepsilon_{n+1} = \varepsilon_n + \hbar\omega, \qquad q_{n+1,\,n} = q_{n,\,n+1}, \qquad q_{-1,\,0} = 0,$$

obtaining

$$\frac{d\varrho_q}{dq} = -\frac{2a\omega}{\hbar}(1 - e^{-\hbar\omega/T}) \sum_{n=0}^{\infty} q_{n,\,n+1}\psi_n\psi_{n+1}e^{-\varepsilon_n/T}.$$

In an exactly similar manner we can prove that

$$q\varrho_q = a(1 + e^{-\hbar\omega/T}) \sum_{n=0}^{\infty} q_{n,\,n+1}\psi_n\psi_{n+1}e^{-\varepsilon_n/T}.$$

A comparison of the two equations gives

$$\frac{d\varrho_q}{dq} = -\left(\frac{2\omega}{\hbar}\tanh\frac{\hbar\omega}{2T}\right) q\varrho_q,$$

† See *Quantum Mechanics*, §23.

whence

$$\varrho_q = \text{constant}\times\exp\left\{-q^2\frac{\omega}{\hbar}\tanh\frac{\hbar\omega}{2T}\right\}.$$

Determining the constant from the normalisation condition, we finally obtain the formula

$$dw_q = \left(\frac{\omega}{\pi\hbar}\tanh\frac{\hbar\omega}{2T}\right)^{1/2}\exp\left\{-q^2\frac{\omega}{\hbar}\tanh\frac{\hbar\omega}{2T}\right\}dq \tag{30.3}$$

(F. Bloch 1932). Thus in the quantum case also the probabilities of various values of the co-ordinate of an oscillator are distributed according to a law of the form $\exp(-\alpha q^2)$, but the coefficient α differs from that in the classical case. In the limit $\hbar\omega \ll T$, where the quantisation is unimportant, formula (30.3) becomes (30.1), as we should expect.

In the opposite limiting case $\hbar\omega \gg T$, formula (30.3) becomes

$$dw_q = \sqrt{\frac{\omega}{\pi\hbar}}\exp(-q^2\omega/\hbar)\,dq,$$

i.e. the purely quantum probability distribution for the co-ordinate in the ground state of the oscillator.† This corresponds to the fact that when $T \ll \hbar\omega$ the oscillations are hardly excited at all.

The probability distribution for the momentum of the oscillator can be written by analogy with (30.3) without repeating the calculation. The reason is that the problem of quantisation of the oscillator is completely symmetrical as regards the co-ordinate and the momentum, and the oscillator wave functions in the p representation are the same as its ordinary co-ordinate wave functions (q being replaced by p/ω).‡ The required distribution is therefore

$$dw_p = \left(\frac{1}{\pi\hbar\omega}\tanh\frac{\hbar\omega}{2T}\right)^{1/2}\exp\left\{-\frac{p^2}{\hbar\omega}\tanh\frac{\hbar\omega}{2T}\right\}dp. \tag{30.4}$$

In the limit of classical mechanics ($\hbar\omega \ll T$) this becomes the usual Maxwellian distribution:

$$dw_p = (2\pi T)^{-1/2}e^{-p^2/2T}\,dp. \tag{30.5}$$

§31. The free energy in the Gibbs distribution

According to formula (7.9) the entropy of a body can be calculated as the mean logarithm of its distribution function:

$$S = -\overline{\log w_n}.$$

Substituting the Gibbs distribution (28.3) gives

$$S = -\log A + \bar{E}/T,$$

† This is the squared modulus of the wave function of the ground state of the oscillator.
‡ See *Quantum Mechanics*, §23, Problem 1.

whence $\log A = (\bar{E}-TS)/T$. But the mean energy $\bar{E}$ is just what is meant by the term "energy" in thermodynamics; hence $\bar{E}-TS = F$ and $\log A = F/T$, i.e. the normalisation constant of the distribution is directly related to the free energy of the body.

Thus the Gibbs distribution may be written in the form

$$w_n = e^{(F-E_n)/T}, \tag{31.1}$$

and this is the form most frequently used. The same method gives in the classical case, using (7.12), the expression

$$\varrho = (2\pi\hbar)^{-s} e^{[F-E(p,q)]/T}. \tag{31.2}$$

The normalisation condition for the distribution (31.1) is

$$\sum_n w_n = e^{F/T} \sum_n e^{-E_n/T} = 1,$$

whence

$$e^{-F/T} = \sum_n e^{-E_n/T},$$

or, taking logarithms,

$$F = -T \log \sum_n e^{-E_n/T}. \tag{31.3}$$

This formula is fundamental in thermodynamic applications of the Gibbs distribution. It affords, in principle, the possibility of calculating the thermodynamic functions for any body whose energy spectrum is known.

The sum in the logarithm in (31.3) is usually called the *partition function* (or *sum over states*). It is just the trace of the operator $\exp(-\hat{H}/T)$, where $\hat{H}$ is the Hamiltonian of the body†:

$$Z \equiv \sum_n e^{-E_n/T} = \operatorname{tr} \exp(-\hat{H}/T). \tag{31.4}$$

This notation has the advantage that any complete set of wave functions may be used in order to calculate the trace.

A similar formula in classical statistics is obtained from the normalisation condition for the distribution (31.2). First of all, however, we must take account of the following fact, which was unimportant so long as we were discussing the distribution function as such and not relating the normalisation coefficient to a particular quantitative property of the body, viz. its free energy. If, for example, two identical atoms change places, then afterwards the microstate of the body is represented by a different phase point, obtained from the original one by replacing the co-ordinates and momenta of one atom by those of the other. On the other hand, since the interchanged atoms are identical, the two states of the body are physically identical. Thus a number of points in phase space correspond to one physical microstate of the body. In integrating the distribution (31.2), however, each state must of course be

† In accordance with the general rules, $\exp(-\hat{H}/T)$ denotes an operator whose eigenfunctions are the same as those of the operator $\hat{H}$ and whose eigenvalues are $e^{-E_n/T}$.

taken only once.† In other words, we must integrate only over those regions of phase space which correspond to physically different states of the body. This will be denoted by a prime to the integral sign.

Thus we have the formula

$$F = -T \log \int' e^{-E(p,q)/T}\, d\Gamma; \tag{31.5}$$

here and in all similar cases below, $d\Gamma$ denotes the volume element in phase space divided by $(2\pi\hbar)^s$:

$$d\Gamma = dp\, dq/(2\pi\hbar)^s. \tag{31.6}$$

Thus the partition function in the quantum formula (31.3) becomes an *integral over states*. As already mentioned in §29, the classical energy $E(p, q)$ can always be written as the sum of the kinetic energy $K(p)$ and the potential energy $U(q)$. The kinetic energy is a quadratic function of the momenta, and the integration with respect to the latter can be effected in a general form. The problem of calculating the partition function therefore actually reduces to that of integrating the function $e^{-U(q)/T}$ with respect to the co-ordinates.

In the practical calculation of the partition function it is usually convenient to extend the region of integration and include an appropriate correction factor. For example, let us consider a gas of N identical atoms. Then we can integrate with respect to the co-ordinates of each atom separately, extending the integration over the whole volume occupied by the gas; but the result must be divided by the number of possible permutations of N atoms, which is $N!$. In other words, the integral $\int'$ can be replaced by the integral over all phase space, divided by $N!$:

$$\int' \dots d\Gamma = \frac{1}{N!}\int \dots d\Gamma. \tag{31.7}$$

Similarly, it is convenient to extend the region of integration for a gas consisting of N identical molecules: the integration with respect to the co-ordinates of each molecule as a whole (i.e. the co-ordinates of its centre of mass) is carried out independently over the whole volume, whilst that with respect to the co-ordinates of the atoms within the molecule is carried out over the "volume" belonging to each molecule (i.e. over a small region in which there is an appreciable probability of finding the atoms forming the molecule). Then the integral must again be divided by $N!$.

† This becomes particularly evident if we consider the classical partition function (integral over states) as the limit of the quantum partition function. In the latter the summation is over all the different quantum states, and there is no problem (remembering that, because of the principle of symmetry of wave functions in quantum mechanics, the quantum state is unaffected by interchanges of identical particles).

From the purely classical viewpoint the need for this interpretation of the statistical integration arises because otherwise the statistical weight would no longer be multiplicative, and so the entropy and the other thermodynamic quantities would no longer be additive.

PROBLEMS

Problem 1. The potential energy of the interaction between the particles in a body is a homogeneous function of degree n in their co-ordinates. Using similarity arguments, determine the form of the free energy of such a body in classical statistics.

Solution. In the partition function

$$Z = \int' e^{-[K(p)+U(q)]/T}\, d\Gamma,$$

we replace each q by λq and each p by $\lambda^{n/2}p$, where λ is an arbitrary constant. If at the same time we replace T by $\lambda^n T$, the integrand is unchanged, but the limits of integration with respect to the co-ordinates are altered: the linear size of the region of integration is multiplied by $1/\lambda$, and so the volume is multiplied by $1/\lambda^3$. In order to restore the limits of integration, we must therefore at the same time replace V by $\lambda^3 V$. The result of these changes is to multiply the integral by $\lambda^{3N(1+n/2)}$ because of the change of variables in $d\Gamma$ ($s = 3N$ co-ordinates and the same number of momenta, N being the number of particles in the body). Thus we conclude that the substitutions $V \to \lambda^3 V$, $T \to \lambda^n T$ give

$$Z \to \lambda^{3N(1+n/2)} Z.$$

The most general form of function $Z(V, T)$ having this property is

$$Z = T^{3N(1/2+1/n)} f(VT^{-3/n}),$$

where f is an arbitrary function of one variable.

Hence we find for the free energy an expression of the form

$$F = -3(\tfrac{1}{2}+1/n)NT \log T + NT\phi(VT^{-3/n}/N), \tag{1}$$

which involves only one unknown function of one variable; the number N is included in the second term in (1) so that F shall have the necessary property of additivity.

Problem 2. Derive the virial theorem for a macroscopic body for which the potential energy of interaction of the particles is a homogeneous function of degree n in their co-ordinates.

Solution. Following the derivation of the virial theorem in mechanics†, we calculate the time derivative of the sum $\sum \mathbf{r}\cdot\mathbf{p}$, where $\mathbf{r}$ and $\mathbf{p}$ are the radius vectors and momenta of the particles in the body. Since $\dot{\mathbf{r}} = \partial K(p)/\partial \mathbf{p}$ and $K(p)$ is a homogeneous function of degree two in the momenta, we have

$$\frac{d}{dt}\sum \mathbf{r}\cdot\mathbf{p} = \sum \mathbf{p}\cdot\frac{\partial K(p)}{\partial \mathbf{p}} + \sum \mathbf{r}\cdot\dot{\mathbf{p}} = 2K(p) + \sum \mathbf{r}\cdot\dot{\mathbf{p}}.$$

The particles in the body execute a motion in a finite region of space with velocities which do not become infinite.The quantity $\sum \mathbf{r}\cdot\mathbf{p}$ is therefore bounded and the mean value of its time derivative is zero, so that

$$2K + \overline{\sum \mathbf{r}\cdot\dot{\mathbf{p}}} = 0,$$

where $K = \overline{K(p)}$. The derivatives $\dot{\mathbf{p}}$ are determined by the forces acting on the particles in the body. In summing over all particles we must take into account not only the forces of interaction between the particles but also the forces exerted on the surface of the body by surrounding bodies:

$$\overline{\sum \mathbf{r}\cdot\dot{\mathbf{p}}} = -\overline{\sum \mathbf{r}\cdot\frac{\partial U(q)}{\partial \mathbf{r}}} - P\oint \mathbf{r}\cdot d\mathbf{f} = -nU - 3PV;$$

the surface integral is transformed to a volume integral and we use the fact that div $\mathbf{r} = 3$. Thus we have $2K - nU - 3PV = 0$ or, in terms of the total energy $E = U + K$,

$$(n+2)K = nE + 3PV. \tag{2}$$

† See *Mechanics*, §10.

This is the required theorem. It is valid in both classical and quantum theory. In the classical case, the mean kinetic energy $K = 3NT/2$, and (2) gives

$$E+(3/n)PV = 3(\tfrac{1}{2}+1/n)NT. \tag{3}$$

This formula could also be derived from the expression (1) for the free energy (Problem 1).

When the particles interact by Coulomb's law ($n = -1$), we have from (2)

$$K = -E+3PV.$$

This is the limiting case of the relativistic relation†

$$E-3PV = \sum mc^2 \sqrt{(1-v^2/c^2)},$$

in which the energy E includes the rest energy of the particles in the body.

§32. Thermodynamic perturbation theory

In the actual calculation of thermodynamic quantities there occur cases where the energy $E(p, q)$ of a body contains relatively small terms which may be neglected to a first approximation. These may be, for instance, the potential energy of the particles of the body in an external field. The conditions under which such terms may be regarded as small are discussed below.

In these cases a kind of "perturbation theory" may be constructed for the calculation of the thermodynamic quantities (R. E. Peierls 1932). We shall first show how this is to be done when the classical Gibbs distribution is applicable.

We write the energy $E(p, q)$ in the form

$$E(p, q) = E_0(p, q)+V(p, q), \tag{32.1}$$

where V represents the small terms. To calculate the free energy of the body, we put

$$\begin{aligned} e^{-F/T} &= \int' e^{-[E_0(p,q)+V(p,q)]/T}\,d\Gamma \\ &\cong \int' e^{-E_0/T}\left(1-\frac{V}{T}+\frac{V^2}{2T^2}\right)d\Gamma; \end{aligned} \tag{32.2}$$

in the expansion in powers of V we shall always omit terms above the second order, in order to calculate the corrections only to the first and second orders of approximation. Taking logarithms and again expanding in series, we have to the same accuracy

$$\begin{aligned} F = F_0 &+ \int' \left(V-\frac{V^2}{2T}\right) e^{[F_0-E_0(p,q)]/T}\,d\Gamma+ \\ &+\frac{1}{2T}\left[\int' V e^{[F_0-E_0(p,q)]/T}\,d\Gamma\right]^2, \end{aligned}$$

where F_0 denotes the "unperturbed" free energy, calculated for $V = 0$.

† See *The Classical Theory of Fields*, §35.

The resulting integrals are the mean values of the corresponding quantities over the "unperturbed" Gibbs distribution. Denoting this averaging by a bar and noticing that $\overline{V^2}-\overline{V}^2 = \overline{(V-\overline{V})^2}$, we have finally

$$F = F_0+\overline{V}-\frac{1}{2T}\overline{(V-\overline{V})^2}. \tag{32.3}$$

Thus the first-order correction to the free energy is just the mean value of the energy perturbation V. The second-order correction is always negative, and is determined by the mean square of the deviation of V from its mean value. In particular, if the mean value $\overline{V}$ is zero, the perturbation reduces the free energy.

A comparison of the terms of the second and first orders in (32.3) enables us to ascertain the condition for this perturbation method to be applicable. Here it must be remembered that both the mean value $\overline{V}$ and the mean square $\overline{(V-\overline{V})^2}$ are roughly proportional to the number of particles; cf. the discussion in §2 concerning r.m.s. fluctuations of the thermodynamic quantities for macroscopic bodies. We can therefore formulate the desired condition by requiring that the perturbation energy per particle should be small in comparison with T (or with kT, if the temperature is measured in degrees).†

Let us now carry out the corresponding calculations for the quantum case. Instead of (32.1) we must now use the analogous expression for the Hamiltonian operator:

$$\hat{H} = \hat{H}_0+\hat{V}.$$

According to the quantum perturbation theory, the energy levels of the perturbed system are given, correct to the second-order terms, by‡

$$E_n = E_n^{(0)}+V_{nn}+\sum_m{}' \frac{|V_{nm}|^2}{E_n^{(0)}-E_m^{(0)}}, \tag{32.4}$$

where the $E_n^{(0)}$ are the unperturbed energy levels (assumed non-degenerate); the prime to the sum signifies that the term with $m = n$ must be omitted. This expression is to be substituted in the formula

$$e^{-F/T} = \sum_n e^{-E_n/T}$$

† In expanding the integrand in (32.2) we have, strictly speaking, expanded in terms of a quantity V/T, which is proportional to the number of particles and is therefore certainly not small, but the further expansion of the logarithm causes the large terms to cancel, and so a series in powers of a small quantity is obtained.

‡ See *Quantum Mechanics*, §38.

and expanded in the same way as above. We thus easily obtain

$$F = F_0+\sum_n V_{nn}w_n+\sum_n\sum_m{}' \frac{|V_{nm}|^2 w_n}{E_n^{(0)}-E_m^{(0)}} - \frac{1}{2T}\sum_n V_{nn}{}^2 w_n+\frac{1}{2T}(\sum_n V_{nn}w_n)^2, \tag{32.5}$$

where $w_n = \exp[(F_0-E_n^{(0)})/T]$ is the unperturbed Gibbs distribution.

The diagonal matrix element V_{nn} is just the mean value of the perturbation energy V in the given (nth) quantum state. The sum

$$\sum_n V_{nn}w_n$$

is therefore the value of V averaged both over the quantum state of the body and over the ("unperturbed") statistical distribution with respect to the various quantum states. We denote this averaging by a bar and find that the correction to the free energy in the first-order approximation is $\overline{V}$, formally the same as the classical result above.

Formula (32.5) may be rewritten as

$$F = F_0+\overline{V}_{nn}-\tfrac{1}{2}\sum_n\sum_m{}' \frac{|V_{nm}|^2(w_m-w_n)}{E_n^{(0)}-E_m^{(0)}}-\frac{1}{2T}\overline{(V_{nn}-\overline{V}_{nn})^2}. \tag{32.6}$$

All the second-order terms in this expression are negative, since w_m-w_n has the same sign as $E_n^{(0)}-E_m^{(0)}$. Thus the correction to the free energy in the second-order approximation is negative in the quantum case also.

As in the classical case, the condition for this method to be applicable is that the perturbation energy per particle should be small compared with T. On the other hand, the condition for the applicability of the ordinary quantum perturbation theory (leading to the expression (32.4) for E_n) is, as we know, that the matrix elements of the perturbation should be small compared with the separations of the corresponding energy levels; roughly speaking, the perturbation energy must be small compared with the separations of the energy levels between which allowed transitions can take place.†

These two conditions are not the same, since the temperature is unrelated to the energy levels of the body. It may happen that the perturbation energy is small compared with T, but is not small, or indeed is even large, compared with the significant separations between energy levels. In such cases the "perturbation theory" for thermodynamic quantities, i.e. formula (32.6), will be applicable while the perturbation theory for the energy levels themselves, i.e. formula (32.4), is not; that is, the limits of convergence of the expansion represented by formula (32.6) may be wider than those of (32.4), from which the former expansion has been derived.

† These are in general the transitions in which the states of only a small number of particles in the body are changed.

The converse case is, of course, also possible (at sufficiently low temperatures).

Formula (32.6) is considerably simplified if not only the perturbation energy but also the differences between energy levels are small in comparison with T. Expanding the difference $w_m - w_n$ in (32.6) in powers of $(E_n^{(0)} - E_m^{(0)})/T$, we find in this case

$$F = F_0 + \bar{V}_{nn} - \frac{1}{2T}\{\sum_m{}' \overline{|V_{nm}|^2 + (V_{nn} - \bar{V}_{nn})^2}\}.$$

The rule of matrix multiplication gives

$$\sum_m{}' |V_{nm}|^2 + V_{nn}{}^2 = \sum_m |V_{nm}|^2 = \sum_m V_{nm}V_{mn} = (V^2)_{nn},$$

and we obtain an expression which is formally exactly the same as formula (32.3). Thus in this case the quantum formula is in formal agreement with the classical formula.

§33. Expansion in powers of ħ

Formula (31.5) is essentially the first and principal term in an expansion of the quantum formula (31.3) for the free energy in powers of $\hbar$ in the quasi-classical case. It is of considerable interest to derive the next non-vanishing term in this expansion (WIGNER, UHLENBECK and GROPPER 1932).

The problem of calculating the free energy amounts to that of calculating the partition function. For this purpose we use the fact that the latter is the trace of the operator $\exp(-\beta\hat{H})$ (see (31.4)), with the notation $\beta = 1/T$ in order to simplify the writing of the involved expressions which appear below. The trace of an operator may be calculated by means of any complete set of orthonormal wave functions. For these it is convenient to use the wave functions of free motion of a system of N non-interacting particles in a large but finite volume V. These functions are

$$\psi_p = \frac{1}{\sqrt{V^N}} \exp\,[(i/\hbar)\sum_i p_i q_i], \tag{33.1}$$

where the q_i are the Cartesian co-ordinates of the particles and the p_i the corresponding momenta, labelled by the suffix i, which takes the values 1, 2, . . ., s, where $s = 3N$ is the number of degrees of freedom of the system of N particles.

The subsequent calculations apply equally to systems containing identical particles (atoms) and to those where the particles are different. In order to allow in a general manner for a possible difference between the particles, we shall add to the particle mass a suffix indicating the degree of freedom: m_i. Of course the three m_i corresponding to any one particle are always equal.

The existence of identical particles in a body means that, in the quantum theory, exchange effects must be taken into account. This means, first of all, that the wave functions (33.1) must be made symmetrical or antisymmetrical in the particle co-ordinates, depending on the statistics obeyed by the particles. It is found, however, that this effect leads only to exponentially small terms in the free energy, and so is of no interest. Secondly, the identity of particles in quantum mechanics affects the manner in which the summation over different values of the particle momenta must be carried out. We shall meet this later, for example in calculating statistical sums for an ideal quantum gas. The effect produces a term of the third order in $\hbar$ in the free energy (as shown later) and so again does not affect the terms of order $\hbar^2$ which we shall calculate here. Thus the exchange effects can be ignored in the calculation.

In each of the wave functions (33.1) the momenta p_i have definite constant values. The possible values of each p_i form a dense discrete set (the distances between neighbouring values being inversely proportional to the linear dimensions of the volume occupied by the system†). The summation of the matrix elements $\exp(-\beta\hat{H})_{pp}$ with respect to all possible values of the momenta may therefore be replaced by integration with respect to $p(\mathrm{d}p = \mathrm{d}p_1\,\mathrm{d}p_2 \ldots \mathrm{d}p_s)$, bearing in mind that the number of quantum states "belonging" to the volume $V^N\,\mathrm{d}p$ of phase space (all values of the co-ordinates of each particle in the volume V and values of the momenta in $\mathrm{d}p$) is

$$V^N\,\mathrm{d}p/(2\pi\hbar)^s.$$

We shall use the notation

$$I = \exp\left[-(i/\hbar)\sum_i p_i q_i\right]\exp(-\beta\hat{H})\exp\left[(i/\hbar)\sum_i p_i q_i\right]. \tag{33.2}$$

The required matrix elements are obtained by integrating with respect to all the co-ordinates:

$$\exp(-\beta\hat{H})_{pp} = \frac{1}{V^N}\int I\,\mathrm{d}q. \tag{33.3}$$

The partition function is then obtained by integration with respect to the momenta.

Altogether, therefore, we must integrate I over all phase space, or more precisely over those of its regions which correspond to physically different states of the body, as explained in §31. This is again denoted by a prime to the integral sign:

$$Z \equiv \sum_n e^{-\beta E_n} = \int' I\,\mathrm{d}\Gamma. \tag{33.4}$$

† See *Quantum Mechanics*, §22.

Let us first calculate I by means of the following procedure. We take the derivative

$$\frac{\partial I}{\partial \beta} = -\exp[-(i/\hbar)\sum p_i q_i]\hat{H}\{\exp[(i/\hbar)\sum p_i q_i]I\},$$

and expand the right-hand side, using the explicit expression for the Hamiltonian of the body:

$$\hat{H} = \sum_i \frac{\hat{p}_i^2}{2m_i} + U = -\tfrac{1}{2}\hbar^2 \sum_i \frac{1}{m_i}\frac{\partial^2}{\partial q_i^2} + U, \tag{33.5}$$

where $U = U(q_1, q_2, \ldots, q_s)$ is the potential energy of interaction between all particles in the body. By means of (33.5) we obtain after a straightforward calculation the following equation for I:

$$\frac{\partial I}{\partial \beta} = -E(p, q)I + \sum_i \frac{\hbar^2}{2m_i}\left(\frac{2i}{\hbar}p_i\frac{\partial I}{\partial q_i} + \frac{\partial^2 I}{\partial q_i^2}\right),$$

where

$$E(p, q) = \sum_i \frac{p_i^2}{2m_i} + U \tag{33.6}$$

is the usual classical expression for the energy of the body.

This equation is to be solved with the obvious condition that $I = 1$ when $\beta = 0$. The substitution

$$I = e^{-\beta E(p,q)}\chi \tag{33.7}$$

gives

$$\frac{\partial \chi}{\partial \beta} = \sum_i \frac{\hbar^2}{2m_i}\left[-\frac{2i\beta p_i}{\hbar}\frac{\partial U}{\partial q_i}\chi + \frac{2ip_i}{\hbar}\frac{\partial \chi}{\partial q_i} - \beta\chi\frac{\partial^2 U}{\partial q_i^2} + \beta^2\chi\left(\frac{\partial U}{\partial q_i}\right)^2 - \right.$$
$$\left. -2\beta\frac{\partial \chi}{\partial q_i}\frac{\partial U}{\partial q_i} + \frac{\partial^2 \chi}{\partial q_i^2}\right] \tag{33.8}$$

with the boundary condition $\chi = 1$ for $\beta = 0$.

In order to obtain an expansion in powers of $\hbar$, we solve equation (33.8) by successive approximations, putting

$$\chi = 1 + \hbar\chi_1 + \hbar^2\chi_2 + \ldots, \tag{33.9}$$

with $\chi_1 = 0$, $\chi_2 = 0, \ldots$ for $\beta = 0$. Substituting this expansion in equation (33.8) and separating terms in different powers of $\hbar$, we obtain the equations

$$\frac{\partial \chi_1}{\partial \beta} = -i\beta\sum_i \frac{p_i}{m_i}\frac{\partial U}{\partial q_i},$$

$$\frac{\partial \chi_2}{\partial \beta} = \sum_i \frac{1}{2m_i}\left[-2i\beta p_i\frac{\partial U}{\partial q_i}\chi_1 + 2ip_i\frac{\partial \chi_1}{\partial q_i} - \beta\frac{\partial^2 U}{\partial q_i^2} + \beta^2\left(\frac{\partial U}{\partial q_i}\right)^2\right].$$

The first equation gives χ_1, and then the second equation gives χ_2. A simple calculation leads to the results

$$\begin{aligned}
\chi_1 &= -\frac{1}{2} i\beta^2 \sum_i \frac{p_i}{m_i} \frac{\partial U}{\partial q_i}, \\
\chi_2 &= -\frac{1}{8}\beta^4 \left(\sum_i \frac{p_i}{m_i}\frac{\partial U}{\partial q_i}\right)^2 + \frac{1}{6}\beta^3 \sum_i \sum_k \frac{p_i}{m_i}\frac{p_k}{m_k}\frac{\partial^2 U}{\partial q_i \partial q_k} + \\
&\quad + \frac{1}{6}\beta^3 \sum_i \frac{1}{m_i}\left(\frac{\partial U}{\partial q_i}\right)^2 - \frac{1}{4}\beta^2 \sum_i \frac{1}{m_i}\frac{\partial^2 U}{\partial q_i^2}.
\end{aligned} \tag{33.10}$$

The required partition function (33.4) is

$$Z = \int' (1 + \hbar\chi_1 + \hbar^2\chi_2) e^{-\beta E(p,q)} \, d\Gamma. \tag{33.11}$$

The term of the first order in $\hbar$ in this integral is easily seen to be zero, since the integrand $\chi_1 e^{-\beta E(p,q)}$ in that term is an odd function of the momenta ($E(p, q)$ being quadratic in the momenta and χ_1, by (33.10), linear), and so the result on integrating with respect to momenta is zero. Thus we can write (33.11) as

$$Z = (1 + \hbar^2 \overline{\chi_2}) \int' e^{-\beta E(p,q)} \, d\Gamma,$$

where $\overline{\chi_2}$ is the value of χ_2 averaged over the classical Gibbs distribution:

$$\overline{\chi_2} = \frac{\int' \chi_2 e^{-\beta E(p,q)} \, d\Gamma}{\int' e^{-\beta E(p,q)} \, d\Gamma}.$$

Substituting this expression for the partition function in formula (31.3), we have for the free energy

$$F = F_{\text{cl}} - \frac{1}{\beta} \log(1 + \hbar^2 \overline{\chi_2}),$$

or, to the same accuracy,

$$F = F_{\text{cl}} - \hbar^2 \overline{\chi_2}/\beta. \tag{33.12}$$

Here F_{cl} denotes the expression for the free energy in classical statistics (formula (31.5)).

Thus the next term after the classical expression in the expansion of the free energy is of the second order in $\hbar$. This is not accidental: in equation (33.8), solved here by the method of successive approximations, the quantum constant appears only as $i\hbar$, and so the resulting expansion is one in powers of $i\hbar$; but the free energy, being a real quantity, can contain only powers of $i\hbar$ which are real. Thus this expansion of the free energy (ignoring exchange effects) is an expansion in even powers of $\hbar$.

It remains to calculate the mean value $\overline{\chi_2}$. We have seen in §29 that in classical statistics the probability distributions for the co-ordinates and momenta are independent. The averaging over momenta and over co-ordinates can therefore be made separately.

The mean value of the product of two different momenta is clearly zero: $\overline{p_i p_k} = \overline{p_i}\cdot\overline{p_k} = 0$. The mean value of the square p_i^2 is m_i/β. We can therefore write

$$\overline{p_i p_k} = (m_i/\beta)\delta_{ik},$$

where $\delta_{ik} = 1$ for $i = k$ and 0 for $i \neq k$. Having averaged with respect to momenta by means of this formula, we obtain

$$\overline{\chi_2} = \frac{\beta^3}{24}\sum_i \frac{1}{m_i}\overline{\left(\frac{\partial U}{\partial q_i}\right)^2} - \frac{\beta^2}{12}\sum_i \frac{1}{m_i}\overline{\frac{\partial^2 U}{\partial q_i^2}}. \tag{33.13}$$

The two terms here may be combined, since the mean values are related by the formula

$$\overline{\frac{\partial^2 U}{\partial q_i^2}} = \beta\overline{\left(\frac{\partial U}{\partial q_i}\right)^2}. \tag{33.14}$$

This is easily seen by noticing that

$$\int \frac{\partial^2 U}{\partial q_i^2} e^{-\beta U}\, dq_i = \frac{\partial U}{\partial q_i} e^{-\beta U} + \beta \int \left(\frac{\partial U}{\partial q_i}\right)^2 e^{-\beta U}\, dq_i.$$

The first term on the right-hand side gives only a surface effect in $\overline{\partial^2 U/\partial q_i^2}$, and since the body is macroscopic this effect may be neglected in comparison with the second term.

Substituting the resulting expression for $\overline{\chi_2}$ in formula (33.12), and replacing β by $1/T$, we find the following final expression for the free energy:

$$F = F_{\text{cl}} + \frac{\hbar^2}{24T^2}\sum_i \frac{1}{m_i}\overline{\left(\frac{\partial U}{\partial q_i}\right)^2}. \tag{33.15}$$

We see that the correction to the classical value is always positive, and is determined by the mean squares of the forces acting on the particles. This correction decreases with increasing particle mass and increasing temperature.

According to the above discussion, the next term in the expansion given here would be of the fourth order. This enables us to calculate quite independently the term of order $\hbar^3$ which occurs in the free energy because of the peculiarities of the summation over momenta resulting from the identity of particles in quantum mechanics. The term in question is formally the same as the correction term which appears in a similar calculation for an ideal gas, and is given by formula (55.14):

$$F^{(3)} = \pm\frac{\pi^{3/2}}{2g}\frac{N^2\hbar^3}{VT^{1/2}m^{3/2}} \tag{33.16}$$

for a body consisting of N identical particles. The upper sign applies for Fermi statistics and the lower sign for Bose statistics; g is the total degree of degeneracy with respect to the directions of the electron and nuclear angular momenta.

From these formulae we can also obtain the correction terms in the probability distribution functions for the co-ordinates and momenta of the atoms of the body. According to the general results in §5, the momentum probability distribution is given by the integral of I with respect to q (see (5.10)):

$$dw_p = \text{constant} \times dp \int I \, dq.$$

The term $\chi_1 e^{-\beta E(p,q)}$ in I contains a total derivative with respect to the co-ordinates, and the integral of it gives a surface effect which can be neglected. Thus we have

$$dw_p = \text{constant} \times \exp\Big(-\beta \sum_i p_i^2/2m_i\Big) dp \int (1+\hbar^2 \chi_2) e^{-\beta U} dq.$$

The third and fourth terms in the expression (33.10) for χ_2 give a small constant (not involving the momenta) on integration, and this can be neglected in the same approximation. Taking out also the factor $\int e^{-\beta U} dq$ and including it in the constant coefficient, we have

$$dw_p = \text{constant} \times \exp\Big(-\beta \sum_i p_i^2/2m_i\Big)\Big[1 - \hbar^2 \frac{\beta^4}{8} \sum_i \sum_k \frac{p_i p_k}{m_i m_k} \overline{\frac{\partial U}{\partial q_i} \frac{\partial U}{\partial q_k}} + $$
$$+ \hbar^2 \frac{\beta^3}{6} \sum_i \sum_k \frac{p_i p_k}{m_i m_k} \overline{\frac{\partial^2 U}{\partial q_i \partial q_k}}\Big] dp.$$

The mean values which appear here are related by

$$\overline{\frac{\partial^2 U}{\partial q_i \partial q_k}} = \beta \overline{\frac{\partial U}{\partial q_i} \frac{\partial U}{\partial q_k}},$$

similarly to (33.14). Hence

$$dw_p = \text{constant} \times \exp\Big(-\beta \sum_i p_i^2/2m_i\Big)\Big[1 + \frac{\hbar^2 \beta^4}{24} \sum_i \sum_k \frac{p_i p_k}{m_i m_k} \overline{\frac{\partial U}{\partial q_i} \frac{\partial U}{\partial q_k}}\Big] dp. \tag{33.17}$$

This expression can be conveniently rewritten in the following final form:

$$dw_p = \text{constant} \times \exp\Big\{-\frac{1}{T}\Big[\sum_i \frac{p_i^2}{2m_i} - \frac{\hbar^2}{24 T^3} \sum_i \sum_k \frac{p_i p_k}{m_i m_k} \overline{\frac{\partial U}{\partial q_i} \frac{\partial U}{\partial q_k}}\Big]\Big\} dp, \tag{33.18}$$

the bracket in (33.17) being replaced by an exponential function to the same degree of accuracy.

Thus we see that the correction to the classical distribution function for the momenta is equivalent to adding to the kinetic energy in the exponent an expression quadratic in the momenta, with coefficients depending on the law of interaction between the particles in the body.

If it is desired to find the probability distribution for any one momentum p_i, then (33.17) must be integrated with respect to all the other momenta. All the terms involving the squares p_k^2 $(k \neq i)$ will then give constants negligible compared with unity, while the terms containing products of different momenta give zero. The result is, again in exponential form,

$$dw_{p_i} = \text{constant} \times \exp\left\{-\frac{p_i^2}{2m_i T}\left[1 - \frac{\hbar^2}{12T^3 m_i}\overline{\left(\frac{\partial U}{\partial q_i}\right)^2}\right]\right\} dp_i. \quad (33.19)$$

Thus the distribution obtained differs from the Maxwellian only in that the true temperature T is replaced by a somewhat higher "effective temperature":

$$dw_{p_i} = \text{constant} \times \exp\{-p_i^2/2m_i T_{\text{eff}}\}\, dp_i,$$

where

$$T_{\text{eff}} = T + \frac{\hbar^2}{12T^2 m_i}\overline{\left(\frac{\partial U}{\partial q_i}\right)^2}.$$

Similarly we can calculate the corrected co-ordinate distribution function by integrating I with respect to the momenta:

$$dw_q = \text{constant} \times dq \int I\, dp.$$

The same calculations as led to (33.13) give

$$dw_q = \text{constant} \times e^{-\beta U}\left[1 + \frac{\hbar^2\beta^3}{24}\sum_i \frac{1}{m_i}\left(\frac{\partial U}{\partial q_i}\right)^2 - \frac{\hbar^2\beta^2}{12}\sum_i \frac{1}{m_i}\frac{\partial^2 U}{\partial q_i^2}\right] dq,$$

or, in exponential form,

$$dw_q = \text{constant} \times \exp\left\{-\frac{1}{T}\left[U - \frac{\hbar^2}{24T^2}\sum_i \frac{1}{m_i}\left(\frac{\partial U}{\partial q_i}\right)^2 + \frac{\hbar^2}{12T}\sum_i \frac{1}{m_i}\frac{\partial^2 U}{\partial q_i^2}\right]\right\} dq. \quad (33.20)$$

§34. The Gibbs distribution for rotating bodies

The problem of the thermodynamic relations for rotating bodies has already been considered in §26. Let us now see how the Gibbs distribution is to be formulated for rotating bodies. This will complete the investigation of their statistical properties. As regards the uniform translational motion, GALILEO's relativity principle shows that, as already mentioned in §26, this motion has only a trivial effect on the statistical properties and so needs no special consideration.

In a system of co-ordinates rotating with the body, the usual Gibbs distribution is valid; in classical statistics,

$$\varrho = (2\pi\hbar)^{-s} e^{[F' - E'(p,q)]/T}, \quad (34.1)$$

where $E'(p, q)$ is the energy of the body in this system, as a function of the co-ordinates and momenta of its particles, and F' the free energy in the same system (which, of course, is not the same as the free energy of the body when at rest). The energy $E'(p, q)$ is related to the energy $E(p, q)$ in a fixed system by

$$E'(p, q) = E(p, q) - \boldsymbol{\Omega}\cdot\mathbf{M}(p, q), \tag{34.2}$$

where $\boldsymbol{\Omega}$ is the angular velocity of rotation and $\mathbf{M}(p, q)$ the angular momentum of the body (see §26). Substituting (34.2) in (34.1), we find the Gibbs distribution for a rotating body in the form†

$$\varrho = (2\pi\hbar)^{-s} e^{[F' - E(p, q) + \boldsymbol{\Omega}\cdot\mathbf{M}(p, q)]/T}. \tag{34.3}$$

In classical statistics the Gibbs distribution for a rotating body can also be represented in another form. To obtain this, we use the following expression for the energy of the body in the rotating co-ordinate system:

$$E' = \sum \tfrac{1}{2}mv'^2 - \tfrac{1}{2}\sum m(\boldsymbol{\Omega}\times\mathbf{r})^2 + U, \tag{34.4}$$

where the $\mathbf{v}'$ are the velocities of the particles relative to the rotating system, and the $\mathbf{r}$ their radius vectors.‡ Denoting by

$$E_0(\mathbf{v}', \mathbf{r}) = \sum \tfrac{1}{2} mv'^2 + U \tag{34.5}$$

the part of the energy which is independent of $\boldsymbol{\Omega}$, we obtain the Gibbs distribution in the form

$$\varrho = (2\pi\hbar)^{-s} \exp\left\{\frac{1}{T}\left[F' - E_0(\mathbf{v}', \mathbf{r}) + \tfrac{1}{2}\sum m(\boldsymbol{\Omega}\times\mathbf{r})^2\right]\right\}.$$

The function ϱ determines the probability corresponding to the element of phase space $dx_1\, dy_1\, dz_1 \ldots dp'_{1x}\, dp'_{1y}\, dp'_{1z} \ldots$, where $\mathbf{p}' = m\mathbf{v}' + m\boldsymbol{\Omega}\times\mathbf{r}$.‡ Since, in obtaining the differentials of the momenta, we must regard the co-ordinates as constant, $d\mathbf{p}' = m\, d\mathbf{v}'$, and the probability distribution expressed in terms of the co-ordinates and velocities of the particle is

$$dw = C \exp\left\{\frac{F'}{T} - \frac{1}{T}\left[E_0(\mathbf{v}', \mathbf{r}) - \sum \tfrac{1}{2} m(\boldsymbol{\Omega}\times\mathbf{r})^2\right]\right\} \times$$
$$\times dx_1\, dy_1\, dz_1 \ldots dv'_{1x}\, dv'_{1y}\, dv'_{1z} \ldots, \tag{34.6}$$

where C denotes for brevity the factor $(2\pi\hbar)^{-s}$ together with the product of the particle masses which appears when we go from the momentum differentials to the velocity differentials.

For a body at rest we have

$$dw = Ce^{[F - E_0(\mathbf{v}, \mathbf{r})]/T}\, dx_1\, dy_1\, dz_1 \ldots dv_{1x}\, dv_{1y}\, dv_{1z} \ldots, \tag{34.7}$$

† The distribution (34.3), like the ordinary Gibbs distribution, is fully in agreement with the result (4.2) derived in §4 from Liouville's theorem: the logarithm of the distribution function is a linear function of the energy and angular momentum of the body.

‡ See *Mechanics*, §39.

with the same expression (34.5) for $E_0(\mathbf{v}, \mathbf{r})$, now a function of the velocities in the fixed co-ordinate system. Thus we see that the Gibbs distribution for the co-ordinates and velocities for a rotating body differs from that for a body at rest only by the additional potential energy $-\frac{1}{2}\sum m(\boldsymbol{\Omega}\times\mathbf{r})^2$. In other words, as regards the statistical properties of the body, the rotation is equivalent to the existence of an external field corresponding to the centrifugal force. The statistical properties are not affected by the Coriolis force.

It should be emphasised, however, that this last result applies only to classical statistics. In the quantum case the expression

$$\hat{w} = \exp\left[(F' - H + \boldsymbol{\Omega}\cdot\mathbf{M})/T\right] \tag{34.8}$$

gives the statistical operator corresponding to (34.3) for a rotating body. Formally we can reduce this operator to a form analogous to (34.6), the velocities $\mathbf{v}'$ being replaced by the operators $\hat{\mathbf{v}}' = \hat{\mathbf{p}}'/m - \boldsymbol{\Omega}\times\mathbf{r}$, but the components of this vector operator do not commute, unlike those of the operator $\hat{\mathbf{v}}$ in the fixed system. The statistical operators corresponding to the expressions (34.6) and (34.7) will therefore in general be markedly different from each other, quite apart from the fact that one of them contains the centrifugal energy.

§35. The Gibbs distribution for a variable number of particles

So far we have always tacitly assumed that the number of particles in a body is some given constant, and have deliberately passed over the fact that in reality particles may be exchanged between different subsystems. In other words, the number N of particles in a subsystem will necessarily fluctuate about its mean value. In order to formulate precisely what we mean by the number of particles, we shall use the term *subsystem* to refer to a part of the system which is enclosed in a fixed volume. Then N will denote the number of particles within that volume.†

Thus the problem arises of generalising the Gibbs distribution to bodies with a variable number of particles. Here we shall write the formulae for bodies consisting of identical particles; the further generalisation to systems containing different particles is obvious (§86).

The distribution function now depends not only on the energy of the quantum state but also on the number N of particles in the body, and the energy levels E_{nN} are of course themselves different for different N (as indicated by the suffix N). The probability that the body contains N particles and is in the nth state will be denoted by w_{nN}.

†In deriving the Gibbs distribution in §28 we have in essence already understood subsystems in this sense; in going from (28.2) to (28.3) we differentiated the entropy whilst regarding the volume of the body (and therefore of the medium) as constant.

The form of this function can be determined in exactly the same way as the function w_n in §28. The only difference is that the entropy of the medium is now a function not only of its energy E' but also of the number N' of particles in it: $S' = S'(E', N')$. Writing $E' = E^{(0)} - E_{nN}$ and $N' = N^{(0)} - N$ (where N is the number of particles in the body, and $N^{(0)}$ the given total number of particles in the entire closed system, which is large compared with N), we have in accordance with (28.2)

$$w_{nN} = \text{constant} \times \exp\{S'(E^{(0)} - E_{nN}, N^{(0)} - N)\};$$

the quantity $\Delta E'$ is regarded as constant, as in §28.

Next, we expand S' in powers of E_{nN} and N, again taking only the linear terms. Equation (24.5), in the form

$$\mathrm{d}S = \frac{\mathrm{d}E}{T} + \frac{P}{T}\mathrm{d}V - \frac{\mu}{T}\mathrm{d}N,$$

shows that $(\partial S/\partial E)_{V,N} = 1/T$, $(\partial S/\partial N)_{E,V} = -\mu/T$. Hence

$$S'(E^{(0)} - E_{nN}, N^{(0)} - N) \cong S'(E^{(0)}, N^{(0)}) - \frac{E_{nN}}{T} + \frac{\mu N}{T},$$

the chemical potential μ (and the temperature) being the same for the body and the medium, from the conditions of equilibrium.

Thus we obtain for the distribution function the expression

$$w_{nN} = A e^{(\mu N - E_{nN})/T}. \tag{35.1}$$

The normalisation constant A can be expressed in terms of the thermodynamic quantities in the same way as in §31. The entropy of the body is

$$S = -\overline{\log w_{nN}} = -\log A - \frac{\mu\bar{N}}{T} + \frac{\bar{E}}{T},$$

and so

$$T\log A = \bar{E} - TS - \mu\bar{N}.$$

But $\bar{E} - TS = F$, and the difference $F - \mu\bar{N}$ is the thermodynamic potential Ω. Thus $T\log A = \Omega$, and (35.1) may be rewritten as

$$w_{nN} = e^{(\Omega + \mu N - E_{nN})/T}. \tag{35.2}$$

This is the final formula for the Gibbs distribution for a variable number of particles.

The normalisation condition for the distribution (35.2) requires that the result of summing the w_{nN} first over all quantum states (for a given N) and then over all values of N should be equal to unity:

$$\sum_N \sum_n w_{nN} = e^{\Omega/T} \sum_N \left(e^{\mu N/T} \sum_n e^{-E_{nN}/T}\right) = 1.$$

Hence we obtain the following expression for the thermodynamic potential Ω:

$$\Omega = -T\log \sum_N \left[e^{\mu N/T} \sum_n e^{-E_{nN}/T}\right]. \tag{35.3}$$

This formula together with (31.3) can be used to alculate the thermodynamic quantities for specific bodies. Formula (31.3) gives the free energy of the body as a function of T, N and V, and (35.3) gives the potential Ω as a function of T, μ and V.

In classical statistics the probability distribution has the form

$$dw_N = \varrho_N \, dp^{(N)} \, dq^{(N)},$$

where

$$\varrho_N = (2\pi\hbar)^{-s} e^{[\Omega + \mu N - E_N(p,\, q)]/T}. \tag{35.4}$$

The variable N is written as a subscript to the distribution function, and the same letter is written as a superscript to the element of phase volume in order to emphasise that a different phase space (of $2s$ dimensions) corresponds to each value of N. The formula for Ω correspondingly becomes

$$\Omega = -T \log \{\sum_N e^{\mu N/T} \int' e^{-E_N(p,q)/T} \, d\Gamma_N\}. \tag{35.5}$$

Finally, we may say a few words concerning the relation between the Gibbs distribution (35.2) for a variable number of particles derived here and the previous distribution (31.1). First of all, it is clear that, for the determination of all the statistical properties of the body except the fluctuations of the total number of particles in it, these two distributions are entirely equivalent. On neglecting the fluctuations of the number N, we obtain $\Omega + \mu N = F$, and the distribution (35.2) is identical with (31.1).

The relation between the distributions (31.1) and (35.2) is to a certain extent analogous to that between the microcanonical and canonical distributions. The description of a system by means of the microcanonical distribution is equivalent to neglecting the fluctuations of its total energy; the canonical distribution in its usual form (31.1) takes into account these fluctuations. The latter form in turn neglects the fluctuations in the number of particles, and may be said to be "microcanonical with respect to the number of particles"; the distribution (35.2) is "canonical" with respect to both the energy and the number of particles.

Thus all three distributions, the microcanonical and the two forms of the Gibbs distribution, are in principle suitable for determining the thermodynamic properties of the body. The only difference from this point of view lies in the degree of mathematical convenience. In practice the microcanonical distribution is the least convenient and is never used for this purpose. The Gibbs distribution for a variable number of particles is usually the most convenient.

§36. The derivation of the thermodynamic relations from the Gibbs distribution

The Gibbs distribution plays a fundamental part throughout statistical physics. We shall therefore give here another justification of it. This distribution has essentially been derived in §4 and 6 directly from LIOUVILLE'S theorem. We have seen that the application of LIOUVILLE'S theorem (together with considerations of the multiplicativity of distribution functions for subsystems) enables us to deduce that the logarithm of the distribution function of a subsystem must be a linear function of its energy:

$$\log w_n = \alpha + \beta E_n, \tag{36.1}$$

the coefficients β being the same for all subsystems in a given closed system (see (6.4), and the corresponding relation (4.5) for the classical case). Hence

$$w_n = e^{\alpha + \beta E_n};$$

using the purely formal notation $\beta = -1/T$, $\alpha = F/T$, we have an expression of the same form as the Gibbs distribution (31.1). It remains to show that the fundamental thermodynamic relations can be derived from the Gibbs distribution itself, i.e. in a purely statistical manner.

We have already seen that the quantity β, and therefore T, must be the same for all parts of a system in equilibrium. It is also evident that $\beta < 0$, i.e. $T > 0$, since otherwise the normalisation sum $\sum w_n$ must diverge: owing to the presence of the kinetic energy of the particles, the energy E_n can take arbitrarily large values. All these properties agree with the fundamental properties of the thermodynamic temperature.

To derive a quantitative relation, we start from the normalisation condition

$$\sum_n e^{(F-E_n)/T} = 1.$$

We differentiate this equation, regarding the left-hand side as a function of T and of various quantities $\lambda_1, \lambda_2, \ldots$ which represent the external conditions to which the body considered is subject; these quantities may, for example, determine the shape and size of the volume occupied by the body. The energy levels E_n depend on $\lambda_1, \lambda_2, \ldots$ as parameters.

Differentiation gives

$$\sum_n \frac{w_n}{T}\left[\mathrm{d}F - \frac{\partial E_n}{\partial \lambda}\,\mathrm{d}\lambda - \frac{F-E_n}{T}\,\mathrm{d}T\right] = 0,$$

where for simplicity only one external parameter is used. Hence

$$\mathrm{d}F \sum_n w_n = \mathrm{d}\lambda \sum_n w_n \frac{\partial E_n}{\partial \lambda} + \frac{\mathrm{d}T}{T}\Big(F - \sum_n w_n E_n\Big).$$

On the left-hand side $\sum w_n = 1$, and on the right-hand side

$$\sum_n w_n E_n = \bar{E}, \quad \sum_n w_n \frac{\partial E_n}{\partial \lambda} = \overline{\frac{\partial E_n}{\partial \lambda}}.$$

Using also the formulae $F-\overline{E} = -TS$ and†

$$\overline{\partial E_n/\partial\lambda} = \overline{\partial\hat{H}/\partial\lambda}, \tag{36.2}$$

we have finally

$$\mathrm{d}F = -S\,\mathrm{d}T + \overline{\partial\hat{H}/\partial\lambda}\cdot\mathrm{d}\lambda.$$

This is the general form for the differential of the free energy.

In the same way we can derive the Gibbs distribution for a variable number of particles. If the number of particles is regarded as a dynamical variable, it is clear that it will be an "integral of the motion", and additive, for a closed system. We must therefore write

$$\log w_{nN} = \alpha + \beta E_n + \gamma N, \tag{36.3}$$

where γ, like β, must be the same for all parts of a system in equilibrium. Putting $\alpha = \Omega/T$, $\beta = -1/T$, $\gamma = \mu/T$, we obtain a distribution of the form (35.2), and then by the same method as above we can deduce an expression for the differential of the potential Ω.

† If the Hamiltonian $\hat{H}$ (and therefore its eigenvalues E_n) depends on a parameter λ, then

$$\partial E_n/\partial\lambda = (\partial\hat{H}/\partial\lambda)_{nn};$$

see *Quantum Mechanics*, §11, Problem. On statistical averaging this gives (36.2).

CHAPTER IV

IDEAL GASES

§37. The Boltzmann distribution

ONE of the most important subjects of study in statistical physics is an *ideal gas*. By this is meant a gas in which the interaction between the particles (molecules) is so weak as to be negligible. Physically, this approximation may be allowable either because the interaction of the particles is small whatever the distances between them or because the gas is sufficiently rarefied. In the latter case, which is the more important, the rarefaction of the gas results in its molecules' being almost always at considerable distances apart, such that the interaction forces are quite small.

The absence of interaction between the molecules enables the quantum-mechanics problem of determining the energy levels E_n of the gas as a whole to be reduced to that of determining the energy levels of a single molecule. These levels will be denoted by ε_k, the suffix k representing the set of quantum numbers which define the state of the molecule. The energies E_n are then given by the sums of the energies of the various molecules.

It must be remembered, however, that, even when there is no direct force interaction, quantum mechanics gives a peculiar mutual effect of particles resulting from their identity (called the *exchange effect*). For example, if the particles "obey Fermi statistics", this effect has the result that no more than one particle can be in each quantum state at one time†; a similar effect but in a different form occurs for particles which "obey Bose statistics".

Let n_k be the number of particles in a gas which are in the kth quantum state; the numbers n_k are sometimes called the *occupation numbers* of the various quantum states. Let us consider the problem of calculating the mean values $\overline{n_k}$ of these numbers, and take in particular the extremely important case where for all k

$$\overline{n_k} \ll 1. \tag{37.1}$$

Physically this case corresponds to a sufficiently rarefied gas. We shall later

† It should be emphasised that, when speaking of the quantum state of an individual particle, we shall always refer to states which are fully determined by a set of values of all the quantum numbers (including the orientation of the angular momentum of the particle, if any). These should not be confused with the quantum energy levels; several different quantum states correspond to a given energy level if the latter is degenerate.

establish a criterion which ensures the fulfilment of this condition, but it may be mentioned immediately that it is in practice satisfied for all ordinary molecular or atomic gases. The condition would be violated only at such high densities that the matter concerned certainly could not be regarded as an ideal gas.

The condition $\overline{n_k} \ll 1$ for the mean occupation numbers signifies that in fact not more than one particle is in each quantum state at any instant. Consequently, we may neglect not only the direct forces of interaction of the particles but also their indirect quantum interactions mentioned above. This in turn enables us to apply the Gibbs distribution formula to the individual molecules. For the Gibbs distribution has been derived for bodies which are relatively small, but at the same time macroscopic, parts of large closed systems. The macroscopic nature of these bodies made it possible to regard them as quasi-closed, i.e. to neglect to some extent their interaction with other parts of the system. In the case under consideration the separate molecules of the gas are quasi-closed, although they are certainly not macroscopic bodies.

Applying the Gibbs distribution formula to the gas molecules, we can say that the probability that a molecule is in the kth state is proportional to $e^{-\varepsilon_k/T}$, and therefore so is the mean number $\overline{n_k}$ of molecules in that state, i.e.

$$\overline{n_k} = ae^{-\varepsilon_k/T}, \tag{37.2}$$

where a is a constant given by the normalisation condition

$$\sum_k \overline{n_k} = N \tag{37.3}$$

(N being the total number of particles in the gas). The distribution of molecules of an ideal gas among the various states that is given by formula (37.2) is called the *Boltzmann distribution*; it was discovered by BOLTZMANN for classical statistics in 1877.

The constant coefficient in (37.2) can be expressed in terms of the thermodynamic quantities for the gas. To do this we shall give another derivation of the formula, based on the application of the Gibbs distribution to the assembly of all particles in the gas that are in a given quantum state. We are able to do this (even if the numbers n_k are not small) since there is no direct force of interaction between these particles and the remainder (or between any of the particles in an ideal gas), and the quantum exchange effects occur only for particles in the same state. Putting $E = n_k\varepsilon_k$, $N = n_k$ and adding the suffix k to Ω in the general formula for the Gibbs distribution for a variable number of particles (35.2), we find the probability distribution for various values of n_k as

$$w_{n_k} = e^{[\Omega_k + n_k(\mu-\varepsilon_k)]/T}. \tag{37.4}$$

In particular, $w_0 = e^{\Omega_k/T}$ is the probability that there are no particles in the state concerned. In the case of interest here, for which $\overline{n_k} \ll 1$, the probability w_0 is almost unity, and so in the expression $w_1 = e^{(\Omega_k+\mu-\varepsilon_k)/T}$ for the probability of finding one particle in the kth state we can put $e^{\Omega_k/T} = 1$ to within terms of a higher order of smallness. Then $w_1 = e^{(\mu-\varepsilon_k)/T}$. The probabilities of values $n_k > 1$ must be taken as zero in the same approximation. Hence

$$\overline{n_k} = \sum_{n_k} w_{n_k} n_k = w_1 \cdot 1,$$

and we have the Boltzmann distribution in the form

$$\overline{n_k} = e^{(\mu-\varepsilon_k)/T}. \qquad (37.5)$$

Thus the coefficient in (37.2) is expressed in terms of the chemical potential of the gas.

§38. The Boltzmann distribution in classical statistics

If the motion of gas molecules (and of the atoms in them) were subject to classical mechanics, we could use, instead of the distribution over quantum states, the distribution of molecules in phase space, i.e. over momenta and co-ordinates. Let dN be the mean number of molecules "contained" in a volume element of phase space of the molecule, $dp\,dq = dp_1 \ldots dp_r\, dq_1 \ldots dq_r$ (r being the number of degrees of freedom of the molecule). We may write this as

$$dN = n(p, q)\, d\tau, \qquad d\tau = dp\, dq/(2\pi\hbar)^r \qquad (38.1)$$

and call $n(p, q)$ the "density in phase space" (although $d\tau$ differs by a factor $(2\pi\hbar)^{-r}$ from the volume element in phase space). We then have, instead of (37.5),

$$n(p, q) = e^{[\mu-\varepsilon(p, q)]/T}, \qquad (38.2)$$

where $\varepsilon(p, q)$ is the energy of the molecule as a function of the co-ordinates and momenta of its atoms.

Usually, however, it is not the entire motion of the molecule which is quasi-classical, but only the motion corresponding to some of its degrees of freedom. In particular, in a gas which is not in an external field, the translational motion of molecules is always quasi-classical. The kinetic energy of the translational motion then appears in the energy ε_k of the molecule as an independent term, while the remaining part of the energy does not involve the co-ordinates x, y, z and momenta p_x, p_y, p_z of the centre of mass of the molecule. This enables us to separate from the general formula for the Boltzmann distribution a factor which gives the distribution of the gas molecules with respect to these variables. The distribution of the molecules in the volume occupied by the gas is clearly just a uniform distribution, and we obtain for the number of molecules per unit volume with momenta (of the translational motion) in given intervals dp_x, dp_y, dp_z the Maxwellian distribution:

$$dN_{\mathbf{p}} = \frac{N}{V(2\pi mT)^{3/2}} \exp\left[-(p_x^2+p_y^2+p_z^2)/2mT\right] dp_x\, dp_y\, dp_z, \qquad (38.3)$$

$$dN_{\mathbf{v}} = \frac{N}{V}\frac{m^{3/2}}{(2\pi T)^{3/2}} \exp\left[-m(v_x^2+v_y^2+v_z^2)/2T\right] dv_x\, dv_y\, dv_z \qquad (38.4)$$

(m being the mass of a molecule), normalised to N/V particles per unit volume.

Let us next consider a gas in an external field, in which the potential energy of a molecule depends only on the co-ordinates of its centre of mass: $u = u(x, y, z)$ (for example, a gravitational field). If, as always occurs in practice, the translational motion in this field is quasi-classical, then $u(x, y, z)$ appears in the energy of the molecule as an independent term. The Maxwellian distribution for the velocities of the molecules remains unchanged, of course, while the distribution for the centre of mass is given by the formula

$$dN_{\mathbf{r}} = n_0 e^{-u(x,y,z)/T}\, dV. \qquad (38.5)$$

This formula gives the number of molecules in an element of volume $dV = dx\, dy\, dz$; the quantity

$$n(\mathbf{r}) = n_0 e^{-u(x,y,z)/T} \qquad (38.6)$$

is the number density of the particles. The constant n_0 is the density at points where $u = 0$. Formula (38.6) is called *Boltzmann's formula.*

In particular, in a uniform gravitational field along the z-axis, $u = mgz$, and the gas density distribution is given by the *barometric formula*

$$n(z) = n_0 e^{-mgz/T}, \qquad (38.7)$$

where n_0 is the density at the level $z = 0$.

At large distances from the Earth, its gravitational field must be described by the exact Newtonian expression, the potential energy u vanishing at infinity. According to formula (38.6) the gas density should remain finite and not zero at infinity, but a finite quantity of gas cannot be distributed in an infinite volume with a density which is nowhere zero. This means that in a gravitational field a gas (such as the atmosphere) cannot be in equilibrium and must be continuously dissipated into space.

PROBLEMS

PROBLEM 1. Find the density of gas in a cylinder of radius R and length l rotating about its axis with angular velocity Ω, there being a total of N molecules in the cylinder.

SOLUTION. It has been mentioned in §34 that the rotation of a body as a whole is equivalent to the presence of an external field with potential energy $-\frac{1}{2}m\Omega^2r^2$ (where r is the distance from the axis of rotation). The gas density is therefore

$$n(r) = Ae^{m\Omega^2r^2/2T}.$$

Normalisation gives

$$n(r) = \frac{Nm\Omega^2 e^{m\Omega^2 r^2/2T}}{2\pi Tl(e^{m\Omega^2 r^2/2T}-1)}.$$

PROBLEM 2. Find the momentum distribution of particles for a relativistic ideal gas.

SOLUTION. The energy of a relativistic particle is given in terms of its momentum by $\varepsilon = c\sqrt{(m^2c^2+p^2)}$, where c is the velocity of light. The normalised momentum distribution is

$$dN_{\mathbf{p}} = \frac{N}{V}\frac{\exp\{-c\sqrt{(m^2c^2+p^2)}/T\}}{2(T/mc^2)^2K_1(mc^2/T)+(T/mc^2)K_0(mc^2/T)}\frac{dp_x\,dp_y\,dp_z}{4\pi(mc)^3},$$

where K_0 and K_1 are Macdonald functions (Hankel functions of imaginary argument). In the calculation of the normalisation integral the following formulae are used:

$$\int_0^\infty e^{-z\cosh t}\sinh^2 t\,dt = K_1(z)/z,$$

$$K_1'(z) = -K_1(z)/z - K_0(z).$$

§39. Molecular collisions

The molecules of a gas enclosed in a vessel collide with its walls as they move. Let us calculate the mean number of impacts between the molecules of a gas and a unit area of the wall per unit time.

We take an element of surface area of the vessel wall and define a co-ordinate system with the z-axis perpendicular to this element, which may then be written as $dx\,dy$. Of the molecules in the gas, those which reach the vessel wall in unit time, i.e. collide with it, are just those whose z co-ordinate does not exceed the component v_z of their velocity along that axis (which, of course, must also be directed towards the wall, not away from it).

The number $d\nu_{\mathbf{v}}$ of collisions of molecules per unit time (and per unit area of the wall surface), in which the velocity components are in given intervals dv_x, dv_y, dv_z is therefore obtained by multiplying the distribution (38.4) by the volume of a cylinder of unit base area and height v_z:

$$d\nu_{\mathbf{v}} = \frac{N}{V}\left(\frac{m}{2\pi T}\right)^{3/2}\exp[-m(v_x^2+v_y^2+v_z^2)/2T]\times v_z\,dv_x\,dv_y\,dv_z. \quad (39.1)$$

From this we easily find the total number ν of impacts of gas molecules on unit area of the vessel wall per unit time. To do so, we integrate (39.1) over all velocities v_z from 0 to ∞ and over v_x and v_y from $-\infty$ to ∞; integration over v_z from $-\infty$ to 0 is not required, since when $v_z < 0$ the molecule is travelling away from the wall, and so does not collide with it. Hence

$$\nu = \frac{N}{V}\sqrt{\frac{T}{2\pi m}} = \frac{P}{\sqrt{(2\pi mT)}}; \quad (39.2)$$

here we have expressed the density of the gas in terms of its pressure by means of CLAPEYRON's equation.

Formula (39.1) may be written in spherical polar co-ordinates in "velocity space", using instead of v_x, v_y, v_z the absolute magnitude of the velocity and the polar angles θ and ϕ which define its direction. Taking the polar axis along the z-axis, we have $v_z = v\cos\theta$ and

$$d\nu_v = \frac{N}{V}\left(\frac{m}{2\pi T}\right)^{3/2} e^{-mv^2/2T}v^3 \sin\theta \cos\theta \, d\theta \, d\phi \, dv. \tag{39.3}$$

Let us now consider collisions between gas molecules. To do this, we must first find the velocity distribution of the molecules (the term velocity everywhere referring to the velocity of the centre of mass) relative to one another. We take any one gas molecule and consider the motion of all the other molecules relative to it, i.e. consider for each molecule not its absolute velocity v (relative to the walls of the vessel) but its velocity v' relative to some other molecule. That is, instead of dealing with individual molecules, we always consider the relative motion of a pair of molecules, ignoring the motion of their common centre of mass.

We know from mechanics that the energy of the relative motion of two particles of masses m_1 and m_2 is $\frac{1}{2}m'v'^2$, where $m' = m_1m_2/(m_1+m_2)$ is their "reduced mass" and v' their relative velocity. The relative-velocity distribution of the molecules of an ideal gas therefore has the same form as the absolute-velocity distribution, except that m is replaced by the reduced mass m'. Since all the molecules are alike, $m' = \frac{1}{2}m$, and the number of molecules per unit volume with a velocity relative to the selected molecule between v' and $v'+dv'$ is

$$dN_{v'} = \frac{N}{V}\frac{\pi}{2}\left(\frac{m}{\pi T}\right)^{3/2} e^{-mv'^2/4T}v'^2 \, dv'. \tag{39.4}$$

A collision between molecules may be accompanied by various processes: deflection (scattering) through a certain angle, dissociation into atoms, and so on. The processes which occur in collisions are usually described by their *cross-sections*. The cross-section for a particular process which occurs in collisions between a given particle and others is the ratio of the probability of such a collision per unit time to the particle flux density (the latter being the number of such particles per unit volume multiplied by their velocity). The number of collisions (per unit time) between this and other particles which are accompanied by a certain process with cross-section σ is therefore

$$\nu' = \frac{N}{V}\frac{\pi}{2}\left(\frac{m}{\pi T}\right)^{3/2} \int_0^\infty e^{-mv'^2/4T}\sigma v'^3 \, dv'. \tag{39.5}$$

The total number of such collisions per unit time throughout the volume of the gas is obviously $\nu'N/2$.

PROBLEMS

PROBLEM 1. Find the number of impacts of gas molecules on unit area of the wall per unit time for which the angle between the direction of the velocity of the molecule and the normal to the surface lies between θ and $\theta + d\theta$.

SOLUTION.

$$d\nu_\theta = \frac{N}{V}\left(\frac{2T}{m\pi}\right)^{1/2} \sin\theta \cos\theta \, d\theta.$$

PROBLEM 2. Find the number of impacts of gas molecules on unit area of the wall per unit time for which the absolute magnitude of the velocity lies between v and $v + dv$.

SOLUTION.

$$d\nu_v = \frac{N}{V}\pi\left(\frac{m}{2\pi T}\right)^{3/2} e^{-mv^2/2T} v^3 \, dv.$$

PROBLEM 3. Find the total kinetic energy E_{kin} of the gas molecules striking unit area of the wall per unit time.

SOLUTION.

$$E_{\text{kin}} = \frac{N}{V}\sqrt{\frac{2T^3}{m\pi}} = P\sqrt{\frac{2T}{m\pi}}.$$

PROBLEM 4. Find the number of collisions between one molecule and the rest per unit time, assuming the molecules to be rigid spheres of radius r.

SOLUTION. The cross-section for collisions between molecules is then $\sigma = \pi(2r)^2 = 4\pi r^2$ (since a collision occurs whenever two molecules pass at a distance less than $2r$). Substitution in (39.5) gives

$$\nu = 16r^2\sqrt{\frac{\pi T}{m}}\frac{N}{V} = 16r^2\sqrt{\frac{\pi}{mT}}P.$$

§40. Ideal gases not in equilibrium

The Boltzmann distribution can also be derived, in a quite different manner, directly from the condition of maximum entropy of the gas as a whole, regarded as a closed system. This derivation is of considerable interest in itself, since it is based on a method whereby the entropy of gas in any non-equilibrium macroscopic state may be calculated.

Any macroscopic state of an ideal gas may be described as follows. Let us distribute all the quantum states of an individual particle of the gas among groups each containing neighbouring states (which, in particular, have neighbouring energy values), both the number of states in each group and the number of particles in these states being still very large. Let the groups of states be numbered $j = 1, 2, \ldots$, and let G_j be the number of states in group j, and N_j the number of particles in these states. Then the set of numbers N_j will completely describe the macroscopic state of the gas.

The problem of calculating the entropy of the gas reduces to that of determining the statistical weight $\Delta\Gamma$ of a given macroscopic state, i.e. the number of microscopic ways in which this state can be realised. Regarding each group of N_j particles as an independent system and denoting its statistical weight by $\Delta\Gamma_j$, we can write

$$\Delta\Gamma = \prod_j \Delta\Gamma_j. \tag{40.1}$$

Thus the problem reduces to that of calculating the $\Delta\Gamma_j$.

In Boltzmann statistics the mean occupation numbers of all quantum states are small in comparison with unity. This means that the numbers of particles N_j must be small compared with the numbers of states G_j ($N_j \ll G_j$), but of course themselves still large. As has been explained in §37, the smallness of the mean occupation numbers enables us to suppose that all the particles are entirely independently distributed among the various states. Placing each of the N_j particles in one of the G_j states, we obtain altogether $G_j^{N_j}$ possible distributions, but among these the distributions which differ only by a permutation of particles are identical, since the particles themselves are identical. The number of permutations of N_j particles is $N_j!$, and so the statistical weight of the distribution of N_j particles among G_j states is

$$\Delta\Gamma_j = G_j^{N_j}/N_j!. \tag{40.2}$$

The entropy of the gas is calculated as the logarithm of the statistical weight:

$$S = \log \Delta\Gamma = \sum \log \Delta\Gamma_j.$$

Substitution of (40.2) gives

$$S = \sum_j (N_j \log G_j - \log N_j!).$$

Since the numbers N_j are large, we can use the approximate formula† $\log N_j! = N_j \log (N_j/e)$, obtaining

$$S = \sum_j N_j \log (eG_j/N_j). \tag{40.3}$$

This formula gives the solution of the problem, determining the entropy of an ideal gas in any macroscopic state defined by the set of numbers N_j. It may be rewritten by using the mean numbers $\bar{n}_j$ of particles in each of the quantum states in group j:

$$\bar{n}_j = N_j/G_j. \tag{40.4}$$

Then

$$S = \sum_j G_j \bar{n}_j \log (e/\bar{n}_j). \tag{40.5}$$

If the motion of the particles is quasi-classical, then in this formula we can change to the particle distribution in phase space. Let the phase space of a particle be divided into regions $\Delta p^{(j)} \Delta q^{(j)}$, each of which is small but

† When N is large, the sum $\log N! = \log 1 + \log 2 + \ldots + \log N$ may be approximately replaced by the integral

$$\int_0^N \log x \cdot dx = N \log (N/e).$$

nevertheless contains a large number of particles. The numbers of quantum states "belonging" to these regions are

$$G_j = \Delta p^{(j)}\,\Delta q^{(j)}/(2\pi\hbar)^r = \Delta\tau^{(j)}, \tag{40.6}$$

where r is the number of degrees of freedom of the particle; the numbers of particles in these states may be written as $N_j = n(p, q)\,\Delta\tau^{(j)}$, where $n(p, q)$ is the particle density distribution in phase space. We substitute these expressions in (40.5), and use the fact that the regions $\Delta\tau^{(j)}$ are small in size and large in number to replace the summation over j by integration over the whole phase space of the particle:

$$S = \int n \log (e/n)\, d\tau. \tag{40.7}$$

In a state of equilibrium, the entropy must be a maximum (as applied to the ideal gas, this statement is sometimes called *Boltzmann's H theorem*). We shall show how this condition may be used to find the distribution function for the gas particles in a state of statistical equilibrium. The problem is to find $\overline{n_j}$ such that the sum (40.5) has the maximum value possible under the subsidiary conditions

$$\sum_j N_j = \sum_j G_j\overline{n_j} = N,$$
$$\sum_j \varepsilon_j N_j = \sum_j \varepsilon_j G_j \overline{n_j} = E,$$

which express the constancy of the total number of particles N and of the total energy E of the gas. Following the usual method of LAGRANGE's undetermined multipliers, we have to equate to zero the derivatives

$$\partial(S+\alpha N+\beta E)/\partial\overline{n_j} = 0, \tag{40.8}$$

where α and β are constants. Effecting the differentiation, we find

$$G_j(-\log \overline{n_j}+\alpha+\beta\varepsilon_j) = 0,$$

whence $\log \overline{n_j} = \alpha+\beta\varepsilon_j$, or

$$\overline{n_j} = e^{\alpha+\beta\varepsilon_j}.$$

This is just the Boltzmann distribution, the constants α and β being given in terms of T and μ by $\alpha = \mu/T$, $\beta = -1/T$.†

§41. The free energy of an ideal Boltzmann gas

We may use the general formula (31.3)

$$F = -T \log \sum_n e^{-E_n/T} \tag{41.1}$$

to calculate the free energy of an ideal gas which obeys Boltzmann statistics.

† These values of α and β could have been foreseen: equations (40.8) can be written as a relation between differentials, $dS+\alpha dN+\beta dE = 0$, which must be the same as the differential of the internal energy at constant volume, $dE = TdS+\mu dN$.

Writing the energy E_n as a sum of energies ε_k, we can reduce the summation over all states of a gas to summation over all states of an individual molecule. Each state of the gas is defined by a set of N values of ε_k (where N is the number of molecules in the gas), which in the Boltzmann case may be regarded as all different (there being not more than one molecule in each molecular state). Writing $e^{-E_n/T}$ as a product of factors $e^{-\varepsilon_k/T}$ for each molecule and summing independently over all states of each molecule, we should obtain

$$\left(\sum_k e^{-\varepsilon_k/T}\right)^N. \tag{41.2}$$

The set of possible values of ε_k is the same for each molecule of the gas, and so the sums $\sum e^{-\varepsilon_k/T}$ are also the same.

The following point must be borne in mind, however. Each set of N different values ε_k which differs only in the distribution of the identical gas molecules over the levels ε_k corresponds to the same quantum state of the gas. But in the statistical sum in formula (41.1) each state must be included only once.† We must therefore again divide the expression (41.2) by the number of possible permutations of N molecules, i.e. by $N!$‡. Thus

$$\sum_n e^{-E_n/T} = \frac{1}{N!}\left(\sum_k e^{-\varepsilon_k/T}\right)^N. \tag{41.3}$$

Substitution of this expression in (41.1) gives

$$F = -TN \log \sum_k e^{-\varepsilon_k/T} + T \log N!.$$

Since N is very large, we can use the formula

$$\log N! \cong N \log (N/e);$$

see the first footnote to §40. This gives the formula

$$F = -NT \log \left[(e/N) \sum_k e^{-\varepsilon_k/T}\right], \tag{41.4}$$

which enables us to calculate the free energy of any gas consisting of identical particles obeying Boltzmann statistics.

In classical statistics, formula (41.4) must be written in the form

$$F = -NT \log \left[(e/N) \int e^{-\varepsilon(p,\, q)/T}\, d\tau\right]; \tag{41.5}$$

the integration is taken over the phase space of the molecule, and $d\tau$ is defined by (38.1).

† See the second footnote to §31.

‡ Here it is important that in Boltzmann statistics the terms containing the same ε_k in (41.2) are of negligible significance.

§42. The equation of state of an ideal gas

It has already been mentioned in §38 that the translational motion of the molecules in a gas is always quasi-classical; the energy of a molecule may be written in the form

$$\varepsilon_k(p_x, p_y, p_z) = (p_x^2+p_y^2+p_z^2)/2m+\varepsilon'_k, \tag{42.1}$$

where the first term is the kinetic energy of the translational motion, and ε'_k denotes the energy levels corresponding to the rotation and internal state of the molecule; ε'_k is independent of the velocities and co-ordinates of the centre of mass of the molecule (assuming that there is no external field).

The partition function in the logarithm in formula (41.4) must now be replaced by the expression

$$\sum_k \frac{1}{(2\pi\hbar)^3} e^{-\varepsilon'_k/T} \int\limits_V \int\limits_{-\infty}^{\infty} \int\limits_{-\infty}^{\infty} \int\limits_{-\infty}^{\infty} \exp[-(p_x^2+p_y^2+p_z^2)/2mT]\, \mathrm{d}p_x\, \mathrm{d}p_y\, \mathrm{d}p_z\, \mathrm{d}V$$
$$= V(mT/2\pi\hbar^2)^{3/2} \sum_k e^{-\varepsilon'_k/T}; \tag{42.2}$$

the integration over V ($\mathrm{d}V = \mathrm{d}x\, \mathrm{d}y\, \mathrm{d}z$) is over the whole volume of the gas. For the free energy we obtain

$$F = -NT \log \left[\frac{eV}{N}\left(\frac{mT}{2\pi\hbar^2}\right)^{3/2} \sum_k e^{-\varepsilon'_k/T}\right]. \tag{42.3}$$

The sum in (42.3) cannot, of course, be calculated in a general form without any assumptions as to the properties of the molecules, but an important fact is that it depends only on the temperature. The dependence of the free energy on the volume is therefore entirely determined by formula (42.3), and so we can derive from it various important general results concerning the properties of an ideal gas (which is not in an external field).

Separating in (42.3) the term containing the volume, we may write this formula as

$$F = -NT \log (eV/N)+Nf(T), \tag{42.4}$$

where $f(T)$ is some function of the temperature. Hence the pressure of the gas is $P = -\partial F/\partial V = NT/V$, or

$$PV = NT. \tag{42.5}$$

Thus we have the familiar equation of state of an ideal gas. If the temperature is measured in degrees, then†

$$PV = NkT. \tag{42.5a}$$

† For a gram-molecule of gas ($N = 6.023\times10^{23}$ = AVOGADRO's number), the product $R = Nk$ is called the *gas constant*: $R = 8.314\times10^7$ erg/deg.

Knowing F, we can find the other thermodynamic quantities also. For example, the thermodynamic potential is

$$\Phi = -NT \log (eV/N) + Nf(T) + PV.$$

Substituting $V = NT/P$ according to (42.5) (since Φ must be expressed as a function of P and T) and using a new function of temperature $\chi(T) = f(T) - T \log T$, we obtain

$$\Phi = NT \log P + N\chi(T). \tag{42.6}$$

The entropy is defined as

$$S = -\partial F/\partial T = N \log (eV/N) - Nf'(T), \tag{42.7}$$

or, as a function of P and T,

$$S = -\partial \Phi/\partial T = -N \log P - N\chi'(T). \tag{42.8}$$

Finally, the energy is

$$E = F + TS = Nf(T) - NTf'(T). \tag{42.9}$$

We see that the energy is a function only of the temperature of the gas (and the same is true of the heat function $W = E + PV = E + NT$). This is evident *a priori*: since the molecules of an ideal gas are assumed not to interact, the change in their mean distance apart when the total volume of the gas varies cannot affect its energy.

As well as E and W, the specific heats $C_v = (\partial E/\partial T)_V$ and $C_p = (\partial W/\partial T)_P$ are functions only of the temperature. In what follows it will be convenient to use the specific heats per molecule, denoted by lower-case letter c:

$$C_v = Nc_v, \qquad C_p = Nc_p. \tag{42.10}$$

Since for an ideal gas $W - E = NT$, the difference $c_p - c_v$ has a fixed value:

$$c_p - c_v = 1 \tag{42.11}$$

(in ordinary units, $c_p - c_v = k$).†

PROBLEMS

PROBLEM 1. Find the work done on an ideal gas in an isothermal change of volume from V_1 to V_2 (or of pressure from P_1 to P_2).

SOLUTION. The required work R is equal to the change in the free energy of the gas, and from (42.4) we have

$$R = F_2 - F_1 = NT \log (V_1/V_2) = NT \log (P_2/P_1).$$

The quantity of heat absorbed in this process is

$$Q = T(S_2 - S_1) = NT \log (V_2/V_1).$$

The latter result also follows directly from the fact that $R + Q$ is the change of energy and is equal to zero for an isothermal process in an ideal gas.

†Since the specific heat is the derivative of the energy (quantity of heat) with respect to temperature, C must be replaced by C/k in the formulae when ordinary units (degrees) are used.

PROBLEM 2. Two vessels contain two identical ideal gases at the same temperature T and with equal numbers of particles N but at different pressures P_1 and P_2. The vessels are then connected. Find the change in entropy.

SOLUTION. Before the vessels are connected, the entropy of the two gases is equal to the sum of their entropies, $S_0 = -N\log(P_1P_2) - 2N\chi'(T)$. After the connection, the temperature of the gases remains the same (as follows from the conservation of energy for the two gases). The pressure is given by the relation

$$\frac{1}{P} = \frac{V_1+V_2}{2NT} = \tfrac{1}{2}\left(\frac{1}{P_1}+\frac{1}{P_2}\right).$$

The entropy is now

$$S = 2N\log\frac{P_1+P_2}{2P_1P_2} - 2N\chi'(T).$$

The change in entropy is therefore

$$\Delta S = N\log\frac{(P_1+P_2)^2}{4P_1P_2}.$$

PROBLEM 3. Find the energy of an ideal gas in a cylindrical vessel of radius R and height h rotating about its axis with angular velocity Ω.

SOLUTION. According to §34, the rotation is equivalent to the presence of an external "centrifugal" field with potential energy $u = -\frac{1}{2}m\Omega^2r^2$ (r being the distance of a particle from the axis of rotation).

When an external field is present, the integrand in (42.2) contains an extra factor $e^{-u/T}$, and so in the argument of the logarithm in (42.3) the volume V is replaced by the integral $\int e^{-u/T}\,dV$. Thus

$$F = F_0 - NT\log\frac{1}{V}\int e^{-u/T}\,dV,$$

where F_0 is the free energy of the gas in the absence of the external field.

In the present case this formula for the free energy becomes (in a rotating co-ordinate system)

$$\begin{aligned} F' &= F_0 - NT\log\frac{1}{\pi R^2h}\int_0^h\int_0^R e^{m\Omega^2r^2/2T}\,2\pi r\,dr\,dz \\ &= F_0 - NT\log\left[\frac{2T}{m\Omega^2R^2}\left(e^{m\Omega^2R^2/2T}-1\right)\right]. \end{aligned}$$

The angular momentum of the gas is

$$\begin{aligned} M &= -\partial F'/\partial\Omega \\ &= -\frac{2NT}{\Omega} + \frac{NmR^2\Omega}{1-e^{-m\Omega^2R^2/2T}}. \end{aligned}$$

The energy in a system rotating with the body is

$$E' = F' - T\,\partial F'/\partial T = E_0 - \frac{Nm\Omega^2R^2}{2(1-e^{-m\Omega^2R^2/2T})} + NT,$$

and in a fixed system of co-ordinates (see (26.5))

$$E = E' + M\Omega = E_0 + \frac{Nm\Omega^2R^2}{2(1-e^{-m\Omega^2R^2/2T})} - NT,$$

where E_0 is the energy of the gas at rest.

§43. Ideal gases with constant specific heat

We shall see later that in many important cases the specific heat of a gas is constant, independent of temperature, over a greater or smaller temperature interval. For this reason we shall now calculate in a general form the thermodynamic quantities for such a gas.

On differentiating the expression (42.9) for the energy, we find that the function $f(T)$ is related to the specific heat c_v by $-Tf''(T) = c_v$. Integration of this relation gives

$$f(T) = -c_v T \log T - \zeta T + \varepsilon_0,$$

where ζ and ε_0 are constants. Substitution in (42.4) gives for the free energy the final expression

$$F = N\varepsilon_0 - NT \log (eV/N) - Nc_v T \log T - N\zeta T. \tag{43.1}$$

The constant ζ is called the *chemical constant* of the gas. For the energy we have

$$E = N\varepsilon_0 + Nc_v T, \tag{43.2}$$

a linear function of the temperature.

The thermodynamic potential Φ of the gas is obtained by adding to (43.2) the quantity $PV = NT$, the volume of the gas being expressed in terms of the pressure and the temperature. The result is

$$\Phi = N\varepsilon_0 + NT \log P - Nc_p T \log T - N\zeta T. \tag{43.3}$$

The heat function $W = E + PV$ is

$$W = N\varepsilon_0 + Nc_p T. \tag{43.4}$$

Finally, differentiating (43.1) and (43.3) with respect to temperature, we obtain the entropy in terms of T and V and of T and P respectively:

$$S = N \log (eV/N) + Nc_v \log T + (\zeta + c_v)N, \tag{43.5}$$

$$S = -N \log P + Nc_p \log T + (\zeta + c_p)N. \tag{43.6}$$

From these expressions for the entropy we can, in particular, derive immediately a relation (called the *Poisson adiabatic*) between the volume, temperature and pressure of an ideal gas (of constant specific heat) undergoing adiabatic expansion or compression. Since the entropy remains constant in an adiabatic process, we have from (43.6) $-N \log P + Nc_p \log T =$ constant, whence $T^{c_p}/P =$ constant or, using (42.11),

$$T^{\gamma} P^{1-\gamma} = \text{constant}, \tag{43.7}$$

where γ denotes the constant ratio

$$\gamma = c_p/c_v. \tag{43.8}$$

Using also the equation of state $PV = NT$, we obtain relations between T and V, and P and V:

$$TV^{\gamma-1} = \text{constant}, \qquad PV^{\gamma} = \text{constant}. \tag{43.9}$$

PROBLEMS

Problem 1. Two identical ideal gases at the same pressure P and containing the same number of particles N but at different temperatures T_1 and T_2 are in vessels with volumes V_1 and V_2. The vessels are then connected. Find the change in entropy.

Solution. Before the vessels are connected, the entropy of the two gases, equal to the sum of their entropies, is by (43.6) $S_0 = -2N \log P + Nc_p \log (T_1 T_2)$.† After the connection, the temperatures of the gases become equal. The sum of the energies of the two gases remains constant. Using the expression (43.2) for the energy, we find $T = \frac{1}{2}(T_1 + T_2)$, where T is the final temperature.

After the connection, the gas contains $2N$ particles and occupies a volume $V_1 + V_2 = N(T_1 + T_2)/P$. Its pressure is then $2NT/(V_1 + V_2) = P$, i.e. the same as before. The entropy is

$$S = -2N \log P + 2Nc_p \log (\tfrac{1}{2}T_1 + \tfrac{1}{2}T_2),$$

and the change in entropy is

$$\begin{aligned} \Delta S &= S - S_0 \\ &= Nc_p \log \frac{(T_1 + T_2)^2}{4T_1 T_2}. \end{aligned}$$

Problem 2. Find the work done on an ideal gas in adiabatic compression.

Solution. In an adiabatic process the quantity of heat $Q = 0$, and so $R = E_2 - E_1$, where $E_2 - E_1$ is the change in energy during the process. According to (43.2) $R = Nc_v(T_2 - T_1)$, where T_1 and T_2 are the gas temperatures before and after the process. R can be expressed in terms of the initial and final volumes V_1 and V_2 by means of the relation (43.10):

$$\begin{aligned} R &= Nc_v T_1[(V_1/V_2)^{\gamma-1} - 1] \\ &= Nc_v T_2[1 - (V_2/V_1)^{\gamma-1}]. \end{aligned}$$

Problem 3. Find the quantity of heat gained by a gas in an *isochoric* process, i.e. one which occurs at constant volume.

Solution. Since in this case the work $R = 0$, we have

$$Q = E_2 - E_1 = Nc_v(T_2 - T_1).$$

Problem 4. Find the work done and quantity of heat gained in an *isobaric* process, i.e. one which occurs at constant pressure.

Solution. At constant pressure

$$R = -P(V_2 - V_1), \qquad Q = W_2 - W_1,$$

whence

$$R = N(T_1 - T_2), \qquad Q = Nc_p(T_2 - T_1).$$

Problem 5. Find the work done on a gas and the quantity of heat which it gains in compression from volume V_1 to V_2 in accordance with the equation $PV^n = a$ (a *polytropic* process).

Solution. The work is

$$R = -\int_{V_1}^{V_2} P \, dV = \frac{a}{n-1}(V_2^{1-n} - V_1^{1-n}).$$

Since the sum of the quantity of heat gained and the work done is equal to the total change in energy, we have $Q = Nc_v(T_2 - T_1) - R$, and since $T = PV/N = (a/N)V^{1-n}$,

$$Q = a\left(c_v + \frac{1}{1-n}\right)\left(V_2^{1-n} - V_1^{1-n}\right).$$

† We everywhere omit the constant terms in the entropy and energy which are unimportant in the solution of problems.

PROBLEM 6. Find the work done on an ideal gas and the quantity of heat which it gains on going through a cyclic process (i.e. one in which it returns to its initial state at the end of the process), consisting of two isochoric and two isobaric processes: the gas goes from a state with pressure and volume P_1, V_1 to states P_1, V_2; P_2, V_2; P_2, V_1; P_1, V_1 again.

SOLUTION. The change in energy in a cyclic process is zero, since the initial and final states are the same. The work done and the quantity of heat gained in such a process are therefore the same with opposite signs ($R = -Q$). In order to find R in the present case, we note that in isochoric processes the work done is zero, and for the two isobaric processes it is respectively $-P_1(V_2-V_1)$ and $-P_2(V_1-V_2)$. Thus $R = (V_2-V_1)(P_2-P_1)$.

PROBLEM 7. The same as Problem 6, but for a cyclic process consisting of two isochoric and two isothermal processes, the successive volumes and temperatures of the gas being V_1, T_1; V_1, T_2; V_2, T_2; V_2, T_1; V_1, T_1.

SOLUTION.

$$R = (T_2-T_1)N \log (V_1/V_2).$$

PROBLEM 8. The same as Problem 6, but for a cyclic process consisting of two isothermal and two adiabatic processes, the successive entropies, temperatures and pressures being S_1, T_1, P_1; S_1, T_2; S_2, T_2, P_2; S_2, T_1; S_1, T_1, P_1.

SOLUTION.

$$Q = (T_2-T_1)(S_2-S_1)$$
$$= (T_2-T_1)[N \log (P_1/P_2)+Nc_p \log (T_2/T_1)].$$

PROBLEM 9. The same as Problem 6, but for a cyclic process consisting of two isobaric and two isothermal processes, the successive states being P_1, T_1; P_1, T_2; P_2, T_2; P_2, T_1; P_1, T_1.

SOLUTION. The work done on the gas in the isobaric processes is (see Problem 4) $N(T_1-T_2)$ and $N(T_2-T_1)$, and that in the isothermal processes is $NT_2 \log (P_2/P_1)$ and $NT_1 \log (P_1/P_2)$. The sum of these is $R = N(T_2-T_1) \log (P_2/P_1)$.

PROBLEM 10. The same as Problem 6, but for a cyclic process consisting of two isobaric and two adiabatic processes, the successive states being P_1, S_1, T_1; P_1, S_2; P_2, S_2, T_2; P_2, S_1; P_1, S_1, T_1.

SOLUTION. The temperature in the second state is $T_2(P_2/P_1)^{(1-\gamma)/\gamma}$, and in the fourth state $T_1(P_1/P_2)^{(1-\gamma)/\gamma}$; these are obtained from T_1 and T_2 by means of (43.7). The quantity of heat gained by the gas in adiabatic processes is zero, and in the isobaric processes it is (see Problem 4)

$$Nc_p[T_2(P_2/P_1)^{(1-\gamma)/\gamma}-T_1] \text{ and}$$
$$Nc_p[T_1(P_1/P_2)^{(1-\gamma)/\gamma}-T_2].$$

Hence

$$Q = Nc_pT_1[(P_1/P_2)^{(1-\gamma)/\gamma}-1]+Nc_pT_2[(P_2/P_1)^{(1-\gamma)/\gamma}-1].$$

PROBLEM 11. The same as Problem 6, but for a cyclic process consisting of two isochoric and two adiabatic processes, the successive states being V_1, S_1, T_1; V_1, S_2; V_2, S_2, T_2; V_2, S_1; V_1, S_1, T_1.

SOLUTION. Using the result of Problem 2, we find

$$R = Nc_vT_2[1-(V_2/V_1)^{\gamma-1}]+Nc_vT_1[1-(V_1/V_2)^{\gamma-1}].$$

PROBLEM 12. Determine the maximum work that can be obtained by connecting vessels containing two identical ideal gases at the same temperature T_0 and with equal numbers of particles N but having different volumes V_1 and V_2.

SOLUTION. The maximum work is done if the process occurs reversibly (i.e. if the entropy remains constant), and is equal to the difference between the energies before and after the process (§19). Before the connection of the vessels, the entropy of the two gases is equal to the sum of their entropies, i.e. by (43.5),

$$S_0 = N \log (e^2 V_1V_2/N^2)+2Nc_v \log T_0.$$

After the connection we have a gas consisting of $2N$ particles occupying a volume V_1+V_2 at some temperature T. Its entropy is

$$S = 2N \log [e(V_1+V_2)/2N]+2Nc_v \log T.$$

Equating S_0 and S gives the temperature T:

$$T = T_0 \left[\frac{4V_1V_2}{(V_1+V_2)^2}\right]^{(\gamma-1)/2}.$$

The energy of the two gases before and after the connection is $E_0 = 2Nc_vT_0$ and $E = 2Nc_vT$ respectively. The maximum work is therefore

$$R_{\max} = E_0-E = 2Nc_v(T_0-T) = 2Nc_vT_0\left[1-\left(\frac{4V_1V_2}{(V_1+V_2)^2}\right)^{(\gamma-1)/2}\right].$$

Problem 13. The same as Problem 12, but for gases with the same pressure P_0 and different temperatures T_1 and T_2 before the connection of the vessels.

Solution. We have similarly

$$R_{\max} = Nc_v\left\{T_1+T_2-2^\gamma\sqrt{(T_1T_2)}\left[\frac{T_1T_2}{(T_1+T_2)^2}\right]^{(\gamma-1)/2}\right\}.$$

Problem 14. Find the minimum work that must be done on an ideal gas in order to compress it from pressure P_1 to P_2 at a constant temperature equal to that of the surrounding medium ($T = T_0$).

Solution. According to (20.2) the minimum work is $R_{\min} = (E_2-E_1) -T_0(S_2-S_1)+P_0(V_2-V_1)$, where the suffixes 1 and 2 refer to the gas before and after compression. In the present case the energy E is unchanged (since the temperature is constant), i.e. $E_2-E_1 = 0$. Using (43.6), we find the change of entropy corresponding to the change of pressure from P_1 to P_2: $S_2-S_1 = N\log(P_1/P_2)$, and the change of volume is $V_2-V_1 = NT_0(1/P_2-1/P_1)$. Hence

$$R_{\min} = NT_0\left[\log\frac{P_2}{P_1}+P_0\left(\frac{1}{P_2}-\frac{1}{P_1}\right)\right].$$

Problem 15. Determine the maximum work which can be obtained from an ideal gas cooled from temperature T to the temperature of the medium T_0 at constant volume.

Solution. From the general formula (20.3),

$$R_{\max} = Nc_v(T-T_0)+Nc_vT_0\log(T_0/T).$$

Problem 16. The same as Problem 15, but for a gas cooled from temperature T to the temperature of the medium T_0 and at the same time expanding from pressure P to the pressure of the medium P_0.

Solution.

$$R_{\max} = Nc_v(T-T_0)+NT_0\log(P/P_0)+Nc_pT_0\log(T_0/T)+N(TP_0/P-T_0).$$

Problem 17. Gas at temperature T_0 flows from a large thermally insulated reservoir into an empty thermally insulated vessel, the gas pressure in the reservoir remaining constant. Find the change in the gas temperature.

Solution. The energy E of the gas in the vessel consists of the energy E_0 which it had in the reservoir and the work done on it to "expel" it from the reservoir. Since the state of the gas in the reservoir may be regarded as steady, we have the condition $W_0 = E$ (cf. §18). The gas temperature in the vessel is therefore $T = \gamma T_0$.

§44. The law of equipartition

Before going on to calculate in detail the thermodynamic quantities for gases, making allowance for the various quantum effects, it is useful to consider the same problem from the point of view of purely classical statistics.

We shall see later when and to what extent the results obtained are applicable to actual gases.

A molecule is a configuration of atoms executing small oscillations about certain equilibrium positions which correspond to minimum potential energy of their interaction. The potential energy is then of the form

$$u = \varepsilon_0 + \sum_{i,k=1}^{r_{\text{vib}}} a_{ik} q_i q_k,$$

where ε_0 is the potential energy of interaction of the atoms when they are all in their equilibrium positions; the second term is a quadratic function of the co-ordinates which give the deviations of the atoms from their equilibrium positions. The number r_{vib} of co-ordinates in this function is the number of vibrational degrees of freedom of the molecule.

This number can be determined from the number n of atoms in the molecule. A molecule containing n atoms has a total of $3n$ degrees of freedom. Three of these correspond to the translational motion of the molecule as a whole, and three to its rotation as a whole. If all the atoms are collinear (and in particular for a diatomic molecule) there are only two rotational degrees of freedom. Thus a non-linear molecule of n atoms has $3n-6$ vibrational degrees of freedom, and a linear one has $3n-5$. For $n = 1$ there are, of course, no vibrational degrees of freedom, since all three degrees of freedom of an atom correspond to translational motion.

The total energy ε of the molecule is the sum of the potential and kinetic energies. The latter is a quadratic function of all the momenta, and these are equal in number to the full $3n$ degrees of freedom of the molecule. The energy ε is therefore of the form $\varepsilon = \varepsilon_0 + f_{\text{II}}(p, q)$, where $f_{\text{II}}(p, q)$ is a quadratic function of the momenta and co-ordinates; the total number of variables in this function is $l = 6n-6$ (for a non-linear molecule) or $6n-5$ (for a linear molecule); in a monatomic gas, $l = 3$, since the co-ordinates do not appear at all in the expression for the energy.

Substituting this expression for the energy in (41.5) we have

$$F = -NT \log \frac{e \cdot e^{-\varepsilon_0/T}}{N} \int e^{-f_{\text{II}}(p,q)/T} \, d\tau.$$

In order to find the dependence on temperature of the integral in this formula, we substitute $p = p'\sqrt{T}$, $q = q'\sqrt{T}$ for all the l variables on which the function $f_{\text{II}}(p, q)$ depends. Since this function is quadratic, $f_{\text{II}}(p, q) = Tf_{\text{II}}(p', q')$, and T cancels in the exponent of the integrand. The transformation of the differentials of these variables in $d\tau$ gives a factor $T^{l/2}$, which can be taken outside the integral. The integration over the vibrational co-ordinates q is taken over the range of values corresponding to vibrations in which the atoms remain within the molecule. However, since the integrand diminishes rapidly with increasing q, the integration may be extended to the whole range from $-\infty$ to ∞, as well as over all the momenta. The above-mentioned change of variables then leaves the limits of integration unaltered, and the whole integral is a constant independent of temperature. Using also the fact that the integration with respect to the co-ordinates of the centre of mass of the molecule gives the volume V occupied by the gas, we obtain for the free energy an expression of the form

$$F = -NT \log (AVe^{-\varepsilon_0/T}T^{l/2}/N),$$

where A is a constant. Expanding the logarithm, we have an expression of exactly the type (43.1) with a constant specific heat equal to

$$c_v = \tfrac{1}{2} l. \tag{44.1}$$

The specific heat $c_p = c_v + 1$ is accordingly

$$c_p = \tfrac{1}{2}(l+2). \tag{44.2}$$

Thus we see that a purely classical ideal gas must have a constant specific heat. Formula (44.1) enables us to state the following rule. Each variable in the energy $\varepsilon(p, q)$ of the molecule gives an equal contribution of $\frac{1}{2}$ to the specific heat c_v of the gas ($\frac{1}{2}k$ in ordinary units) or, what is the same thing, an equal contribution of $\frac{1}{2}T$ to its energy. This is called the *law of equipartition*.

Since for the translational and rotational degrees of freedom the energy $\varepsilon(p, q)$ contains only the corresponding momenta, we can say that each of these degrees of freedom gives a contribution of $\frac{1}{2}$ to the specific heat. Each vibrational degree of freedom corresponds to two variables (co-ordinate and momentum) in the energy $\varepsilon(p, q)$, and its contribution to the specific heat is 1.

For the model considered here it is easy to find a general formula for the energy distribution of the gas molecules. For convenience we shall measure the energy of a molecule from the value ε_0, i.e. omit this constant from the expression for $\varepsilon(p, q)$. Let us consider the volume in the phase space of the molecule whose points correspond to values of $\varepsilon(p, q)$ not exceeding a given value ε, i.e. determine the integral $\tau(\varepsilon) = \int d\tau$ taken over the region $\varepsilon(p, q) \leqslant \varepsilon$. According to the foregoing discussion, $\varepsilon(p, q)$ is a quadratic function of l variables. We replace those l quantities p, q on which the energy $\varepsilon(p, q)$ depends by new variables $p' = p/\sqrt{\varepsilon}$, $q' = q/\sqrt{\varepsilon}$. Then the condition $\varepsilon(p, q) \leqslant \varepsilon$ becomes $\varepsilon'(p', q') \leqslant 1$, and $\int d\tau$ becomes $\varepsilon^{l/2}\int d\tau'$. The integral $\int d\tau'$ is clearly independent of ε, and so $\tau = \text{constant}\times\varepsilon^{l/2}$, whence $d\tau(\varepsilon) = \text{constant}\times\varepsilon^{\frac{1}{2}l-1}\, d\varepsilon$ and the energy probability distribution is

$$dw_\varepsilon = Ae^{-\varepsilon/T}\varepsilon^{\frac{1}{2}l-1}\, d\varepsilon.$$

Determining A from the normalisation condition, we find

$$dw_\varepsilon = \frac{1}{T^{l/2}\Gamma(\frac{1}{2}l)} e^{-\varepsilon/T} \varepsilon^{\frac{1}{2}l-1}\, d\varepsilon. \tag{44.3}$$

PROBLEM

Find the specific heat of an ideal gas in the extreme relativistic case, where the energy of a particle is related to its momentum by $\varepsilon = cp$, c being the velocity of light.

SOLUTION. According to (41.5) we have

$$F = -NT \log \frac{eV}{N(2\pi\hbar)^3} \int_0^\infty e^{-cp/T} \cdot 4\pi p^2\, dp$$

or, after carrying out the integration,

$$F = -NT \log (AVT^3/N),$$

where A is a constant. The specific heat is therefore $c_v = 3$, which is twice the value for a non-relativistic monatomic gas.

§45. Monatomic ideal gases

The complete calculation of the free energy (and therefore of the other thermodynamic quantities) for an ideal gas requires a calculation of the specific form of the partition function in the logarithm in (42.3),

$$Z = \sum_k e^{-\varepsilon'_k/T}.$$

Here ε'_k are the energy levels of the atom or molecule (the kinetic energy of the translational motion of the particle being excluded). If the summation is taken only over all the different energy levels, it must be remembered that a level may be degenerate, and in this case the corresponding term must appear in the sum over all states as many times as the degree of degeneracy. Let this be g_k. The degree of degeneracy of the level is often called in this connection its *statistical weight*. Omitting for brevity the prime in ε'_k, we can write the partition function concerned in the form

$$Z = \sum_k g_k e^{-\varepsilon_k/T}. \tag{45.1}$$

The free energy of the gas is

$$F = -NT \log \left[\frac{eV}{N} \left(\frac{mT}{2\pi\hbar^2} \right)^{3/2} Z \right]. \tag{45.2}$$

Turning now to the consideration of monatomic gases, we must first of all make the following important comment. As the gas temperature increases, so does the number of atoms in excited states, including the states of the continuous spectrum, which correspond to ionisation of the atom. When the temperature is not too high, the relative number of ionised atoms in the gas is negligible, but the gas is almost completely ionised at temperatures T of the order of the ionisation energy I_{ion}, and not only for $T \gg I_{\text{ion}}$ (see §106). Thus a non-ionised gas can reasonably be considered only at temperatures such that $T \ll I_{\text{ion}}$.†

The atomic terms (neglecting their fine structure) are so situated that the separation between the ground state and the first excited level is comparable with the ionisation energy. At temperatures $T \ll I_{\text{ion}}$, the gas will therefore be practically free not only of ionised atoms but also of excited atoms, and so all the atoms may be regarded as being in the ground state.

Let us first consider the simplest case, that of atoms which in their ground state have neither orbital angular momentum nor spin ($L = S = 0$), such as the atoms of the inert gases. The ground state is not degenerate, and the partition function reduces to a single term, $Z = e^{-\varepsilon_0/T}$. For monatomic gases it is customary to put $\varepsilon_0 = 0$, i.e. to measure the energy from the ground state of the atom, so that $Z = 1$. Expanding the logarithm in (45.2) as a sum of logarithms, we obtain for the free energy an expression of the type (43.1), with constant specific heat

$$c_v = 3/2 \tag{45.3}$$

and chemical constant

$$\zeta = \frac{3}{2}\log\frac{m}{2\pi\hbar^2}. \tag{45.4}$$

This value of the specific heat is due entirely to the translational degrees of freedom of the atom ($\frac{1}{2}$ for each degree of freedom); it will be remembered that the translational motion of the gas particles is always quasi-classical. The "electronic degrees of freedom" under these conditions (no excited atoms in the gas) have, of course, no effect on the thermodynamic quantities.‡

These expressions enable us to deduce a criterion for the validity of Boltzmann statistics. In this statistics it is assumed that

$$\overline{n_k} = e^{(\mu-\varepsilon_k)/T} \ll 1$$

† For different atoms the temperature I_{ion}/k lies between 5×10^4 degrees (alkali metal atoms) and 28×10^4 degrees (helium).

‡ The "electronic part" of the thermodynamic quantities, naturally, can never be treated classically. In this connection we may note the fact (which in essence has been tacitly assumed already) that in classical statistics the atoms must be regarded as particles without internal structure. The impossibility of applying to effects within the atom a statistics based on classical mechanics is further shown by the absurd result obtained on substituting the interaction energy between the electrons and the atomic nucleus in the classical distribution formulae. This energy is of the form $-a/r$, where r is the distance of the electron from the nucleus and a is a constant. The substitution would give a factor $e^{a/rT}$ in the distribution, which becomes infinite for $r = 0$. This would mean that all the electrons would have to "fall" into the nucleus in thermal equilibrium.

(see (37.1)). It is clearly sufficient to require the fulfilment of the condition

$$e^{\mu/T} \ll 1.$$

For the chemical potential $\mu = \Phi/N$ we have from (43.3), with c_v and ζ given by (45.3) and (45.4),

$$\mu = T \log \left[\frac{P}{T^{5/2}} \left(\frac{2\pi\hbar^2}{m} \right)^{3/2} \right]$$

$$= T \log \left[\frac{N}{V} \left(\frac{2\pi\hbar^2}{mT} \right)^{3/2} \right]. \tag{45.5}$$

Thus we obtain the condition

$$(N/V)(\hbar^2/mT)^{3/2} \ll 1. \tag{45.6}$$

For a given temperature, this condition requires that the gas should be sufficiently rarefied. Substitution of numerical values shows that in practice, for any atomic (or molecular) gas, this condition can be violated only at densities where the interaction of the particles becomes important and the gas can in any case no longer be regarded as ideal.

It is useful to note the following intuitive interpretation of the above condition. Since the majority of atoms have energies of the order of T, and therefore momenta of the order of $\sqrt{(mT)}$, we can say that all the atoms occupy in phase space a volume of the order of $V(mT)^{3/2}$, corresponding to $\sim V(mT)^{3/2}/\hbar^3$ quantum states. In the Boltzmann case this number must be large compared with the number N of particles, and hence we have (45.6).

Finally, we may make the following comment. The formulae derived in this section appear at first sight to contradict NERNST's theorem, since neither the entropy nor the specific heat is zero at $T = 0$. However, it must be remembered that, under the conditions for which NERNST's theorem is stated, all actual gases condense at sufficiently low temperatures. For NERNST's theorem requires that the entropy of a body should tend to zero at $T = 0$ for a fixed value of its volume. But as $T \to 0$ the saturated vapour pressure of all substances becomes arbitrarily small, so that a fixed finite quantity of matter in a fixed finite volume cannot remain gaseous as $T \to 0$.

If we consider a model of a gas, possible in principle, which consists of mutually repulsive particles, then, although such a gas will never condense, at sufficiently low temperatures Boltzmann statistics ceases to be valid, and the application of Fermi or Bose statistics leads, as we shall see later, to expressions which are in agreement with NERNST's theorem.

§46. Monatomic gases. The effect of the electronic angular momentum

If only one of the angular momenta L and S is non-zero in the ground state of the atom, then this state again has no fine structure. In practice the absence of fine structure of the ground state is always due to a zero orbital angular momentum; the spin S is sometimes not zero (for example, atoms in the vapour of alkali metals).

A level with spin S is $(2S+1)$-fold degenerate. The only difference as compared with the case discussed in §45 is that the partition function Z is now $2S+1$ instead of 1, and so the chemical constant (45.4) is increased by the quantity†

$$\zeta_S = \log(2S+1). \tag{46.1}$$

If the ground state of an atom has a fine structure, it must be remembered that the intervals in this structure may generally be comparable with T, and so all the components of the fine structure of the ground state must be taken into account in the partition function.

The fine-structure components differ in the value of the total angular momentum of the atom (with given orbital angular momentum L and spin S). Let these levels, measured from the lowest of them, be denoted by ε_J. Each level with a given J is $(2J+1)$-fold degenerate with respect to orientations of the total angular momentum.‡ The partition function therefore becomes

$$Z = \sum_J (2J+1)e^{-\varepsilon_J/T}; \tag{46.2}$$

the summation is taken over all possible values of J for the given L and S. We obtain for the free energy

$$F = -NT\log\left[\frac{eV}{N}\left(\frac{mT}{2\pi\hbar^2}\right)^{3/2}\sum_J(2J+1)e^{-\varepsilon_J/T}\right]. \tag{46.3}$$

This expression becomes considerably simpler in two limiting cases. Let us assume that the temperature is so high that T is large in comparison with with all the fine-structure intervals: $T \gg \varepsilon_J$. Then we can put $e^{-\varepsilon_J/T} \cong 1$, and Z becomes simply the total number of fine-structure components $(2S+1)(2L+1)$. The expression for the free energy involves the constant specific heat $c_v = 3/2$ as before, and the quantity

$$\zeta_{SL} = \log[(2S+1)(2L+1)] \tag{46.4}$$

is added to the chemical constant (45.4).

Similar expressions for the thermodynamic quantities (with a different ζ) are obtained in the opposite limiting case where T is small compared with the

† We may write out for reference the formula for the chemical potential of a monatomic ideal gas with statistical weight (degree of degeneracy) of the ground state g:

$$\mu = T\log\left[\frac{P}{gT^{5/2}}\left(\frac{2\pi\hbar^2}{m}\right)^{3/2}\right] = T\log\left[\frac{N}{gV}\left(\frac{2\pi\hbar^2}{mT}\right)^{3/2}\right]. \tag{46.1a}$$

This applies also to a Boltzmann gas of elementary particles; for instance, in an electron gas $g = 2$.

‡ We assume that Russell–Saunders coupling is valid in the atom; see *Quantum Mechanics*, §72.

fine-structure intervals.† In this case all terms may be neglected in the sum (46.2) except the one with $\varepsilon_J = 0$ (the lowest component of the fine structure, i.e. the ground state of the atom). In consequence the quantity added to the chemical constant (45.4) is

$$\zeta_J = \log(2J+1), \tag{46.5}$$

where J is the total angular momentum of the atom in the ground state.

Thus, when the ground state of the atom has fine structure, the specific heat of the gas at sufficiently low and sufficiently high temperatures has the same constant value, but in the intermediate range it depends on the temperature and passes through a maximum. It must be borne in mind, however, that for gases concerned in practice (heavy-metal vapours, atomic oxygen, etc.) only the range of high temperatures, where the specific heat becomes constant, is of importance.

So far we have ignored the possibility that the atom has a non-zero nuclear spin i. The existence of such a spin causes the *hyperfine splitting* of atomic levels. The intervals in this structure are, however, so small that they may be neglected in comparison with T at all temperatures where the gas remains a gas.‡ In calculating the partition function, the energy differences between the hyperfine multiplet components may be entirely neglected, and the splitting need be taken into account only as increasing the degree of degeneracy of each level (and therefore the sum Z) by a factor $2i+1$. Accordingly, the free energy contains an additional "nuclear" term

$$F_{\text{nuc}} = -NT\log(2i+1). \tag{46.6}$$

This term does not affect the specific heat of the gas (the corresponding energy $E_{\text{nuc}} = 0$) and simply changes the entropy by $S_{\text{nuc}} = N\log(2i+1)$, and the chemical constant by $\zeta_{\text{nuc}} = \log(2i+1)$.

Because the interaction between the nuclear spin and the electron shells is extremely weak, the "nuclear" part of the thermodynamic quantities usually plays no part in the various thermal processes and does not appear in the equations. We shall therefore omit these terms, as is usually done; in other words, we shall measure the entropy not from zero but from the value S_{nuc} due to the nuclear spins.

§47. Diatomic gases with molecules of unlike atoms. Rotation of molecules

Turning now to the calculation of the thermodynamic quantities for a diatomic gas, we may point out first of all that, just as monatomic gases can

† As examples, the quantities ε_J/k for the components of the triplet ground state of the oxygen atom are 230° and 320°; for those of the quintet ground state of the iron atom they are between 600° and 1400°, and for the doublet ground state of the chlorine atom 1300°.

‡ The temperatures corresponding to the hyperfine structure intervals of various atoms range from 0.1° to 1.5°.

reasonably be considered only for temperatures T which are small compared with the ionisation energy, a diatomic gas can be regarded as such only if T is small compared with the dissociation energy of the molecule.† This in turn means that only the lowest electronic state of the molecule need be retained in the partition function.

Let us begin with the most important case, where the gas molecule in the lowest electronic state has neither spin nor orbital angular momentum about the axis ($S = 0$, $\Lambda = 0$); such an electronic state has, of course, no fine structure. We must also distinguish molecules composed of unlike atoms (including different isotopes of the same element) from those composed of like atoms, since the latter case has certain specific properties. In the present section we shall assume that the molecule consists of unlike atoms.

The energy level of a diatomic molecule is, to a certain approximation, the sum of three independent parts: the electron energy (which includes also the energy of the Coulomb interaction of the nuclei in their equilibrium position and will be measured from the sum of the energies of the separated atoms), the rotational energy, and the vibrational energy of the nuclei within the molecule. For a singlet electronic state, these levels may be written‡

$$\varepsilon_{vK} = \varepsilon_0 + \hbar\omega(v+\tfrac{1}{2}) + \hbar^2 K(K+1)/2I, \qquad (47.1)$$

where ε_0 is the electron energy, $\hbar\omega$ the vibrational quantum, v the vibrational quantum number, K the rotational quantum number (angular momentum of the molecule), $I = m'r_0^2$ the moment of inertia of the molecule ($m' = m_1m_2/(m_1+m_2)$ is the reduced mass of the two atoms and r_0 the equilibrium value of the distance between the nuclei).

When the expression (47.1) is substituted in the partition function, the latter is resolved into three independent factors:

$$Z = e^{-\varepsilon_0/T} Z_{\mathrm{rot}} Z_{\mathrm{vib}}, \qquad (47.2)$$

where the "rotational" and "vibrational" sums are defined by

$$Z_{\mathrm{rot}} = \sum_{K=0}^{\infty} (2K+1) e^{-\hbar^2 K(K+1)/2TI}, \qquad (47.3)$$

$$Z_{\mathrm{vib}} = \sum_{v=0}^{\infty} e^{-\hbar\omega(v+\frac{1}{2})/T}, \qquad (47.4)$$

the factor $2K+1$ in Z_{rot} taking account of the degeneracy of the rotational levels with respect to the orientations of the angular momentum **K**. Accordingly, the free energy is the sum of three parts:

$$F = -NT \log\left[\frac{eV}{N}\left(\frac{mT}{2\pi\hbar^2}\right)^{3/2}\right] + F_{\mathrm{rot}} + F_{\mathrm{vib}} + N\varepsilon_0, \qquad (47.5)$$

† As examples, the temperatures I_{diss}/k for some diatomic molecules are H_2 52,000°, N_2 85,000°, O_2 59,000°, Cl_2 29,000°, NO 61,000°, CO 98,000°.

‡ See *Quantum Mechanics*, §82.

where $m = m_1+m_2$ is the mass of the molecule. The first term may be called the *translational part* F_{tr} (since it arises from the degrees of freedom of the translational motion of the molecules), and

$$F_{rot} = -NT \log Z_{rot}, \qquad F_{vib} = -NT \log Z_{vib} \tag{47.6}$$

the *rotational* and *vibrational* parts. The translational part is always given by a formula of the type (43.1) with a constant specific heat $c_{tr} = 3/2$ and chemical constant

$$\zeta_{tr} = \frac{3}{2} \log \frac{m}{2\pi\hbar^2}. \tag{47.7}$$

The total specific heat of the gas is the sum of several terms:

$$\begin{aligned} c_v &= c_{tr}+c_{rot}+c_{vib}, \\ c_p &= c_{tr}+c_{rot}+c_{vib}+1, \end{aligned} \tag{47.8}$$

which arise respectively from the thermal excitation of the translational motion of the molecule, its rotation and the vibrations of atoms within the molecule.

Let us next calculate the rotational free energy. If the temperature is so high that $T \gg \hbar^2/2I$ (i.e. the "rotational quantum" $\hbar^2/2I$ is small compared with T)†, then the terms with large K are the most important in the sum (47.3). For large values of K, the rotation of the molecule is quasi-classical. In this case, therefore, the partition function Z_{rot} can be replaced by the corresponding classical integral:

$$Z_{rot} = \int e^{-\varepsilon(\mathbf{M})/T} \, d\tau_{rot}, \tag{47.9}$$

where $\varepsilon(\mathbf{M})$ is the classical expression for the kinetic energy of rotation as a function of the angular momentum $\mathbf{M}$. Using a system of co-ordinates ξ, η, ζ rotating with the molecule, with the ζ-axis along the axis of the molecule, and bearing in mind that a diatomic molecule has two rotational degrees of freedom and the rotational angular momentum of a linear mechanical system is perpendicular to its axis, we can write

$$\varepsilon(\mathbf{M}) = (M_\xi^2+M_\eta^2)/2I.$$

The element $d\tau_{rot}$ is the product of the differentials dM_ξ, dM_η and the differentials $d\phi_\xi$, $d\phi_\eta$ of the "generalised co-ordinates" corresponding to M_ξ, M_η (i.e. the infinitesimal angles of rotation about the ξ and η axes), divided by $(2\pi\hbar)^2$.‡ The product of two infinitesimal angles of rotation about the ξ and η axes is just the element of solid angle do_ζ for the direction of the third axis

† In practice this condition is always satisfied for all gases except the two isotopes of hydrogen. As examples, the values of $\hbar^2/2kI$ are: H_2 85.4°, D_2 43°, HD 64°, N_2 2.9°, O_2 2.1°, Cl_2 0.36°, NO 2.4°, HCl 15.2°.

‡ It must be remembered that this notation is to some extent arbitrary, since $d\phi_\xi$ and $d\phi_\eta$ are not total differentials of any function of the position of the axes.

ζ, and integration over the solid angle gives 4π. Thus†

$$Z_{\text{rot}} = \frac{4\pi}{(2\pi\hbar)^2} \int_{-\infty}^{\infty} \int_{-\infty}^{\infty} \exp\left[-\frac{1}{2TI}(M_\xi^2+M_\eta^2)\right] \mathrm{d}M_\xi\, \mathrm{d}M_\eta = 2IT/\hbar^2.$$

Hence the free energy is

$$F_{\text{rot}} = -NT \log T - NT \log (2I/\hbar^2). \tag{47.10}$$

Thus, at the relatively high temperatures under consideration, the rotational part of the specific heat is a constant, $c_{\text{rot}} = 1$, in accordance with the general results of the classical treatment in §44 ($\frac{1}{2}$ for each rotational degree of freedom). The rotational part of the chemical constant is $\zeta_{\text{rot}} = \log (2I/\hbar^2)$. We shall see below that there is a considerable range of temperatures over which the condition $T \gg \hbar^2/2I$ holds and at the same time the vibrational part of the free energy, and therefore the vibrational part of the specific heat, are zero. Over this range the specific heat of a diatomic gas $c_v = c_{\text{tr}}+c_{\text{rot}}$, i.e.

$$c_v \doteq 5/2, \qquad c_p = 7/2, \tag{47.11}$$

and the chemical constant $\zeta = \zeta_{\text{tr}}+\zeta_{\text{rot}}$:

$$\zeta = \log [(2I/\hbar^5)(m/2\pi)^{3/2}]. \tag{47.12}$$

In the opposite limiting case of low temperatures, $T \ll \hbar^2/2I$, it is sufficient to retain the first two terms of the sum:

$$Z_{\text{rot}} = 1+3e^{-\hbar^2/IT},$$

and for the free energy we have in the same approximation

$$F_{\text{rot}} = -3NTe^{-\hbar^2/IT}. \tag{47.13}$$

Hence the entropy is

$$S_{\text{rot}} = \frac{3N\hbar^2}{IT} e^{-\hbar^2/IT}(1+IT/\hbar^2) \tag{47.14}$$

and the specific heat is

$$c_{\text{rot}} = 3N(\hbar^2/IT)^2 e^{-\hbar^2/IT}. \tag{47.15}$$

Thus the rotational entropy and specific heat of the gas tend to zero essentially exponentially as $T \to 0$. At low temperatures, therefore, a diatomic gas behaves like a monatomic one; both the specific heat and the chemical constant have the same values as in a monatomic gas of particles of mass m.

† This value of Z_{rot} can also be derived in another way: assuming that the numbers K in the sum (47.3) are large and replacing the summation by integration with respect to K, we have

$$Z_{\text{rot}} \cong \int_0^\infty 2Ke^{-K^2\hbar^2/2IT}\, \mathrm{d}K = 2TI/\hbar^2.$$

In the general case of arbitrary temperatures the sum Z_{rot} must be calculated numerically. Fig. 3 shows c_{rot} as a function of $2TI/\hbar^2$. The rotational specific heat has a maximum of 1.1 at $T = 0.81(\hbar^2/2I)$, and then tends asymptotically to the classical value 1.†

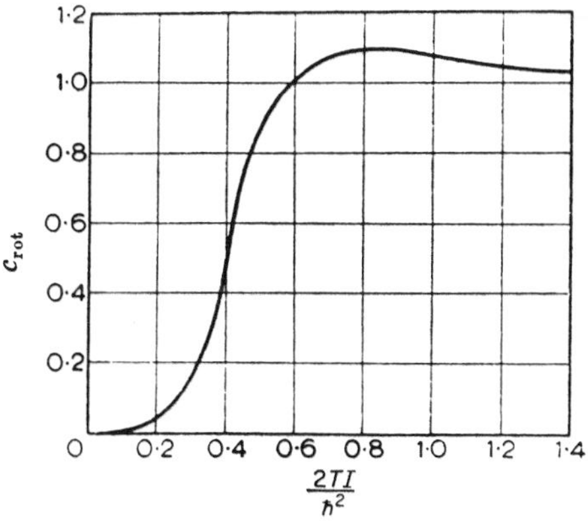

FIG. 3

§48. Diatomic gases with molecules of like atoms. Rotation of molecules

Diatomic molecules consisting of like atoms have certain specific properties which necessitate changes in some of the formulae derived in §47.

First of all, let us consider the limiting case of high temperatures, where a classical treatment is possible. Since the two nuclei are identical, two opposite positions of the axis of the molecule (differing only in that the two nuclei are interchanged) now correspond to the same physical state of the molecule. The classical partition function (47.9) must therefore be halved, and so the chemical constant becomes

$$\zeta_{rot} = \log (I/\hbar^2); \qquad (48.1)$$

accordingly the factor 2 disappears from the argument of the logarithm in the sum $\zeta_{tr}+\zeta_{rot}$ (47.12).

More important changes are needed at temperatures where the quantum treatment has to be used. Since in practice the entire problem is of interest only in its application to the two isotopes of hydrogen (H_2 and D_2), we shall consider these gases in what follows. The requirement of quantum-mechanical

† An asymptotic expansion of the thermodynamic quantities for large values of $2TI/\hbar^2$ may be obtained. The first two terms of the expansion for the specific heat are

$$c_{rot} = 1 + \frac{1}{45}\left(\frac{\hbar^2}{2TI}\right)^2.$$

It must be remembered, however, that this expansion gives only a poor approximation to the function $c_{rot}(T)$.

symmetry in the nuclei† results in the electronic state ${}^1\Sigma_g^+$(the ground state of the hydrogen molecule) having rotational levels of different nuclear spin degeneracy for even and odd values of K: levels with even and odd K respectively occur only for even and odd total spin of the two nuclei, and have relative degrees of degeneracy $g_g = i/(2i+1)$, $g_u = (i+1)/(2i+1)$ for a half-integral spin i of the nuclei, and $g_g = (i+1)/(2i+1)$, $g_u = i/(2i+1)$ for integral i. For hydrogen there is an accepted terminology whereby the molecules in states of greater nuclear statistical weight are called *orthohydrogen* molecules, and those in states of smaller statistical weight are called *parahydrogen* molecules. Thus for the H_2 and D_2 molecules the statistical weights are

$$H_2\,(i=\tfrac{1}{2})\begin{cases}\text{ortho } g_u = \tfrac{3}{4},\\ \text{para }\; g_g = \tfrac{1}{4},\end{cases}\qquad D_2\,(i=1)\begin{cases}\text{ortho } g_g = \tfrac{2}{3},\\ \text{para }\; g_u = \tfrac{1}{3}.\end{cases}$$

The suffix g denotes that the molecule has an even total nuclear spin (0 for H_2, 0 or 2 for D_2) and even rotational angular momenta K; the suffix u signifies odd total nuclear spins (1 for H_2 and D_2) and odd values of K.

Whereas, in molecules with unlike nuclei, the nuclear degrees of degeneracy of all the rotational levels are the same, and so the allowance for this degeneracy simply gives an unimportant change in the chemical constant, here it causes a change in the form of the partition function, which must now be written‡

$$Z_{\text{rot}} = g_g Z_g + g_u Z_u, \tag{48.2}$$

where

$$\begin{aligned} Z_g &= \sum_{K=0,\,2,\,\ldots} (2K+1)e^{-\hbar^2 K(K+1)/2IT},\\ Z_u &= \sum_{K=1,\,3,\,\ldots} (2K+1)e^{-\hbar^2 K(K+1)/2IT}. \end{aligned} \tag{48.3}$$

Similarly the free energy becomes

$$F_{\text{rot}} = -NT\log(g_g Z_g + g_u Z_u), \tag{48.4}$$

and the remaining thermodynamic quantities are likewise changed. At high temperatures,

$$Z_g \cong Z_u \cong \tfrac{1}{2}Z_{\text{rot}} = TI/\hbar^2,$$

so that the previous classical expression is obtained for the free energy, as it should be.

As $T \to 0$ the sum Z_g tends to unity and Z_u tends exponentially to zero; at low temperatures, therefore, the gas behaves as if monatomic (the specific heat $c_{\text{rot}} = 0$) and the chemical constant simply contains a "nuclear part" $\zeta_{\text{nuc}} = \log g_g$.

† See *Quantum Mechanics*, §86.

‡ The normalisation of the nuclear statistical weights which we use (such that $g_g + g_u = 1$) signifies that the entropy is measured from $\log(2i+1)^2$, in accordance with the condition stated at the end of §46.

The above formulae relate, of course, to a gas in complete thermal equilibrium. In such a gas the ratio of the numbers of molecules of parahydrogen and orthohydrogen is a definite function of temperature, which from the Boltzmann distribution is

$$x_{H_2} = N_{\text{ortho}-H_2}/N_{\text{para}-H_2} = g_u Z_u/g_g Z_g = 3Z_u/Z_g,$$
$$1/x_{D_2} = N_{\text{ortho}-D_2}/N_{\text{para}-D_2} = g_g Z_g/g_u Z_u = 2Z_g/Z_u. \quad (48.5)$$

As the temperature varies from 0 to ∞, the ratio x_{H_2} varies from 0 to 3, and x_{D_2} from 0 to $\frac{1}{2}$ (at $T = 0$ all the molecules are, of course, in the state with the lowest value of K, namely $K = 0$, corresponding to pure para-H_2 and ortho-D_2).

It must be borne in mind, however, that the probability of a change in the total nuclear spin in a collision between molecules is very small. The molecules of orthohydrogen and parahydrogen consequently behave practically as different modifications of hydrogen and are not† converted into each other. In practice, therefore, we are concerned not with a gas in equilibrium but with a non-equilibrium mixture of the ortho and para modifications, the relative amounts of which have given constant values.‡ The free energy of such a mixture is equal to the sum of the free energies of the two components.

In particular, for $x = \infty$ (pure ortho-H_2 or para-D_2) we have

$$F_{\text{rot}} = -NT \log (g_u Z_u).$$

At low temperatures ($\hbar^2/2IT \gg 1$) only the first term in the sum need be retained in Z_u, so that $Z_u = 3e^{-\hbar^2/IT}$, and the free energy is

$$F_{\text{rot}} = N\hbar^2/I - NT \log (3g_u).$$

This means that the gas will behave as if monatomic ($c_{\text{rot}} = 0$), the chemical constant including an additional term $\log (3g_u)$, and the energy a constant term $N\hbar^2/I$, corresponding to the rotational energy of all the molecules, with $K = 1$.

§49. Diatomic gases. Vibrations of atoms

The vibrational part of the thermodynamic quantities for a gas becomes important at considerably higher temperatures than the rotational part, because the intervals in the vibrational structure of the terms are large compared with those in the rotational structure.||

† In the absence of suitable catalysts.

‡ For an ordinary gas which has been at room temperature for a considerable time the ratios are $x_{H_2} = 3$, $x_{D_2} = \frac{1}{2}$.

|| As examples, the values of $\hbar\omega/k$ for some diatomic gases are H_2 6100°, N_2 3340°, O_2 2230°, NO 2690°, HCl 4140°.

We shall suppose, however, that the temperature is not large enough to excite the very high vibrational levels. Then the vibrations are small, and therefore harmonic, and the energy levels are given by the usual expression $\hbar\omega(v+\frac{1}{2})$ as in (47.4).

The calculation of the vibrational partition function Z_{vib} (47.4) is elementary. Owing to the very rapid convergence of the series, the summation may be formally extended to $v = \infty$. We shall measure the energy of the molecule from the lowest vibrational level ($v = 0$), i.e. include $\frac{1}{2}\hbar\omega$ in the constant ε_0 in (47.1). Then

$$Z_{\text{vib}} = \sum_{v=0}^{\infty} e^{-\hbar\omega v/T} = 1/(1-e^{-\hbar\omega/T}),$$

and hence the free energy is

$$F_{\text{vib}} = NT \log(1-e^{-\hbar\omega/T}), \tag{49.1}$$

the entropy

$$S_{\text{vib}} = -N\log(1-e^{-\hbar\omega/T})+N\hbar\omega/T(e^{\hbar\omega/T}-1), \tag{49.2}$$

the energy

$$E_{\text{vib}} = N\hbar\omega/(e^{\hbar\omega/T}-1), \tag{49.3}$$

and the specific heat

$$c_{\text{vib}} = \left(\frac{\hbar\omega}{T}\right)^2 \frac{e^{\hbar\omega/T}}{(e^{\hbar\omega/T}-1)^2}. \tag{49.4}$$

Fig. 4 shows c_{vib} as a function of $T/\hbar\omega$.

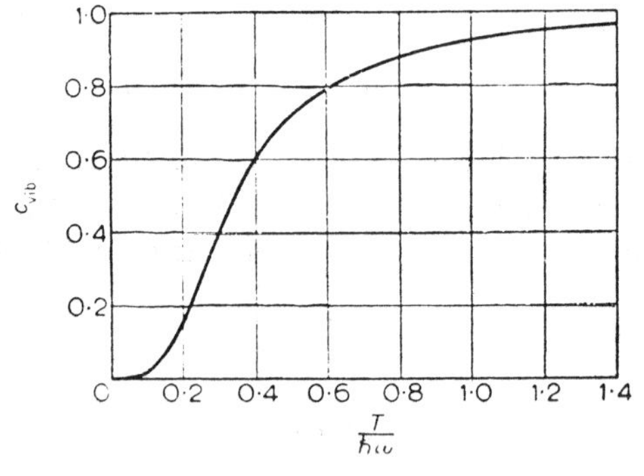

FIG. 4

At low temperatures ($\hbar\omega \gg T$) all these quantities tend exponentially to zero:

$$\begin{aligned} F_{\text{vib}} &= -NTe^{-\hbar\omega/T}, \\ c_{\text{vib}} &= (\hbar\omega/T)^2 e^{-\hbar\omega/T}. \end{aligned} \tag{49.5}$$

At high temperatures ($\hbar\omega \ll T$) we have

$$F_{\text{vib}} = -NT\log T+NT\log(\hbar\omega)-N\cdot\tfrac{1}{2}\hbar\omega, \tag{49.6}$$

corresponding to a constant specific heat $c_{\text{vib}} = 1$† and a chemical constant $\zeta_{\text{vib}} = -\log(\hbar\omega)$. Adding these to the values (47.11), (47.12), we find that at temperatures $T \gg \hbar\omega$ the total specific heat of a diatomic gas is‡

$$c_v = 7/2, \qquad c_p = 9/2, \tag{49.7}$$

and the chemical constant is

$$\zeta = \log\left[\frac{(2)I}{\omega\hbar^6}\left(\frac{m}{2\pi}\right)^{3/2}\right]; \tag{49.8}$$

the factor (2) must be omitted for molecules consisting of like atoms. The first two terms in the expansion of E_{vib} are

$$E_{\text{vib}} = NT - \tfrac{1}{2}N\hbar\omega. \tag{49.9}$$

The constant term $-\frac{1}{2}N\hbar\omega$ appears here because the energy is measured from the lowest quantum level (i.e. from the energy of the "zero-point" vibrations), whereas the classical energy would have to be measured from the minimum of the potential energy.

The expression (49.6) for the free energy can also be derived classically, of course, since for $T \gg \hbar\omega$ the important quantum numbers v are the large ones, where the motion is quasi-classical. The classical energy of small oscillations of frequency ω is

$$\varepsilon_{\text{vib}}(p, q) = \frac{p^2}{2m'} + \tfrac{1}{2}m'\omega^2 q^2,$$

where m' is the reduced mass. The integration with this expression for ε gives for the partition function

$$Z_{\text{vib}} = \frac{1}{2\pi\hbar}\int_{-\infty}^{\infty}\int_{-\infty}^{\infty} e^{-\varepsilon_{\text{vib}}/T}\, dp\, dq = T/\hbar\omega, \tag{49.10}$$

which corresponds to (49.6)||; owing to the rapid convergence of the integral, the integration with respect to q may be taken from $-\infty$ to ∞.

At sufficiently high temperatures, when vibrations with large v are excited, the anharmonicity of the vibrations and their interaction with the rotation of the molecule may become important. These effects are in principle of the same order of magnitude. Since v is large, the corresponding correction to the thermodynamic quantities may be determined classically.

Let us consider a molecule as a mechanical system of two particles interacting in accordance with the law $U(r)$ in a co-ordinate system in which their centre of mass is at rest. The energy (Hamiltonian) which gives a precise

† Again in accordance with the classical results of §44.

‡ As Fig. 4 shows, c_{vib} actually approaches its limiting value of 1 when $T \cong \hbar\omega$; for $T/\hbar\omega = 1$, $c_{\text{vib}} = 0.93$. As a practical condition for the applicability of the classical expressions we may write $T \gg \hbar\omega/3$.

|| The same result is obtained on replacing the summation over v by an integration.

classical description of the rotation and vibrations of the system is the sum of the kinetic energy (the energy of a particle with the reduced mass m') and the potential energy $U(r)$. The partition function, after integration over the momenta, reduces to an integral over the co-ordinates: $\int e^{-U(r)/T}\,dV$, and after integration over the angles (in spherical polar co-ordinates) there remains the integral

$$\int_0^\infty e^{-U(r)/T} r^2\,dr.$$

The approximation corresponding to independent harmonic vibrations and rotation of the molecule is obtained by putting $U(r) = U_0+\frac{1}{2}m'\omega^2(r-r_0)^2$ and, in the integration, replacing the slowly varying factor r^2 by r_0^2, where r_0 is the equilibrium distance between the two particles: $U_0 = U(r_0)$. In order to take into account the anharmonicity of the vibrations and their interaction with the rotation we now write

$$U(r) = U_0+\tfrac{1}{2}m'\omega^2 r_0^2(\xi^2-\alpha\xi^3+\beta\xi^4), \tag{49.11}$$

where $\xi = r/r_0-1$, and α and β are constants†, and then expand the whole integrand in powers of ξ, separating the factor $\exp\{-(U_0+\frac{1}{2}m'\omega^2 r_0^2\xi^2)/T\}$. In the expansion, only those terms need be retained which after integration give the highest and next highest powers of the temperature; the integration over ξ is taken from $-\infty$ to ∞. The zero-order term in the expansion gives the usual value of the partition function, and the remaining terms give the required correction. Omitting the calculations, we shall state the final result for the correction to the free energy:

$$F_{\text{anh}} = -NT^2\frac{1}{2I\omega^2}\left[1+3\alpha-\frac{3}{2}\beta+\frac{15}{8}\alpha^2\right]. \tag{49.12}$$

Thus the anharmonicity of the vibrations and their interaction with the rotation give a correction to the free energy which is proportional to the square of the temperature. Accordingly the specific heat has a further term proportional to the first power of the temperature.

§50. Diatomic gases. The effect of the electronic angular momentum

Some types of molecule, though not many, have a non-zero orbital angular momentum or spin in their electronic ground state.

The presence of a non-zero orbital angular momentum Λ causes a twofold degeneracy of the electronic term, corresponding to the two possible directions

† These constants can be expressed in terms of the spectroscopic constants of the molecule; see *Quantum Mechanics*, §82.

of this angular momentum with respect to the axis of the molecule.† This affects the thermodynamic quantities: because of the doubling of the partition function, a quantity

$$\zeta_\Lambda = \log 2 \tag{50.1}$$

is added to the chemical constant.

The presence of a non-zero spin S causes a splitting into $2S+1$ levels, but the intervals in this fine structure are so small (when $\Lambda = 0$) that they can always be neglected in calculating the thermodynamic quantities. The presence of the spin simply increases the degree of degeneracy of each level by a factor $2S+1$, and so the chemical constant is increased by

$$\zeta_S = \log(2S+1). \tag{50.2}$$

The fine structure which occurs when $S \neq 0$, $\Lambda \neq 0$ requires special consideration. Here the fine-structure intervals may reach values which have to be taken into account in calculating the thermodynamic quantities. We shall derive the formulae for the case of a doublet electron term.‡ Each component of the electron doublet has its vibrational and rotational structure, the parameters of which may be regarded as the same for each component. The partition function (47.2) therefore contains a further factor

$$Z_{\mathrm{el}} = g_0 + g_1 e^{-\Delta/T},$$

where g_0, g_1 are the degrees of degeneracy of the components of the doublet, and Δ their separation. The free energy must accordingly contain an "electronic part"

$$F_{\mathrm{el}} = -NT\log(g_0 + g_1 e^{-\Delta/T}). \tag{50.3}$$

We may also give the "electronic" specific heat which must be added to the other parts of the specific heat:

$$c_{\mathrm{el}} = \frac{(\Delta/T)^2}{[1+(g_0/g_1)e^{\Delta/T}][1+(g_1/g_0)e^{-\Delta/T}]}. \tag{50.4}$$

In the limits $T \to 0$ and $T \to \infty$, c_{el} is of course zero, and it has a maximum at some temperature $T \sim \Delta$.

† Strictly speaking, the term is split into two levels (*Λ-doubling*), but the separation between these is so small that it may be entirely neglected here.

‡ This case occurs for NO; the electronic ground state of the NO molecule is the doublet $\Pi_{1/2,\,3/2}$ with width $\Delta = 178°$. Each component of the doublet is doubly degenerate.

An unusual case occurs for oxygen. The electronic ground state of the O_2 molecule is a very narrow triplet $^3\Sigma$, the width of which may be neglected, but it happens by chance that the next (excited) state $^1\Delta$ (doubly degenerate) is relatively near, at $\Delta = 11{,}300°$, and at high temperatures it may be excited, with a consequent effect on the thermodynamic quantities.

PROBLEM

Determine the correction to the free energy for oxygen due to the first excited electronic state of the O_2 molecule (see the last footnote). The temperature is large compared with the vibrational quantum, but small compared with the distance Δ between the ground state $^3\Sigma$ and the excited state $^1\Delta$.

SOLUTION. The partition function is

$$Z = 3\frac{T}{\hbar\omega}\frac{TI}{\hbar^2}+2e^{-\Delta/T}\frac{T}{\hbar\omega}\frac{TI'}{\hbar^2},$$

where the two terms on the right are the partition functions for the ground and excited states, each of which is the product of electronic, vibrational and rotational factors. The required correction to the free energy is therefore

$$F_{^1\Delta} = -NT\log\left(1+\frac{2\omega r_0'^2}{3\omega' r_0^2}e^{-\Delta/T}\right) \cong -NT\cdot\frac{2\omega r_0'^2}{3\omega' r_0^2}e^{-\Delta/T},$$

where ω, r_0, ω', r_0' are the frequencies and equilibrium distances between the nuclei in the ground and excited electronic states.

§51. Polyatomic gases

The free energy of a polyatomic gas, like that of a diatomic gas, can be written as the sum of translational, rotational and vibrational parts. The translational part, as before, is characterised by values of the specific heat and chemical constant

$$c_{\text{tr}} = 3/2, \qquad \zeta_{\text{tr}} = (3/2)\log(m/2\pi\hbar^2). \tag{51.1}$$

Owing to the large moments of inertia of polyatomic molecules (and the corresponding smallness of their rotational quanta) their rotation may always be treated classically.† The polyatomic molecule has three rotational degrees of freedom and three principal moments of inertia I_1, I_2, I_3, which are in general different; its kinetic energy of rotation is therefore

$$\varepsilon_{\text{rot}} = \frac{M_\xi^2}{2I_1}+\frac{M_\eta^2}{2I_2}+\frac{M_\zeta^2}{2I_3}, \tag{51.2}$$

where ξ, η, ζ are co-ordinates in a rotating system whose axes coincide with the principal axes of inertia of the molecule; for the present we disregard the special case of molecules consisting of collinear atoms. This expression is to be substituted in the partition function

$$Z_{\text{rot}} = \int' e^{-\varepsilon_{\text{rot}}/T}\,d\tau_{\text{rot}}, \tag{51.3}$$

where

$$d\tau_{\text{rot}} = \frac{1}{(2\pi\hbar)^3}\,dM_\xi\,dM_\eta\,dM_\zeta\,d\phi_\xi\,d\phi_\eta\,d\phi_\zeta,$$

† Rotation quantisation effects would be observable only in methane CH_4, where they should occur at temperatures of about 50°K; see the Problem at the end of this section.

and the prime denotes, as usual, that the integration is to be taken only over the physically different orientations of the molecule.

If the molecule has axes of symmetry, rotations about these axes leave the molecule unchanged, and amount to an interchange of identical atoms. It is clear that the number of physically indistinguishable orientations of the molecule is equal to the number of possible different rotations about the axes of symmetry, including a rotation through 360° (the identical transformation). Denoting this number† by σ, we can take the integration in (51.3) simply over all orientations and divide by σ.

In the product $d\phi_\xi \, d\phi_\eta \, d\phi_\zeta$ of three infinitesimal angles of rotation, $d\phi_\xi \, d\phi_\eta$ may be regarded as an element do_ζ of solid angle for directions of the ζ-axis. The integration over o_ζ is independent of that over rotations $d\phi_\zeta$ about the ζ-axis, and gives 4π. The integration over ϕ_ζ gives a further 2π. Integrating also over M_ξ, M_η, M_ζ from $-\infty$ to ∞, we finally have

$$Z_{\text{rot}} = \frac{8\pi^2}{\sigma(2\pi\hbar)^3}(2\pi T)^{3/2}(I_1 I_2 I_3)^{1/2} = (2T)^{3/2}(\pi I_1 I_2 I_3)^{1/2}/\sigma\hbar^3$$

Hence the free energy is

$$F = -\frac{3}{2}NT\log T - NT\log\frac{(8\pi I_1 I_2 I_3)^{1/2}}{\sigma\hbar^3}. \tag{51.4}$$

Thus we have for the rotational specific heat, in accordance with §44,

$$c_{\text{rot}} = 3/2, \tag{51.5}$$

and the chemical constant is

$$\zeta_{\text{rot}} = \log\frac{(8\pi I_1 I_2 I_3)^{1/2}}{\sigma\hbar^3}. \tag{51.6}$$

For a *linear molecule*, i.e. one where all the atoms are collinear, there are, as in the diatomic molecule, only two rotational degrees of freedom and one moment of inertia I. The rotational specific heat and the chemical constant are, as in a diatomic gas,

$$c_{\text{rot}} = k, \qquad \zeta_{\text{rot}} = \log(2I/\sigma\hbar^2), \tag{51.7}$$

where $\sigma = 1$ for an asymmetric molecule (such as NNO) and $\sigma = 2$ for a molecule symmetrical about its midpoint (such as OCO).

The vibrational part of the free energy of a polyatomic gas is calculated in a similar way to that for a diatomic gas, given above. The only difference is that a polyatomic molecule has not one but several vibrational degrees of freedom: a non-linear molecule of n atoms clearly has $r_{\text{vib}} = 3n-6$ vibrational degrees of freedom, while for a linear molecule of n atoms $r_{\text{vib}} = 3n-5$ (see §44). The number of vibrational degrees of freedom determines

† For instance, in H_2O (an isosceles triangle) $\sigma = 2$, in NH_3 (an equilateral triangular pyramid) $\sigma = 3$, in CH_4 (a tetrahedron) $\sigma = 12$, and in C_6H_6 (a regular hexagon) $\sigma = 12$.

the number of *normal modes of vibration* of the molecule, to each of which there corresponds a frequency ω_α (the suffix α numbering the normal modes). It must be remembered that some of the frequencies ω_α may be equal, in which case the frequency concerned is said to be *degenerate*.

In the harmonic approximation, where the vibrations are assumed small (only temperatures for which this is so will be considered), all the normal modes are independent, and the vibrational energy is the sum of the energies of the individual modes. The vibrational partition function therefore falls into a product of partition functions of the individual modes, and the free energy F_{vib} is a sum of expressions of the type (49.1):

$$F_{\text{vib}} = NT\sum_{\alpha} \log\left(1-e^{-\hbar\omega_\alpha/T}\right). \tag{51.8}$$

Each frequency appears in this sum a number of times equal to its degeneracy. Similar sums are obtained for the vibrational parts of the other thermodynamic quantities.

Each of the normal modes gives, in its own classical limit ($T \gg \hbar\omega_\alpha$), a contribution $c_{\text{vib}}^{(\alpha)} = 1$ to the specific heat; for T greater than the greatest $\hbar\omega_\alpha$ we should obtain

$$c_{\text{vib}} = r_{\text{vib}}. \tag{51.9}$$

In practice, however, this limit is not reached, since polyatomic molecules usually decompose at considerably lower temperatures.

The various frequencies ω_α for a polyatomic molecule generally range over a very wide interval. As the temperature increases, the various normal modes successively contribute to the specific heat. In consequence the specific heat of polyatomic gases may often be regarded as approximately constant over fairly wide intervals of temperature.

We may mention the possibility of a curious change from vibration to rotation, an instance of which is afforded by the ethane molecule C_2H_6. This molecule consists of two CH_3 groups at a certain distance apart and oriented in a certain way to each other. One of the normal vibrations of the molecule is a "torsional" vibration, in which one of the CH_3 groups is twisted relative to the other. As the energy of the vibrations increases, their amplitude increases and ultimately, at sufficiently high temperatures, the vibration becomes a free rotation. The contribution of this degree of freedom to the specific heat, which is approximately 1 when the vibrations are fully excited, therefore begins to decrease as the temperature increases further, approaching asymptotically the value $\frac{1}{2}$ typical of a rotation.

Finally, it may be mentioned that, if the molecule has a non-zero spin S (for example, the molecules NO_2 and ClO_2), the chemical constant includes a term

$$\zeta_S = \log(2S+1). \tag{51.10}$$

PROBLEM

Determine the rotational partition function for methane at low temperatures.

SOLUTION. As already mentioned in the first footnote to this section, a quantum calculation of Z_{rot} for methane is required at sufficiently low temperatures.

The CH_4 molecule is a tetrahedron of the spherical-top type, and so its rotational levels are $\hbar^2 J(J+1)/2I$, where I is the common value of the three principal moments of inertia, and J the rotational quantum number. Since the spin i of the H nucleus is $\frac{1}{2}$, and that of the C^{12} nucleus is zero, the total nuclear spin of the CH_4 molecule may be 0, 1 or 2, the corresponding nuclear statistical weights being 1, 3 or 5.† For any given value of J there are definite numbers of states corresponding to values of the total nuclear spin. The following table gives these numbers for the first five values of J.

Nuclear spin	0	1	2
$J = 0$	–	–	1
1	–	1	–
2	2	1	–
3	–	2	1
4	2	2	1

The value of the sum Z_{rot} which is obtained by taking into account the total degree of degeneracy with respect to orientations of the rotational angular momentum and nuclear spin must be divided by 16 if the entropy is to be measured from the value $\log(2i+1)^4 = \log 16$ (cf. the second footnote to §48). The result is

$$Z_{rot} = \frac{5}{16}+\frac{9}{16}e^{-\hbar^2/IT}+\frac{25}{16}e^{-3\hbar^2/IT}+\frac{77}{16}e^{-6\hbar^2/IT}+\frac{117}{16}e^{-10\hbar^2/IT}+\dots.$$

† See *Quantum Mechanics*, §105, Problem 5.

CHAPTER V

THE FERMI AND BOSE DISTRIBUTIONS

§52. The Fermi distribution

IF THE temperature of an ideal gas (at a given density) is sufficiently low, Boltzmann statistics becomes inapplicable, and a different statistics must be devised, in which the mean occupation numbers of the various quantum states of particles are not assumed small.

This statistics, however, differs according to the type of wave functions by which the gas is described when regarded as a system of N identical particles. These functions must be either antisymmetrical or symmetrical with respect to interchanges of any pair of particles, the former case occurring for particles with half-integral spin, and the latter case for those with integral spin.

For a system of particles described by antisymmetrical wave functions, *Pauli's principle* applies: in each quantum state there cannot simultaneously be more than one particle. The statistics based on this principle is called *Fermi statistics*, or Fermi–Dirac statistics.†

As in §37, we shall apply the Gibbs distribution to the set of all particles in the gas which are in a given quantum state; as already mentioned in §37, this may be done even if there is an exchange interaction between the particles. We again denote by Ω_k the thermodynamic potential of this set of particles; by the general formula (35.3),

$$\Omega_k = -T \log \sum_{n_k} \left(e^{(\mu-\varepsilon_k)/T}\right)^{n_k}, \tag{52.1}$$

since the energy of n_k particles in the kth state is just $n_k \varepsilon_k$. According to PAULI's principle, the occupation numbers of each state can take only the values 0 and 1. Hence

$$\Omega_k = -T \log \left(1 + e^{(\mu-\varepsilon_k)/T}\right).$$

Since the mean number of particles in the system is equal to minus the derivative of the potential Ω with respect to the chemical potential μ, the required mean number of particles in the kth quantum state is here obtained

† It was proposed by FERMI for electrons, and its relation to quantum mechanics was elucidated by DIRAC (1926).

as the derivative

$$\bar{n}_k = -\frac{\partial \Omega_k}{\partial \mu} = \frac{e^{(\mu-\varepsilon_k)T}}{1+e^{(\mu-\varepsilon_k)/T}},$$

or finally

$$\bar{n}_k = \frac{1}{e^{(\varepsilon_k-\mu)/T}+1}. \tag{52.2}$$

This is the distribution function for an ideal gas obeying Fermi statistics, which for brevity will be called a *Fermi gas.* When $e^{(\mu-\varepsilon_k)/T} \ll 1$ it tends to the Boltzmann distribution function, as it should.†

The Fermi distribution is normalised by the condition

$$\sum_k \frac{1}{e^{(\varepsilon_k-\mu)/T}+1} = N, \tag{52.3}$$

where N is the total number of particles in the gas. This equation implicitly determines the chemical potential as a function of T and N.

The thermodynamic potential Ω of the gas as a whole is obtained by summation of Ω_k over all quantum states:

$$\Omega = -T\sum_k \log\left(1+e^{(\mu-\varepsilon_k)/T}\right). \tag{52.4}$$

§53. The Bose distribution

Let us now consider the statistics obeyed by an ideal gas consisting of particles described by symmetrical wave functions, namely *Bose statistics* or Bose–Einstein statistics.‡

The occupation numbers of the quantum states when the wave functions are symmetrical are unrestricted and can take any values. The distribution function may be derived as in §52; we put

$$\Omega_k = -T \log \sum_{n_k=0}^{\infty} \left(e^{(\mu-\varepsilon_k)/T}\right)^{n_k}.$$

This geometric progression is convergent only if $e^{(\mu-\varepsilon_k)/T} < 1$. Since this condition must be satisfied for all ε_k, including $\varepsilon_k = 0$, it is clear that we must certainly have

$$\mu < 0. \tag{53.1}$$

† In Boltzmann statistics, the expression (52.1) must be expanded in powers of the small quantity $e^{(\mu-\varepsilon_k)/T}$; the first term of the expansion is

$$\Omega_k = -Te^{(\mu-\varepsilon_k)/T},$$

whence differentiation with respect to μ again gives the Boltzmann distribution formula.

‡ This was introduced by Bose for light quanta, and generalised by Einstein (1924).

Thus in Bose statistics the chemical potential is always negative. In this connection it may be recalled that in Boltzmann statistics the chemical potential is always negative, and large in absolute value; in Fermi statistics, μ may be either negative or positive.

Summation of the geometric progression gives

$$\Omega_k = T \log\left(1 - e^{(\mu-\varepsilon_k)/T}\right).$$

Hence we find the mean occupation numbers $\bar{n}_k = -\partial\Omega_k/\partial\mu$:

$$\bar{n}_k = \frac{1}{e^{(\varepsilon_k-\mu)/T} - 1}. \qquad (53.2)$$

This is the distribution function for an ideal gas which obeys Bose statistics (or, as we shall call it for brevity, a *Bose gas*). It differs from the Fermi distribution function in the sign of unity in the denominator. Like that function, it tends of course to the Boltzmann distribution function when $e^{(\mu-\varepsilon_k)/T} \ll 1$. The total number of particles in the gas is given by the formula

$$N = \sum_k \frac{1}{e^{(\varepsilon_k-\mu)/T} - 1}, \qquad (53.3)$$

and the thermodynamic potential Ω of the gas as a whole is obtained by summation of Ω_k over all quantum states:

$$\Omega = T\sum_k \log\left(1 - e^{(\mu-\varepsilon_k)/T}\right). \qquad (53.4)$$

§54. Fermi and Bose gases not in equilibrium

As in §40, we can calculate the entropy also for Fermi and Bose gases not in equilibrium, and again derive the Fermi and Bose distribution functions from the condition that the entropy is a maximum.

In the Fermi case there can be no more than one particle in each quantum state, but the numbers N_j are not small, and are in general of the same order of magnitude as the numbers G_j. (The notation is as in §40.)

The number of possible ways of distributing N_j identical particles among G_j states with not more than one particle in each is just the number of ways of selecting N_j of the G_j states, i.e. the number of combinations of G_j things N_j at a time. Thus

$$\Delta\Gamma_j = G_j!/N_j!\,(G_j - N_j)!. \qquad (54.1)$$

Taking the logarithm of this expression and using for the logarithm of each factorial the formula $\log N! = N \log(N/e)$, we find

$$S = \sum_j \{G_j \log G_j - N_j \log N_j - (G_j - N_j)\log(G_j - N_j)\}. \qquad (54.2)$$

Again using the mean occupation numbers of the quantum states, $\overline{n_j} = N_j/G_j$, we finally have the following expression for the entropy of a Fermi gas not in equilibrium:

$$S = -\sum_j G_j[\overline{n_j} \log \overline{n_j} + (1-\overline{n_j}) \log (1-\overline{n_j})]. \qquad (54.3)$$

From the condition for this expression to be a maximum according to (40.8) we easily find that the equilibrium distribution is given by the formula

$$\overline{n_j} = 1/(e^{\alpha+\beta\varepsilon_j}+1),$$

which is the Fermi distribution, as it should be.

Finally, for Bose statistics, each quantum state may contain any number of particles, so that the statistical weight $\Delta\Gamma_j$ is the total number of ways of distributing N_j particles among G_j states. This number is†

$$\Delta\Gamma_j = (G_j+N_j-1)!/(G_j-1)!N_j!. \qquad (54.4)$$

Taking the logarithm of this expression and neglecting unity in comparison with the very large numbers G_j+N_j and G_j, we obtain

$$S = \sum_j \{(G_j+N_j) \log (G_j+N_j) - N_j \log N_j - G_j \log G_j\}. \qquad (54.5)$$

In terms of the numbers $\overline{n_j}$ we can write the entropy of a Bose gas not in equilibrium as

$$S = \sum_j G_j[(1+\overline{n_j}) \log (1+\overline{n_j}) - \overline{n_j} \log \overline{n_j}]. \qquad (54.6)$$

It is easily seen that the condition for this expression to be a maximum in fact gives the Bose distribution.

The two formulae (54.2) and (54.5) for the entropy naturally tend, in the limiting case $N_j \ll G_j$, to the Boltzmann formula (40.3), and the statistical weights (54.1) and (54.4) for Fermi and Bose statistics tend to the Boltzmann expression (40.2); to see this, we must put $G_j! \cong (G_j-N_j)!G_j^{N_j}$, $(G_j+N_j-1)! \cong (G_j-1)!G_j^{N_j}$. It must be remembered, however, that, in going to the limit, terms of order N_j^2/G_j are neglected in the statistical weights (as may easily be

†The problem is to find the number of ways of distributing N_j identical balls among G_j urns. Let us imagine the balls as a line of N_j points, and number the urns; let us then imagine the latter to be separated by G_j-1 vertical strokes placed at intervals along the line of points. For example, the diagram

$$\cdot|\cdot\cdot\cdot||\cdot\cdot\cdot\cdot|\cdot\cdot$$

represents ten balls distributed among five urns: one in the first, three in the second, none in the third, four in the fourth and two in the fifth. The total number of places (occupied by points and strokes) in the line is G_j+N_j-1. The required number of distributions of the balls among the urns is the number of ways of choosing G_j-1 positions for the strokes, i.e. the number of combinations of N_j+G_j-1 things G_j-1 at a time, and this gives the result (54.4).

verified), and these terms are not in general small; but when the logarithm is taken these terms give a correction to the entropy which is of the relatively small order N_j/G_j.

Finally, we shall give a formula for the entropy of a Bose gas in the important limiting case where the number of particles in each quantum state is large (so that $N_j \gg G_j$, $\overline{n_j} \gg 1$). We know from quantum mechanics that this case corresponds to the classical wave picture of the field. The statistical weight (54.4) becomes

$$\Delta\Gamma_j = N_j^{G_j-1}/(G_j-1)! \tag{54.7}$$

and the entropy is

$$S = \sum_j G_j \log(eN_j/G_j). \tag{54.8}$$

We shall make use of this formula in §65.

§55. Fermi and Bose gases of elementary particles

Let us consider a gas consisting of elementary particles, or of particles which under certain conditions may be regarded as elementary. As has already been mentioned, the Fermi or Bose distribution need not be used for ordinary atomic or molecular gases, since these gases are in practice always described with sufficient accuracy by the Boltzmann distribution.

All the formulae derived in the present section are exactly similar in form for both Fermi and Bose statistics, differing only as regards one sign. The upper sign will always correspond to Fermi statistics and the lower sign to Bose statistics.

The energy of an elementary particle is just the kinetic energy of its translational motion, which is always quasi-classical. We therefore have

$$\varepsilon = (p_x^2+p_y^2+p_z^2)/2m, \tag{55.1}$$

and in the distribution function we make the usual change to the distribution in the phase space of the particle. Here it must be borne in mind that, for a given value of the momentum, the state of the particle still depends on the orientation of its spin. Hence the number of particles in a volume element $dp_x\,dp_y\,dp_z\,dV$ in phase space is found by multiplying the distribution (52.2) or (53.2) by

$$g\,d\tau = g\,dp_x\,dp_y\,dp_z\,dV/(2\pi\hbar)^3,$$

where $g = 2S+1$ (S being the spin of the particle), giving

$$dN = \frac{g\,d\tau}{e^{(\varepsilon-\mu)/T}\pm 1}. \tag{55.2}$$

Integrating over V (which simply involves replacing dV by the total volume

V of the gas) we find the distribution for the components p_x, p_y, p_z of the particle momentum; using spherical polar co-ordinates in momentum space and integrating over angles, we find the distribution for the absolute magnitude of the momentum:

$$dN_p = \frac{gVp^2\,dp}{2\pi^2\hbar^3(e^{(\varepsilon-\mu)/T}\pm 1)}, \tag{55.3}$$

where $\varepsilon = p^2/2m$, or the energy distribution

$$dN_\varepsilon = \frac{gVm^{3/2}}{2^{1/2}\pi^2\hbar^3}\frac{\sqrt{\varepsilon}\,d\varepsilon}{e^{(\varepsilon-\mu)/T}\pm 1}. \tag{55.4}$$

These formulae take the place of the classical Maxwellian distribution.

Integrating (55.4) with respect to ε, we obtain the total number of particles in the gas:

$$N = \frac{gVm^{3/2}}{2^{1/2}\pi^2\hbar^3}\int_0^\infty \frac{\sqrt{\varepsilon}\,d\varepsilon}{e^{(\varepsilon-\mu)/T}\pm 1}.$$

In terms of a new variable of integration $z = \varepsilon/T$, this equation can be written

$$\frac{N}{V} = \frac{g(mT)^{3/2}}{2^{1/2}\pi^2\hbar^3}\int_0^\infty \frac{\sqrt{z}\,dz}{e^{z-\mu/T}\pm 1}. \tag{55.5}$$

This formula implicitly determines the chemical potential μ of the gas as a function of its temperature T and density N/V.

With the same change from summation to integration in formulae (52.4), (53.4), we find for the potential Ω the expression

$$\Omega = \mp\frac{VgTm^{3/2}}{2^{1/2}\pi^2\hbar^3}\int_0^\infty \sqrt{\varepsilon}\log(1\pm e^{(\mu-\varepsilon)/T})\,d\varepsilon.$$

Integration by parts gives

$$\Omega = -\frac{2}{3}\frac{gVm^{3/2}}{2^{1/2}\pi^2\hbar^3}\int_0^\infty \frac{\varepsilon^{3/2}\,d\varepsilon}{e^{(\varepsilon-\mu)/T}\pm 1}. \tag{55.6}$$

This expression is the same, apart from the factor $-\frac{2}{3}$, as the total energy of the gas,

$$E = \int_0^\infty \varepsilon\,dN_\varepsilon = \frac{gVm^{3/2}}{2^{1/2}\pi^2\hbar^3}\int_0^\infty \frac{\varepsilon^{3/2}\,d\varepsilon}{e^{(\varepsilon-\mu)/T}\pm 1}. \tag{55.7}$$

Since $\Omega = -PV$, we have therefore

$$PV = \tfrac{2}{3}E. \tag{55.8}$$

This result is exact, and so must hold good in the limiting case of a Boltzmann gas also; and in fact, on substituting the Boltzmann value $E = 3NT/2$, we obtain CLAPEYRON's equation.

From formula (55.6), substituting $\varepsilon/T = z$, we obtain

$$\Omega = -PV = VT^{5/2}f(\mu/T), \tag{55.9}$$

where f is a function of a single variable, i.e. Ω/V is a homogeneous function of order 5/2 in μ and T.† Hence

$$\frac{S}{V} = -\frac{1}{V}\left(\frac{\partial\Omega}{\partial T}\right)_{V,\mu} \quad \text{and} \quad \frac{N}{V} = -\frac{1}{V}\left(\frac{\partial\Omega}{\partial\mu}\right)_{T,V}$$

are homogeneous functions of order 3/2 in μ and T, and their ratio S/N is a homogeneous function of order zero, i.e. $S/N = \phi(\mu/T)$. Hence we see that in an adiabatic process (S = constant) the ratio μ/T remains constant, and since $N/VT^{3/2}$ is also a function of μ/T only we have

$$VT^{3/2} = \text{constant}. \tag{55.10}$$

Then (55.9) shows that

$$PV^{5/3} = \text{constant}, \tag{55.11}$$

and also $T^{5/2}/P$ = constant. These equations are the same as that of the Poisson adiabatic (43.9) for an ordinary monatomic gas, but it must be emphasised that the exponents in (55.10), (55.11) are here unrelated to the ratio of specific heats, since the relations $c_p/c_v = 5/3$ and $c_p - c_v = 1$ are not valid.

Formula (55.6), in the form

$$P = \frac{g \cdot 2^{1/2}m^{3/2}T^{5/2}}{3\pi^2\hbar^3}\int_0^\infty \frac{z^{3/2}\,dz}{e^{z-\mu/T}\pm 1}, \tag{55.12}$$

together with (55.5) determines the equation of state of the gas (in parametric form, with parameter μ), i.e. the relation between P, V and T. In the limiting case of a Boltzmann gas ($e^{\mu/T} \ll 1$) these formulae give CLAPEYRON's equation, as they should. We shall show this by means of a calculation which also gives the first correction term in the expansion in the equation of state.

For $e^{\mu/T} \ll 1$ we expand the integrand in (55.12) as a series of powers of

† If the energy is calculated from (55.9) as

$$E = N\mu + TS - PV = -\mu\partial\Omega/\partial\mu - T\,\partial\Omega/\partial T + \Omega,$$

we again obtain (55.8).

$e^{\mu/T-z}$ and, retaining only the first two terms, obtain

$$\int_0^\infty \frac{z^{3/2}\,dz}{e^{z-\mu/T}\pm 1} \cong \int_0^\infty z^{3/2}e^{\mu/T-z}(1\mp e^{\mu/T-z})\,dz$$

$$= \tfrac{3}{4}\sqrt{\pi}e^{\mu/T}(1\mp \frac{1}{2^{5/2}}e^{\mu/T}).$$

Substitution in (55.12) gives

$$\Omega = -PV = -\frac{gVm^{3/2}T^{5/2}}{(2\pi)^{3/2}\hbar^3}e^{\mu/T}(1\mp\frac{1}{2^{5/2}}e^{\mu/T}).$$

If only the first term of the expansion is retained, we obtain precisely the Boltzmann value of the chemical potential of a monatomic gas (formula (45.5), where $g = 1$). The next term gives the required correction, so that we can put

$$\Omega = \Omega_{\mathrm{Bol}} \pm \frac{gVm^{3/2}T^{5/2}}{16\pi^{3/2}\hbar^3}e^{2\mu/T}. \tag{55.13}$$

But the small additions to all the thermodynamic potentials (expressed in terms of the appropriate variables; see (24.16)) are the same. Hence, expressing the correction term in Ω in terms of T and V (which can be done to the same accuracy by means of the Boltzmann expressions), we obtain immediately the correction to the free energy:

$$F = F_{\mathrm{Bol}} \pm \frac{\pi^{3/2}}{2g}\cdot\frac{N^2\hbar^3}{VT^{1/2}m^{3/2}}. \tag{55.14}$$

Finally, differentiating with respect to volume, we obtain the required equation of state:

$$PV = NT\left[1\pm\frac{\pi^{3/2}}{2g}\frac{N\hbar^3}{V(mT)^{3/2}}\right] \tag{55.15}$$

The condition for the correction term in this formula to be small is naturally the same as the condition (45.6) for Boltzmann statistics to be applicable. Thus we see that the deviations of an ideal gas from classical properties, occurring when the temperature is lowered at constant density (the gas then being said to become *degenerate*), cause in Fermi statistics an increase in pressure as compared with its value in an ordinary gas; we may say that in this case the quantum exchange effects lead to the occurrence of an additional effective "repulsion" between the particles.

In Bose statistics, on the other hand, the value of the gas pressure changes in the opposite direction, becoming less than the classical value; we may say that here there is an effective "attraction" between the particles.

§56. A degenerate electron gas

The study of the properties of a Fermi gas at sufficiently low temperatures is of fundamental significance. As we shall see below, the temperatures concerned may in practice be very high in other respects.

In what follows we shall discuss an electron gas, with a view to the most important applications of Fermi statistics. For electrons, $g = 2$, but we shall avoid substituting this value in the formulae, so that the results will be directly applicable to other cases also.

Let us first consider an electron gas at a temperature of absolute zero (a *completely degenerate* Fermi gas). In such a gas, the electrons will be distributed among the various quantum states so that the total energy of the gas has its least possible value. Since no more than one electron can be in each quantum state, the electrons occupy all states with energies from the least value (zero) to some greatest value which depends on the number of electrons in the gas.

The number of quantum states of translational motion of a particle with absolute magnitude of momentum in the interval from p to $p+dp$ is $4\pi p^2\, dp \cdot V/(2\pi\hbar)^3$. Multiplying this by g, we obtain the total number of quantum states with such momenta:

$$gVp^2\, dp/2\pi^2\hbar^3. \tag{56.1}$$

The number of electrons occupying all states with momenta from zero to some p_0 is therefore

$$N = \frac{gV}{2\pi^2\hbar^3}\int_0^{p_0} p^2\, dp = \frac{gVp_0^3}{6\pi^2\hbar^3},$$

whence the limiting momentum p_0 is given by

$$p_0 = \left(\frac{6\pi^2}{g}\right)^{1/3}\left(\frac{N}{V}\right)^{1/3}\hbar, \tag{56.2}$$

and the limiting energy by

$$\varepsilon_0 = \frac{p_0^2}{2m} = \left(\frac{6\pi^2}{g}\right)^{2/3}\frac{\hbar^2}{2m}\left(\frac{N}{V}\right)^{2/3}. \tag{56.3}$$

This energy has a simple thermodynamic significance. In accordance with the foregoing discussion, the Fermi distribution function over quantum states,

$$\frac{1}{e^{(\varepsilon-\mu)/T}+1}, \tag{56.4}$$

tends to unity as $T \to 0$ for all $\varepsilon < \mu$ and to zero for $\varepsilon > \mu$, as shown by the continuous line in Fig. 5. Hence we see that the chemical potential of the gas at absolute zero is the same as the limiting energy of the electrons:

$$\mu = \varepsilon_0. \tag{56.5}$$

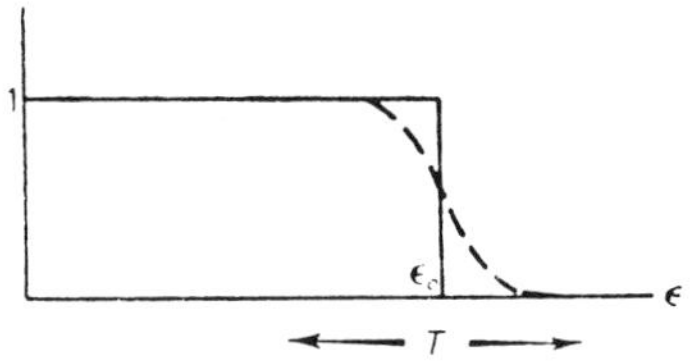

FIG. 5

The total energy of the gas is obtained by multiplying the number of states (56.1) by $p^2/2m$ and integrating over all momenta:

$$E = \frac{gV}{4m\pi^2\hbar^3}\int_0^{p_0} p^4\, dp = \frac{gVp_0^{\ 5}}{20m\pi^2\hbar^3},$$

or, substituting (56.2),

$$E = \frac{3}{10}\left(\frac{6\pi^2}{g}\right)^{2/3}\frac{\hbar^2}{m}\left(\frac{N}{V}\right)^{2/3}N. \tag{56.6}$$

Finally, from the general relation (55.8) we find the equation of state of the gas:

$$P = \frac{1}{5}\left(\frac{6\pi^2}{g}\right)^{2/3}\frac{\hbar^2}{m}\left(\frac{N}{V}\right)^{5/3}. \tag{56.7}$$

Thus the pressure of a Fermi gas at absolute zero is proportional to the 5/3 power of its density.

Formulae (56.6), (56.7) are approximately valid also at temperatures which are sufficiently close to absolute zero (for a given gas density). The condition for them to be applicable (for the gas to be "strongly degenerate") is clearly that T should be small in comparison with the limiting energy ε_0:

$$T \ll (\hbar^2/m)(N/V)^{2/3}. \tag{56.8}$$

This condition is, as we should expect, the opposite of the condition (45.6) for Boltzmann statistics to be valid. The temperature defined by the relation $T_0 \cong \varepsilon_0$ is called the *degeneracy temperature*.

A degenerate electron gas has the peculiar property that it increasingly approaches the "ideal gas" state as its density increases. This is easily seen as follows.

Let us consider a gas consisting of electrons and a corresponding number of positively charged nuclei which balance the charge on the electrons; a gas composed of electrons alone would obviously be entirely unstable, but we have not mentioned the nuclei hitherto, because the assumption of ideal-gas properties means that the presence of the nuclei does not affect the thermodynamic quantities for the electron gas. The energy (per electron) of the Coulomb interaction between the electrons and the nuclei is of the order of Ze^2/a, where Ze is the nuclear charge and $a \sim (ZV/N)^{1/3}$ is the mean distance between the electrons and the nuclei. The condition for an ideal gas is that this energy should be small compared with the mean kinetic energy of the electrons, which in order of magnitude is equal to the limiting energy ε_0. The inequality $Ze^2/a \ll \varepsilon_0$, after the substitution of $a \sim (ZV/N)^{1/3}$ and the expression (56.3) for ε_0, gives the condition

$$N/V \gg (e^2m/\hbar^2)^3Z^2. \tag{56.9}$$

We see that this condition is more nearly met as the density N/V of the gas increases.†

PROBLEM

Determine the number of collisions with a wall in an electron gas at absolute zero (taking $g = 2$).

SOLUTION. The number of electrons per unit volume with momenta in the interval dp at an angle to the normal to the wall in the interval $d\theta$ is $2\cdot 2\pi \sin\theta \, d\theta \, p^2 \, dp/(2\pi\hbar)^3$. The required number of collisions ν (per unit area of wall) is obtained by multiplying by $v\cos\theta$ ($v = p/m$) and integrating with respect to θ from 0 to $\frac{1}{2}\pi$ and with respect to p from 0 to p_0. The result is

$$\nu = \frac{3(3\pi^2)^{1/3}}{16}\frac{\hbar}{m}\left(\frac{N}{V}\right)^{4/3}.$$

§57. The specific heat of a degenerate electron gas

At temperatures which are low compared with the degeneracy temperature T_0, the distribution function (56.4) has the form shown by the broken line in Fig. 5: it is appreciably different from unity or zero only in a narrow range of values of the energy ε close to the limiting energy ε_0. The width of this "transition zone" of the Fermi distribution is of the order of T.

The expressions (56.6), (56.7) are the first terms in the expansions of the corresponding quantities in powers of the small ratio T/T_0. Let us now determine the next terms in the expansions.

Formula (55.6) involves an integral of the form

$$I = \int_0^\infty \frac{f(\varepsilon)\,d\varepsilon}{e^{(\varepsilon-\mu)/T}+1},$$

† The degeneracy temperature corresponding to the electron gas density $(e^2m/\hbar^2)^3Z^2$ is $40Z^{4/3}$ eV $\cong 0.5\times 10^6 Z^{4/3}$ degrees.

where $f(\varepsilon)$ is a function such that the integral converges; in (55.6), $f(\varepsilon) = \varepsilon^{3/2}$. We transform this integral by the substitution $\varepsilon - \mu = Tz$:

$$I = \int_{-\mu/T}^{\infty} \frac{f(\mu+Tz)}{e^z+1} T\,dz$$

$$= T\int_0^{\mu/T} \frac{f(\mu-Tz)}{e^{-z}+1}\,dz + T\int_0^{\infty} \frac{f(\mu+Tz)}{e^z+1}\,dz.$$

In the first integral we put $1/(e^{-z}+1) = 1-1/(e^z+1)$, obtaining

$$I = \int_0^{\mu} f(\varepsilon)\,d\varepsilon - T\int_0^{\mu/T} \frac{f(\mu-Tz)}{e^z+1}\,dz + T\int_0^{\infty} \frac{f(\mu+Tz)}{e^z+1}\,dz.$$

In the second of these integrals we replace the upper limit by infinity, sinc $\mu/T \gg 1$ and the integral is rapidly convergent.† This gives

$$I = \int_0^{\mu} f(\varepsilon)\,d\varepsilon + T\int_0^{\infty} \frac{f(\mu+Tz)-f(\mu-Tz)}{e^z+1}\,dz.$$

We now expand the numerator of the second integrand as a Taylor series of powers of z and integrate term by term:

$$I = \int_0^{\mu} f(\varepsilon)\,d\varepsilon + 2T^2 f'(\mu)\int_0^{\infty} \frac{z\,dz}{e^z+1} +$$

$$+\tfrac{1}{3}T^4 f'''(\mu)\int_0^{\infty} \frac{z^3\,dz}{e^z+1} + \ldots .$$

Substituting the values‡ of the integrals, we have finally

$$I = \int_0^{\mu} f(\varepsilon)\,d\varepsilon + \frac{\pi^2}{6}T^2 f'(\mu) + \frac{7\pi^4}{360}T^4 f'''(\mu) + \ldots . \tag{57.1}$$

† This amounts to neglecting exponentially small terms. It must be remembered that the expansion (57.1) derived below is an asymptotic, not a convergent, series.

‡ Integrals of this type are calculated as follows:

$$\int_0^{\infty} \frac{z^{x-1}\,dz}{e^z+1} = \int_0^{\infty} z^{x-1}e^{-z} \sum_{n=0}^{\infty} (-)^n e^{-nz}\,dz = \Gamma(x) \sum_{n=1}^{\infty} (-)^{n+1} \frac{1}{n^x}$$

$$= (1-2^{1-x})\Gamma(x) \sum_{n=1}^{\infty} \frac{1}{n^x},$$

The third term in the expansion is given for reference; it will not be needed here.

Putting $f = \varepsilon^{3/2}$ in formula (57.1) and substituting in (55.6), we obtain the required next term in the expansion of the potential Ω at low temperatures:

$$\Omega = \Omega_0 - VT^2 \frac{g\sqrt{(2\mu)}m^{3/2}}{12\hbar^3}, \tag{57.2}$$

where Ω_0 denotes the value of Ω at absolute zero.

Regarding the second term as a small correction to Ω_0 and expressing μ in it in terms of T and V by means of the "zero-order approximation" (56.5), we can immediately write down an expression for the free energy (according to (24.16)):

$$F = F_0 - \tfrac{1}{2}\beta NT^2(V/N)^{2/3}, \tag{57.3}$$

---

or

$$\int_0^\infty \frac{z^{x-1}\,dz}{e^z+1} = (1-2^{1-x})\Gamma(x)\zeta(x) \qquad (x > 0), \tag{1}$$

where $\zeta(x) = \sum_{n=1}^{\infty} 1/n^x$ is the Riemann zeta function.

For $x = 1$, the expression (1) becomes indeterminate; the value of the integral is

$$\int_0^\infty \frac{dz}{e^z+1} = \log 2. \tag{2}$$

For x an even integer ($= 2n$) the zeta function can be expressed in terms of the Bernoulli numbers B_n:

$$\int_0^\infty \frac{z^{2n-1}\,dz}{e^z+1} = \frac{2^{2n-1}-1}{2n}\pi^{2n}B_n. \tag{3}$$

The following integrals are calculated similarly:

$$\int_0^\infty \frac{z^{x-1}\,dz}{e^z-1} = \Gamma(x)\zeta(x) \qquad (x > 1). \tag{4}$$

For x an even integer ($= 2n$),

$$\int_0^\infty \frac{z^{2n-1}\,dz}{e^z-1} = \frac{(2\pi)^{2n}B_n}{4n}. \tag{5}$$

For reference we shall give the values of the first few Bernoulli numbers and of some zeta functions:

$$B_1 = 1/6, \quad B_2 = 1/30, \quad B_3 = 1/42, \quad B_4 = 1/30;$$
$$\zeta(3/2) = 2.612, \quad \zeta(5/2) = 1.341, \quad \zeta(3) = 1.202,$$
$$\zeta(5) = 1.037, \quad \Gamma(3/2) = \tfrac{1}{2}\sqrt{\pi}, \quad \Gamma(5/2) = \tfrac{3}{4}\sqrt{\pi}.$$

using for brevity the notation

$$\beta = \left(\frac{g\pi}{6}\right)^{2/3} \frac{m}{\hbar^2}. \tag{57.4}$$

Hence we find the entropy

$$S = \beta NT(V/N)^{2/3}, \tag{57.5}$$

specific heat†

$$C = \beta NT(V/N)^{2/3}, \tag{57.6}$$

and energy of the gas:

$$\begin{aligned} E &= E_0 + \tfrac{1}{2}\beta NT^2(V/N)^{2/3} \\ &= E_0[1+0.0713g^{4/3}(mT/\hbar^2)^2(V/N)^{4/3}]. \end{aligned} \tag{57.7}$$

Thus the specific heat of a degenerate Fermi gas at low temperatures is proportional to the temperature.

§58. A relativistic degenerate electron gas

As the gas is compressed, the mean energy of the electrons increases (ε_0 increases); when it becomes comparable with mc^2, relativistic effects begin to be important. Here we shall discuss in detail a completely degenerate extreme relativistic electron gas, the energy of whose particles is large compared with mc^2. In this case the relation between the energy and momentum of a particle is

$$\varepsilon = cp. \tag{58.1}$$

The previous formulae (56.1) and (56.2) give the number of quantum states and hence the limiting momentum p_0. The limiting energy (i.e. the chemical potential of the gas) is now

$$\varepsilon_0 = cp_0 = \left(\frac{6\pi^2}{g}\right)^{1/3} \hbar c \left(\frac{N}{V}\right)^{1/3}. \tag{58.2}$$

The total energy of the gas is

$$E = \frac{gcV}{2\pi^2\hbar^3}\int_0^{p_0} p^3\,dp = \frac{gcp_0^4}{8\pi^2\hbar^3}V,$$

or

$$E = \tfrac{3}{4}\left(\frac{6\pi^2}{g}\right)^{1/3} \hbar cN \left(\frac{N}{V}\right)^{1/3}. \tag{58.3}$$

The gas pressure can be obtained by differentiating the energy with respect to the volume at constant entropy (equal to zero). This gives

$$P = \frac{E}{3V} = \tfrac{1}{4}\left(\frac{6\pi^2}{g}\right)^{1/3} \hbar c \left(\frac{N}{V}\right)^{4/3}. \tag{58.4}$$

† The suffix v or p to the specific heat is omitted, since C_v and C_p are the same in this approximation. We have seen in §23 that, if S tends to zero as T^n when $T \to 0$, the difference $C_p - C_v$ tends to zero as T^{2n+1}, and so in this case $C_p - C_v \sim T^3$.

The pressure of an extreme relativistic electron gas is proportional to the 4/3 power of the density.

It should be mentioned that the relation

$$PV = \tfrac{1}{3}E \tag{58.5}$$

is actually valid for an extreme relativistic gas not only at absolute zero but at all temperatures. This is easily seen by exactly the same method as that used to derive the relation (55.8), with the energy given by $\varepsilon = cp$ instead of $\varepsilon = p^2/2m$. With $\varepsilon = cp$, formula (52.4) leads to the following expression for Ω:

$$\Omega = -\frac{gTV}{2\pi^2c^3\hbar^3}\int_0^\infty \varepsilon^2 \log\left(1+e^{(\mu-\varepsilon)/T}\right) d\varepsilon,$$

or, integrating by parts,

$$\Omega = -\tfrac{1}{3}\frac{gV}{2\pi^2c^3\hbar^3}\int_0^\infty \frac{\varepsilon^3\, d\varepsilon}{e^{(\varepsilon-\mu)/T}+1} = -\tfrac{1}{3}E. \tag{58.6}$$

Thus the limiting value that the pressure of any macroscopic body can have for a given E (see §27) is reached for an extreme relativistic Fermi gas.

Using the variable of integration $z = \varepsilon/T$, we have

$$\Omega = -\frac{gVT^4}{6\pi^2c^3\hbar^3}\int_0^\infty \frac{z^3\, dz}{e^{z-\mu/T}+1}.$$

This shows that

$$\Omega = VT^4 f(\mu/T). \tag{58.7}$$

Hence, as in §55, we find that in an adiabatic process the volume, pressure and temperature of an extreme relativistic Fermi gas are related by

$$PV^{4/3} = \text{constant}, \quad VT^3 = \text{constant}, \quad T^4/P = \text{constant}. \tag{58.8}$$

These are the same as the usual equation of the Poisson adiabatic with $\gamma = 4/3$; but it must be emphasised that γ here is not the ratio of the specific heats of the gas.

PROBLEMS

Problem 1. Determine the number of collisions with a wall in an extreme relativistic completely degenerate electron gas.†

Solution. The calculation is as in §56, Problem; it must be remembered that the electron velocity $v \cong c$. The result is $\nu = \tfrac{1}{4}cN/V$.

Problem 2. Determine the specific heat of a degenerate extreme relativistic electron gas.

† In all the Problems we put $g = 2$.

SOLUTION. Applying the formula (57.1) to the integral in (58.6), we find

$$\Omega = \Omega_0 - \frac{(\mu T)^2}{6(c\hbar)^3} V.$$

Hence the entropy

$$S = \frac{\mu^2}{3(c\hbar)^3} VT = N\frac{(3\pi^2)^{2/3}}{3c\hbar} T\left(\frac{V}{N}\right)^{1/3}$$

and the specific heat

$$C = N\frac{(3\pi^2)^{2/3}}{3c\hbar}\left(\frac{V}{N}\right)^{1/3} T.$$

PROBLEM 3. Determine the equation of state of a relativistic completely degenerate electron gas (the electron energy and momentum being related by $\varepsilon^2 = c^2p^2+m^2c^4$).

SOLUTION. The previous formulae (56.1) and (56.2) give the number of states and the limiting momentum p_0, and the total energy is

$$E = \frac{Vc}{\pi^2\hbar^3}\int_0^{p_0} p^2\sqrt{(m^2c^2+p^2)}\,dp,$$

whence

$$E = \frac{cV}{8\pi^2\hbar^3}\{p_0(2p_0^2+m^2c^2)\sqrt{(p_0^2+m^2c^2)}-(mc)^4\sinh^{-1}(p_0/mc)\}.$$

The pressure $P = -(\partial E/\partial V)_{S=0}$ is

$$P = \frac{c}{8\pi^2\hbar^3}\left\{p_0\left(\frac{2}{3}p_0^2-m^2c^2\right)\sqrt{(p_0^2+m^2c^2)}+(mc)^4\sinh^{-1}(p_0/mc)\right\}.$$

These formulae are conveniently put in parametric form, using as parameter the quantity $\xi = 4\sinh^{-1}(p_0/mc)$. Then

$$N/V = (mc/\hbar)^3\cdot(1/3\pi^2)\sinh^3\tfrac{1}{4}\xi,$$

$$P = (m^4c^5/32\pi^2\hbar^3)(\tfrac{1}{3}\sinh\xi-\frac{8}{3}\sinh\tfrac{1}{2}\xi+\xi),$$

$$E/V = (m^4c^5/32\pi^2\hbar^3)(\sinh\xi-\xi).$$

§59. A degenerate Bose gas

At low temperatures the properties of a Bose gas bear no resemblance to those of a Fermi gas. This is evident from the fact that for a Bose gas the state of lowest energy, occupied by the gas at $T = 0$, must be that with $E = 0$ (all the particles being in the quantum state $\varepsilon = 0$), whereas a Fermi gas has a non-zero energy at absolute zero.

If the temperature of the gas is lowered at constant density N/V, the chemical potential μ given by equation (55.5) (with the lower sign) will increase, i.e. its absolute magnitude will decrease (since μ is negative). It reaches the value $\mu = 0$ at a temperature determined by the equation

$$\frac{N}{V} = \frac{g(mT)^{3/2}}{2^{1/2}\pi^2\hbar^3}\int_0^\infty \frac{\sqrt{z}\,dz}{e^z-1}. \tag{59.1}$$

The integral in (59.1) can be expressed in terms of the zeta function; see the second footnote to §57. Denoting the required temperature by T_0, we obtain

$$T_0 = \frac{3.31}{g^{2/3}} \frac{\hbar^2}{m} \left(\frac{N}{V}\right)^{2/3}. \tag{59.2}$$

For $T < T_0$, equation (55.5) has no negative solutions, whereas in Bose statistics the chemical potential must be negative at all temperatures.

This apparent contradiction arises because under the conditions in question it is not legitimate to go from the summation in formula (53.3) to the integration in (55.5): in this process the first term in the sum (with $\varepsilon_k = 0$) is multiplied by $\sqrt{\varepsilon} = 0$ and so disappears from the sum; but, as the temperature decreases, more and more particles must occupy that state of lowest energy, until at $T = 0$ they are all in it. The mathematical effect of this is that, when the limit $\mu \to 0$ is taken in the sum (53.3), the sum of all the terms in the series except the first tends to a finite limit given by the integral (55.5), but the first term (with $\varepsilon_k = 0$) tends to infinity. Consequently, by letting μ tend not to zero but to some small finite value, we can make this first term in the sum take the desired finite value.

In reality, therefore, the situation for $T < T_0$ is as follows. Particles with energy $\varepsilon > 0$ are distributed according to formula (55.4) with $\mu = 0$:

$$dN_\varepsilon = \frac{gm^{3/2}V}{2^{1/2}\pi^2\hbar^3} \frac{\sqrt{\varepsilon}\, d\varepsilon}{e^{\varepsilon/T} - 1}. \tag{59.3}$$

The total number of particles with energies $\varepsilon > 0$ will thus be

$$N_{\varepsilon>0} = \int dN_\varepsilon = \frac{gV(mT)^{3/2}}{2^{1/2}\pi^2\hbar^3} \int_0^\infty \frac{\sqrt{z}\, dz}{e^z - 1} = N(T/T_0)^{3/2}.$$

The remaining

$$N_{\varepsilon=0} = N[1 - (T/T_0)^{3/2}] \tag{59.4}$$

particles are in the lowest state, i.e. have energy $\varepsilon = 0$.† The energy of the gas for $T < T_0$ is, of course, determined only by the particles with $\varepsilon > 0$; putting $\mu = 0$ in (55.7), we have

$$E = \frac{gV(mT)^{3/2}T}{2^{1/2}\pi^2\hbar^3} \int_0^\infty \frac{z^{3/2}\, dz}{e^z - 1}.$$

This integral reduces to $\zeta(5/2)$ (see the second footnote to §57), and we obtain

$$\begin{aligned} E &= 0.770NT(T/T_0)^{3/2} \\ &= 0.128g(m^{3/2}T^{5/2}/\hbar^3)V. \end{aligned} \tag{59.5}$$

† The steady increase of particles in the state with $\varepsilon = 0$ is often called *Bose–Einstein condensation*. It should be emphasised that this refers only to "condensation" in momentum space; no condensation actually occurs in the gas, of course.

The specific heat is therefore

$$C_v = 5E/2T, \tag{59.6}$$

i.e. is proportional to $T^{3/2}$. Integration of the specific heat gives the entropy:

$$S = 5E/3T, \tag{59.7}$$

and the free energy is

$$F = E - TS = -\tfrac{2}{3}E. \tag{59.8}$$

This is obvious, since for $\mu = 0$

$$F = \Phi - PV = N\mu + \Omega = \Omega.$$

The pressure is

$$P = -(\partial F/\partial V)_T = 0.0851 g m^{3/2} T^{5/2}/\hbar^3. \tag{59.9}$$

We see that for $T < T_0$ the pressure is proportional to $T^{5/2}$ and is independent of the volume. This is the natural consequence of the fact that particles in a state with $\varepsilon = 0$ have no momentum and make no contribution to the pressure.

At the point $T = T_0$ itself, all the above-mentioned thermodynamic quantities are continuous, but it may be shown that the derivative of the specific heat with respect to temperature is discontinuous there (see the Problem). The curve of the specific heat itself as a function of temperature has a change in slope at $T = T_0$, and the specific heat has its maximum value there (equal to $1.28 \times 3N/2$).

PROBLEM

Determine the discontinuity of the derivative $(\partial C_v/\partial T)_V$ at $T = T_0$.

SOLUTION. To solve this problem we must determine the energy of the gas for small positive $T - T_0$. The equation (55.5) is identical with

$$N = N_0(T) + \frac{gVm^{3/2}}{2^{1/2}\pi^2\hbar^3} \int_0^\infty \left[\frac{1}{e^{(\varepsilon-\mu)/T}-1} - \frac{1}{e^{\varepsilon/T}-1}\right] \sqrt{\varepsilon}\, d\varepsilon,$$

where $N_0(T)$ is given by (59.1). Expanding the integrand and using the fact that μ is small near the point $T = T_0$, and therefore the important part of the integral arises from the region where ε is small, we find that the integral is equal to

$$T\mu \int_0^\infty \frac{d\varepsilon}{\sqrt{\varepsilon}(\varepsilon + |\mu|)} = -\pi T \sqrt{|\mu|}.$$

Substituting this value and then expressing μ in terms of $N - N_0$, we have

$$-\mu = \frac{2\pi^2\hbar^6}{g^2 m^3}\left(\frac{N_0 - N}{TV}\right)^2.$$

To the same accuracy we can write

$$\frac{\partial E}{\partial \mu} = -\frac{3}{2}\frac{\partial \Omega}{\partial \mu} = \frac{3}{2}N \cong \frac{3}{2}N_0,$$

whence

$$E = E_0 + \frac{3}{2} N_0 \mu = E_0 - \frac{3\pi^2 \hbar^6}{g^2 m^3} N_0 \left(\frac{N_0 - N}{TV}\right)^2,$$

where $E_0 = E_0(T)$ denotes the energy for $\mu = 0$, i.e. the function (59.5). The second derivative of the second term with respect to temperature will clearly give the required discontinuity. The result of the calculation is

$$\Delta \left(\frac{\partial C_v}{\partial T}\right)_V = -\frac{6\pi^2 \hbar^6}{g^2 m^3 V^2} \left[N_0 \left(\frac{1}{T} \frac{\partial N_0}{\partial T}\right)^2\right]_{T=T_0} = -3.66\, N/T_0.$$

The value of the derivative $(\partial C_v/\partial T)_V$ for $T = T_0 - 0$ is, from (59.5), $+2.89\, N/T_0$, and for $T = T_0 + 0$ it is therefore $-0.77\, N/T_0$.

§60. Black-body radiation

The most important application of Bose statistics relates to electromagnetic radiation which is in thermal equilibrium—called *black-body radiation*. Such radiation may be regarded as a "gas" consisting of photons. The linearity of the equations of electrodynamics expresses the fact that photons do not interact with one another (the principle of superposition for the electromagnetic field), so that the "photon gas" is an ideal gas. Because the angular momentum of the photons is integral, this gas obeys Bose statistics.

If the radiation is not in a vacuum but in a material medium, the condition for an ideal photon gas requires also that the interaction between radiation and matter should be small. This condition is satisfied in gases throughout the radiation spectrum except for frequencies in the neighbourhood of absorption lines of the material, but at high densities of matter it may be violated except at very high temperatures.

It should be remembered that at least a small amount of matter must be present if thermal equilibrium is to be reached in the radiation, since the interaction between the photons themselves may be regarded as completely absent.† The mechanism by which equilibrium can be established consists in the absorption and emission of photons by matter. This results in a very important specific property of the photon gas: the number of photons N in it is variable, and not a given constant as in ordinary gas. Thus N itself must be determined from the conditions of thermal equilibrium. From the condition that the free energy of the gas should be a minimum (for given T and V), we obtain as one of the necessary conditions $\partial F/\partial N = 0$. Since $(\partial F/\partial N)_{T,V} = \mu$, this gives

$$\mu = 0, \tag{60.1}$$

i.e. the chemical potential of the photon gas is zero.

† Apart from the entirely negligible interaction which is due to the possible production of virtual electron-positron pairs.

The distribution of photons among the various quantum states with energies $\varepsilon_k = \hbar\omega_k$, where the ω_k are the eigenfrequencies of the radiation in a given volume V, is therefore given by formula (53.2) with $\mu = 0$:

$$\overline{n_k} = 1/(e^{\hbar\omega_k/T}-1). \tag{60.2}$$

This is called *Planck's distribution.*

Assuming that the volume is sufficiently large, we can make the usual change† from the discrete to the continuous distribution of eigenfrequencies of the radiation. The number of modes of oscillation for which the components of the wave vector **f** lie in the intervals $\mathrm{d}f_x, \mathrm{d}f_y, \mathrm{d}f_z$ is $V\,\mathrm{d}f_x\,\mathrm{d}f_y\,\mathrm{d}f_z/(2\pi)^3$, and the number of modes for which the absolute magnitude of the wave vector lies in the range $\mathrm{d}f$ is correspondingly $V\cdot 4\pi f^2\,\mathrm{d}f/(2\pi)^3$. Using the frequency $\omega = cf$ and multiplying by 2 (for the two independent directions of polarisation of the oscillations), we obtain the number of quantum states of photons with frequencies between ω and $\omega+\mathrm{d}\omega$:

$$V\omega^2\,\mathrm{d}\omega/\pi^2c^3. \tag{60.3}$$

Multiplying the distribution (60.2) by this quantity, we find the number of photons in this frequency interval:

$$\mathrm{d}N_\omega = \frac{V}{\pi^2c^3}\frac{\omega^2\,\mathrm{d}\omega}{e^{\hbar\omega/T}-1}, \tag{60.4}$$

and a further multiplication by $\hbar\omega$ gives the radiation energy in this segment of the spectrum:

$$\mathrm{d}E_\omega = \frac{V\hbar}{\pi^2c^3}\frac{\omega^3\,\mathrm{d}\omega}{e^{\hbar\omega/T}-1}. \tag{60.5}$$

This formula for the spectral energy distribution of black-body radiation is called *Planck's formula* (1900). In terms of the wavelength $\lambda = 2\pi c/\omega$, it becomes

$$\mathrm{d}E_\lambda = \frac{16\pi^2c\hbar V}{\lambda^5}\frac{\mathrm{d}\lambda}{e^{2\pi\hbar c/T\lambda}-1}. \tag{60.6}$$

At low frequencies ($\hbar\omega \ll T$), formula (60.5) gives

$$\mathrm{d}E_\omega = V(T/\pi^2c^3)\omega^2\,\mathrm{d}\omega. \tag{60.7}$$

This is the *Rayleigh–Jeans formula*. It should be noticed that formula (60.7) does not contain the quantum constant $\hbar$, and can be derived by multiplying by T the "number of modes" (60.3); in this sense it corresponds to classical statistics, in which an energy T must correspond to each "vibrational degree of freedom"— the law of equipartition (§44).

† See *The Classical Theory of Fields*, §52.

In the opposite limiting case of high frequencies ($\hbar\omega \gg T$), formula (60.5) becomes

$$dE_\omega = V(\hbar/\pi^2c^3)\omega^3 e^{-\hbar\omega/T}\, d\omega. \tag{60.8}$$

This is *Wien's formula.*

Fig. 6 shows a graph of the function $x^3/(e^x-1)$, corresponding to the distribution (60.5).

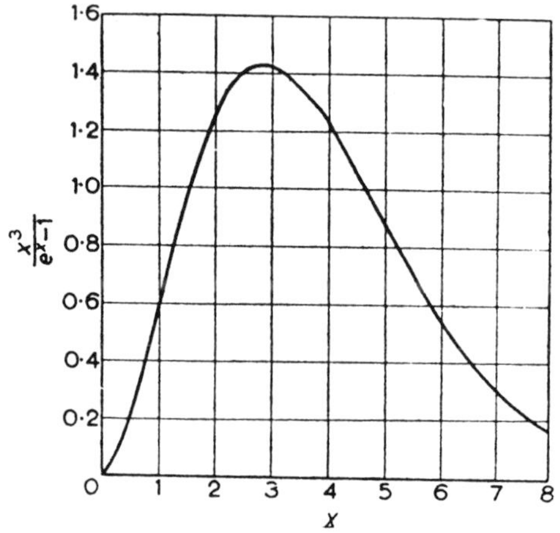

FIG. 6

The "density" of the spectral frequency distribution of the energy of black-body radiation, $dE_\omega/d\omega$, has a maximum at a frequency ω_m given by

$$\hbar\omega_m/T = 2.822. \tag{60.9}$$

Thus, when the temperature rises, the position of the maximum of the distribution is displaced towards higher frequencies in proportion to T (the *displacement law*).†

Let us calculate the thermodynamic quantities for black-body radiation. For $\mu = 0$, the free energy is the same as Ω (since $F = \Phi - PV = N\mu + \Omega$). According to formula (53.4), in which we put $\mu = 0$ and change in the usual way (by means of (60.3)) from summation to integration, we obtain

$$F = T\frac{V}{\pi^2c^3}\int_0^\infty \omega^2 \log\left(1-e^{-\hbar\omega/T}\right) d\omega. \tag{60.10}$$

With the new variable of integration $x = \hbar\omega/T$, integration by parts gives

$$F = -V\frac{T^4}{3\pi^2\hbar^3c^3}\int_0^\infty \frac{x^3\, dx}{e^x-1}.$$

† The wavelength distribution "density" $dE_\lambda/d\lambda$ also has a maximum, but at a different value of the corresponding ratio: $2\pi\hbar c/T\lambda_m = 4.965$. Thus the maximum λ_m of the wavelength distribution is displaced in inverse proportion to the temperature.

The integral is equal to $\pi^4/15$ (see the second footnote to §57). Thus

$$F = -V\cdot\pi^2T^4/45(\hbar c)^3$$
$$= -4\sigma VT^4/3c. \qquad (60.11)$$

If T is measured in degrees, the coefficient σ (called the *Stefan–Boltzmann constant*) is

$$\sigma = \pi^2k^4/60\hbar^3c^2$$
$$= 5.67\times10^{-5}\ \text{g/sec}^3\ \text{deg}^4. \qquad (60.12)$$

The entropy is

$$S = -\partial F/\partial T = 16\sigma VT^3/3c, \qquad (60.13)$$

and is proportional to the cube of the temperature. The total radiation energy $E = F+TS$ is

$$E = 4\sigma VT^4/c = -3F. \qquad (60.14)$$

This expression could, of course, be derived also by direct integration of the distribution (60.5). Thus the total energy of black-body radiation is proportional to the fourth power of the temperature. This is *Boltzmann's law*.

For the specific heat of the radiation $C_v = (\partial E/\partial T)_V$ we have

$$C_v = 16\sigma T^3V/c. \qquad (60.15)$$

Finally, the pressure is

$$P = -(\partial F/\partial V)_T = 4\sigma T^4/3c, \qquad (60.16)$$

$$PV = \tfrac{1}{3}E. \qquad (60.17)$$

Thus for a photon gas the same limiting value of the pressure is obtained as for an extreme relativistic electron gas (§58); this is as it should be, since the relation (60.17) is a direct consequence of the linear relation ($\varepsilon = cp$) between the energy and momentum of a particle.

The total number of photons in black-body radiation is

$$N = \frac{V}{\pi^2c^3}\int_0^\infty \frac{\omega^2\,d\omega}{e^{\hbar\omega/T}-1} = \frac{VT^3}{\pi^2c^3\hbar^3}\int_0^\infty \frac{x^2\,dx}{e^x-1}.$$

The integral can be expressed in terms of $\zeta(3)$; see the second footnote to §57. Thus

$$N = \frac{2\zeta(3)}{\pi^2}\left(\frac{T}{\hbar c}\right)^3 V = 0.244\left(\frac{T}{\hbar c}\right)^3 V. \qquad 60.18)$$

In an adiabatic expansion (or compression) of the photon gas, the volume and temperature are related by VT^3 = constant. From (60.16), the pressure and volume are then related by $PV^{4/3}$ = constant. A comparison with (58.8) shows that the equation of the adiabatic for a photon gas coincides (as we should expect) with that for an extreme relativistic electron gas.

Let us consider a body in thermal equilibrium with black-body radiation around it. The body continually reflects and absorbs photons incident on it, and at the same time emits new ones, and in equilibrium all these processes balance in such a way that the distribution of photons in frequency and direction remains unchanged on the average.

Owing to the complete isotropy of the black-body radiation, each volume element emits a flux of energy uniformly in all directions. We use the notation

$$e_0(\omega) = \frac{1}{4\pi V}\frac{\mathrm{d}E_\omega}{\mathrm{d}\omega} = \frac{\hbar\omega^3}{4\pi^3c^3(e^{\hbar\omega/T}-1)} \tag{60.19}$$

for the "spectral density" of black-body radiation per unit volume and unit solid angle. Then the energy flux density with frequencies in the interval $\mathrm{d}\omega$ leaving each point and entering the solid angle element $\mathrm{d}o$ is $ce_0(\omega)\,\mathrm{d}o\,\mathrm{d}\omega$. The radiation energy (with frequencies in $\mathrm{d}\omega$) incident in unit time on unit area of the surface of the body at an angle θ to the normal is therefore $ce_0(\omega)\times \cos\theta\,\mathrm{d}o\,\mathrm{d}\omega$, $\mathrm{d}o = 2\pi\sin\theta\,\mathrm{d}\theta$.

Let $A(\omega, \theta)$ denote the "absorbing power" of the body as a function of the frequency and direction of incidence of the radiation; this quantity is defined as the fraction of the radiation energy incident on the surface of the body, in the given frequency interval, which is absorbed by the body, not including the radiation (if any) which passes through the body. Then the quantity of radiation absorbed per unit time and surface area will be

$$ce_0(\omega)A(\omega, \theta)\cos\theta\,\mathrm{d}o\,\mathrm{d}\omega. \tag{60.20}$$

Let us assume that the body does not scatter radiation and is not fluorescent, i.e. that the reflection occurs without change in the angle θ or in the frequency. We shall also suppose that the radiation does not pass through the body; in other words, all radiation not reflected is completely absorbed. Then the quantity of radiation (60.20) must be balanced by the radiation emitted by the body itself in the same directions at the same frequencies. Denoting by $J(\omega, \theta)\,\mathrm{d}\omega\,\mathrm{d}o$ the intensity of emission from unit area of the surface and equating it to the absorbed energy, we obtain

$$J(\omega, \theta) = ce_0(\omega)A(\omega, \theta)\cos\theta. \tag{60.21}$$

The functions $J(\omega, \theta)$ and $A(\omega, \theta)$ are, of course, different for different bodies, but we see that their ratio is independent of the properties of the body and is a universal function of frequency and direction:

$$J(\omega, \theta)/A(\omega, \theta) = ce_0(\omega)\cos\theta,$$

which is determined by the energy distribution in the black-body radiation spectrum (at a temperature equal to that of the body). This is *Kirchhoff's law*.

If the body scatters radiation, KIRCHHOFF's law can be formulated only in a more restricted way. Since in this case reflection occurs with a change in the angle θ, we can derive from the condition of equilibrium only the requirement that the radiation (of a given frequency) absorbed from all directions should be equal to the total emission from the body in all directions:

$$\int J(\omega, \theta)\, do = ce_0(\omega) \int A(\omega, \theta) \cos\theta\, do. \qquad (60.22)$$

The angle θ also changes, in general, when radiation can pass through the body (because of refraction on entering and leaving the body). In this case the relation (60.22) must be integrated over the entire surface of the body; the functions $A(\omega, \theta)$ and $J(\omega, \theta)$ now depend not only on the material of the body but also on its shape and on the point considered on its surface.

Finally, when there is scattering with change of frequency (fluorescence), KIRCHHOFF's law applies only to the integrals over both direction and frequency of the radiation:

$$\iint J(\omega, \theta)\, do\, d\omega = c \iint e_0(\omega) A(\omega, \theta) \cos\theta\, do\, d\omega. \qquad (60.23)$$

A body which completely absorbs all radiation incident on it is called a *black body*.† For such a body, $A(\omega, \theta) = 1$ by definition, and its emissive power is entirely determined by the function

$$J_0(\omega, \theta) = ce_0(\omega) \cos\theta, \qquad (60.24)$$

which is the same for all black bodies. It may be noted that the intensity of emission from a black body is a very simple function of direction, being proportional to the cosine of the angle to the normal to the surface of the body. The total intensity of emission from a black body, J_0, is obtained by integrating (60.24) over all frequencies and over a hemisphere:

$$J_0 = c \int_0^\infty e_0(\omega)\, d\omega \int^{\pi/2} 2\pi \cos\theta \sin\theta\, d\theta = cE/4V,$$

where E is given by (60.14). Thus

$$J_0 = \sigma T^4, \qquad (60.25)$$

† Such a body can be realised in the form of a cavity with highly absorbing internal walls and a small aperture. Any ray entering through the aperture can return to it and leave the cavity only after repeated reflection from the walls of the cavity. When the aperture is sufficiently small, therefore, the cavity will absorb practically all the radiation incident on the aperture, and so the surface of the aperture will be a black body.

i.e. the total intensity of emission from a black body is proportional to the fourth power of its temperature.

Finally, let us consider radiation not in thermal equilibrium, having a non-equilibrium spectral or directional distribution. Let $e(\omega, \mathbf{n})\,d\omega\,do$ be the volume density of this radiation in the frequency interval $d\omega$ and with the direction $\mathbf{n}$ of the wave vector lying in the solid-angle element do. We can use the concept of the temperature of the radiation in each small interval of frequency and direction, defined as the temperature for which the density $e(\omega, \mathbf{n})$ is equal to that given by PLANCK's formula, i.e. $e(\omega, \mathbf{n}) = e_0(\omega)$. Denoting this temperature by $T_{\omega, \mathbf{n}}$, we have

$$T_{\omega, \mathbf{n}} = \frac{\hbar\omega}{\log\left\{1+\dfrac{\hbar\omega^3}{4\pi^3c^3}\cdot\dfrac{1}{e(\omega, \mathbf{n})}\right\}}. \tag{60.26}$$

Let us imagine a black body emitting into a surrounding vacuum. The radiation is propagated freely along straight lines and will not be in thermal equilibrium outside the body; it is by no means isotropic, as equilibrium radiation must be. Since the photons are propagated in a vacuum and do not interact with one another, we are in a position to apply LIOUVILLE's theorem rigorously to the photon distribution function in the corresponding phase space of co-ordinates and wave-vector components.† According to this theorem, the distribution function remains constant along the phase trajectories. But the distribution function is, apart from a factor dependent on frequency, the same as the volume density of radiation of a given frequency and direction, $e(\omega, \mathbf{n}, \mathbf{r})$. Since the radiation frequency is also constant during propagation, we have the following important result: in every solid-angle element where radiation is propagated (from a given point in space) the radiation density $e(\omega, \mathbf{n}, \mathbf{r})$ is equal to the density within the emitting black body, i.e. to the black-body radiation density $e_0(\omega)$. Whereas, however, for equilibrium radiation the density exists for all directions, here it exists only for a certain interval of directions.

Defining the temperature of non-equilibrium radiation by (60.26), we can express the result differently by saying that the temperature $T_{\omega, \mathbf{n}}$ is equal to the temperature T of the emitting black body for all directions in which radiation is being propagated (at any given point in space). If the radiation temperature is defined from the density averaged over all directions, however, it is of course less than the temperature of the black body.

All these consequences of LIOUVILLE's theorem remain fully valid when reflecting mirrors and refracting lenses are present, provided, of course, that the conditions for geometrical optics to be applicable are still satisfied. By

† When considering the limiting case of geometrical optics, we can speak of co-ordinates of a photon.

means of lenses or mirrors the radiation can be focused, i.e. the range of directions from which rays reach a given point in space can be enlarged. This may increase the mean radiation temperature at the point considered, but the foregoing discussion shows that there is no means of raising it above the temperature of the black body which emitted the radiation.

CHAPTER VI

THE CONDENSED STATE

§61. Solids at low temperatures

SOLIDS form another suitable topic for the application of statistical methods of calculating the thermodynamic quantities. A characteristic property of solids is that the atoms in them execute only small vibrations about certain equilibrium positions, the crystal lattice sites. The configuration of the lattice sites which corresponds to thermal equilibrium of the body is preferred, i.e. distinguished from all other possible distributions, and must therefore be regular. In other words, a solid in thermal equilibrium must be *crystalline*.

As well as crystals, there exist in Nature also *amorphous* solids, in which the atoms vibrate about randomly situated points. Such bodies are thermodynamically metastable, and must ultimately become crystalline. In practice, however, the relaxation times are so long that amorphous bodies behave as if stable for an almost unlimited time. All the following calculations apply equally to both crystalline and amorphous substances. The only difference is that, since amorphous bodies are not in equilibrium, NERNST's theorem does not apply to them, and as $T \to 0$ their entropy tends to a non-zero value. Consequently, for amorphous bodies the formula (61.7) derived below for the entropy has to be augmented by some constant S_0 (and the free energy by a corresponding term $-TS_0$); we shall omit this unimportant constant, which, in particular, does not affect the specific heats of a body.

The "residual" entropy, which does not vanish as $T \to 0$, may also be observed in crystalline solids, because of what is called *ordering* of crystals. If the number of crystal lattice sites at which atoms of a given kind can be situated is equal to the number of such atoms, there will be one atom near each site; that is, the probability of finding an atom (of the kind in question) in the neighbourhood of each site is equal to unity. Such crystals are said to be *completely ordered*. There are also, however, crystals in which the atoms may be not only at their "own" positions (i.e. those which they occupy in complete ordering) but also at certain "other" positions. In that case the number of sites that may be occupied by an atom of the given kind is greater than the number of such atoms, and the probability of finding atoms of this kind at either the old or the new sites will not be unity.

For example, solid carbon monoxide is a molecular crystal, in which the

CO molecule can have two opposite orientations differing by interchange of the two atoms; the number of sites that may be occupied by carbon (or oxygen) atoms is here equal to twice the number of these atoms.

In a state of complete thermodynamic equilibrium at absolute zero, any crystal must be completely ordered, and the atoms of each kind must occupy entirely definite positions. However, because the processes of lattice rearrangement are slow, especially at low temperatures, a crystal which is incompletely ordered at a high temperature may in practice remain so even at very low temperatures. This "freezing" of the disorder leads to the existence of a constant residual term in the entropy of the crystal. For instance, in the example of the CO crystal mentioned above, if the CO molecules have the two orientations with equal probability, the residual entropy will be $S_0 = \log 2$.

According to classical mechanics, all the atoms are at rest at absolute zero, and the potential energy of their interaction must be a minimum in equilibrium. At sufficiently low temperatures, therefore, the atoms must always execute only small vibrations, i.e. all bodies must be solid. In reality, however, quantum effects may bring about exceptions to this rule. One such is liquid helium, the only substance which remains liquid at absolute zero (at pressures below 25 atmospheres); all other substances solidify well before quantum effects become important.†

We may note that for a body to be solid its temperature must be sufficiently low. The quantity T must certainly be small in comparison with the energy of interaction of the atoms (in practice, all solids melt or decompose at higher temperatures). From this it results that the vibrations of atoms in a solid about their equilibrium positions are always small.

Let N be the number of molecules in the body, and ν the number of atoms in each molecule. Then the number of atoms is $N\nu$. Of the total number of degrees of freedom $3N\nu$, three correspond to translational and three to rotational motion of the body as a whole. The number of vibrational degrees of freedom is therefore $3N\nu-6$, but since $3N\nu$ is extremely large we can, of course, neglect 6 and assume that the number of vibrational degrees of freedom is just $3N\nu$.

It should be emphasised that in discussing solids we shall entirely ignore the "internal" (electronic) degrees of freedom of the atoms. Hence, if these degrees of freedom are important (as they may be, for example, in metals), the following formulae will relate only to the "lattice" part of the thermodynamic quantities for the solid, which is due to the vibrations of the atoms. In order to obtain the total values of these quantities, the "electronic" part (see §69) must be added to the "lattice" part.

† Quantum effects become important when the de Broglie wavelength corresponding to the thermal motion of the atoms becomes comparable with the distances between atoms. In liquid helium this occurs at 2–3°K.

In mechanical terms, a system with $3N\nu$ vibrational degrees of freedom may be regarded as an assembly of $3N\nu$ independent oscillators, each corresponding to one normal mode of vibration. The thermodynamic quantities relating to one vibrational degree of freedom have already been calculated in §49. From the formulae there we can immediately write down the free energy of the solid as†

$$F = N\varepsilon_0 + T\sum_{\alpha} \log\left(1 - e^{-\hbar\omega_\alpha/T}\right). \tag{61.1}$$

The summation is over all $3N\nu$ normal vibrations, which are labelled by the suffix α. We have added to the sum over vibrations a term $N\varepsilon_0$ which represents the energy of interaction between all the atoms in the body in their equilibrium positions (more precisely, when executing their "zero-point" vibrations); this energy is obviously proportional to the number N of molecules in the body, so that ε_0 is the energy per molecule. It must be remembered that ε_0 is, in general, not constant, but a function of the density (or specific volume) of the body: when the volume changes, so do the distances between the atoms, and therefore the energy of their interaction. For a given volume, however, ε_0 does not depend on the temperature: $\varepsilon_0 = \varepsilon_0(V/N)$.

The remaining thermodynamic quantities can be derived in the usual way from the free energy.

Let us now consider the limiting case of low temperatures. For small T, only the terms with low frequencies ($\hbar\omega_\alpha \sim T$) are of importance in the sum over α. But vibrations with low frequencies are just ordinary *sound waves*, whose wavelength is related to the frequency by $\lambda \sim u/\omega$, where u is the velocity of sound. In sound waves the wavelength is large in comparison with the lattice constant ($\lambda \gg a$), and so $\omega \ll u/a$. In other words, if the vibrations can be regarded as sound waves, the temperature must satisfy a condition which may be written in the form

$$T \ll \hbar u/a. \tag{61.2}$$

Let us assume that the body is isotropic (an amorphous solid). In an isotropic solid, longitudinal sound waves can be propagated (with velocity u_l, say), and so can transverse waves with two independent directions of polarisation and equal velocities of propagation (u_t, say). The frequency of these waves is linearly related to the absolute magnitude of the wave vector $\mathbf{k}$ by $\omega = u_l k$ or $\omega = u_t k$.

The number of vibrational modes in the spectrum of sound waves with absolute magnitude of the wave vector lying in the interval dk and with a given polarisation is $V\cdot 4\pi k^2\, dk/(2\pi)^3$, where V is the volume of the body. Putting for one of the three independent polarisations $k = \omega/u_l$ and for the

† Quantised vibrations were first used by Einstein (1907) to calculate the thermodynamic quantities for a solid.

other two $k = \omega/u_t$, we find that the interval $d\omega$ contains altogether

$$V\frac{\omega^2\,d\omega}{2\pi^2}\left(\frac{1}{u_l^3}+\frac{2}{u_t^3}\right) \tag{61.3}$$

vibrations.

A mean velocity of sound $\bar{u}$ can be defined according to the formula

$$\frac{3}{\bar{u}^3} = \frac{2}{u_t^3}+\frac{1}{u_l^3}.$$

Then the expression (61.3) becomes

$$V\cdot 3\omega^2\,d\omega/2\pi^2\bar{u}^3. \tag{61.4}$$

In this form it is applicable not only to isotropic bodies but also to crystals, where $\bar{u} = \bar{u}(V/N)$ must be understood as the velocity of propagation of sound in the crystal, averaged in a certain way. The determination of the averaging procedure requires the solution of the problem (which belongs to the theory of elasticity) of the propagation of sound in a crystal of given symmetry.

By means of (61.4) we can change from the summation in (61.1) to integration, obtaining

$$F = N\varepsilon_0+T\frac{3V}{2\pi^2\bar{u}^3}\int\limits_0^\infty \log\left(1-e^{-\hbar\omega/T}\right)\omega^2\,d\omega; \tag{61.5}$$

because of the rapid convergence of the integral when T is small, the integration can be taken from 0 to ∞. This expression (apart from the term $N\varepsilon_0$) differs from the formula (60.10) for the free energy of black-body radiation only in that the velocity of light c is replaced by the velocity of sound $\bar{u}$ and a factor 3/2 appears. This resemblance is not surprising, since the frequency of sound vibrations is related to their wave number by the same type of linear formula as is valid for photons. The integers v_α in the energy levels $\sum v_\alpha\hbar\omega_\alpha$ of a system of sound oscillators may be regarded as "occupation numbers" of the various "quantum states" with energies $\varepsilon_\alpha = \hbar\omega_\alpha$, the values of these numbers being arbitrary (as in Bose statistics). The appearance of the extra factor 3/2 in (61.5) is due to the fact that sound vibrations have three possible directions of polarisation instead of two as for photons.

Thus, without having to repeat the calculations, we can use the expression (60.11) derived in §60 for the free energy of black-body radiation, if c is replaced by $\bar{u}$ and a factor 3/2 included. The free energy of a solid is therefore

$$F = N\varepsilon_0 - V\cdot\pi^2T^4/30(\hbar\bar{u})^3; \tag{61.6}$$

the entropy is

$$S = V\cdot 2\pi^2T^3/15(\hbar\bar{u})^3, \tag{61.7}$$

the energy

$$E = N\varepsilon_0 + V\cdot\pi^2T^4/10(\hbar\bar{u})^3, \tag{61.8}$$

and the specific heat

$$C = 2\pi^2T^3V/5(\hbar\bar{u})^3. \tag{61.9}$$

Thus the specific heat of a solid at low temperatures is proportional to the cube of the temperature† (DEBYE 1912). We write the specific heat as C simply (not distinguishing C_v and C_p), since at low temperatures the difference $C_p - C_v$ is a quantity of a higher order of smallness than the specific heat itself (see § 23; here $S \sim T^3$ and so $C_p - C_v \sim T^7$).

For solids having a simple crystal lattice (elements and simple compounds) the T^3 law for the specific heat does in fact begin to hold at temperatures of the order of tens of degrees, but for bodies with a complex lattice this law may be expected to be satisfactorily obeyed only at much lower temperatures.

§62. Solids at high temperatures

Let us now turn to the opposite limiting case of high temperatures (of order $T \gg \hbar u/a$, where a is the lattice constant). In this case we can put $1 - e^{-\hbar\omega_\alpha/T} \cong \hbar\omega_\alpha/T$, and formula (61.1) becomes

$$F = N\varepsilon_0 + T\sum_\alpha \log(\hbar\omega_\alpha/T). \tag{62.1}$$

The sum over α contains altogether $3N\nu$ terms. We define the "geometric mean" frequency $\bar{\omega}$ by

$$\log \bar{\omega} = \frac{1}{3N\nu}\sum_\alpha \log \omega_\alpha. \tag{62.2}$$

Then the free energy of the solid is given by

$$F = N\varepsilon_0 - 3N\nu T \log T + 3N\nu T \log \hbar\bar{\omega}. \tag{62.3}$$

The mean frequency $\bar{\omega}$, like $\bar{u}$, is a function of the density, $\bar{\omega}(V/N)$.

From (62.3) we find the energy of the body, $E = F - T\partial F/\partial T$:

$$E = N\varepsilon_0 + 3N\nu T. \tag{62.4}$$

The case of high temperatures corresponds to the classical treatment of the vibrations of the atoms; it is therefore clear why formula (62.4) accords exactly with the law of equipartition (§44): apart from the constant $N\varepsilon_0$, an energy T corresponds to each of the $3N\nu$ vibrational degrees of freedom.

For the specific heat we have

$$C = Nc = 3N\nu, \tag{62.5}$$

where $c = 3\nu$ is the specific heat per molecule. We again write the specific heat as C simply, since in solids the difference between C_p and C_v is always negligible (see the end of §64).

† It may be recalled that when "electronic degrees of freedom" are present these formulae give only the "lattice" part of the thermodynamic quantities. However, even when there is an "electronic part" (as in metals) it begins to affect the specific heat, for example, only at temperatures of a few degrees.

Thus at sufficiently high temperatures the specific heat of a solid is constant and depends only on the number of atoms in the body. In particular, the specific heat per atom ($\nu = 1$) must be the same for different elements and equal to 3 (in ordinary units, $3k$); this is *Dulong and Petit's law*. At ordinary temperatures this law is well satisfied for many elements. Formula (62.5) is valid at high temperatures for a number of simple compounds also, but for more complex compounds it gives a limiting value of the specific heat which in general is not reached before the substance melts or decomposes.

Substituting (62.5) in (62.3) and (62.4), we can write the free energy and energy of a solid as

$$F = N\varepsilon_0 - NcT \log T + NcT \log \hbar\bar{\omega}, \tag{62.6}$$

$$E = N\varepsilon_0 + NcT. \tag{62.7}$$

The entropy $S = -\partial F/\partial T$ is

$$S = Nc \log T - Nc \log (\hbar\bar{\omega}/e). \tag{62.8}$$

Formula (62.1) can also, of course, be derived directly from classical statistics, using the general formula (31.5)

$$F = -T \log \int' e^{-E(p,q)/T}\, d\Gamma. \tag{62.9}$$

For a solid, the integration over the co-ordinates in this integral is carried out as follows. Each atom is regarded as being situated near a particular lattice site, and the integration over its co-ordinates is taken only over a small neighbourhood of that site. It is clear that all the points in the region of integration thus defined will correspond to physically different microstates, and no additional factor is needed in the integral.†

We substitute in (62.9) the energy expressed in terms of the co-ordinates and momenta of the normal modes:

$$E(p, q) = \tfrac{1}{2} \sum_{\alpha} (p_\alpha^2 + \omega_\alpha^2 q_\alpha^2), \tag{62.10}$$

and write $d\Gamma$ in the form

$$d\Gamma = \frac{1}{(2\pi\hbar)^{3N\nu}} \prod_{\alpha} dp_\alpha\, dq_\alpha.$$

Then the integral becomes a product of $3N\nu$ integrals, all of the form

$$\int_{-\infty}^{\infty}\int_{-\infty}^{\infty} \exp\{-(p_\alpha^2 + \omega_\alpha^2 q_\alpha^2)/2T\} dp_\alpha\, dq_\alpha = 2\pi T/\omega_\alpha,$$

leading to formula (62.1); because of the rapid convergence of the integral, the integration over q_α may be extended from $-\infty$ to ∞.

† Whereas it was for a gas, where the integration over the co-ordinates of each particle was taken over the whole volume (cf. the end of §31).

At sufficiently high temperatures (provided that the solid does not melt or decompose) the effects of anharmonic vibrations of the atoms may become appreciable. The nature of these effects as regards the thermodynamic quantities for the body may be investigated as follows; cf. the similar calculations for gases in §49. Taking into account the terms following the quadratic terms in the expansion of the potential energy of the vibrations in powers of q_α, we have

$$E(p, q) = f_2(p, q)+f_3(q)+f_4(q)+\ldots,$$

where $f_2(p, q)$ denotes the harmonic expression (62.10) (a quadratic form in q_α and p_α), and $f_3(q)$, $f_4(q)$, ... are forms homogeneous in all the co-ordinates q_α, of degree three, four, etc. Substituting in the partition function in (62.9) $q_\alpha = q_\alpha'/\sqrt{T}$, $p_\alpha = p_\alpha'/\sqrt{T}$, we obtain

$$Z = \int' e^{-E(p,q)/T}\, d\Gamma$$

$$= T^{3N\nu}\int' \exp\{-f_2(p', q')-\sqrt{T}f_3(q')-Tf_4(q')-\ldots\}\, d\Gamma.$$

We see that, when the integrand is expanded in powers of the temperature, all odd powers of $\sqrt{T}$ are multiplied by odd functions of the co-ordinates, which give zero on integration over the co-ordinates. Hence Z is a series $Z = Z_0+TZ_1+T^2Z_2+\ldots$ which contains only integral powers of the temperature. On substitution in (62.9), the first correction term to the free energy will accordingly be of the form

$$F_{\text{anh}} = AT^2, \tag{62.11}$$

i.e. proportional to the square of the temperature. In the specific heat it gives a correction† proportional to the temperature itself. It should be emphasised that the expansion under discussion here is essentially one in powers of the ratio T/ε_0, which is always small, and not, of course, in powers of the ratio $T/\hbar\bar{\omega}$, which in the present case is large.

PROBLEMS

PROBLEM 1. Determine the maximum work which can be obtained from two identical solid bodies at temperatures T_1 and T_2 when their temperatures are made equal.

SOLUTION. The solution is entirely similar to that in §43, Problem 12, and gives

$$|R|_{\max} = Nc(\sqrt{T_1}-\sqrt{T_2})^2.$$

PROBLEM 2. Determine the maximum work which can be obtained from a solid when it is cooled from a temperature T to the temperature T_0 of the medium (at constant volume).

SOLUTION. From formula (20.3) we have

$$|R|_{\max} = Nc(T-T_0)+NcT_0 \log(T_0/T).$$

† This correction is usually negative (corresponding to positive A in (62.11)).

§63. Debye's interpolation formula

Thus in both the limiting cases of low and high temperatures it is possible to make a sufficiently complete calculation of the thermodynamic quantities for a solid. In the intermediate temperature range, such a calculation is impossible, since the sum over frequencies in (61.1) depends considerably on the actual frequency distribution over the whole spectrum of vibrations of the body concerned.

It is therefore of interest to construct a single interpolation formula giving the correct values of the thermodynamic quantities in the two limiting cases. More than one such formula can be found, of course, but we should expect that a reasonable interpolation formula will give at least a qualitatively correct description of the behaviour of the body throughout the intermediate range.

The form of the thermodynamic quantities for a solid at low temperatures is given by the distribution (61.4) of the frequencies in the vibration spectrum. At high temperatures it is important that all the $3N\nu$ vibrations are excited. To construct the required interpolation formula, therefore, it is reasonable to start from a model in which the law (61.4) (which in reality is valid only at low frequencies) governs the frequency distribution over the whole vibration spectrum, the spectrum beginning at $\omega = 0$ and terminating at some finite frequency ω_m determined by the condition that the total number of vibrations is equal to the correct value $3N\nu$:

$$\frac{3V}{2\pi^2\bar{u}^3}\int_0^{\omega_m}\omega^2\,d\omega = \frac{V\omega_m^{\ 3}}{2\pi^2\bar{u}^3} = 3N\nu,$$

whence

$$\omega_m = \bar{u}(6\pi^2 N\nu/V)^{1/3}. \tag{63.1}$$

Thus the frequency distribution in this model is given by the formula

$$9N\nu\omega^2\,d\omega/\omega_m^{\ 3} \quad (\omega \leqslant \omega_m) \tag{63.2}$$

for the number of vibrations with frequencies in the interval $d\omega$ (here $\bar{u}$ has been expressed in terms of ω_m).

Changing from the sum in (61.1) to an integral, we now have

$$F = N\varepsilon_0 + T\cdot\frac{9N\nu}{\omega_m^{\ 3}}\int_0^{\omega_m}\omega^2\log\left(1-e^{-\hbar\omega/T}\right)d\omega.$$

The *Debye temperature* or *characteristic temperature* Θ of the body is defined by

$$\Theta = \hbar\omega_m \tag{63.3}$$

(and is, of course, dependent on the density of the body). Then

$$F = N\varepsilon_0 + 9N\nu T(T/\Theta)^3 \int_0^{\Theta/T} z^2 \log(1-e^{-z})\,dz. \tag{63.4}$$

Integrating by parts and using the *Debye function*

$$D(x) = \frac{3}{x^3}\int_0^x \frac{z^3\,dz}{e^z-1}, \tag{63.5}$$

we can rewrite this formula as

$$F = N\varepsilon_0 + N\nu T[3\log(1-e^{-\Theta/T}) - D(\Theta/T)]. \tag{63.6}$$

Hence the energy $E = F - T\partial F/\partial T$ is

$$E = N\varepsilon_0 + 3N\nu T D(\Theta/T) \tag{63.7}$$

and the specific heat is

$$C = 3N\nu\{D(\Theta/T) - (\Theta/T)D'(\Theta/T)\}. \tag{63.8}$$

Fig. 7 shows a graph of $C/3N\nu$ as a function of T/Θ.

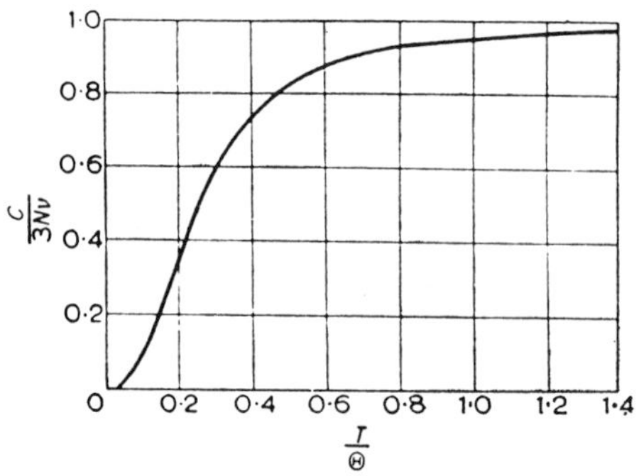

FIG. 7

Formulae (63.6)–(63.8) are the required interpolation formulae for the thermodynamic quantities for a solid (DEBYE 1912).

It is easy to see that in both limiting cases these formulae in fact give the correct results. For $T \ll \Theta$ (low temperatures) the argument Θ/T of the Debye function is large. In a first approximation we can replace x by ∞ in the upper limit of the integral in the definition (63.5) of $D(x)$; the resulting definite integral is $\pi^4/15$, and so†

$$D(x) \cong \pi^4/5x^3 \qquad (x \gg 1).$$

† Replacing $\int_0^x$ by $\int_0^\infty - \int_x^\infty$, expanding $(e^z-1)^{-1}$ in the second integrand in powers of

Substituting this in (63.8), we obtain

$$C = (12N\nu\pi^4/5)(T/\Theta)^3, \tag{63.9}$$

which is the same as (61.9). At high temperatures ($T \gg \Theta$) the argument of the Debye function is small; for $x \ll 1$ we have $D(x) \cong 1$ to a first approximation,† and (63.8) gives $C = 3N\nu$, again in full agreement with the previous result (62.5).‡

It is useful to point out that the actual form of the function $D(x)$ is such that the criterion of applicability of the limiting expressions for the specific heat is the relative magnitude of T and $\frac{1}{4}\Theta$: the specific heat may be regarded as constant for $T \gg \frac{1}{4}\Theta$ and proportional to T^3 for $T \ll \frac{1}{4}\Theta$.||

According to DEBYE's formula, the specific heat is some universal function of the ratio Θ/T. In other words, according to this formula, the specific heats of bodies must be the same if the bodies are in *corresponding states*, i.e. have the same value of Θ/T.

DEBYE's formula gives a good description of the variation of specific heat with temperature (as far as can be expected from an interpolation formula) only for certain substances with simple crystal lattices: most of the elements, and some simple compounds such as the halides. It is inapplicable in practice to substances of more complex structure; this is quite reasonable, since in such substances the vibration spectrum is extremely complicated.

In particular, DEBYE's formula is totally inapplicable to highly anisotropic crystals. Such crystals may have a "layer" or "chain" structure, and the potential energy of interaction of the atoms within each "layer" or "chain" is then considerably greater than the binding energy between different layers or chains. Accordingly the vibrational spectrum will be described by not one but several characteristic temperatures, of different orders of magnitude. The T^3 law for the specific heat will be valid only at temperatures which are small compared with the smallest of the characteristic temperatures; in the intermediate ranges, new limiting laws apply.⁺

---

e^{-z}, and integrating term by term, we find that, for $x \gg 1$,

$$D(x) = \frac{\pi^4}{5x^3} - 3e^{-x}\{1+O(1/x)\}.$$

The value given in the text is therefore correct to within exponentially small terms.

† For $x \ll 1$ a direct expansion of the integrand in powers of x and integration term by term gives

$$D(x) = 1 - \frac{3}{8}x + \frac{1}{20}x^2 - \ldots.$$

‡ The specific heat at high temperatures accurate to the next term in the expansion is

$$C = 3N\nu\left\{1 - \frac{1}{20}(\Theta/T)^2\right\}.$$

|| As examples, the values of Θ for a number of substances, derived from their specific-heat values, are Pb 90°, Ag 210°, Al 400°, KBr 180°, NaCl 280°. For diamond, Θ is particularly large, ~ 2000°

⁺ See I. M. LIFSHITZ, *Zhurnal éksperimental'noĭ i teoreticheskoĭ fiziki* **22**, 471, 1952.

§64. Thermal expansion of solids

The term proportional to T^4 in the free energy at low temperatures (61.6) can be regarded as a small correction to $F_0 = N\varepsilon_0(V/N)$. The small correction to the free energy (for given V and T) is equal to the small correction to the thermodynamic potential Φ (for given P and T; see (15.12)). We can therefore write immediately

$$\Phi = \Phi_0(P) - \pi^2 T^4 V_0(P)/30(\hbar\bar{u})^3. \tag{64.1}$$

Here $\Phi_0(P)$ is the temperature-independent part of the thermodynamic potential, $V_0(P)$ the volume expressed as a function of pressure by means of the relations $P = -\partial F_0/\partial V = -N\,\mathrm{d}\varepsilon_0/\mathrm{d}V$, and $\bar{u} = \bar{u}(P)$ is the mean velocity of sound, expressed in terms of the pressure by means of the same relations. The dependence of the volume of the body on the temperature is given by $V = \partial\Phi/\partial P$:

$$V = V_0(P) - \frac{\pi^2 T^4}{30\hbar^3}\frac{\mathrm{d}}{\mathrm{d}P}\left(\frac{V_0}{\bar{u}^3}\right). \tag{64.2}$$

The thermal expansion coefficient $\alpha = (1/V)(\partial V/\partial T)_P$ is

$$\alpha = -\frac{2\pi^2 T^3}{15\hbar^3 V_0}\frac{\mathrm{d}}{\mathrm{d}P}\left(\frac{V_0}{\bar{u}^3}\right). \tag{64.3}$$

We see that at low temperatures α is proportional to the cube of the temperature. This result is already obvious from NERNST's theorem (§23) together with the T^3 law for the specific heat.

Similarly, at high temperatures we can consider the second and third terms in (62.6) as small corrections to the first term (we must always have $T \ll \varepsilon_0$ if the body is solid), and obtain

$$\Phi = \Phi_0(P) - NcT\log T + NcT\log\hbar\bar{\omega}(P), \tag{64.4}$$

whence

$$V = V_0(P) + (NcT/\bar{\omega})\,\mathrm{d}\bar{\omega}/\mathrm{d}P. \tag{64.5}$$

The thermal expansion coefficient is

$$\alpha = (Nc/V_0\bar{\omega})\,\mathrm{d}\bar{\omega}/\mathrm{d}P, \tag{64.6}$$

and is independent of the temperature.

When the pressure increases, the atoms in a solid come closer together, and the amplitude of their vibrations (for a given energy) decreases, i.e. the frequency increases. Thus $\mathrm{d}\bar{\omega}/\mathrm{d}P > 0$, so that $\alpha > 0$, and solids expand when the temperature rises. Similar considerations show that the coefficient α given by formula (64.3) is also positive.

Finally, we can make use of the law of corresponding states given at the end of §63. The statement that the specific heat is a function only of the ratio T/Θ is equivalent to saying that the thermodynamic potential, for example, is of the form

$$\Phi = \Phi_0(P) + \Theta f(T/\Theta). \tag{64.7}$$

The volume is

$$V = V_0(P) + (\mathrm{d}\Theta/\mathrm{d}P)[f - (T/\Theta)f'],$$

and the thermal expansion coefficient is

$$\alpha = -(T/V_0\Theta^2)(\mathrm{d}\Theta/\mathrm{d}P)f''.$$

Similarly, we find the heat function $W = \Phi - T\partial\Phi/\partial T$ and the specific heat $C = \partial W/\partial T$:

$$C = -(T/\Theta)f''.$$

Taking the ratio of the two expressions for α and C, we obtain

$$\frac{\alpha}{C} = \frac{1}{\Theta V_0(P)}\frac{\mathrm{d}\Theta}{\mathrm{d}P}. \tag{64.8}$$

Thus, within the limits of validity of the law of corresponding states, the ratio of the thermal expansion coefficient to the specific heat of a solid is independent of temperature (*Grüneisen's law*).

It has already been mentioned that in solids the difference between the specific heats C_p and C_v is very slight. At low temperatures this is a general consequence of NERNST's theorem, which applies to all bodies. At high temperatures we have, using the thermodynamic relation (16.9),

$$C_p - C_v = -T\frac{[(\partial V/\partial T)_P]^2}{(\partial V/\partial P)_T} = -T\frac{\alpha^2 V_0^2}{\mathrm{d}V_0/\mathrm{d}P},$$

where $\alpha = \alpha(P)$ is the thermal expansion coefficient (64.6). We see that the difference $C_p - C_v$ is proportional to T; essentially this means that its expansion in powers of T/ε_0 begins with a first-order term, whereas that of the specific heat itself begins with a zero-order (constant) term. Hence it follows that in solids $C_p - C_v \ll C$ at high temperatures also.

§65. Phonons

In the foregoing sections we have treated the thermal motion of the atoms in a solid as a set of small normal vibrations of the crystal lattice. Let us now examine the properties of these vibrations in more detail.

Each unit cell of the crystal generally contains several atoms. Thus each atom can be specified by stating which unit cell contains it and giving the number of the atom in that cell. The position of the unit cell can be defined by the radius vector $\mathbf{r}_s$ of any particular vertex of the cell; $\mathbf{r}_s$ takes values given by

$$\mathbf{r}_s = s_1\mathbf{a}_1 + s_2\mathbf{a}_2 + s_3\mathbf{a}_3, \tag{65.1}$$

where s_1, s_2, s_3 are integers, and $\mathbf{a}_1$, $\mathbf{a}_2$, $\mathbf{a}_3$ are the basic lattice vectors (i.e. the lengths of the sides of the unit cell).

Let the displacements of the atoms in the vibrations be denoted by $u_n^{\mathbf{s}}$, where the index **s** gives the number of the cell and the suffix n gives both the number of the atom in the cell and the co-ordinate axis (x, y, z) along which the displacement is considered; n therefore takes altogether $3r$ values, where r is the number of atoms in the cell.

The vibrations occur under the influence of the forces exerted on each atom by the remaining atoms in the lattice. These forces are functions of the displacements, and since the latter are small the forces may be expanded in powers of the $u_n^{\mathbf{s}}$, retaining only the linear terms. This expansion does not contain zero-order terms, since for $u_n^{\mathbf{s}} = 0$ all the atoms are in equilibrium and the forces acting on them must vanish. Thus the equations of motion of the atoms in the lattice are of the form

$$\ddot{u}_n^{\mathbf{s}} = -\sum_{n', \mathbf{s}'} \lambda_{nn'}{}^{\mathbf{s}'-\mathbf{s}} u_{n'}{}^{\mathbf{s}'}. \tag{65.2}$$

The constant coefficients λ depend only on the differences $\mathbf{s}' - \mathbf{s}$, since the interaction forces between atoms can clearly depend only on the relative position of the lattice cells, not on their absolute position in space.†

We shall seek solutions of equations (65.2) in the form of a "monochromatic plane wave"

$$u_n^{\mathbf{s}} = u_n \exp\,[i(\mathbf{k}\cdot\mathbf{r}_{\mathbf{s}} - \omega t)]. \tag{65.3}$$

The (complex) amplitude u_n depends only on the suffix n, i.e. is different for different atoms in the same cell but not for equivalent atoms in different cells.

Substitution of (65.3) in (65.2) gives

$$\omega^2 u_n e^{i\mathbf{k}\cdot\mathbf{r}_{\mathbf{s}}} = \sum_{n', \mathbf{s}'} \lambda_{nn'}{}^{\mathbf{s}'-\mathbf{s}} u_{n'} e^{i\mathbf{k}\cdot\mathbf{r}_{\mathbf{s}'}}.$$

Dividing both sides of this equation by $e^{i\mathbf{k}\cdot\mathbf{r}_{\mathbf{s}}}$, defining the vector $\mathbf{r}_{\boldsymbol{\sigma}} = \mathbf{r}_{\mathbf{s}'} - \mathbf{r}_{\mathbf{s}}$ with suffix $\boldsymbol{\sigma} = \mathbf{s}' - \mathbf{s}$ and changing from summation over $\mathbf{s}'$ to summation over $\boldsymbol{\sigma}$, we have

$$\sum_{n', \boldsymbol{\sigma}} \lambda_{nn'}{}^{\boldsymbol{\sigma}} e^{i\mathbf{k}\cdot\mathbf{r}_{\boldsymbol{\sigma}}} u_{n'} - \omega^2 u_n = 0. \tag{65.4}$$

This set of linear homogeneous equations for the amplitudes has non-zero solutions only if the determinant is equal to zero:

$$\Big|\sum_{\boldsymbol{\sigma}} \lambda_{nn'}{}^{\boldsymbol{\sigma}} e^{i\mathbf{k}\cdot\mathbf{r}_{\boldsymbol{\sigma}}} - \omega^2 \delta_{nn'}\Big| = 0. \tag{65.5}$$

Since the suffixes n, n' each take $3r$ values, the order of the determinant is $3r$, so that (65.5) is an algebraic equation in ω^2 of degree $3r$.

† The coefficients λ are connected by certain relations which express the fact that a simple displacement or rotation of the lattice as a whole causes no forces to act on the atoms. We shall not pause to write out these relations here.

Each of the $3r$ solutions of this equation determines the frequency ω as a function of the wave vector $\mathbf{k}$, usually called the *dispersion relation*. Thus for any given value of the wave vector the frequency can in general take $3r$ different values. In other words, we can say that the frequency is a many-valued function of the wave vector, with $3r$ branches: $\omega = \omega_n(\mathbf{k})$, where the suffix n labels the values of the frequency for a given $\mathbf{k}$. Geometrically, the functional relation $\omega = \omega(k_x, k_y, k_z)$ is represented by a four-dimensional hypersurface. The different branches of the function correspond to different sheets of this hypersurface.

The hypersurface $\omega = \omega(k_x, k_y, k_z)$ may intersect itself, i.e. its sheets need not be completely separate. Such intersections may occur both for "accidental" values of $\mathbf{k}$ and for values which are distinguished by the symmetry of their position with respect to the reciprocal lattice.

In the former case the intersections can occur only in a manifold of one (not two) dimensions, i.e. in a line.† The existence of such intersections could be theoretically predicted only by actually solving the equations of motion of the atoms in a particular lattice.

Intersections resulting from the symmetry of the crystal can be treated by the methods of group theory. We shall not pause to discuss this question here, but simply mention that in this case various types of intersection are possible, not only of two but of more than two hypersurfaces.‡

Among these $3r$ branches there must be some which for wavelengths large compared with the distances between atoms correspond to ordinary elastic (i.e. sound) waves in the crystal. It is known from the theory of elasticity that waves of three types can be propagated in a crystal regarded as a continuous medium; these types differ as regards the dependence of ω on $\mathbf{k}$, but for all three types ω is a homogeneous function of the first order in the components of the vector $\mathbf{k}$, and vanishes when $\mathbf{k} = 0$. Thus the $3r$ branches of the functions $\omega(\mathbf{k})$ must include three for which the frequency vanishes with $\mathbf{k}$, and for small $\mathbf{k}$ is a homogeneous function of the first order in the components of $\mathbf{k}$, i.e. is of the form

$$\omega = \alpha(\mathbf{n})k, \tag{65.6}$$

where $\alpha(\mathbf{n})$ is some function of the direction of the vector $\mathbf{k}$ ($\mathbf{n}$ being a unit vector in the direction of $\mathbf{k}$). These three types of wave are called *elastic* or *acoustic*; they are characterised by the fact that the lattice vibrates as a whole, as a continuous medium. In the limiting case of infinite wavelength, these vibrations become a simple parallel displacement of the entire lattice.

† Cf. *Quantum Mechanics*, §79.

‡ See L. P. Bouckaert, R. Smoluchowski and E. Wigner, *Physical Review* **50**, 58, 1936; F. Hund, *Zeitschrift für Physik* **99**, 119, 1936; C. Herring, *Physical Review* **52**, 361, 365, 1937.

In the remaining $3(r-1)$ types of wave the frequency does not vanish when $\mathbf{k} = 0$, but tends to a constant limit as $\mathbf{k} \to 0$. These are called *optical vibrations* of the lattice. In this case the atoms in a given cell are in relative motion, and in the limit $\mathbf{k} = 0$ the centre of mass of the cell remains fixed. It is evident that if each cell contains only one atom there can be no optical vibrations.

We may note that the $3r-3$ limiting frequencies (for $\mathbf{k} = 0$) of the optical vibrations need not all be different from one another. When the crystal has certain symmetry properties, the limiting frequencies of some of the branches of the optical vibrations may coincide (i.e. the self-intersections of the hypersurface $\omega = \omega(\mathbf{k})$ may pass through the point $\mathbf{k} = 0$). If the limiting frequency for one of the optical branches does not coincide with that for another branch, then the frequency $\omega(\mathbf{k})$ for that branch can be expanded (near $\mathbf{k} = 0$) in powers of the components k_i of the vector $\mathbf{k}$. It is easy to see that this expansion can contain only even powers of the k_i. For, by the symmetry of the mechanical equations of motion under time reversal, if the propagation of a wave (65.3) is possible, then so is that of a similar wave in the opposite direction. But such a change of direction is equivalent to a change in the sign of $\mathbf{k}$. Accordingly, the functions $\omega(\mathbf{k})$ must be even: $\omega(-\mathbf{k}) = \omega(\mathbf{k})$, and this proves the above statement. Thus in the present case the dependence of the frequency of optical vibrations on the wave vector near $\mathbf{k} = 0$ has the form

$$\omega = \omega_0 + \sum_{i,\, k=x,\, y,\, z} \alpha_{ik} k_i k_k, \tag{65.7}$$

where ω_0 is the limiting frequency and the α_{ik} are certain constants.

If, however, the limiting frequencies of several branches coincide, the frequencies $\omega(\mathbf{k})$ for these branches can not in general be expanded in powers of $\mathbf{k}$, since the point $\mathbf{k} = 0$ is a singular one (branch point) for them, and near such a point a function cannot be expanded in series. We can say only that, near $\mathbf{k} = 0$, the difference $\omega - \omega_0$ will be a homogeneous function of the k_i of either the first or the second order (depending on the symmetry of the crystal).

The wave vector $\mathbf{k}$ of the lattice vibrations has the following important property. The vector $\mathbf{k}$ appears in the expression (65.3) only through the exponential factor $e^{i\mathbf{k}\cdot\mathbf{r}_s}$. But this factor is unchanged when $\mathbf{k}$ is augmented by any vector of the form $2\pi\mathbf{b}$, $\mathbf{b} = p_1\mathbf{b}_1 + p_2\mathbf{b}_2 + p_3\mathbf{b}_3$, where $\mathbf{b}$ is any vector of the reciprocal lattice (see §135), $\mathbf{b}_1$, $\mathbf{b}_2$, $\mathbf{b}_3$ are the basic vectors of the reciprocal lattice, and p_1, p_2, p_3 are integers. This means that the wave vector of the lattice vibrations is defined only to within an arbitrary vector of the reciprocal lattice, multiplied by 2π.

Thus in each branch of the function $\omega(\mathbf{k})$ it is sufficient to consider values of the wave vector which lie in some particular finite interval: if the co-ordinate axes (in general oblique) are taken along the three basic vectors of the reciprocal lattice, it is sufficient to consider values of the three components of the wave vector in the intervals

$$-\pi b_1 \leqslant k_x \leqslant \pi b_1, \quad -\pi b_2 \leqslant k_y \leqslant \pi b_2, \quad -\pi b_3 \leqslant k_z \leqslant \pi b_3. \qquad (65.8)$$

In other words, for the vector $\mathbf{k}/2\pi$ we must take all its possible values lying in one reciprocal lattice cell. This applies, of course, to both acoustic and optical vibrations.

Concerning the whole of the foregoing discussion, it must be emphasised once more that this has related only to the "harmonic approximation", in which only those terms are taken into account, in the potential energy of the vibrating particles, which are quadratic in the displacements of the atoms. It is only in this approximation that the various monochromatic waves (65.3) do not "interact" but are freely propagated through the lattice. When the subsequent "anharmonic" terms are taken into account, various processes of scattering of these waves by one another appear.

Moreover, it is assumed that the lattice is perfectly periodic, but it must be borne in mind that the perfect periodicity is to some extent perturbed, even without allowing for possible impurities and other lattice defects, if the crystal contains randomly distributed atoms of different isotopes. This perturbation, however, is comparatively small if the relative difference of atomic weights of the isotopes is small or if the abundance of one isotope greatly exceeds those of the others. In such cases, which are those usually encountered, the above description remains valid in a first approximation, and in subsequent approximations there occur various processes of "scattering" of waves by "inhomogeneities" in the lattice.

Let us now consider the quantum treatment of lattice vibrations.

Instead of the waves (65.3), in which the atoms have definite displacements at any instant, quantum theory uses the concept of *sound quanta* or *phonons*, which are "quasi-particles" propagated through the lattice, with definite energies and directions of motion. Since the energy of an oscillator in quantum mechanics is an integral multiple of $\hbar\omega$ (where ω is the frequency of the classical wave), the phonon energy ε is related to the frequency ω by

$$\varepsilon = \hbar\omega, \qquad (65.9)$$

in the same way as for light quanta or photons. The wave vector $\mathbf{k}$ determines the *quasi-momentum* $\mathbf{p}$ of the phonon:

$$\mathbf{p} = \hbar\mathbf{k}. \qquad (65.10)$$

This is a quantity in many ways analogous to the ordinary momentum, but there is an important difference between them due to the fact that the quasi-momentum is defined only to within an arbitrary additive constant vector of the form $2\pi\hbar\mathbf{b}$; values of $\mathbf{p}$ differing by such a quantity are physically equivalent.

The velocity of a phonon is given by the group velocity of the corresponding classical waves, $\mathbf{v} = \partial\omega/\partial\mathbf{k}$. This formula may also be written

$$\mathbf{v} = \partial\varepsilon(\mathbf{p})/\partial\mathbf{p}, \tag{65.11}$$

which is exactly analogous to the usual relation between the energy, momentum and velocity of particles.

The whole of the discussion above concerning the relation between the frequency and the wave vector for waves is entirely applicable to the relation between the energy and the quasi-momentum for phonons. In particular, the function $\varepsilon = \varepsilon(\mathbf{p})$ in general has $3r$ different branches. These include three "kinds" of phonons for which, at sufficiently small values of the quasi-momentum, the energy ε is a homogeneous function of the first order in the components of $\mathbf{p}$. The velocity of such phonons for small $\mathbf{p}$ has a value which depends only on the direction of $\mathbf{p}$ and not on its magnitude. This velocity is clearly just the corresponding velocity of sound in the crystal.

The free propagation of the waves (65.3) in the "harmonic" approximation corresponds, in the quantum description, to the free motion of phonons which do not interact at all, i.e. do not "collide" with one another. In subsequent approximations, various processes of elastic and inelastic collisions between phonons appear. These collisions provide the mechanism whereby thermal equilibrium is established in the "phonon gas", i.e. whereby an equilibrium thermal motion is established in the lattice.

In the collision of two (or more) phonons, the laws of conservation of energy and of quasi-momentum must be obeyed. The latter law, however, requires the conservation of the sum of the quasi-momenta of the colliding phonons only to within an arbitrary additive vector of the form $2\pi\hbar\mathbf{b}$, as a result of the non-uniqueness of the quasi-momentum itself. Thus the quasi-momenta of the two phonons before the collision ($\mathbf{p}_1$, $\mathbf{p}_2$) and after the collision ($\mathbf{p}_1'$, $\mathbf{p}_2'$) must be related by

$$\mathbf{p}_1+\mathbf{p}_2 = \mathbf{p}_1'+\mathbf{p}_2'+2\pi\hbar\mathbf{b}. \tag{65.12}$$

Any number of identical phonons may be created simultaneously in the lattice. That is, any number of phonons may be in each of the phonon quantum states. This means that the phonon gas obeys Bose statistics. Since, furthermore, the total number of "particles" in this gas is not given and is itself determined by the equilibrium conditions, its chemical potential is zero (see §60). The mean number of phonons in a given quantum state (with quasi-momentum $\mathbf{p}$ and energy ε) is determined in thermal equilibrium by PLANCK's function:

$$\overline{n_\mathbf{p}} = 1/(e^{\varepsilon(\mathbf{p})/T}-1). \tag{65.13}$$

It may be noted that at high temperatures ($T \gg \varepsilon$) this expression becomes

$$\overline{n_\mathbf{p}} = T/\varepsilon, \tag{65.14}$$

i.e. the number of phonons in a given state is proportional to the temperature.

The concept of phonons can be used to describe the non-equilibrium states of a solid in the same way as for an ideal gas. Any non-equilibrium macroscopic state of a solid is defined by some non-equilibrium distribution of phonons among their quantum states. The entropy of a body in such a state can be calculated by means of the formulae derived in §54 (for a Bose gas). In particular, when there are many phonons in each state, the entropy is

$$S = \sum_j G_j \log (eN_j/G_j),$$

where N_j is the number of phonons in a group of G_j neighbouring states (see (54.8)). This case corresponds to high temperatures ($T \gg \Theta$).

We can rewrite this formula in an integral form corresponding to the classical wave picture of thermal vibrations. The number of phonon states (of each of $3r$ "kinds") which "correspond" to the interval $dk_x\, dk_y\, dk_z$ of values of the wave vector and the space volume element dV is

$$d\tau = \frac{dp_x\, dp_y\, dp_z\, dV}{(2\pi\hbar)^3} = \frac{dk_x\, dk_y\, dk_z\, dV}{(2\pi)^3}.$$

Let $U(\mathbf{r}, \mathbf{k})\, d\tau$ be the energy of thermal vibrations with wave vectors in $dk_x\, dk_y\, dk_z$ in the space volume dV. The corresponding number of phonons is $U(\mathbf{r}, \mathbf{k})\, d\tau/\hbar\omega(\mathbf{k})$. Substituting these expressions for G_j and N_j and changing to integration, we have the following formula for the entropy of a solid with a given non-equilibrium distribution of energy in the spectrum of thermal vibrations:

$$S = \sum \int \log [eU(\mathbf{r}, \mathbf{k})/\hbar\omega(\mathbf{k})]\, d\tau. \qquad (65.15)$$

The summation is over the $3r$ branches of the function $\omega(\mathbf{k})$.

§66. Quantum liquids with Bose-type spectrum

Unlike gases and solids, liquids do not allow a calculation in a general form of the thermodynamic quantities or even of their dependence on temperature. The reason lies in the existence of a strong interaction between the molecules of the liquid while at the same time we do not have the smallness of the vibrations which makes the thermal motion in solids especially simple. The strength of the interaction between molecules makes it necessary to know the precise law of interaction in order to calculate the thermodynamic quantities, and this law is different for different liquids.

A general theoretical treatment is, however, possible for liquids at temperatures near absolute zero.† This problem is of considerable fundamental

† The results given in §§66–68 are due to L. Landau.

interest, although in practice there is in Nature only one substance, namely helium, which can remain liquid down to absolute zero. In this connection it will be recalled (see §61) that according to classical mechanics all bodies should be solid at absolute zero, but helium, owing to the unusually weak interaction between its atoms, remains liquid down to temperatures at which quantum effects become important (a *quantum liquid*), after which it need not solidify.

The calculation of the thermodynamic quantities requires a knowledge of the energy level spectrum of the body in question. It should be emphasised that, for a system of strongly interacting particles such as a quantum liquid, we can speak only of levels corresponding to stationary quantum states of the whole liquid, and never to states of the individual atoms. In calculating the partition function at temperatures close to absolute zero, only the energy levels of low excitation in the liquid need be taken into account, i.e. the levels which lie not too far above the ground state.

The following point is fundamental to the whole of the subsequent discussion. Any low excited state of a macroscopic body can be regarded in quantum mechanics as an assembly of *elementary excitations*, which behave as *quasi-particles* moving in the volume occupied by the body and having definite energies and momenta. So long as the number of elementary excitations is sufficiently small, they do not "interact" with one another (i.e. their energies are additive), and so the assembly of them may be regarded as an ideal gas. It should again be emphasised that the concept of elementary excitations arises as a means of quantum description of the collective motion of the atoms in a liquid, and they cannot in any way be identified with the individual atoms or molecules.

An example of quasi-particles is given by the phonons discussed in §65, which describe states of a crystal whose atoms are executing small vibrations about equilibrium positions.

One possible type of energy spectrum of low excited states of a quantum liquid (a "Bose-type" spectrum) is characterised by the fact that the elementary excitations may appear and disappear singly. But the angular momentum of any quantum system (here the whole liquid) can change only by an integer. Thus the elementary excitations appearing singly must have integral angular momentum and therefore obey Bose statistics. Any liquid consisting of atoms which obey Bose statistics must have a spectrum of this type.

A very important property of quasi-particles is their dispersion relation, i.e. the relation between their energy and momentum. In a spectrum of the type under consideration, the elementary excitations with small momenta p (i.e. large wavelengths $\hbar/p$) correspond to ordinary sound waves in the liquid, i.e. are phonons. This means that the energy of such elementary excitations is a linear function of the momentum:

$$\varepsilon = up, \tag{66.1}$$

where u is the velocity of sound in the liquid. It must be emphasised that the momentum of an elementary excitation in the liquid is the true momentum, and not the quasi-momentum as for a phonon in the periodic crystal lattice of a solid.

As the momentum increases, the function $\varepsilon(p)$ ceases to be linear, of course; its subsequent form depends on the particular law of interaction between the molecules of the liquid, and therefore cannot be determined in a general form. It must be remembered that, for sufficiently large momenta, the function $\varepsilon(p)$ will not exist, since elementary excitations with too large momenta are unstable and decompose into several excitations with smaller momenta (and energies).†

If the function $\varepsilon(p)$ for small p is known, we can calculate the thermodynamic quantities for the liquid at temperatures close to absolute zero such that practically all the elementary excitations in the liquid have low energies, i.e. are phonons. The corresponding formulae can be written down without special calculation by making direct use of the expressions derived in §61 for the thermodynamic quantities in a solid at low temperatures. The only difference is that, instead of the three possible directions of polarisation of sound waves in a solid (one longitudinal and two transverse), in a liquid there exists only one (longitudinal), and so all the expressions for the thermodynamic quantities are to be divided by three. For example, the free energy of liquid is

$$F = F_0 - V\cdot\pi^2 T^4/90(\hbar u)^3, \tag{66.2}$$

where F_0 is the free energy of the liquid at absolute zero. The energy of the liquid is

$$E = E_0 + V\cdot\pi^2 T^4/30(\hbar u)^3, \tag{66.3}$$

and the specific heat

$$C = V\cdot 2\pi^2 T^3/15(\hbar u)^3; \tag{66.4}$$

this is proportional to the cube of the temperature.

Liquid helium (the isotope He^4) has an energy spectrum of this type. An analysis of experimental values of its thermodynamic quantities shows that they can be fully represented by the type of dispersion relation for elementary excitations shown in Fig. 8: after an initial linear rise, the function $\varepsilon(p)$ reaches a maximum, then decreases and passes through a minimum at a certain value p_0 of the momentum.‡ In thermal equilibrium, most of the

† The properties of the spectrum near the "termination point" of the curve $\varepsilon = \varepsilon(p)$ have been examined by L. P. PITAEVSKIĬ (*Soviet Physics JETP* **9**, 830, 1959).

‡ A qualitative theory of this type of spectrum has been given by R. P. FEYNMAN, *Physical Review* **94**, 262, 1954; see also L. P. PITAEVSKIĬ, *Soviet Physics JETP* **4**, 439, 1957.

elementary excitations in the liquid have energies in regions near the minima of the function $\varepsilon(p)$, i.e. in the region of small ε (near $\varepsilon = 0$) and in the region of $\varepsilon(p_0)$. These regions are therefore of particular importance. Near the point $p = p_0$, the function $\varepsilon(p)$ can be expanded in powers of the difference $p - p_0$. There is no linear term in the expansion, and we have as far as the second-order terms

$$\varepsilon = \Delta + (p-p_0)^2/2\mu, \tag{66.5}$$

where $\Delta = \varepsilon(p_0)$ and μ are constants. Quasi-particles of this type are called *rotons*.†

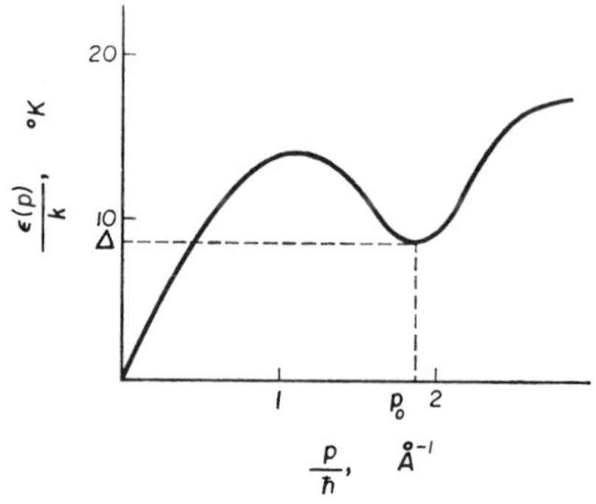

FIG. 8

The empirical values of the constants Δ, p_0 and μ are $\Delta = 8.5°\text{K}$, $p_0/\hbar = 1.9 \times 10^{-8}\ \text{cm}^{-1}$, $\mu = 0.16\ m_{\text{He}}$ (where m_{He} is the mass of the He^4 atom).

Since the roton energy always includes the quantity Δ, which is large compared with T at temperatures sufficiently low for a "roton gas" to be considered, this gas may be described by the Boltzmann distribution instead of the Bose distribution. Accordingly, to calculate the "roton" part of the thermodynamic quantities for liquid helium we start from formula (41.5):

$$F = -NT \log \frac{eV}{N(2\pi\hbar)^3} \int e^{-\varepsilon/T}\, d^3p.$$

The number N of particles in the roton gas is not a given number, but is itself determined by the condition of minimum free energy. Equating $\partial F/\partial N$ to zero, we find for the number of rotons

$$N_r = \frac{V}{(2\pi\hbar)^3} \int e^{-\varepsilon/T}\, d^3p.$$

† The special name does not, of course, connote any fundamental difference between phonons and rotons, which simply correspond to different parts of the same curve, and there is a continuous transition from one to the other.

The corresponding free energy is

$$F_r = -\frac{VT}{(2\pi\hbar)^3}\int e^{-\varepsilon/T}\,d^3p.$$

The expression (66.5) is to be substituted in these formulae. Since $p_0^2 \gg \mu T$, in integrating with respect to p we can take the factor p^2 outside the integral and replace it by p_0^2 with sufficient accuracy. In integrating the exponential we can extend the range of integration from $-\infty$ to ∞. The result is

$$N_r = \frac{2(\mu T)^{1/2}p_0^2 V}{(2\pi)^{3/2}\hbar^3}\,e^{-\Delta/T},$$
$$F_r = -TN_r. \tag{66.6}$$

Hence the roton contributions to the entropy and the specific heat are respectively

$$S_r = N_r\left(\frac{3}{2}+\frac{\Delta}{T}\right),$$
$$C_r = N_r\left[\frac{3}{4}+\frac{\Delta}{T}+\left(\frac{\Delta}{T}\right)^2\right]. \tag{66.7}$$

We see that the temperature dependence of the roton part of the thermodynamic quantities is essentially exponential. At sufficiently low temperatures (below about 0.8°K for liquid helium) the roton part is therefore less than the phonon part, while at high temperatures the position is reversed and the roton contribution is greater than that of the phonons.

§67. Superfluidity

A quantum liquid with an energy spectrum of the type described above must possess a remarkable property, known as *superfluidity*: the property of flowing through narrow capillaries or slits without exhibiting viscosity.† Let us first consider a liquid at absolute zero, at which temperature the liquid is in its ground state.

Let us consider a liquid flowing along a capillary at a constant velocity **v**. Because of the friction against the walls of the tube and the friction within the liquid itself, the presence of viscosity would have the effect that the kinetic energy of the liquid would be dissipated and the flow would gradually become slower.

It will be more convenient to discuss the flow in a co-ordinate system moving with the liquid. In such a system the liquid (e.g. helium) is at rest, and the walls of the capillary move with velocity $-\mathbf{v}$. When viscosity is present, the liquid at rest must also begin to move. It is physically evident that the entrainment of the liquid by the walls of the tube cannot initiate

† This phenomenon was discovered in liquid helium (helium II) by P. L. Kapitsa (1938).

movement of the liquid as a whole. The motion must arise from a gradual excitation of internal motions, that is, from the appearance of elementary excitations in the liquid.

Let us suppose that a single elementary excitation appears in the liquid, with momentum **p** and energy $\varepsilon(\mathbf{p})$. Then the energy E_0 of the liquid (in the co-ordinate system in which it was originally at rest) is equal to the energy ε of the excitation, and its momentum $\mathbf{P}_0$ is equal to **p**. Let us now return to the co-ordinate system in which the capillary is at rest. According to the familiar formulae of mechanics for the transformation of energy and momentum, we obtain for the energy E and momentum **P** of the liquid in this system

$$E = E_0+\mathbf{P}_0\cdot\mathbf{v}+\tfrac{1}{2}Mv^2, \qquad \mathbf{P} = \mathbf{P}_0+M\mathbf{v}, \tag{67.1}$$

where M is the mass of the liquid. Substituting ε and **p** for E_0 and $\mathbf{P}_0$, we have

$$E = \varepsilon+\mathbf{p}\cdot\mathbf{v}+\tfrac{1}{2}Mv^2.$$

The term $\frac{1}{2}Mv^2$ is the original kinetic energy of the flowing liquid; the expression $\varepsilon+\mathbf{p}\cdot\mathbf{v}$ is the change in energy due to the appearance of the excitation. This change must be negative, since the energy of the moving liquid must decrease: $\varepsilon+\mathbf{p}\cdot\mathbf{v} < 0$.

For a given value of **p**, the quantity on the left-hand side of this inequality is a minimum when **p** and **v** are antiparallel; thus we must always have $\varepsilon-pv < 0$, or

$$v > \varepsilon/p. \tag{67.2}$$

This inequality must be satisfied for at least some values of the momentum p of the elementary excitation. Hence the final condition for the occurrence of excitations to be possible in the liquid as it moves along the capillary is obtained by finding the minimum of ε/p. Geometrically, the ratio ε/p is the slope of the line drawn from the origin (in the $p\varepsilon$-plane) to some point on the curve $\varepsilon = \varepsilon(p)$. Its minimum value is clearly given by the point at which the line from the origin is a tangent to the curve. If this minimum is not zero, then, for velocities of flow below a certain value, excitations cannot appear in the liquid. This means that the flow will not become slower, i.e. that the liquid exhibits the phenomenon of superfluidity.

The condition just derived for the presence of superfluidity is essentially equivalent to the requirement that the curve $\varepsilon = \varepsilon(p)$ should not touch the axis of abscissae at the origin (ignoring the unlikely possibility that it touches this axis at some other point). Thus any spectrum in which sufficiently small excitations are phonons will lead to superfluidity.

Let us now consider the same liquid at a temperature other than absolute zero (but close to it). In this case the liquid contains excitations, and is not in the ground state. The arguments given above remain valid, since they made no direct use of the fact that the liquid was originally in the ground state.

The motion of the liquid relative to the walls of the tube when the above condition is satisfied still cannot cause any new elementary excitations to appear in it. It is, however, necessary to elucidate the effect of excitations already present in the liquid.

To do this, we calculate as follows. Let us imagine that the "quasi-particle gas" moves as a whole with respect to the liquid, with a translational velocity **v**. The distribution function for the gas moving as a whole is obtained from the distribution function $n(\varepsilon)$ for the gas at rest by replacing the energy ε of a particle by $\varepsilon-\mathbf{p}\cdot\mathbf{v}$, where **p** is the momentum of the particle.† Hence the total momentum of the gas per unit volume is

$$\mathbf{P} = \int \mathbf{p}n(\varepsilon-\mathbf{p}\cdot\mathbf{v})\, d^3p.$$

Let us assume that the velocity **v** is small, and expand the integrand in powers of $\mathbf{p}\cdot\mathbf{v}$. The zero-order term gives zero on integration over the directions of the vector **p**, leaving

$$\mathbf{P} = -\int \mathbf{p}(\mathbf{p}\cdot\mathbf{v})\, \frac{dn(\varepsilon)}{d\varepsilon}\, d^3p.$$

Integrating over the directions of the vector **p** gives

$$\mathbf{P} = -\mathbf{v}\frac{4}{3}\pi \int_0^\infty p^4 \frac{dn(\varepsilon)}{d\varepsilon}\, dp. \tag{67.3}$$

For phonons‡, $\varepsilon = up$, and integration by parts gives

$$\mathbf{P} = -\mathbf{v}\frac{4\pi}{3u}\int_0^\infty p^4 \frac{dn(p)}{dp}\, dp$$

$$= \mathbf{v}\,\frac{16\pi}{3u}\int_0^\infty p^3 n(p)\, dp.$$

† For an ordinary gas this is a direct consequence of GALILEO's relativity principle, and is proved by a simple change of co-ordinates, but in the present case such arguments cannot be applied directly, since the "quasi-particle gas" is moving not in a vacuum but "through the liquid". Nevertheless, the statement remains valid, as can be seen from the following argument. Let the excitation gas be moving relative to the liquid with velocity **v**. Let us take a co-ordinate system in which the gas is at rest as a whole, and the liquid is accordingly moving with velocity $-\mathbf{v}$ (system K). According to the transformation formula (67.1), the energy E of the liquid in the system K is related to the energy in a system K_0 where the liquid is at rest by $E = E_0-\mathbf{P}_0\cdot\mathbf{v}+\frac{1}{2}Mv^2$. Let an elementary excitation of energy $\varepsilon(p)$ in K_0 arise in the liquid. Then the additional energy of the liquid in K is $\varepsilon-\mathbf{p}\cdot\mathbf{v}$, and this proves the statement.

‡ For phonons, the function $n(\varepsilon)$ is the Bose distribution function with chemical potential zero. Hence $n(\varepsilon-\mathbf{p}\cdot\mathbf{v})$ is proportional to $1/(e^{(\varepsilon-\mathbf{p}\cdot\mathbf{v})/T}-1)$. It should be noted that the superfluidity condition $v < \varepsilon/p$ is precisely the condition for this latter expression to be positive and finite for all energies.

But the integral

$$\int_0^\infty upn(p)\cdot 4\pi p^2 \, \mathrm{d}p = \int \varepsilon n(\varepsilon) \, \mathrm{d}^3p$$

is just the energy E_{ph} of the phonon gas per unit volume, so that we have finally

$$\mathbf{P} = \mathbf{v}\cdot 4E_{\mathrm{ph}}/3u^2. \tag{67.4}$$

First of all, we see that the motion of the quasi-particle gas is accompanied by a transfer of mass: the effective mass per unit volume of the gas is determined by the proportionality coefficient between the momentum **P** and the velocity **v** in (67.3) or (67.4). On the other hand, in the flow of a liquid along a capillary (say) there is nothing to prevent the "particles" of this gas from colliding with the walls of the tube and exchanging momentum with them. In consequence the excitation gas will be slowed down, like any ordinary gas flowing along a capillary.

Thus we have the following fundamental result. At non-zero temperatures, part of the mass of the liquid will behave as a normal viscous liquid which "sticks" as it moves along the walls of the vessel; the remaining part of the mass will behave as a superfluid without viscosity. Here it is very important that there is "no friction" between these two parts of the mass of the liquid as they pass "through one another", that is, there is no transfer of momentum from one to the other. For the existence of such motion of one part of the mass of the liquid relative to the other has been derived by the consideration of statistical equilibrium in a uniformly moving excitation gas. But if any relative motion can occur in a state of thermal equilibrium, it is not accompanied by friction.

It should be emphasised that the treatment of the liquid as a "mixture" of normal and superfluid "parts" is simply a form of words convenient for the description of the phenomena occurring in a quantum liquid. Like any description of quantum effects in classical terms, it is not entirely adequate. In reality we should say that in a quantum liquid there can exist simultaneously two motions, each of which has a corresponding "effective mass" such that the sum of these two masses is equal to the actual total mass of the liquid. One of these motions is "normal", i.e. has the same properties as that of an ordinary viscous liquid; the other is "superfluid". The two motions occur without transfer of momentum from one to the other. It is particularly emphasised that no distinction is made between "superfluid" and "normal" among the actual particles of the liquid. In a certain sense we can speak of superfluid and normal masses of liquid, but this does not mean that the liquid can really be separated into two such parts.†

†Concerning the hydrodynamic properties of a superfluid liquid see *Fluid Mechanics*, Chapter XVI.

Formula (67.4) determines the normal part of the mass of the liquid at temperatures so low that all the elementary excitations may be regarded as phonons. Substituting the expression given by (66.3) for the energy of the phonon gas, we find for the normal part ϱ_n of the density of the liquid

$$\varrho_n = 2\pi^2 T^4/45\hbar^3 u^5. \tag{67.5}$$

To calculate the roton part of ϱ_n we note that, since the rotons can be described by a Boltzmann distribution, $\partial n/\partial \varepsilon = -n/T$, and from (67.3)

$$(\varrho_n)_r = \frac{4\pi}{3T(2\pi\hbar)^3}\int p^4 n \, dp = \frac{1}{3T}\int \frac{p^2 n}{(2\pi\hbar)^3} d^3p = \frac{\overline{p^2}}{3T}\frac{N_r}{V}.$$

Since $p_0^2 \gg \mu T$, we can put $\overline{p^2} = p_0^2$ with sufficient accuracy; also substituting N_r from (66.6), we have finally

$$(\varrho_n)_r = p_0^2 N_r/3TV = \frac{2\mu^{1/2}p_0^4}{3(2\pi)^{3/2}T^{1/2}\hbar^3} e^{-\Delta/T}. \tag{67.6}$$

The roton part of ϱ_n is comparable with the phonon part at about 0.6°K and predominates at higher temperatures.

As the temperature increases, an increasing fraction of the mass of the liquid becomes normal. At the point where the whole of the liquid mass becomes normal, the property of superfluidity disappears entirely. This is called the *λ-point* of the liquid (2.19°K for helium) and is a phase transition point of the second kind.

The part of the curve $\varrho_n(T)$ near the λ-point cannot be calculated exactly, of course. But because of the very rapid increase of ϱ_n given by (67.6) we may expect that the temperature of the λ-point can be approximately obtained by putting $\varrho_n/\varrho = 1$ and using that formula. Such a calculation gives a value 2.8°K, in fair agreement with the true value.

A phase transition of the second kind always involves the appearance or disappearance of some qualitative property (see §137). In the case of the λ-point of liquid helium, this change can be macroscopically described as the appearance or disappearance of the superfluid component of the liquid. From the deeper microscopic viewpoint the change concerns certain properties of the density matrix of the atoms in the co-ordinate representation. This matrix $\varrho(\mathbf{r}', \mathbf{r})$ is defined as the integral

$$\varrho(\mathbf{r}', \mathbf{r}) = \int \Psi^*(\mathbf{r}', q)\Psi(\mathbf{r}, q)\, dq,$$

where $\Psi(\mathbf{r}, q)$ is the wave function of the body, $\mathbf{r}$ denoting the radius vector of one particle and q the set of co-ordinates of all the other particles, the integration being over the latter. For an isotropic body (a liquid) the density matrix depends only on the co-ordinate differences $|\mathbf{r}'-\mathbf{r}|$. For ordinary liquids the value of $\varrho(\mathbf{r}', \mathbf{r})$ tends to zero when the distance $\mathbf{r}'-\mathbf{r}$ increases indefinitely, but for a superfluid liquid the limit is not zero.

The Fourier components of the density matrix, i.e. the integrals of the form

$$\int \varrho(\mathbf{r}', \mathbf{r}) e^{i\mathbf{k}\cdot(\mathbf{r}'-\mathbf{r})} \, d^3(\mathbf{r}'-\mathbf{r}), \tag{67.7}$$

are the same, apart from a constant factor, as

$$\int \left| \int \Psi(\mathbf{r}, q) e^{i\mathbf{k}\cdot\mathbf{r}} \, dV \right|^2 dq,$$

i.e. they determine the probability distribution for the various values of the momentum $\mathbf{p} = \hbar\mathbf{k}$ of the particle. If $\varrho(\mathbf{r}', \mathbf{r}) \to 0$ when $|\mathbf{r}'-\mathbf{r}| \to \infty$, then the probability density (in $\mathbf{p}$-space) remains finite as $\mathbf{p} \to 0$, but if $\varrho(\mathbf{r}', \mathbf{r})$ has a finite value ϱ_∞ at infinity, the value of the integral (67.7) tends to infinity as $\mathbf{p} \to 0$, the integral being equal to $(2\pi)^3\varrho_\infty\delta(\mathbf{k})$. This corresponds to a finite probability of zero momentum of the particle; it may be noted in passing that ϱ_∞, which determines this probability, must be positive.

Thus this property of the density matrix is equivalent to the statement that in a superfluid liquid, unlike a non-superfluid, a finite number of particles have zero momentum. However, to avoid misunderstanding we must emphasise that these particles cannot be identified with the "superfluid part" of the liquid. Apart from the fact that such an identification could have no justification, its incorrectness is seen from the fact that at absolute zero the whole of the liquid becomes superfluid, whereas by no means all its particles have zero momentum.†

§68. Quantum liquids with Fermi-type spectrum

It has already been noted in §66 that any quantum liquid composed of particles with integral spin must have a Bose-type spectrum. A liquid composed of particles with half-integral spin, on the other hand, can also have a spectrum of another type, which may be called a *Fermi-type spectrum*; the liquid helium isotope He^3 is of this kind. It must be emphasised, however, that a spectrum of this type cannot be a universal property of liquids consisting of particles with half-integral spin. The type of spectrum depends also on the specific nature of the interaction between atoms. This is clear from the following simple consideration: if the interaction is such that it causes the atoms to tend to associate in pairs, then in the limit we obtain a

† All these properties of the particle distribution function are evident consequences of the model discussed below (§78) of a slightly non-ideal Bose gas at temperatures close to zero, when it is certainly "superfluid".

molecular liquid consisting of molecules with integral spin and therefore having a Bose-type spectrum.

The energy spectrum of a Fermi-type quantum liquid has a structure which is to some extent similar to that of an ideal Fermi gas. The ground state of the latter corresponds to the case where all the quantum states of individual particles with momenta from zero to some p_0 are occupied. The excited states of the gas occur when a particle goes from a state in the occupied band to a state with $p > p_0$.

In a liquid, of course, there are no quantum states for individual particles, but to construct a spectrum of this type we start from the assumption that the classification of energy levels remains unchanged when the interaction between the atoms is gradually "switched on", i.e. as we go from the gas to the liquid. In this classification the gas particles become the elementary excitations, whose number is equal to the number of atoms and which obey Fermi statistics.

Each of these quasi-particles has a definite momentum (we shall return later to the question of the validity of this assumption). Let $n(\mathbf{p})$ be their momentum distribution function. The above principle of classification consists in supposing that, if this function is given, the energy E of the liquid is uniquely determined and that the ground state corresponds to a distribution function in which all quasi-particle states with absolute magnitudes of momentum lying in a certain restricted interval are occupied. In the simplest but most natural case this interval extends, as in the gas, from zero to a certain limiting value p_0, forming a sphere in momentum space (called the *Fermi sphere*).† In other words, the ground state corresponds to a "step function" for the quasi-particle distribution, which ends abruptly at $p = p_0$. The value of p_0 is related to the density of the liquid (number of particles per unit volume) by the same formula as for a gas.

It is most important to emphasise that the total energy E of the liquid is not simply the sum of the energies ε of the quasi-particles. In other words, E is a functional of the distribution function in a general form, and does not reduce to the integral $\int n\varepsilon \, d\tau$ (as it does for a gas, where the quasi-particles are the same as the actual particles).

Since the primary concept is E, the question arises how the energy ε of a quasi-particle should be defined.

We shall normalise the distribution function by the condition

$$\int n \, d\tau = N/V, \tag{68.1}$$

† In principle it might be possible for a "cavity" to appear within this sphere during the gradual transition from a gas to a liquid, so that the ground state would correspond to the occupation of all states with absolute values of momentum lying in the interval between two finite non-zero values.

where N is the number of particles in the liquid of volume V, and $d\tau$ here denotes $d^3p/(2\pi\hbar)^3$; this condition will later be made more precise. The change in E due to an infinitesimal change in the distribution function may be written as

$$\delta E/V = \int \varepsilon\, \delta n\, d\tau. \tag{68.2}$$

The quantity ε is the functional derivative of the energy with respect to the distribution function, and corresponds to the change in the energy of the system when a single quasi-particle of momentum $\mathbf{p}$ is added. This quantity plays the part of the Hamiltonian of a quasi-particle in the field of the other particles. It is also a functional of the distribution function, i.e. the form of the function $\varepsilon(\mathbf{p})$ is determined by the distribution of all the quasi-particles in the liquid.

In this connection it may be noted that an elementary excitation in the type of spectrum considered may in a certain sense be treated like an atom in the self-consistent field of other atoms. This self-consistency is, of course, not to be understood in the sense usual in quantum mechanics. Here its nature is more profound; in the Hamiltonian of the atom, not only is allowance made for the effect of the surrounding particles on the potential energy, but the dependence of the kinetic-energy operator on the momentum operator is also modified.

It is easy to see that the quasi-particle distribution function is (in equilibrium) an ordinary Fermi distribution, the particle energy being represented by the quantity ε defined according to (68.2). For, because the energy levels of the liquid and of the ideal Fermi gas are classified in the same manner, the entropy of the liquid is determined by a similar combinatorial expression

$$S = -\int \{n \log n + (1-n) \log (1-n)\}\, d\tau \tag{68.3}$$

to that for a gas (cf. (54.3)). Varying this expression with the additional conditions of constant total number of particles and constant total energy (the variation of the latter is given by formula (68.2)), we obtain the required distribution:

$$n = 1/(e^{(\varepsilon-\mu)/T}+1). \tag{68.4}$$

It must be emphasised, however, that, despite the formal analogy between this expression and the ordinary Fermi distribution, it is not precisely the same, since ε is itself a functional of n, and formula (68.4) is therefore, strictly speaking, a very complicated implicit expression for n.

Hitherto we have ignored the possible spin of the quasi-particles. In reality all the quantities (n, ε, etc.) are not only functions of momentum but also operator functions of the operator (matrix) of the spin $\hat{\mathbf{s}}$ of the quasi-particles.

If the liquid is in thermal equilibrium, it is homogeneous and isotropic; the scalar quantity ε can then depend only on scalar arguments. The operator $\hat{\mathbf{s}}$ can therefore appear only in the form $\hat{\mathbf{s}}^2$ or $(\hat{\mathbf{s}}\cdot\mathbf{p})^2$; the first power of the product $\hat{\mathbf{s}}\cdot\mathbf{p}$ is inadmissible, since the spin vector is axial and this product is therefore a pseudoscalar, not a true scalar. For spin $\frac{1}{2}$ we have $\hat{\mathbf{s}}^2 = \frac{3}{4}$, $(\hat{\mathbf{s}}\cdot\mathbf{p})^2 = \frac{1}{4}\mathbf{p}^2$, so that $\hat{\mathbf{s}}$ does not appear. Thus in this case the energy of a quasi-particle is independent of its spin.

The fact that ε is independent of the spin signifies that all the energy levels of the quasi-particles are doubly degenerate. The statement that a quasi-particle has spin is essentially an expression of the existence of this degeneracy. In this sense we can say that the spin of the quasi-particles in a spectrum of the type considered is always $\frac{1}{2}$, whatever the spin of the actual particles in the liquid. For with any spin s other than $\frac{1}{2}$ the terms of the form $(\hat{\mathbf{s}}\cdot\mathbf{p})^2$ would give a splitting of the $(2s+1)$-fold degenerate levels into $\frac{1}{2}(2s+1)$ doubly degenerate levels. In other words, $\frac{1}{2}(2s+1)$ different branches of the function $\varepsilon(\mathbf{p})$ appear, each corresponding to quasi-particles "with spin $\frac{1}{2}$".

To simplify the formulae, we shall assume in what follows that all quantities are independent of the spin operator. Then the presence of spin $\frac{1}{2}$ can be taken into account by including a factor 2, which we shall incorporate in the definition of $d\tau$:

$$d\tau = 2\, d^3p/(2\pi\hbar)^3.$$

When spin dependence is present, the formulae are modified only in that the integration over phase space must be accompanied by the taking of the trace of matrix functions.

Let us now return to the assumption made above that a definite momentum can be assigned to each quasi-particle. The condition for this assumption to be valid is that the uncertainty of momentum (due to the finite mean free path of the quasi-particle) should be small not only in comparison with the momentum itself but also in comparison with the width of the "transition zone" of the distribution (over which it differs appreciably from a step function). It is easy to see that this condition is satisfied if the distribution $n(\mathbf{p})$ differs from a step function only by a fairly small deviation near the limiting momentum, i.e. near the surface of the Fermi sphere. For, by PAULI's principle, only quasi-particles in the transition zone of the distribution can undergo mutual scattering, and as a result of this scattering they must enter free states in that zone. Hence the collision probability is proportional to the square of the width Δp of the zone. Accordingly, the uncertainty of momentum due to scattering processes is also proportional to $(\Delta p)^2$. It is therefore clear that, when Δp is sufficiently small, the uncertainty in momentum will be small in comparison not only with p_0 but also with Δp.

Thus the method described is valid only for excited states of the liquid which are described by a quasi-particle distribution function differing from a step function only in a small region near the upper limit. In particular, for thermodynamic equilibrium distributions only sufficiently low temperatures (small compared with the degeneracy temperature T_0) are permissible. In these conditions we can, as a first approximation, replace the functional ε in (68.4) by its value calculated for a step-function distribution. Then ε becomes an entirely definite function of the magnitude of the momentum, and formula (68.4) reduces to the ordinary Fermi distribution.

Thus the function $\varepsilon(p)$ has a direct physical significance only in the neighbourhood of the surface of the Fermi sphere. Expanding this function in powers of $p-p_0$ there, we have

$$\Delta\varepsilon = \varepsilon - \mu \cong v_0(p-p_0), \tag{68.5}$$

where

$$v_0 = [\partial\varepsilon/\partial p]_{p=p_0} \tag{68.6}$$

is the "velocity" of the quasi-particles on the Fermi surface.† In an ideal Fermi gas, where the quasi-particles are identical with the actual particles, we have $\varepsilon = p^2/2m$, and so $v_0 = p_0/m$. By analogy we can define for a Fermi liquid the quantity

$$m^* = p_0/v_0, \tag{68.7}$$

called the *effective mass* of the quasi-particles.‡

This quantity determines, in particular, the specific heat of the liquid at low temperatures, which is given by the same formula (57.6) as for a gas, with m replaced by m^*. This follows from the fact that the expression (68.3) for the entropy in terms of the distribution function is the same for a liquid and for a gas, and so is the expression (68.4) for the distribution function in terms of ε, and in calculating the integral (68.3) only the interval of momenta near p_0 is important.

Let $\delta\varepsilon$ denote the change in energy of a quasi-particle caused by a small deviation of the distribution function from step-function form. This quantity must be a linear functional:

$$\delta\varepsilon(\mathbf{p}) = \int f(\mathbf{p}, \mathbf{p}')\, \delta n'\, d\tau'. \tag{68.8}$$

The function $f(\mathbf{p}, \mathbf{p}')$ is the second functional derivative of E, and is therefore symmetrical in the variables $\mathbf{p}$ and $\mathbf{p}'$. It plays an important part in the theory of the Fermi liquid. In the ideal-gas approximation, $f \equiv 0$.

† It may be noted that a spectrum of this type does not admit the appearance of superfluidity. In the discussion in §67 it is now necessary to write $\Delta\varepsilon$ in place of ε, and the inequality (67.2) $v > \Delta\varepsilon/p$ can be satisfied for any v.

‡ For liquid He^3, $p_0/\hbar \cong 0.8\times 10^8$ cm^{-1}, $m^* \cong 2.4 m_{He^3}$. It may also be noted that for liquid He^3 the theory given here is quantitatively valid only for temperatures up to a few tenths of a degree.

The function f depends, in general, not only on the momenta but also on the spins. Whereas the fundamental distribution is isotropic, the function f will in general contain terms of the form $\phi_{ik}(\mathbf{p}, \mathbf{p}')\hat{s}_i\hat{s}_k$. In particular, the exchange interaction of the quasi-particles leads to terms of the form $\phi(\mathbf{p}, \mathbf{p}')\hat{\mathbf{s}}\cdot\hat{\mathbf{s}}'$. Below, however, we shall assume for simplicity that the function f is independent of the spins.

In the absence of an external field, the momentum of the liquid per unit volume is the same as the current density of mass; this follows directly from GALILEO's relativity principle. The velocity of a quasi-particle is $\partial\varepsilon/\partial\mathbf{p}$, and so the quasi-particle current is given by $\int n(\partial\varepsilon/\partial\mathbf{p})\,\mathrm{d}\tau$. Since the number of quasi-particles in the liquid is equal to the number of actual particles, it is clear that, in order to obtain the total transfer of mass by quasi-particles, we must multiply their number current by the mass m of an actual particle. Thus we obtain the equation

$$\int \mathbf{p}n\,\mathrm{d}\tau = \int m(\partial\varepsilon/\partial\mathbf{p})n\,\mathrm{d}\tau. \tag{68.9}$$

Varying both sides of this equation and using (68.8), we have

$$\begin{aligned}\int \mathbf{p}\,\delta n\,\mathrm{d}\tau &= m\int\frac{\partial\varepsilon}{\partial\mathbf{p}}\,\delta n\,\mathrm{d}\tau + m\iint\frac{\partial f(\mathbf{p}, \mathbf{p}')}{\partial\mathbf{p}}\,n\,\delta n'\,\mathrm{d}\tau\,\mathrm{d}\tau' \\ &= m\int\frac{\partial\varepsilon}{\partial\mathbf{p}}\,\delta n\,\mathrm{d}\tau - m\iint f(\mathbf{p}, \mathbf{p}')\frac{\partial n'}{\partial\mathbf{p}'}\,\delta n\,\mathrm{d}\tau\,\mathrm{d}\tau';\end{aligned}$$

in the second integral on the right we have renamed the variables of integration and integrated by parts. Since δn is arbitrary, this gives

$$\frac{\mathbf{p}}{m} = \frac{\partial\varepsilon}{\partial\mathbf{p}} - \int f\frac{\partial n'}{\partial\mathbf{p}'}\,\mathrm{d}\tau'. \tag{68.10}$$

Let us now apply this relation to momenta near the boundary of the Fermi distribution, at the same time replacing the distribution function by a step function. Then the energy ε is a function of momentum for which the expression (68.5) may be used, and the derivative $\partial n/\partial\mathbf{p}$ is essentially a delta function: $\partial n/\partial\mathbf{p} = -(\mathbf{p}/p)\delta(p-p_0)$. This enables us to integrate over the magnitude of the momentum in (68.10):

$$\int f\frac{\partial n'}{\partial\mathbf{p}'}\frac{2p'^2\,\mathrm{d}p'\,\mathrm{d}o'}{(2\pi\hbar)^3} = -\frac{2p_0}{(2\pi\hbar)^3}\int f\mathbf{p}_0'\,\mathrm{d}o'.$$

In the function $f(\mathbf{p}, \mathbf{p}')$ both arguments are taken to have magnitude p_0, so that f actually depends only on the angle θ between $\mathbf{p}_0$ and $\mathbf{p}_0'$. Substituting this result in (68.10), multiplying both sides by $\mathbf{p}_0$ and then dividing by p_0^2, we obtain the following relation between the actual mass of the particles and the effective mass of the quasi-particles:

$$\frac{1}{m} = \frac{1}{m^*} + \frac{p_0}{2(2\pi\hbar)^3} \cdot 4 \int f \cos\theta \, do'. \tag{68.11}$$

Finally, let us calculate the compressibility of a Fermi liquid (at absolute zero) or, what is the same thing, the velocity of sound in it, which is the square root of the compressibility.† The density of the liquid is $\varrho = mN/V$, and the square of the velocity of sound is

$$u^2 = \frac{\partial P}{\partial(mN/V)} = -\frac{V^2}{mN}\frac{\partial P}{\partial V}.$$

For $T = 0$, $S = 0$ also, and so it is not necessary to distinguish the isothermal and adiabatic compressibilities. To calculate this derivative we may express it in terms of the derivative of the chemical potential. Since the latter depends on N and V only through the ratio N/V, we have

$$\frac{\partial \mu}{\partial N} = -\frac{V}{N}\frac{\partial \mu}{\partial V} = -\frac{V^2}{N^2}\frac{\partial P}{\partial V};$$

for T = constant $= 0$, $d\mu = -V\,dP/N$. Thus

$$u^2 = \frac{N}{m}\frac{\partial \mu}{\partial N}. \tag{68.12}$$

Since $\mu = \varepsilon(p_0) \equiv \varepsilon_0$, the change $\delta\mu$ resulting from a change in the number of particles by δN is

$$\delta\mu = \int f \,\delta n' \, d\tau' + \frac{\partial \varepsilon}{\partial p_0}\delta p_0. \tag{68.13}$$

The second term appears because a change in the total number of particles also affects the value of the limiting momentum, δN and δp_0 being related by

$$2\cdot 4\pi p_0^2 \delta p_0 V/(2\pi\hbar)^3 = \delta N.$$

Since $\delta n'$ is appreciably different from zero only when $p \cong p_0$, we can write for the integral in (68.13)

$$\int f \,\delta n' \, d\tau' \cong \int f \, do' \int \delta n' \, d\tau'/4\pi = \int f \, do' \, \delta N/4\pi V.$$

Substituting this in (68.13) and putting $\partial\varepsilon_0/\partial p_0 = p_0/m^*$, we obtain

$$\frac{\partial \mu}{\partial N} = \frac{1}{4\pi V}\int f \, do' + \frac{(2\pi\hbar)^3}{8\pi p_0 m^* V}.$$

† It must be remembered, however, that in practice ordinary sound could not be propagated in a Fermi liquid at absolute zero, since its viscosity increases without limit as $T \to 0$.

Finally, substituting m^* from (68.11) and multiplying by $N/m = 2\cdot 4\pi p_0^3 V/3(2\pi\hbar)^3 m$, we have

$$u^2 = \frac{p_0^2}{3m^2} + \frac{1}{6m}\left(\frac{p_0}{2\pi\hbar}\right)^3 \cdot 4 \int f(1-\cos\theta)\, do'. \tag{68.14}$$

If the function f depends on the spins of both particles, the factor 4 before the integrals in (68.11) and (68.14) must be replaced by the trace taken over both spin variables.

The above type of energy spectrum for a Fermi liquid may be unstable in certain conditions. The liquid then enters a state characterised by a spectrum of a different type (with an "energy gap"), in which it has the property of superfluidity. This is caused by the existence of forces of attraction between the atoms of the liquid at sufficiently low temperatures. The nature of the resulting spectrum will be discussed in §80, using the model of a Fermi gas with a weak attraction between particles.†

§69. The electronic spectra of metals

The concept of elementary excitations is also needed in order to describe the electronic spectra of solids. The electron shells of the atoms in a crystal interact strongly with one another, and so it is not possible to speak of the energy levels of individual atoms, but only of levels for the assembly of electron shells of all the atoms in the whole body. The nature of the electronic spectrum is different for different types of solid.

An "electron liquid" in a normal (not superconducting) metal has a spectrum of the Fermi type discussed in §68. Such a spectrum has, as we have seen, a structure similar to that of the spectrum of an ideal Fermi gas. In the present case, however, we are concerned with electrons in an external electric field created by the nuclei of the atoms (which we regard as fixed in their equilibrium positions at the lattice sites). We must therefore ascertain first of all the properties which an "ideal gas" of electrons would have in such a field. This problem is equivalent to that of the behaviour of a single electron in an external field periodic in space; the latter problem was first considered by F. Bloch (1929). The periodicity of the field means that it is unchanged in a parallel displacement by any vector of the form $\mathbf{a} = s_1\mathbf{a}_1 + s_2\mathbf{a}_2 + s_3\mathbf{a}_3$, where $\mathbf{a}_1$, $\mathbf{a}_2$, $\mathbf{a}_3$ are the basic lattice vectors:

$$U(\mathbf{r}+\mathbf{a}) = U(\mathbf{r}). \tag{69.1}$$

Hence the Schrödinger's equation which describes the motion of an electron in such a field is also invariant under any transformation $\mathbf{r} \to \mathbf{r}+\mathbf{a}$.

† When the temperature is sufficiently low this effect must ultimately occur in liquid He^3 (as shown by L. P. Pitaevskiĭ, 1959). The reason is that in the interaction of neutral atoms there is always a range of distances (large compared with atomic dimensions) where there is attraction, called the van der Waals attraction; see *Quantum Mechanics*, §89.

From this it follows that, if $\psi(\mathbf{r})$ is the wave function of a stationary state, then $\psi(\mathbf{r}+\mathbf{a})$ is also a solution of SCHRÖDINGER's equation, and describes the same state of the electron. This means that the two functions must be the same apart from a constant factor: $\psi(\mathbf{r}+\mathbf{a}) = \text{constant}\times\psi(\mathbf{r})$. It is evident that the constant must be of unit modulus, since otherwise the wave function would tend to infinity on repeating the displacement through $\mathbf{a}$ (or $-\mathbf{a}$) an unlimited number of times. The general form of a function having this property is

$$\psi_{n\mathbf{k}}(\mathbf{r}) = e^{i\mathbf{k}\cdot\mathbf{r}}u_{n\mathbf{k}}(\mathbf{r}), \tag{69.2}$$

where $\mathbf{k}$ is an arbitrary (real) constant vector and $u_{n\mathbf{k}}$ a periodic function:

$$u_{n\mathbf{k}}(\mathbf{r}+\mathbf{a}) = u_{n\mathbf{k}}(\mathbf{r}). \tag{69.3}$$

For a given value of $\mathbf{k}$, SCHRÖDINGER's equation has in general an infinity of different solutions, corresponding to an infinite discrete set of different values of the electron energy $\varepsilon(\mathbf{k})$, and labelled by the suffix n in $\psi_{n\mathbf{k}}$. A similar suffix (frequently called the *energy band number*) must be added to the various branches of the function $\varepsilon = \varepsilon_n(\mathbf{k})$.†

All the functions $\psi_{n\mathbf{k}}$ with different n or $\mathbf{k}$ are, of course, orthogonal. In particular, the orthogonality of the functions $u_{n\mathbf{k}}$ follows from that of the $\psi_{n\mathbf{k}}$ with different n and the same $\mathbf{k}$, and because of their periodicity the ntegration need be taken only over the volume v of one unit lattice cell. With the appropriate normalisation,

$$\int u_{n'\mathbf{k}}{}^{*}u_{n\mathbf{k}}\,dv = \delta_{nn'}. \tag{69.4}$$

The significance of the vector $\mathbf{k}$ is that its value determines the behaviour of the wave function under translation, this function being multiplied by $e^{i\mathbf{k}\cdot\mathbf{a}}$ when $\mathbf{r}\to\mathbf{r}+\mathbf{a}$:

$$\psi_{n\mathbf{k}}(\mathbf{r}+\mathbf{a}) = e^{i\mathbf{k}\cdot\mathbf{a}}\psi_{n\mathbf{k}}(\mathbf{r}). \tag{69.5}$$

Hence it follows immediately that the value of $\mathbf{k}$ is by definition non-unique: values differing by a vector of the form $2\pi\mathbf{b}$ (where $\mathbf{b}$ denotes, as in §65, any vector of the reciprocal lattice) lead to the same behaviour of the wave function under translation (since the factor $e^{i(\mathbf{k}+2\pi\mathbf{b})\cdot\mathbf{a}} = e^{i\mathbf{k}\cdot\mathbf{a}}$). In other words, such values of $\mathbf{k}$ are physically equivalent; they correspond to the same state of the electron, i.e. the same wave function. We may say that the functions $\psi_{n\mathbf{k}}$ are periodic (with the periods of the reciprocal lattice) with respect to the suffix $\mathbf{k}$:

$$\psi_{n,\,\mathbf{k}+2\pi\mathbf{b}}(\mathbf{r}) = \psi_{n\mathbf{k}}(\mathbf{r}). \tag{69.6}$$

† The general properties of the multi-sheet hypersurface $\varepsilon = \varepsilon_n(k_x, k_y, k_z)$ resulting from the symmetry of the lattice are the same as those of the phonon hypersurface $\omega = \omega(k_x, k_y, k_z)$ mentioned in §65.

The energy is likewise periodic:

$$\varepsilon_n(\mathbf{k}+2\pi\mathbf{b}) = \varepsilon_n(\mathbf{k}). \tag{69.7}$$

The functions (69.2) bear a certain similarity to the wave functions of the free electron, $\psi = \text{constant}\times e^{i\mathbf{p}\cdot\mathbf{r}/\hbar}$, the conserved momentum being represented by the constant vector $\mathbf{p} = \hbar\mathbf{k}$. As with the phonon, we are again led to the concept of the "quasi-momentum" of an electron in a periodic field. It should be emphasised that there is no actual conserved momentum in this case, since there is no law of conservation of momentum in an external field. Nevertheless, it is worth noting that an electron in a periodic field is still characterised by a certain constant vector.†

All the physically different values of the vector $\mathbf{k}/2\pi$ lie in a single unit cell of the reciprocal lattice. The "volume" of this cell is $1/v$, where v is the volume of a unit cell in the crystal lattice itself (see §135). On the other hand, the volume of $\mathbf{k}$-space, divided by $(2\pi)^3$, gives the number of corresponding states "belonging" to a finite (unit) volume of the body. Thus the number of states (per unit volume of the crystal) included in each energy band is $1/v$, i.e. equal to the number of unit cells.

Let us next consider two electrons in a periodic field. Considering them together as one system with wave function $\psi(\mathbf{r}_1, \mathbf{r}_2)$, we find that, under a parallel displacement ($\mathbf{r}_1 \to \mathbf{r}_1+\mathbf{a}$, $\mathbf{r}_2 \to \mathbf{r}_2+\mathbf{a}$), this function must be multiplied by a factor of the form $e^{i\mathbf{k}\cdot\mathbf{a}}$, where $\mathbf{k}$ may be called the quasi-momentum of the system. On the other hand, when the distance between the electrons is large, $\psi(\mathbf{r}_1, \mathbf{r}_2)$ reduces to the product of the wave functions of the individual electrons and is multiplied by $e^{i\mathbf{k}_1\cdot\mathbf{a}}e^{i\mathbf{k}_2\cdot\mathbf{a}}$ in the translation.

The equation $e^{i\mathbf{k}\cdot\mathbf{a}} = e^{i(\mathbf{k}_1+\mathbf{k}_2)\cdot\mathbf{a}}$ shows that

$$\mathbf{k} = \mathbf{k}_1+\mathbf{k}_2+2\pi\mathbf{b}.$$

† In a stationary state with a given quasi-momentum $\hbar\mathbf{k}$, the true momentum can take an infinite number of values of the form $\hbar(\mathbf{k}+2\pi\mathbf{b})$ with different probabilities. This follows from the fact that the expansion of a periodic function in Fourier series has the form

$$u_{n\mathbf{k}} = \sum_{\mathbf{b}} a_{n\mathbf{k}\mathbf{b}} e^{2\pi i\mathbf{b}\cdot\mathbf{r}},$$

and so the expansion of $\psi_{n\mathbf{k}}$ (69.2) in plane waves is

$$\psi_{n\mathbf{k}} = \sum_{\mathbf{b}} a_{n\mathbf{k}\mathbf{b}} e^{i(\mathbf{k}+2\pi\mathbf{b})\cdot\mathbf{r}}.$$

The property (69.6) signifies that the coefficients in this expansion must depend on $\mathbf{k}$ and $\mathbf{b}$ only through the sum $\mathbf{k}+2\pi\mathbf{b}$, so that we can write

$$\psi_{n\mathbf{k}} = \sum_{\mathbf{b}} a_{n,\,\mathbf{k}+2\pi\mathbf{b}} e^{i(\mathbf{k}+2\pi\mathbf{b})\cdot\mathbf{r}}. \tag{69.2a}$$

The two properties (69.5) and (69.6) appear explicitly from this representation of the wave function. Equating (69.2) and (69.2a) and integrating over the volume of the unit cell, we obtain

$$a_{n\mathbf{k}} = \frac{1}{v}\int u_{n\mathbf{k}}(\mathbf{r})\, dv.$$

Hence, in particular, it follows that, in a collision of two electrons moving in a periodic field, the sum of their quasi-momenta is conserved apart from a vector of the reciprocal lattice:

$$\mathbf{k}_1+\mathbf{k}_2 = \mathbf{k}_1'+\mathbf{k}_2'+2\pi\mathbf{b}.$$

A further analogy between the quasi-momentum and the true momentum is seen when we determine the mean velocity of the electron. To calculate this, we must know the velocity operator $\hat{\mathbf{v}} = \dot{\hat{\mathbf{r}}}$ in the $\mathbf{k}$ representation. The operators in this representation act on the coefficients $c_{n\mathbf{k}}$ in an expansion of an arbitrary wave function ψ in terms of the eigenfunctions $\psi_{n\mathbf{k}}$ (69.2):

$$\psi = \sum_n \int c_{n\mathbf{k}}\psi_{n\mathbf{k}}\, d^3k. \tag{69.8}$$

Let us first find the operator $\hat{\mathbf{r}}$. We have identically

$$\hat{\mathbf{r}}\psi = \sum_n \int c_{n\mathbf{k}}\mathbf{r}\psi_{n\mathbf{k}}\, d^3k = \sum_n \int c_{n\mathbf{k}}\left(-i\frac{\partial\psi_{n\mathbf{k}}}{\partial\mathbf{k}}+ie^{i\mathbf{k}\cdot\mathbf{r}}\frac{\partial u_{n\mathbf{k}}}{\partial\mathbf{k}}\right) d^3k.$$

In the first term we integrate by parts, while in the second term we expand the function $\partial u_{n\mathbf{k}}/\partial\mathbf{k}$ (which, like $u_{n\mathbf{k}}$ itself, is periodic) in terms of the set of mutually orthogonal functions $u_{n\mathbf{k}}$ with the same $\mathbf{k}$, writing the expansion in the form

$$\frac{\partial u_{n\mathbf{k}}}{\partial\mathbf{k}} \sum_n \Omega^{n\mathbf{k}}_{m\mathbf{k}} u_{m\mathbf{k}}. \tag{69.9}$$

Then we obtain

$$\begin{aligned}\hat{\mathbf{r}}\psi &= \sum_n \int \psi_{n\mathbf{k}} i\frac{\partial c_{n\mathbf{k}}}{\partial\mathbf{k}}\, d^3k + i\sum_{n,\,m}\int c_{n\mathbf{k}}\Omega^{n\mathbf{k}}_{m\mathbf{k}}\psi_{m\mathbf{k}}\, d^3k \\ &= \sum_n \int \left\{ i\frac{\partial c_{n\mathbf{k}}}{\partial\mathbf{k}} + i\sum_m \Omega^{m\mathbf{k}}_{n\mathbf{k}} c_{m\mathbf{k}} \right\} \psi_{n\mathbf{k}}\, d^3k.\end{aligned}$$

On the other hand, from the definition of the operator $\hat{\mathbf{r}}$ we must have

$$\hat{\mathbf{r}}\psi = \sum_n \int (\hat{\mathbf{r}}c_{n\mathbf{k}})\psi_{n\mathbf{k}}\, d^3k.$$

Comparing this with the expression obtained above, we find

$$\hat{\mathbf{r}} = i\partial/\partial\mathbf{k}+i\hat{\Omega}, \tag{69.10}$$

where the operator $\hat{\Omega}$ is specified by its matrix $\Omega^{m\mathbf{k}}_{n\mathbf{k}}$. It is important to notice that this matrix is diagonal with respect to the suffix $\mathbf{k}$.

The velocity operator is obtained, according to the general rules, by commuting the operator $\hat{\mathbf{r}}$ with the Hamiltonian. In the $\mathbf{k}$ representation, the Hamiltonian is just the energy $\varepsilon(\mathbf{k})$ expressed as a function of $\mathbf{k}$. Hence we have

$$\hat{\mathbf{v}} = (i/\hbar)(\varepsilon\hat{\mathbf{r}} - \hat{\mathbf{r}}\,\varepsilon)$$

$$= -\frac{1}{\hbar}\left(\varepsilon\frac{\partial}{\partial\mathbf{k}} - \frac{\partial}{\partial\mathbf{k}}\varepsilon\right) - \frac{1}{\hbar}(\varepsilon\hat{\Omega} - \hat{\Omega}\varepsilon),$$

or

$$\hat{\mathbf{v}} = \frac{1}{\hbar}\frac{\partial\varepsilon(\mathbf{k})}{\partial\mathbf{k}} + i\hat{\dot{\Omega}}. \tag{69.11}$$

The matrix elements of $\dot{\Omega}$ are related to those of Ω by

$$\dot{\Omega}^{n\mathbf{k}}_{m\mathbf{k}} = \frac{1}{\hbar}[\varepsilon_n(\mathbf{k}) - \varepsilon_m(\mathbf{k})]\Omega^{n\mathbf{k}}_{m\mathbf{k}}.$$

We see from this that $\dot{\Omega}^{n\mathbf{k}}_{n\mathbf{k}} = 0$, i.e. $\dot{\Omega}$ has no elements diagonal in the band number.

The mean value of the velocity is equal to the diagonal matrix element of the operator (69.11). Using the above result we have

$$\mathbf{v} = \partial\varepsilon(\mathbf{k})/\hbar\partial\mathbf{k}, \tag{69.12}$$

in complete analogy to the usual classical relation.

Thus we have elucidated the main fundamental differences between the nature of the classification of states for a free electron and an electron in a periodic field. For the former, the particle energy is uniquely determined by its momentum, which can take a continuous and unlimited range of values. For the latter, the continuous parameter is represented by the quasi-momentum, and all its physically non-equivalent values lie in a finite interval—the reciprocal lattice cell. In addition to this continuous parameter, the electron energy depends also on a discrete quantum number, the band number.† In each band the energy takes values in a certain finite interval; it is important to note that different bands may partly overlap (though retaining their "individuality", of course, since a different dispersion relation $\varepsilon = \varepsilon_n(\mathbf{k})$ corresponds to each band).

All these properties apply to the classification of levels in the spectrum of an electron Fermi liquid in a metal, the particles (electrons) being represented by quasi-particles. An important characteristic of this spectrum for any particular metal is the shape and position of the limiting Fermi surface in **k**-space (the reciprocal lattice). This surface may have various and in general complicated forms. It may be singly or multiply connected, closed or open. The latter means that the Fermi surface may enclose certain finite regions of the reciprocal lattice cell, or it may cut off a part of the cell volume bounded by the faces of the cell. If we imagine the Fermi surface to be continued in a periodic manner throughout the reciprocal lattice, each cell will contain

† The dependence on the spin projection (in a non-magnetic metal) is very slight, and we shall ignore it.

similar closed regions, but the open surfaces will run continuously throughout the reciprocal lattice. We shall not pause here to make a detailed study of the topological properties of Fermi surfaces, which affect the kinetic properties of the metal rather than its thermodynamic properties.

It is important to note that the Fermi surface, being a surface of equal energy $\varepsilon_n(\mathbf{k}) = \mu$ (the limiting energy is the same as the chemical potential μ at absolute zero), in general comprises various sheets corresponding to several overlapping bands (the value of μ is, of course, the same for all bands).

In an isotropic "free" Fermi liquid, as discussed in §68, the Fermi surface is a sphere whose radius is determined by the density of the liquid. A similar relation applies to the electron liquid in a metal, but the specific features resulting from the periodicity of the lattice field lead to some changes in the way in which this relation is formulated.

The number of electrons in the metal may conveniently be referred to one unit lattice cell. Let ν be the total number of electrons in the atoms in one cell, and let Δ_{Fn} denote the fraction of the reciprocal lattice cell volume lying "below" the Fermi surface of the nth energy band (i.e. the part of the volume in which $\varepsilon_n(\mathbf{k}) < \mu$). Then the relation

$$\nu - 2l = 2\sum_n \Delta_{Fn} \tag{69.13}$$

holds, where l is some integer and the factor 2 on the right-hand side takes into account, as usual, the two orientations of the quasi-particle spin. The intuitive significance of the particular feature of this relation—the subtraction of an even integer from ν—is clear from the analogy with the spectrum of an ideal Fermi gas (in a periodic field). The number $2l$ corresponds to electrons completely occupying the l lowest bands, so that the position of the limiting energy in the partly occupied bands is determined by the number of electrons which belong only to these bands.

In accordance with the same analogy with the ideal-gas spectrum, we can say that each quasi-particle transports in its motion a charge equal to that of the electron.†

It has already been mentioned in §68 that the function $\varepsilon(\mathbf{p})$ has direct physical significance as the energy of a quasi-particle only in the neighbourhood of the Fermi surface. This neighbourhood also determines the electronic parts of the thermodynamic quantities for the metal (the principal terms in their expansions at temperatures considerably below the degeneracy temperature‡). The difference from an ideal gas arises, in this approximation,

† We may note that on the basis of very general considerations we can certainly exclude the possibility that the effective charge carried by a quasi-particle is not constant but depends on the state of the metal. In an inhomogeneous body this charge would then vary through the body, and this would violate the gauge invariance of the equations of electrodynamics.

‡ For most metals the electron degeneracy temperature is of the order of 10^4 degrees.

only from the different number of states of the quasi-particles near the Fermi surface.†

Let the number of states (per unit volume of the metal) belonging to the energy range $d\varepsilon$ be $\varrho\, d\varepsilon$. The volume element in **p**-space between infinitely close surfaces of equal energy μ and $\mu + d\varepsilon$ is $df\, d\varepsilon/v$, where df is an element of area on the Fermi surface, and v the magnitude of the vector $\mathbf{v} = \partial\varepsilon/\partial\mathbf{p}$ normal to that surface (the quasi-particle "velocity" on the Fermi surface). Hence

$$\varrho = \frac{2}{(2\pi\hbar)^3} \sum_n \int \frac{df_n}{v_n}, \tag{69.14}$$

where n is the band number and the integration is over the whole of the Fermi surface within one reciprocal lattice cell (when the Fermi surface is an open one, the cell faces are not, of course, included in the region of integration).

The quantity (69.14) replaces in the thermodynamic quantities the expression which for a gas of free particles (where the Fermi surface is a sphere of radius p_0) has the form

$$\frac{2}{(2\pi\hbar)^3} \frac{4\pi p_0^2}{p_0/m} = \frac{m p_0}{\pi^2 \hbar^3}.$$

For example, the thermodynamic potential Ω of a metal is (cf. (57.2))

$$\Omega = \Omega_0 - \frac{1}{6}\pi^2 \varrho V T^2; \tag{69.15}$$

the term Ω_0 includes the lattice part of the potential and the contribution from the electrons for $T = 0$. Regarding the second term in (69.15) as a small correction to Ω_0, we can (cf. §§57 and 64) write down a similar formula for the thermodynamic potential Φ:

$$\Phi = \Phi_0 - \frac{1}{6}\pi^2 \varrho V T^2, \tag{69.16}$$

where now ϱ and V are assumed to be expressed in terms of P and T (in the "zero-order approximation").

Determining the entropy from (69.16), and then the specific heat, we find for the electron contribution to the latter

$$C_e = \tfrac{1}{3}\pi^2 \varrho V T. \tag{69.17}$$

The total specific heat of the metal consists of the electron and lattice parts. The latter is proportional to T^3 (for $T \ll \Theta$), so that at sufficiently low temperatures the electron contribution to the specific heat becomes predominant.

For the same reason, the electron contribution to the thermal expansion of the metal also becomes predominant in this range of temperatures.

† To avoid misunderstanding, we should emphasise that formulae (68.11)–(68.14) derived in §68 for a Fermi liquid not in an external field do not, of course, apply to an electron liquid in a metal.

Determining from (69.16) the volume $V = \partial\Phi/\partial P$ and hence the thermal expansion coefficient $\alpha = (1/V)(\partial V/\partial T)_P$, we find

$$\alpha_e = -T\frac{\pi^2}{3V}\frac{\partial}{\partial P}(V\varrho). \tag{69.18}$$

It may be noted that here, as also in the high-temperature range (see §64), the ratio α/C is independent of temperature.

The interaction between electrons and lattice vibrations (phonons) causes no qualitative change in the energy spectrum of a normal metal, and is important only as regards kinetic effects, but it causes the appearance of a certain effective attraction between the electrons, the mechanism of which may be described as an exchange of a phonon in the mutual scattering of two electrons. When this effect is sufficiently large, it may predominate over the Coulomb repulsion of the electrons, and so there may occur at low temperatures the change in the nature of the spectrum mentioned at the end of §68 and the appearance of *superconductivity*.

§70. The electronic spectra of solid dielectrics

The principal feature of the energy spectrum of a dielectric non-paramagnetic crystal (a topic first discussed in 1931, by Ya. I. Frenkel') is that even the first excited level is at a finite distance from the ground state—that is, there is an "energy gap" between the ground level and the spectrum of excited levels. The existence of this gap, which in ordinary dielectrics is of the order of a few electron-volts, means that the "electronic parts" of the thermodynamic quantities are exponentially small (proportional to $e^{-\Delta/T}$, where Δ is the width of the gap).

An elementary excitation in the spectrum under consideration can be intuitively described as an excited state of an individual atom, but this state can not be assigned to any particular atom; it is "collectivised" and is propagated in the crystal in the form of an "excitation wave", as it were jumping from one atom to another. As in other cases, these excitations may be regarded as quasi-particles, here called *excitons*, which have definite energies and quasi-momenta. Like all excitations that can appear singly, excitons have integral angular momenta and obey Bose statistics.

For a given value of the quasi-momentum $\mathbf{p}$, the energy of an exciton can take a discrete series of different values $\varepsilon_n(\mathbf{p})$. The components of the quasi-momentum can take, as we know, a continuous series of values in finite intervals, and for each n the function $\varepsilon_n(\mathbf{p})$ gives a "band" of exciton energy values; different bands may partly overlap. The least possible value of the functions $\varepsilon_n(\mathbf{p})$, i.e. the least possible energy of the exciton, as already mentioned, is not zero.

As well as excitons, excitations of another type may be present in a dielectric. These may be regarded as resulting from the ionisation of individual atoms. Each such ionisation causes the appearance in the dielectric of two independently propagated "particles": an electron and a "hole". The latter is a "lack" of one electron in an atom, and therefore behaves as a positively charged particle. Here again, when speaking of the motion of electrons and holes in a crystal, we really mean of course certain "collective" excited states of the electrons in the dielectric which (unlike the exciton states) are accompanied by the transfer of a negative or positive unit charge.

Electrons and holes have half-integral spin and obey Fermi statistics, but this does not mean that the electron-hole spectrum of a dielectric resembles the Fermi-type spectrum described in §68. A characteristic feature of the latter is the existence of a limiting value p_0 of the momentum, but in the present case there is no such quantity, and an electron and hole appearing simultaneously can have entirely arbitrary quasi-momenta.

The electron and the hole have a Coulomb interaction. The spectrum of eigenvalues of the energy of particles with Coulomb attraction consists of a discrete series of negative levels which become increasingly close together as the value zero is approached, at which a continuous spectrum of positive values begins. In the present case the discrete levels correspond to exciton excitations ("bound" electrons and holes) and the continuous levels to electron-hole excitations. We can therefore say that, for a given value of the quasi-momentum, the possible values of the exciton energy form a discrete series which become increasingly close together as the energy increases and pass into a continuous series of values corresponding to a freely moving electron and hole.

In the foregoing discussion the electron spectrum has been considered in isolation from the motion of the atomic nuclei, which have been assumed fixed at the crystal lattice sites. This assumption is by no means always justified. The interaction of the electrons with the lattice vibrations may be so strong that the above treatment is inadmissible. In a dielectric, the interaction of an electron with the vibrations of the lattice deforms it in the vicinity of the electron, and this deformation, of course, considerably affects also the motion of the electron itself. An electron together with the lattice deformation which it causes is called a *polaron*, a concept first used by S. I. PEKAR (1946).

§71. Negative temperatures

Let us now consider some peculiar effects related to the properties of paramagnetic dielectrics. In such substances the atoms have angular momenta, and therefore magnetic moments, which are more or less freely oriented. The interaction of these moments (magnetic or exchange interaction, depending on their distance apart) brings about a new "magnetic" spectrum which is superposed on the ordinary dielectric spectrum.

This new spectrum lies entirely within a finite interval of energy, of the order of magnitude of the energy of interaction of the magnetic moments of all the atoms of the body, lying at fixed distances apart at the crystal lattice sites; the amount of this energy per atom may be from tenths of a degree to hundreds of degrees. In this respect the magnetic energy spectrum is completely different from the ordinary spectra, which, owing to the presence of the kinetic energy of the particles, extend to arbitrarily high energy values.

Because of this property we can consider an interval of temperatures large compared with the maximum possible interval of energy values per atom. The free energy F_{mag} pertaining to the magnetic part of the spectrum is calculated in exactly the same way as in §32.

Let E_n be the energy levels of the system of interacting moments. Then we have for the required partition function

$$\begin{aligned} Z_{\text{mag}} &= \sum_n e^{-E_n/T} \\ &\cong \sum_n \left(1 - \frac{E_n}{T} + \frac{E_n{}^2}{2T^2}\right). \end{aligned}$$

Here, as in §32, a formal expansion in powers of the quantity E_n/T, which is not in general small, will give (after taking logarithms) an expansion in terms of a small quantity $\sim E_n/NT$, where N is the number of atoms. The total number of levels in the spectrum under consideration is finite and equal to the number of all possible combinations of orientations of the atomic moments; for example, if all the moments are equal, this number is g^N, where g is the number of possible orientations of an individual moment relative to the lattice. Denoting by a bar here the ordinary arithmetic mean, we can write Z_{mag} as

$$Z_{\text{mag}} = g^N\left(1 - \frac{1}{T}\overline{E_n} + \frac{1}{2T^2}\overline{E_n{}^2}\right).$$

Finally, taking logarithms and again expanding in series with the same accuracy, we obtain for the free energy the expression

$$\begin{aligned} F_{\text{mag}} &= -T \log Z_{\text{mag}} \\ &= -NT \log g + \overline{E_n} - \frac{1}{2T}\overline{(E_n - \overline{E_n})^2}. \end{aligned} \qquad (71.1)$$

Hence the entropy is

$$S_{\text{mag}} = N \log g - \frac{1}{2T^2}\overline{(E_n - \overline{E_n})^2}, \qquad (71.2)$$

the energy

$$E_{\text{mag}} = \overline{E}_n - \frac{1}{T}\overline{(E_n - \overline{E}_n)^2}, \tag{71.3}$$

and the specific heat

$$C_{\text{mag}} = \frac{1}{T^2}\overline{(E_n - \overline{E}_n)^2}. \tag{71.4}$$

We shall regard the atomic magnetic moments fixed at the lattice sites and interacting with one another as a single isolated system, ignoring its interaction with the lattice vibrations, which is usually very weak. Formulae (71.1)–(71.4) determine the thermodynamic quantities for this system at high temperatures.

The proof given in §10 that the temperature is positive was based on the condition for the system to be stable with respect to the occurrence of internal macroscopic motions within it. But the system of moments here considered is by its nature incapable of macroscopic motion, and so the previous arguments do not apply to it; nor does the proof based on the normalisation condition for the Gibbs distribution (§36), since in the present case the system has only a finite number of energy levels, themselves finite, and so the normalisation sum converges for any value of T.

Thus we have the interesting result that the system of interacting moments may have either a positive or a negative temperature. Let us examine the properties of the system at various temperatures.

At $T = 0$, the system is in its lowest quantum state, and its entropy is zero. As the temperature increases, the energy and entropy of the system increase monotonically. At $T = \infty$, the energy is $\overline{E}_n$ and the entropy reaches its maximum value $N \log g$; these values correspond to a distribution with equal probability over all quantum states of the system, which is the limit of the Gibbs distribution as $T \to \infty$.

The temperature $T = -\infty$ is physically identical with $T = \infty$; the two values give the same distribution and the same values of the thermodynamic quantities for the system. A further increase in the energy of the system corresponds to an increase in the temperature from $T = -\infty$, with decreasing absolute magnitude since the temperature is negative. The entropy decreases monotonically (Fig. 9).† Finally, at $T = 0-$ the energy reaches its greatest value and the entropy returns to zero, the system then being in its highest quantum state.

Thus the region of negative temperatures lies not "below absolute zero" but "above infinity". In this sense we can say that negative temperatures are "higher" than positive ones. This is in accordance with the fact that, when a

† The curve $S = S(E)$ is symmetrical near its maximum, but in general need not be symmetrical far from the maximum.

system at a negative temperature interacts with one at a positive temperature (i.e. the lattice vibrations), energy must pass from the former to the latter system; this is easily seen by the same method as that used in §9 to discuss the exchange of energy between bodies at different temperatures.

States with negative temperature can be attained in practice in a paramagnetic system of nuclear moments in a crystal where the relaxation time t_2 for

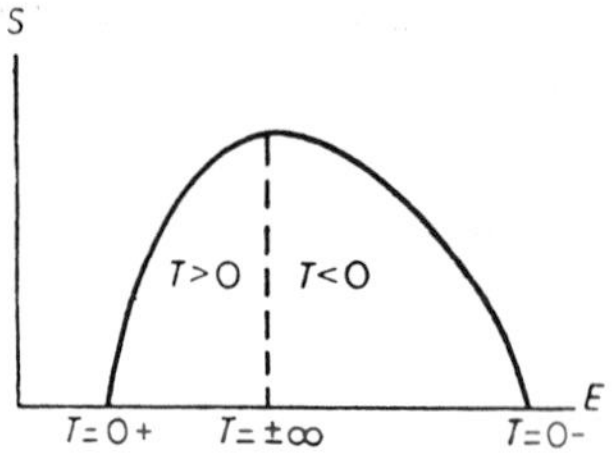

FIG. 9

the interaction between nuclear spins is very small compared with the relaxation time t_1 for the spin-lattice interaction (E. M. PURCELL and R. V. POUND 1951). Let the crystal be magnetised in a strong magnetic field, and let the direction of the field then be reversed so quickly that the spins "cannot follow it". The system is thus in a non-equilibrium state, with an energy which is obviously higher than $\overline{E_n}$. During a time of the order of t_2, the system reaches an equilibrium state with the same energy. If the field is then adiabatically removed, the system remains in the equilibrium state, which will clearly have a negative temperature. The subsequent exchange of energy between the spin system and the lattice, whereby their temperatures are equalised, takes place in a time of the order of t_1.

CHAPTER VII

NON-IDEAL GASES

§72. Deviations of gases from the ideal state

THE equation of state of an ideal gas can often be applied to actual gases with sufficient accuracy. This approximation may, however, be inadequate, and it is then necessary to take account of the deviations of an actual gas from the ideal state which result from the interaction between its component molecules.

Here we shall do this on the assumption that the gas is still so rarefied that triple, quadruple etc. collisions between molecules may be neglected, and their interaction may be assumed to occur only through binary collisions.

To simplify the formulae, let us first consider a monatomic actual gas. The motion of its particles may be treated classically, so that its energy has the form

$$E(p, q) = \sum_{a=1}^{N} \frac{p_a^2}{2m} + U, \tag{72.1}$$

where the first term is the kinetic energy of the N atoms of the gas, and U is the energy of their mutual interaction. In a monatomic gas, U is a function only of the distances between the atoms. The partition function $\int e^{-E(p,q)/T}\,d\Gamma$ becomes the product of the integral over the momenta of the atoms and the integral over their co-ordinates. The latter integral is

$$\int \cdots \int e^{-U/T}\,dV_1 \ldots dV_N,$$

where the integration over each $dV_a = dx_a\,dy_a\,dz_a$ is taken over the whole volume V occupied by the gas. For an ideal gas, $U = 0$, and this integral would be simply V^N. It is therefore clear that, on calculating the free energy from the general formula (31.5), we obtain

$$F = F_{\text{id}} - T \log \frac{1}{V^N} \int \cdots \int e^{-U/T}\,dV_1 \ldots dV_N, \tag{72.2}$$

where F_{id} is the free energy of an ideal gas.

Adding and subtracting unity in the integrand and using the fact that $\int dV_1 \ldots dV_N = V^N$, we can rewrite formula (72.2) as

$$F = F_{\text{id}} - T \log \left\{ \frac{1}{V^N} \int \cdots \int (e^{-U/T} - 1)\,dV_1 \ldots dV_N + 1 \right\}. \tag{72.3}$$

For the subsequent calculations we make use of the following formal device. Let us suppose that the gas is not only sufficiently rarefied but also so small in quantity that not more than one pair of atoms may be assumed to be colliding in the gas at any one time. This assumption does not affect the generality of the resulting formulae, since we know from the additivity of the free energy that it must have the form $F = Nf(T, V/N)$ (see §24), and therefore the formulae deduced for a small quantity of gas are necessarily valid for any quantity.

The interaction between atoms is very small except when the two atoms concerned are very close together, i.e. are almost colliding. The integrand in (72.3) is therefore appreciably different from zero only when some pair of atoms are very close together. According to the above assumption, not more than one pair of atoms can satisfy this condition at any one time, and this pair can be selected from N atoms in $\frac{1}{2}N(N-1)$ ways. Consequently, the integral in (72.3) may be written

$$\tfrac{1}{2}N(N-1)\int \ldots \int (e^{-U_{12}/T}-1)\, dV_1 \ldots dV_N,$$

where U_{12} is the energy of interaction of the two atoms (it does not matter which two, as they are all identical); U_{12} depends only on the co-ordinates of two atoms, and we can therefore integrate over the remaining co-ordinates, obtaining V^{N-2}. We can also, of course, write N^2 instead of $N(N-1)$, since N is very large; substituting the resulting expression in (72.3) in place of the integral, and using the fact that $\log(1+x) \approx x$ for $x \ll 1$, we have†

$$F = F_{\text{id}} - \frac{TN^2}{2V^2}\iint (e^{-U_{12}/T}-1)\, dV_1\, dV_2,$$

where $dV_1\, dV_2$ is the product of differentials of the co-ordinates of the two atoms.

But U_{12} is a function only of the distance between the two atoms, i.e. of the differences of their co-ordinates. Thus, if the co-ordinates of the two atoms are expressed in terms of the co-ordinates of their centre of mass and their relative co-ordinates, U_{12} will depend only on the latter (the product of whose differentials will be denoted by dV). We can therefore integrate with respect to the co-ordinates of the centre of mass, again obtaining the volume V. The final result is

$$F = F_{\text{id}} + N^2TB(T)/V, \tag{72.4}$$

where

$$B(T) = \tfrac{1}{2}\int (1-e^{-U_{12}/T})\, dV. \tag{72.5}$$

† We shall see later that the first term in the argument of the logarithm in (72.3) is proportional to N^2/V. The expansion in question therefore depends on precisely the assumption made above, that not only the density N/V but also the quantity of the gas is small.

From this we find the pressure $P = -\partial F/\partial V$:

$$P = \frac{NT}{V}\left(1+\frac{NB(T)}{V}\right), \tag{72.6}$$

since $P_{\text{id}} = NT/V$. Equation (72.6) is the equation of state of the gas in the approximation considered.

As we know (§15), the changes in the free energy and the thermodynamic potential resulting from small changes in the external conditions or properties of a body are equal, one being taken at constant volume and the other at constant pressure.

If we regard the deviation of a gas from the ideal state as such a change, we can change directly to Φ from (72.4). To do so, we need only express the volume in terms of the pressure in (72.4) by means of the equation of state for an ideal gas, obtaining

$$\Phi = \Phi_{\text{id}}+NBP. \tag{72.7}$$

The volume may hence be expressed as a function of the pressure:

$$V = \frac{NT}{P}+NB. \tag{72.8}$$

The whole of the foregoing discussion applies to monatomic gases. The same formulae remain valid, however, for polyatomic gases also. In this case the potential energy of interaction of the molecules depends not only on their distance apart but also on their relative orientation. If (as almost always happens) the rotation of the molecules may be treated classically, we can say that U_{12} is a function of the co-ordinates of the centres of mass of the molecules and of rotational co-ordinates (angles) which define the spatial orientation of the molecules. It is easy to see that the only difference from the case of a monatomic gas amounts to the fact that $\mathrm{d}V_a$ must be taken as the product of the differentials of all these co-ordinates of the molecule. But the rotational co-ordinates can always be so chosen that the integral $\int \mathrm{d}V_a$ is again equal to the volume V of the gas. For the integration over the co-ordinates of the centre of mass gives this volume V, while the integration over angles gives a constant, and the angles can always be normalised so that this constant is unity. Thus all the formulae derived in this section have the same form for polyatomic gases, the only difference being that in (72.5) $\mathrm{d}V$ is now the product of the differentials of co-ordinates defining both the distance between two molecules and their relative orientation.†

All the above formulae are meaningful, of course, only if the integral (72.5) converges. For this to be so it is certainly necessary that the forces of interaction

† If the particles in the gas have spin, the form of the function U_{12} depends in general on the orientation of the spins. In that case a summation over spin orientations must be added to the integration with respect to $\mathrm{d}V$.

between the molecules should decrease sufficiently rapidly with increasing distance. If U_{12} decreases at large distances according to a power law $\sim r^{-n}$, we must have† $n > 3$.

If this condition is not satisfied, a gas consisting of identical particles can not exist as a homogeneous body. In this case every region of matter will be subject to very large forces exerted by distant parts of the gas. The regions near to and far from the boundary of the volume occupied by the gas will therefore be in quite different conditions, and so the gas is no longer homogeneous.

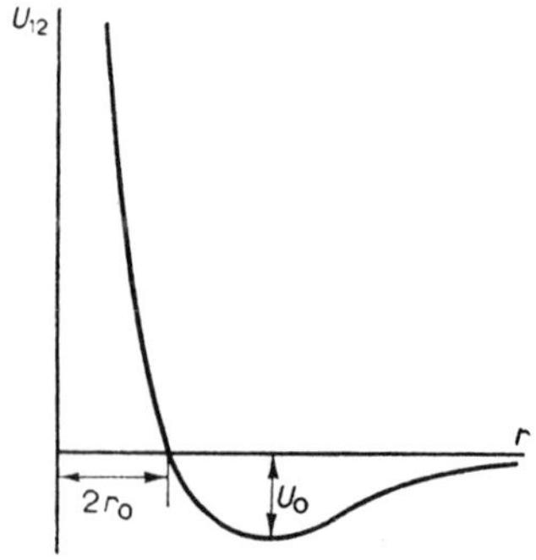

FIG. 10

For monatomic gases the function $U_{12}(r)$ has the form shown in Fig. 10; the abscissa is the distance r between the atoms. At small distances, U_{12} increases with decreasing distance, corresponding to repulsive forces between the atoms; beginning approximately at the place where the curve crosses the abscissa axis, it rises steeply, so that U_{12} rapidly becomes very large, corresponding to the mutual "impenetrability" of the atoms (for which reason the distance r_0 is sometimes called the "radius" of the atom). At large distances, U_{12} increases slowly, approaching zero asymptotically. The increase of U_{12} with distance corresponds to a mutual attraction of the atoms. The minimum of U_{12} corresponds to a "stable" equilibrium. The absolute value U_0 of the energy at this point is usually small, being of the order of the critical temperature of the substance.

For a polyatomic gas, the interaction energy has a similar form, but it cannot, of course, be represented by the curve in Fig. 10, since it is a function of a large number of variables.

This information as to the nature of the function U_{12} is sufficient to determine the sign of $B(T)$ in the limiting cases of high and low temperatures.

† This condition is always satisfied for atomic and molecular gases: the forces of interaction between electrically neutral atoms or molecules (including dipoles), when averaged over the relative orientations of the particles, decrease at large distances as $U_{12} \sim 1/r^6$; see *Quantum Mechanics*, §89.

At high temperatures ($T \gg U_0$) we have $|U_{12}|/T \ll 1$ throughout the region $r > 2r_0$, and the integrand in $B(T)$ (72.5) is almost zero. Thus the value of the integral is mainly determined by the region $r < 2r_0$, where U_{12}/T is large and positive; in this region, therefore, the integrand is positive, and so the integral itself is positive. Thus $B(T)$ is positive at high temperatures.

At low temperatures ($T \ll U_0$), on the other hand, the region $r > 2r_0$ is the important one in the integral, and in this region U_{12}/T is large and negative. At sufficiently low temperatures, therefore, $B(T)$ must be negative, and its temperature dependence is essentially given by the exponential factor $-e^{U_0/T}$.

Being positive at high temperatures and negative at low temperatures, $B(T)$ must pass through zero at some temperature.†

Finally, let us consider a Joule–Thomson process involving a non-ideal gas. The variation of temperature during the process is given by the derivative

$$\left(\frac{\partial T}{\partial P}\right)_W = \frac{1}{C_p}\left[T\left(\frac{\partial V}{\partial T}\right)_P - V\right]; \qquad (72.9)$$

see (18.2). For an ideal gas this derivative is of course zero, but for a gas with the equation of state (72.8) we have

$$\left(\frac{\partial T}{\partial P}\right)_W = \frac{N}{C_p}\left(T\frac{\mathrm{d}B}{\mathrm{d}T} - B\right) = \frac{N}{2C_p}\int\left[e^{-U_{12}/T}\left(1 - \frac{U_{12}}{T}\right) - 1\right]\mathrm{d}V. \qquad (72.10)$$

As in the discussion of $B(T)$, it is easy to see that at high temperatures $(\partial T/\partial P)_W < 0$, i.e. when the gas goes from a higher to a lower pressure in a Joule–Thomson process its temperature rises; at low temperatures, $(\partial T/\partial P)_W > 0$, i.e. the gas temperature falls with the pressure. At a definite temperature for each gas, called the *inversion point*, the Joule–Thomson effect must therefore change sign.‡

PROBLEMS

PROBLEM 1. Determine $B(T)$ for a gas whose particles repel one another according to $U_{12} = \alpha/r^n$ ($n > 3$).

SOLUTION. In (72.5) we put $\mathrm{d}V = 4\pi r^2\,\mathrm{d}r$ and integrate by parts with respect to r from 0 to ∞; the substitution $\alpha r^{-n} = x$ then reduces the integral to a gamma function:

$$B(T) = \frac{2\pi}{3}\left(\frac{\alpha}{T}\right)^{3/n}\Gamma\left(1 - \frac{3}{n}\right).$$

† The temperature T_B for which $B(T_B) = 0$ is called the *Boyle point*. If PV/T is plotted against P for various given T, the isotherm $T = T_B$ has a horizontal tangent as $P \to 0$, and separates isotherms with positive and negative initial slopes; all the isotherms start from the point $PV/T = 1$, $P = 0$.

‡ It will be recalled that we are considering a gas which is only slightly non-ideal, so that the pressure is relatively low. The result that the inversion point is independent of pressure is valid only in this approximation; cf. §74, Problem 4.

PROBLEM 2. The *fugacity* of a gas is the pressure P^* which it would have for given values of the temperature and chemical potential if so rarefied that it could be regarded as an ideal gas. Determine the fugacity of a gas with the thermodynamic potential (72.7).

SOLUTION. The chemical potential of the gas is (with μ_{id} given by (42.6))

$$\mu = \mu_{\text{id}} + BP = T \log P + \chi(T) + BP.$$

Equating this to $T \log P^* + \chi(T)$ by the definition of the fugacity, we have to the same accuracy as that of (72.7)

$$P^* = P\left(1+\frac{BP}{T}\right) = \frac{NT}{V}\left(1+\frac{2NB}{V}\right).$$

§73. Expansion in powers of the density

The equation of state (72.6) derived in §72 consists essentially of the first two terms in an expansion of the pressure in powers of $1/V$:

$$P = \frac{NT}{V}\left(1+\frac{NB(T)}{V}+\frac{N^2C(T)}{V^2}+\dots\right). \tag{73.1}$$

The first term in the expansion corresponds to an ideal gas, i.e. no interaction between molecules. The second term is obtaining by taking into account the interaction between pairs of molecules, and the subsequent terms must involve the interactions between groups of three, four etc. molecules.†

The coefficients B, C,... in the expansion (73.1) are called the second, third etc. *virial coefficients*. To determine these quantities, it is convenient to begin by calculating the potential Ω, not the free energy. Let us again consider a monatomic gas, and start from the general formula (35.5), which for a gas consisting of identical particles becomes

$$e^{-\Omega/T} = \sum_{N=0}^{\infty} \frac{1}{N!} e^{\mu N/T} \int e^{-E_N(p,q)/T} \, d\Gamma_N. \tag{73.2}$$

The factor $1/N!$ is included and the integration is then taken simply over the whole phase space of the system of N particles; cf. (31.7).

In the successive terms of the sum over N, the energy $E_N(p, q)$ has the following forms. For $N = 0$, of course, $E_0(p, q) \equiv 0$. For $N = 1$, it is simply the kinetic energy of one atom:

$$E_1(p, q) = p^2/2m.$$

For $N = 2$ it consists of the kinetic energy of the two atoms and the energy of their interaction:

$$E_2(p, q) = \sum_{a=1}^{2} \frac{p_a^{\,2}}{2m} + U_{12}.$$

† The dimensionless small parameter with respect to which the expansion is made is actually the ratio Nv_0/V of the "volume" v_0 of one molecule to the gas volume per molecule V/N.

Similarly,

$$E_3(p, q) = \sum_{a=1}^{3} \frac{p_a^2}{2m} + U_{123},$$

where U_{123} is the interaction energy of three atoms (which in general is not equal to the sum $U_{12}+U_{13}+U_{23}$), and so on.

We substitute these expressions in (73.2) and use the notation

$$\xi = \frac{e^{\mu/T}}{(2\pi\hbar)^3} \int e^{-p^2/2mT} \, d^3p = \left(\frac{mT}{2\pi\hbar^2}\right)^{3/2} e^{\mu/T}. \tag{73.3}$$

We shall see below that this is simply equal to P_{id}/T, where P_{id} is the pressure of an ideal gas for given T and V. We obtain

$$\Omega = -T \log \left\{1+\xi V+\frac{\xi^2}{2!} \iint e^{-U_{12}/T} \, dV_1 \, dV_2 + \right.$$
$$\left. + \frac{\xi^3}{3!} \iiint e^{-U_{123}/T} \, dV_1 \, dV_2 \, dV_3 + \ldots \right\}.$$

Each of the $U_{12}, U_{123}, \ldots$ is a function only of the distances between the atoms; hence, by using the relative co-ordinates of the atoms (relative to the first atom, say), we reduce the multiplicity of the integrals by one, with a further factor V entering:

$$\Omega = -PV = -T \log \left\{1+\xi V+\frac{\xi^2 V}{2!} \int e^{-U_{12}/T} \, dV_2 + \right.$$
$$\left. + \frac{\xi^3 V}{3!} \iint e^{-U_{123}/T} \, dV_2 \, dV_3 + \ldots \right\}.$$

Finally, we expand this expression in powers of ξ; the resulting series can be written as

$$P = T \sum_{n=1}^{\infty} \frac{J_n}{n!} \xi^n, \tag{73.4}$$

where

$$J_1 = 1, \qquad J_2 = \int (e^{-U_{12}/T} - 1) \, dV_2,$$
$$J_3 = \iint (e^{-U_{123}/T} - e^{-U_{12}/T} - e^{-U_{13}/T} - e^{-U_{23}/T} + 2) \, dV_2 \, dV_3, \tag{73.5}$$

etc. The structure of the integrals J_n is evident; the integrand in J_n is appreciably different from zero only if n atoms are close together, i.e. in a collision of n atoms.

Differentiating (73.4) with respect to μ, we obtain the number of particles in the gas, since $N = -(\partial\Omega/\partial\mu)_{T,V} = V(\partial P/\partial\mu)_{T,V}$. Bearing in mind that by definition (73.3) $\partial\xi/\partial\mu = \xi/T$, we have

$$N = V \sum_{n=1}^{\infty} \frac{J_n}{(n-1)!} \xi^n. \tag{73.6}$$

The two equations (73.4) and (73.6) give in parametric form (with parameter ξ) the relation between P, V and T, i.e. the equation of state of the gas. Eliminating ξ, we can obtain the equation of state in the form of the series (73.1) with any desired number of terms.†

§74. Van der Waals' formula

In gases the interaction between molecules is very weak. As this interaction increases, the properties of the gas differ more and more from those of ideal gases, and finally the gas condenses into a liquid. In the latter, the interaction between molecules is strong, and the properties of this interaction (and therefore those of the liquid) depend considerably on the particular liquid concerned. For this reason it is, as already mentioned, impossible to derive any general formulae giving a quantitative description of the properties of aliq uid.

We can, however, find an interpolation formula which gives a qualitative description of the transition between the liquid and gaseous states. This formula must give the correct results in the two limiting cases. For rarefied gases it must become the formulae valid for ideal gases. When the density increases and the gas approaches the liquid state, it must take account of the finite compressibility of the substance. Such a formula will then give a qualitative description of the behaviour of the gas in the intermediate range also.

To derive such a formula, let us examine in more detail the deviations from the ideal state at high temperatures. As in the preceding sections, we shall first consider a monatomic gas; by the same arguments as used previously, all the resulting formulae will be equally applicable to polyatomic gases.

The type of interaction between gas atoms described in §72 (Fig. 10) enables us to determine the form of the leading terms in the expansion of $B(T)$ in inverse powers of the temperature; here we shall assume that the ratio U_0/T is small:

$$U_0/T \ll 1. \tag{74.1}$$

Since U_{12} depends only on the distance r between the atoms, we write in the integral (72.5) $\mathrm{d}V = 4\pi r^2\,\mathrm{d}r$. Dividing the range of integration with respect to r into two parts, we write

$$\int (1-e^{-U_{12}/T})\,\mathrm{d}V = 4\pi \int_0^{2r_0} (1-e^{-U_{12}/T})r^2\,\mathrm{d}r + 4\pi \int_{2r_0}^{\infty} (1-e^{-U_{12}/T})r^2\,\mathrm{d}r.$$

† In the first approximation, $P = T\xi$, $N = V\xi$, whence $P = NT/V = P_{\mathrm{id}}$. In the second approximation, $P = T\xi(1+\frac{1}{2}J_2\xi)$, $N = V\xi(1+J_2\xi)$; eliminating ξ from these equations (with the same accuracy), we have

$$P = \frac{NT}{V} - \frac{N^2T}{2V^2}J_2,$$

in agreement with (72.6).

For values of r between 0 and $2r_0$, the potential energy U_{12} is in general very large. In the first integral we can therefore neglect $e^{-U_{12}/T}$ in comparison with unity. Then this integral is equal to $2b$, where $b = 16\pi r_0^3/3$. If we regard r_0 as the "radius" of the atom, b is four times its "volume" (for polyatomic gases, of course, the constant b is not equal to four times the "volume" of a molecule).

In the second integral, U_{12} nowhere exceeds U_0 in absolute magnitude (Fig. 10). Thus $-U_{12}/T$ is always small compared with unity, since, even when $U_{12} = -U_0$, $-U_{12}/T = U_0/T \ll 1$. We can therefore expand $e^{-U_{12}/T}$ in powers of U_{12}/T and take only the first two terms. The second integral then becomes

$$\frac{1}{T}\int_{2r_0}^{\infty} 4\pi U_{12} r^2 \,\mathrm{d}r.$$

Since U_{12} is negative throughout the range of integration, the integral itself is negative, and we write it as $-2a/T$, where a is a positive constant.

Thus we find

$$B(T) = b - a/T, \tag{74.2}$$

and substitution of this in (72.4) gives for the free energy of the gas

$$F = F_{\mathrm{id}} + N^2(Tb-a)/V. \tag{74.3}$$

Substitution in (72.7) gives for the thermodynamic potential

$$\Phi = \Phi_{\mathrm{id}} + NP(b-a/T). \tag{74.4}$$

The desired interpolation formula can be obtained from (74.3), which itself does not satisfy the necessary conditions, since it does not take account of the finite compressibility of the substance. In (74.3) we substitute the expression for F_{id} from (42.4); this gives

$$F = Nf(T) - NT\log(e/N) - NT(\log V - Nb/V) - N^2a/V. \tag{74.5}$$

In deriving formula (72.4) for the free energy of a gas we assumed that the gas, though not sufficiently rarefied to be regarded as an ideal gas, nevertheless occupies such a large volume that we can neglect triple and higher-order collisions between molecules, so that the distances between molecules are in general considerably larger than their dimensions. We may say that the gas volume V is always considerably greater than Nb, i.e. $Nb/V \ll 1$; using also the fact that $\log(1+x) \cong x$ for $x \ll 1$, we find

$$\begin{aligned}\log(V-Nb) &= \log V + \log(1-Nb/V)\\ &= \log V - Nb/V.\end{aligned}$$

Hence (74.5) may be written

$$\begin{aligned}F &= Nf(T) - NT\log[e(V-Nb)/N] - N^2a/V\\ &= F_{\mathrm{id}} - NT\log(1-Nb/V) - N^2a/V.\end{aligned} \tag{74.6}$$

In this form the conditions stated above are satisfied, since when V is large the formula for the free energy of an ideal gas is obtained, and when V is small the formula shows that the gas cannot be indefinitely compressed, since the argument of the logarithm becomes negative when $V < Nb$.

If the free energy is known, we can determine the gas pressure:

$$P = -\partial F/\partial V = \frac{NT}{V-Nb} - \frac{N^2 a}{V^2}$$

or

$$\left(P + \frac{N^2 a}{V^2}\right)(V-Nb) = NT. \tag{74.7}$$

This is the required interpolation formula for the equation of state of an actual gas. It is called *van der Waals' equation.*

VAN DER WAALS' formula is, of course, only one of an infinity of possible interpolation formulae which satisfy the requirements stated, and there is no physical reason to select any one of them; VAN DER WAALS' formula is merely the simplest and most convenient.†

From (74.6) we can find the entropy of the gas:

$$S = S_{\mathrm{id}} + N \log(1 - Nb/V), \tag{74.8}$$

and thence the energy $E = F + TS$:

$$E = E_{\mathrm{id}} - N^2 a/V. \tag{74.9}$$

Thus the specific heat $C_v = (\partial E/\partial T)_V$ of a van der Waals gas is equal to that of an ideal gas; it depends only on the temperature and, in particular, may be constant. The specific heat C_p is easily seen (cf. Problem 1) to depend not only on the temperature but also on the volume, and so can not be constant.

The second term in (74.9) corresponds to the energy of interaction of the gas molecules; it is, of course, negative, since on average the forces between molecules are attractive.

PROBLEMS

PROBLEM 1. Find $C_p - C_v$ for a non-ideal gas described by VAN DER WAALS' formula.

SOLUTION. Using formula (16.10) and VAN DER WAALS' equation, we find

$$C_p - C_v = \frac{N}{1 - 2Na(V-Nb)^2/TV^3}.$$

PROBLEM 2. Find the equation of an adiabatic process for a van der Waals gas of constant specific heat C_v.

SOLUTION. Substituting in (74.8) $S_{\mathrm{id}} = N \log V + Nc_v \log T$ (omitting unimportant constants) and putting S = constant, we obtain the relation $(V-Nb)T^{c_v}$ = constant. This differs from the corresponding equation for an ideal gas in that V is replaced by $V-Nb$.

† In actual applications of this formula, the values of the constants a and b must be chosen so as to give the best agreement with experiment. The constant b cannot then be regarded as four times the "molecular volume", even for a monatomic gas.

PROBLEM 3. For a gas of the same kind as in Problem 2, find the change in temperature on expansion into a vacuum from volume V_1 to V_2.

SOLUTION. In an expansion into a vacuum, the energy of the gas remains constant. Thus formula (74.9), with $E_{\text{id}} = Nc_vT$, gives

$$T_2 - T_1 = \frac{Na}{c_v}\left(\frac{1}{V_2} - \frac{1}{V_1}\right).$$

PROBLEM 4. For a van der Waals gas find the temperature dependence of the inversion point for the Joule–Thomson effect.

SOLUTION. The inversion point is determined by the equation $(\partial T/\partial V)_P = T/V$ (see (72.9)). Substitution of T from (74.7) leads to an equation which has to be solved simultaneously with (74.7). Algebraic calculation gives the following dependence of the inversion point on pressure:

$$T_{\text{inv}} = \frac{2a}{9b}(2 \pm \sqrt{(1-3b^2P/a)})^2.$$

For any given pressure $P < a/3b^2$ there are two inversion points, between which the derivative $(\partial T/\partial P)_W$ is positive, while outside this temperature interval it is negative. When $P > a/3b^2$ there are no inversion points and $(\partial T/\partial P)_W < 0$ everywhere.†

§75. Completely ionised gases

The method given above for calculating the thermodynamic quantities for a non-ideal gas is certainly inapplicable to a gas consisting of charged particles with Coulomb interaction, since the integrals which appear in the formulae then diverge. Such gases must therefore be treated separately.

Let us consider a completely ionised gas or *plasma*. The charges on its particles will be denoted by $z_a e$, where the suffix a refers to the different kinds of ion; e is the unit charge and the z_a are positive or negative integers. Also, let n_{a0} be the number of ions of the ath kind per unit volume of the gas. The gas as a whole is, of course, electrically neutral, so that

$$\sum_a z_a n_{a0} = 0. \tag{75.1}$$

We shall suppose that the gas does not deviate greatly from the ideal state. For this to be so it is certainly necessary that the mean energy of the Coulomb interaction between two ions ($\sim(ze)^2/r$, where $r \sim n^{-1/3}$ is the mean distance between ions) should be small compared with the mean kinetic energy of the ions ($\sim T$). Thus we must have $(ze)^2 n^{1/3} \ll T$ or

$$n \ll (T/z^2e^2)^3. \tag{75.2}$$

To calculate the thermodynamic quantities for such a gas we must first determine the change E_{Coul} in the energy of the gas (as compared with the energy of an ideal gas) due to the Coulomb interaction of the particles. As we

† The upper inversion point as $P \to 0$ ($T_{\text{inv}} = 2a/b$) corresponds to the case considered at the end of §72. The lower inversion point for small P may not occur in a gas owing to its condensation into a liquid.

know from electrostatics, the electrical interaction energy of a system of charged particles can be written as half the sum of the products of each charge and the potential of the field at the position of that charge due to all the other charges. In the present case

$$E_{\text{Coul}} = V \cdot \tfrac{1}{2} \sum_a e z_a n_{a0} \phi_a, \tag{75.3}$$

where ϕ_a is the potential of the field acting on an ion of the ath kind due to the other charges. To calculate these potentials we proceed as follows.†

Each ion creates around itself a non-uniformly charged *ion cloud*, which on average is spherically symmetrical. In other words, if we select any particular ion in the gas and consider the density of distribution of the other ions relative to that ion, this density will depend only on the distance r from the centre. Let the density of distribution of ions (of the ath kind) in this ion cloud be denoted by n_a. The potential energy of each ion of the ath kind in the electric field around the ion considered is $z_a e\phi$, where ϕ is the potential of this field. Hence BOLTZMANN's formula (38.6) gives

$$n_a = n_{a0} e^{-z_a e\phi/T}. \tag{75.4}$$

The constant coefficient is put equal to n_{a0}, since at a large distance from the centre (where $\phi \to 0$) the density of the ion cloud must become equal to the mean ion density in the gas.

The potential ϕ of the field in the ion cloud is related to the charge density in it (equal to $\sum e z_a n_a$) by the electrostatic POISSON's equation:

$$\triangle \phi = -4\pi e \sum_a z_a n_a. \tag{75.5}$$

Formulae (75.4) and (75.5) together give the equations of the "self-consistent" electric field of the electrons and ions.

With the above assumption that the interaction of the ions is relatively weak, the energy $e z_a \phi$ is small in comparison with T, and formula (75.4) may be written in the approximate form

$$n_a = n_{a0} - \frac{n_{a0} e z_a}{T} \phi. \tag{75.6}$$

Substituting this in equation (75.5) and using the condition (75.1) for the gas to be neutral as a whole, we obtain

$$\triangle \phi - \varkappa^2 \phi = 0, \tag{75.7}$$

where

$$\varkappa^2 = \frac{4\pi e^2}{T} \sum_a n_{a0} z_a^2. \tag{75.8}$$

The quantity $\varkappa$ has the dimensions of reciprocal length.

† This method was used by DEBYE and HÜCKEL to calculate the thermodynamic quantities for strong electrolytes (1923).

The spherically symmetric solution of equation (75.7) is ϕ = constant× $e^{-\varkappa r}/r$. In the immediate neighbourhood of the centre, the field must become the purely Coulomb field of the charge considered; this charge will be denoted by $z_b e$. In other words, for sufficiently small r we must have $\phi \cong e z_b/r$. This shows that the constant must be taken as $z_b e$, and so the required potential distribution is given by

$$\phi = e z_b e^{-\varkappa r}/r. \tag{75.9}$$

Hence we see, incidentally, that the field becomes very small at distances large compared with $1/\varkappa$. The length $1/\varkappa$ can therefore be regarded as determining the dimensions of the ion cloud due to a given ion; it is also called the *Debye–Hückel length*. All the calculations given here assume, of course, that this length is large in comparison with the mean distances between ions. This condition is clearly identical with (75.2).

Expanding the potential (75.9) in series for small $\varkappa r$, we have

$$\phi = \frac{e z_b}{r} - e z_b \varkappa + \dots .$$

The terms omitted vanish when $r = 0$. The first term is the Coulomb field of the ion itself; the second term is clearly the potential produced by all the other ions in the "cloud" at the point occupied by the ion considered, and is the quantity to be substituted in formula (75.3): $\phi_a = -e z_a \varkappa$.

Thus we have the following expression for the "Coulomb part" of the plasma energy:

$$E_{\text{Coul}} = -\tfrac{1}{2} V \varkappa e^2 \sum_a n_{a0} z_a^2 = -V e^3 \sqrt{\frac{\pi}{T}} \Big(\sum_a n_{a0} z_a^2\Big)^{3/2}, \tag{75.10}$$

or, in terms of the total numbers of different ions in the gas $N_a = n_{a0} V$,

$$E_{\text{Coul}} = -e^3 \sqrt{\frac{\pi}{TV}} \Big(\sum_a N_a z_a^2\Big)^{3/2}. \tag{75.11}$$

This energy is inversely proportional to the square root of the temperature and to that of the volume of the gas.

Integrating the thermodynamic relation $E/T^2 = -(\partial/\partial T)(F/T)$, we can deduce from E_{Coul} the corresponding change in the free energy:

$$F = F_{\text{id}} - \frac{2e^3}{3} \sqrt{\frac{\pi}{TV}} \Big(\sum_a N_a z_a^2\Big)^{3/2}; \tag{75.12}$$

the constant of integration must be taken as zero, since when $T \to \infty$ we must have $F = F_{\text{id}}$. Hence the pressure is

$$P = \frac{T}{V} \sum N_a - \frac{e^3}{3 V^{3/2}} \sqrt{\frac{\pi}{T}} \Big(\sum_a N_a z_a^2\Big)^{3/2}. \tag{75.13}$$

The thermodynamic potential is obtained from F in the same way as in §72 (i.e. by regarding the second term in (75.12) as a small correction to F_{id}):

$$\Phi = \Phi_{\text{id}} - \frac{2e^3}{3T}\left(\frac{\pi P}{\sum N_a}\right)^{1/2} \left(\sum_a N_a z_a^2\right)^{3/2}. \tag{75.14}$$

In the foregoing theory it is assumed that the plasma is far from degeneracy, i.e. obeys Boltzmann statistics. It is possible in principle for conditions to occur such that the electron component of the gas is already degenerate while at the same time the interaction between particles in the gas is unimportant, i.e. the gas is "almost ideal". It has been mentioned at the end of §56 that such a situation occurs at very high densities of a degenerate gas; the "gas of nuclei" may still be far from degeneracy, owing to the large mass of the nuclei. In this case the above calculations are inapplicable. At temperatures of the order of the degeneracy temperature the principal contribution to the correction terms in the thermodynamic quantities (as compared with those of an ideal gas) in a degenerate plasma comes from the exchange part of the electrical interaction of the electrons, which in the classical case is unimportant and which we have ignored. Moreover, in calculating the "self-consistent" interaction of the electrons and ions the motion of the electrons can no longer be regarded as quasi-classical.†

§76. **The method of correlation functions**

The advantage of the Debye–Hückel method described in §75 lies in its simplicity and physical clarity. Its basic drawback, however, is that it cannot be generalised to calculate further approximations with respect to the concentration. We shall therefore also give a brief description of another method (proposed by N. N. Bogolyubov, 1946), which, though more complicated, allows in principle the calculation of further terms in the expansions of the thermodynamic quantities.

This method is based on a consideration of what are called *correlation functions* between the simultaneous positions of several particles at given points in space. The simplest and most important of these is the binary correlation function w_{ab}, which is proportional to the probability of finding two particles (ions) simultaneously at given points $\mathbf{r}_a$ and $\mathbf{r}_b$; the ions a and b may be of either the same or different kinds. Because the gas is isotropic and homogeneous, this function of course depends only on $r = |\mathbf{r}_b - \mathbf{r}_a|$. We choose the

† The calculations for this case have been carried out by A. A. Vedenov, *Soviet Physics JETP* **(36) 9,** 446, 1959.

At sufficiently low temperatures the ordered arrangement of the nuclei in a "crystal lattice" becomes thermodynamically favourable rather than their random motion in the gas. Under these conditions the corrections due to the interaction between the electrons and the nuclei are of a different kind; see the second footnote to §108.

normalisation coefficient in the function w_{ab} such that this function tends to unity as $r \to \infty$. Then $\iint w_{ab}\,\mathrm{d}V_a\,\mathrm{d}V_b = V^2$.

If the function w_{ab} is known, the required energy E_{Coul} can be obtained by integration, using the obvious formula

$$E_{\mathrm{Coul}} = \frac{1}{2V^2}\sum_{a,\,b} N_a N_b \iint u_{ab} w_{ab}\,\mathrm{d}V_a\,\mathrm{d}V_b, \tag{76.1}$$

where the summation is over all the kinds of ions, and u_{ab} is the Coulomb interaction energy of a pair of ions at distance r.

According to the Gibbs distribution formula, the function w_{ab} is given by

$$w_{ab} = \frac{1}{V^{N-2}}\int \exp\left\{\frac{F-F_{\mathrm{id}}-U}{T}\right\}\mathrm{d}V_1\,\mathrm{d}V_2\ldots\mathrm{d}V_{N-2}, \tag{76.2}$$

where U is the Coulomb interaction energy of all the ions, and the integration is over the co-ordinates of all the ions except the two considered. For an approximate calculation of this integral we proceed as follows.

We differentiate equation (76.2) with respect to the co-ordinates of ion b:

$$\frac{\partial w_{ab}}{\partial \mathbf{r}_b} = -\frac{w_{ab}}{T}\frac{\partial u_{ab}}{\partial \mathbf{r}_b} - \frac{1}{TV}\sum_c N_c \int \frac{\partial u_{bc}}{\partial \mathbf{r}_b} w_{abc}\,\mathrm{d}V_c, \tag{76.3}$$

where the summation in the last term is over all the kinds of ions, and w_{abc} is the ternary correlation function, defined by

$$w_{abc} = \frac{1}{V^{N-3}}\int \exp\left\{\frac{F-F_{\mathrm{id}}-U}{T}\right\}\mathrm{d}V_1\,\mathrm{d}V_2\ldots\mathrm{d}V_{N-3}$$

analogously to (76.2).

Assuming the gas sufficiently rarefied and considering only the first-order terms, we can express the ternary correlation function in terms of binary correlations: neglecting the possibility that all three ions are close together, we have $w_{abc} = w_{ab}w_{bc}w_{ac}$. In the same approximation we can suppose that even the pairs of particles are not so close together that the w_{ab} are appreciably different from unity. We define the small quantities

$$\omega_{ab} = w_{ab} - 1 \tag{76.4}$$

and write

$$w_{abc} = \omega_{ab} + \omega_{bc} + \omega_{ac} + 1, \tag{76.5}$$

neglecting the higher powers of the ω_{ab}.

When this expression is substituted in the integral on the right-hand side of (76.3), only the term in ω_{ac} remains; the others give zero identically, because of the isotropy of the gas. In the first term on the right of (76.3) it is sufficient to put $w_{ab} = 1$. Thus

$$\frac{\partial \omega_{ab}}{\partial \mathbf{r}_b} = -\frac{1}{T}\frac{\partial u_{ab}}{\partial \mathbf{r}_b} - \frac{1}{TV}\sum_c N_c \int \omega_{ac}\frac{\partial u_{bc}}{\partial \mathbf{r}_b}\,\mathrm{d}V_c.$$

We now take the divergence of both sides of this equation, using the facts that $u_{ab} = z_a z_b e^2/r$, $\mathbf{r} = \mathbf{r}_b - \mathbf{r}_a$, and the well-known formula $\triangle(1/r) = -4\pi\delta(\mathbf{r})$. The integration is then trivial because of the presence of the delta function, and we have

$$\triangle\,\omega_{ab}(\mathbf{r}) = \frac{4\pi z_a z_b e^2}{T}\,\delta(\mathbf{r}) + \frac{4\pi e^2 z_b}{TV}\sum_c N_c z_c \omega_{ac}(\mathbf{r}). \tag{76.6}$$

The solution of this system of equations can be sought in the form

$$\omega_{ab}(\mathbf{r}) = z_a z_b \omega(\mathbf{r}), \tag{76.7}$$

whereby they are reduced to a single equation

$$\triangle\,\omega(\mathbf{r}) = \frac{4\pi e^2}{T}\,\delta(\mathbf{r}) + \frac{4\pi e^2}{TV}\sum_c N_c z_c^2 \omega(\mathbf{r}). \tag{76.8}$$

This final equation has the same form as equation (75.7) in DEBYE and HÜCKEL's method; the term containing the delta function in (76.8) corresponds to the boundary condition as $r \to 0$ imposed on the function $\phi(\mathbf{r})$ in (75.7). It is easy to see that this gives the same result as before for the energy E_{Coul}.

In the next approximation the calculations become more laborious. In particular, the assumption (76.5) is now insufficient, and ternary correlations which do not reduce to binary ones must be introduced. For these we obtain an equation analogous to (76.3) but involving quaternary correlations; in this (the second) approximation the latter reduce to ternary ones.†

§77. Calculation of the virial coefficient in quantum mechanics

In calculating the virial coefficients in §§72–74 we have used classical statistics, as is practically always justifiable. There is, however, methodological interest in the problem of calculating these coefficients in the quantum case; and such a case may actually occur for helium at sufficiently low temperatures. We shall show how the second virial coefficient may be calculated with allowance for the quantisation of the binary interaction of the gas particles (BETH and UHLENBECK 1937). We shall consider a monatomic gas whose atoms have no electronic angular momentum; and having in mind the case of helium, we shall also suppose for definiteness that the nuclei of the atoms have no spin and that the atoms obey Bose statistics.

In the approximation concerned, it is sufficient to retain only the first three terms in the sum over N in formula (35.3), which determines the potential Ω:

$$\Omega = -T\log\left\{1 + \sum_n e^{(\mu - E_{1n})/T} + \sum_n e^{(2\mu - E_{2n})/T}\right\}. \tag{77.1}$$

† The terms of next higher order in the thermodynamic quantities for a plasma have in fact been calculated (using a different method) by A. A. VEDENOV and A. I. LARKIN, *Soviet Physics JETP* **(36) 9**, 806, 1959.

Here the E_{1n} are the energy levels of a single atom, and the E_{2n} are those of a system of two interacting atoms. Our object is to calculate only those correction terms in the thermodynamic quantities which are due to the direct interaction of the atoms; the corrections due to quantum exchange effects exist even in an ideal gas and are given by formula (55.15), according to which the "exchange" part of the second virial coefficient is (in the case of Bose statistics)

$$B_{\text{exch}} = -\tfrac{1}{2}(\pi\hbar^2/mT)^{3/2}. \qquad (77.2)$$

Thus the problem reduces to the calculation of

$$Z^{(2)} = \sum_n e^{(2\mu - E_{2n})/T},$$

and from this must be subtracted the expression which would be obtained for two non-interacting atoms.

The energy levels E_{2n} consist of the kinetic energy of the motion of the centre of mass of the two atoms ($p^2/4m$, where $\mathbf{p}$ is the momentum of that motion, and m the mass of an atom) and the energy of their relative motion. Let the latter be denoted by ε; this component is given by the energy levels of a particle of mass $\frac{1}{2}m$ (the reduced mass of the two atoms) moving in a central field $U_{12}(r)$, where U_{12} is the potential energy of the interaction of the atoms. The motion of the centre of mass is always quasi-classical; integrating over its co-ordinates and momenta in the usual manner (cf. §42), we obtain

$$Z^{(2)} = Ve^{2\mu/T}(mT/\pi\hbar^2)^{3/2}\sum e^{-\varepsilon/T}.$$

If we denote by Z_{int} the part of $Z^{(2)}$ which depends on the interaction of the particles, we can write Ω in the form

$$\Omega = \Omega_{\text{id}} - TVe^{2\mu/T}(mT/\pi\hbar^2)^{3/2}Z_{\text{int}}.$$

Regarding the second term as a small correction to the first term, and expressing it in terms of T, V and N by means of formula (45.5) for the chemical potential of an ideal gas, we obtain for the free energy the expression

$$F = F_{\text{id}} - T\frac{8N^2}{V}\left(\frac{\pi\hbar^2}{mT}\right)^{3/2} Z_{\text{int}}.$$

Differentiation with respect to V gives the pressure, and the required part of the virial coefficient that is due to the interaction of the atoms is

$$B_{\text{int}}(T) = -8(\pi\hbar^2/mT)^{3/2}Z_{\text{int}}. \qquad (77.3)$$

The spectrum of energy levels ε consists of a discrete spectrum of negative values (corresponding to a finite relative motion of the atoms) and a continuous spectrum of positive values (infinite motion). We denote the former by ε_n; the latter may be written in the form p^2/m, where $\mathbf{p}$ is the momentum of the relative motion of the atoms when the distance between them has become very large. The whole of the sum $\sum e^{|\varepsilon_n|/T}$ over the discrete spectrum appears in Z_{int}; from the integral over the continuous spectrum we must separate the part corresponding to the free motion of non-interacting particles. To do this we proceed as follows.

At large distances r, the wave function of a stationary state with orbital angular momentum l and positive energy p^2/m has the asymptotic form†

$$\psi = \frac{\text{constant}}{r} \times \sin\left(\frac{p}{\hbar} r - \tfrac{1}{2} l\pi + \delta_l\right),$$

where the phase shifts $\delta_l = \delta_l(p)$ depend on the specific form of the field $U_{12}(r)$. Let us formally suppose that the range of variation of the distance r is bounded by some very large but finite value R. Then the momentum p can take only a discrete series of values given by the boundary condition that $\psi = 0$ for $r = R$:

$$\frac{p}{\hbar} R - \tfrac{1}{2} l\pi + \delta_l = s\pi,$$

where s is an integer. For large R these values are very close together and the summation in

$$\sum_{\mathbf{p}} e^{-p^2/mT}$$

may be replaced by an integration. To do so, we multiply the summand (for a given l) by

$$\mathrm{d}s = \frac{1}{\pi}\left(\frac{R}{\hbar} + \frac{\mathrm{d}\delta_l}{\mathrm{d}p}\right) \mathrm{d}p$$

and integrate over p; the result must then be multiplied by $2l+1$ (the degree of degeneracy with respect to orientations of the orbital angular momentum) and summed over l:

$$\sum_{\mathbf{p}} e^{-p^2/mT} = \frac{1}{\pi} \sum_l (2l+1) \int_0^\infty \left(\frac{R}{\hbar} + \frac{\mathrm{d}\delta_l}{\mathrm{d}p}\right) e^{-p^2/mT}\, \mathrm{d}p.$$

For particles obeying Bose statistics and having no spin, the co-ordinate wave functions must be symmetric; this means that only even values of l are possible, and so the summation over l is over all even integers.

In free motion, all the phase shifts $\delta_l = 0$. The expression remaining when $\delta_l = 0$ is therefore the part of the sum which is unrelated to the interaction of the atoms and is to be omitted. Thus we obtain the following expression for Z_{int}:

$$Z_{\text{int}} = \sum_n e^{|\varepsilon_n|/T} + \frac{1}{\pi} \sum_l \int_0^\infty (2l+1) \frac{\mathrm{d}\delta_l}{\mathrm{d}p} e^{-p^2/mT}\, \mathrm{d}p, \tag{77.4}$$

† See *Quantum Mechanics*, §33.

and the virial coefficient $B = B_{\text{exch}} + B_{\text{int}}$ is

$$B(T) = -\tfrac{1}{2}(\pi\hbar^2/mT)^{3/2}(1+16Z_{\text{int}}). \tag{77.5}$$

The phase shifts δ_l determine the scattering amplitude for particles moving in the field $U_{12}(r)$ by means of the formula†

$$f(\theta) = \frac{\hbar}{2ip}\sum_l (2l+1)(e^{2i\delta_l}-1)P_l(\cos\theta),$$

where the P_l are Legendre polynomials, and θ the angle between the directions of incidence and scattering; in the present case the summation is over all even values of l. It is therefore possible to express the integral in (77.4) in terms of the scattering amplitude. By direct substitution of the expression for $f(\theta)$ the following relation may easily be verified:

$$\sum_l (2l+1)\frac{\mathrm{d}\delta_l}{\mathrm{d}p} = \frac{1}{2\hbar}\frac{\mathrm{d}}{\mathrm{d}p}\{p[f(0)+f^*(0)]\} + \frac{i}{4\pi\hbar^2}\int p^2\left(f\frac{\partial f^*}{\partial p} - f^*\frac{\partial f}{\partial p}\right)\mathrm{d}o.$$

The sum on the left-hand side appears in the integrand in (77.4), and on substituting it and integrating by parts in one of the terms we find

$$Z_{\text{int}} = \sum_n e^{|\varepsilon_n|/T} + \frac{1}{\pi\hbar mT}\int\limits_0^\infty p^2 e^{-p^2/mT}[f(0)+f^*(0)]\,\mathrm{d}p +$$
$$+\frac{i}{(2\pi\hbar)^2}\iint p^2 e^{-p^2/mT}\left(f\frac{\partial f^*}{\partial p} - f^*\frac{\partial f}{\partial p}\right)\mathrm{d}p\,\mathrm{d}o. \tag{77.6}$$

If there are discrete levels in the field $U_{12}(r)$, then at sufficiently low temperatures the dependence of $B(T)$ on temperature will be mainly governed by the sum over the discrete levels, which increases exponentially with decreasing T. There may, however, be no discrete levels; then the virial coefficient will vary as a power of the temperature (if we bear in mind that the scattering amplitude tends to a constant limit as $p \to 0$, we easily find that at sufficiently low temperatures B will be determined mainly by the term B_{exch}).

It may be noted that in the case of a weak interaction, when particle collisions can be described by the Born approximation, the scattering amplitude is small and the third term in (77.6), which is quadratic in the amplitude, may be omitted. For weak interaction there are no bound states, and so the first term in (77.6) is also absent. Using the familiar expression for the scattering amplitude $f(0)$ in the Born approximation, which is proportional to $\int U_{12}r^2\,\mathrm{d}r$, it is easy to see that the expression for F agrees exactly with (32.3) (without the quadratic term), as it should in this case.

† See *Quantum Mechanics*, §122. The cross-section for scattering into the solid-angle element do is $|f(\theta)|^2$ do.

PROBLEM

Determine the quantum correction (of the order of $\hbar^2$) in the quasi-classical case in the virial coefficient $B(T)$ for a monatomic gas.

SOLUTION. The correction to the classical free energy is given by formula (33.15). Bearing in mind that in the present case only binary interaction of atoms occurs, and that U_{12} depends only on the distance between atoms, we find

$$B_{qu} = \frac{\pi\hbar^2}{6mT^3}\int_0^\infty \left(\frac{dU_{12}}{dr}\right)^2 e^{-U_{12}/T} r^2\, dr.$$

This expression is the correction to the classical value given by (72.5). It may be noted that $B_{qu} > 0$.

§78. A degenerate "almost ideal" Bose gas

The problem of the thermodynamic properties of an "almost ideal" highly degenerate gas (the case of slight degeneracy having been considered in §77) has no direct physical significance, since the gases which actually exist in Nature condense at temperatures near absolute zero. Nevertheless, because of the considerable methodological interest of this problem, it is useful to discuss it for an imaginary gas whose particles interact in such a way that condensation does not occur.

The condition for a gas to be "only slightly non-ideal" is that the "range of interaction" a of the molecular forces should be small compared with the mean distance between the particles, $l \sim (V/N)^{1/3}$. Together with the condition $a \ll l$, the inequality

$$ka \ll 1 \tag{78.1}$$

will also be satisfied, where $k = p/\hbar$ are the wave numbers of the gas particles. In conditions of strong degeneracy, the existence of such an inequality is evident from dimensional arguments, and it may also be proved directly by estimating the order of magnitude of the particle momenta.†

We shall consider here only binary interactions between particles (again denoting the interaction energy of two particles by U_{12}). Our purpose is to calculate the leading terms in the expansion of the thermodynamic quantities in powers of the ratio a/l by using some form of quantum perturbation theory. The difficulty is that, because of the rapid increase of the interaction energy U at small distances between the particles, perturbation theory (the Born approximation) cannot in fact be directly applied to particle collisions. This difficulty can, however, be circumvented in the following way.

† For a degenerate Fermi gas the order of magnitude of the limiting momentum is given by formula (56.2): $p_0/\hbar \sim (N/V)^{1/3} \ll 1/a$. For a Bose gas we shall see below that the majority of the particles (outside the "condensate") have momenta $p/\hbar \sim \sqrt{(aN/V)}$, for which the inequality (78.1) again holds.

In the Born approximation, the scattering cross-section in a collision of two particles of mass m is given by the squared modulus of the scattering amplitude

$$f = -\frac{m}{4\pi\hbar^2}\int U_{12}e^{-i\mathbf{q}\cdot\mathbf{r}}\,dV,$$

where $\hbar\mathbf{q}$ is the momentum transferred in the collision.† When the condition (78.1) holds, i.e. in "slow" collisions, $\mathbf{q}\cdot\mathbf{r} \ll 1$ throughout the important range of integration, and the amplitude tends to a constant limit, which we here denote by $-a$:‡

$$a = mU_0/4\pi\hbar^2, \qquad U_0 = \int U_{12}(r)\,dV. \tag{78.2}$$

Since this quantity entirely defines the properties of collisions, it must also determine the thermodynamic properties of the gas (when the Born approximation is applicable).

Thus the following procedure can be used. We formally replace the true energy U_{12} by another function with the same value of the scattering amplitude but such as to permit the application of perturbation theory. So long as the final result of the calculations contains U_{12} only in the scattering amplitude (i.e. in any approximation for which this is true), this result will be the same as that which would be obtained for the true interaction.

Let us first calculate the energy spectrum of low excited states of an almost ideal Bose gas. It has been shown by N. N. Bogolyubov (1947) that this may be done by applying perturbation theory in the second quantisation method.

The Hamiltonian of a system of N particles (which we shall assume to have no spin), taking account only of binary interaction between the particles, is written as follows in the second quantisation method:||

$$\hat{H} = \sum_{\mathbf{p}} \frac{p^2}{2m}\hat{a}_{\mathbf{p}}^{+}\hat{a}_{\mathbf{p}} + \tfrac{1}{2}\sum U_{\mathbf{p}_1\mathbf{p}_2}^{\mathbf{p}_1'\mathbf{p}_2'}\hat{a}_{\mathbf{p}_1'}^{+}\hat{a}_{\mathbf{p}_2'}^{+}\hat{a}_{\mathbf{p}_2}\hat{a}_{\mathbf{p}_1}. \tag{78.3}$$

Here $\hat{a}_{\mathbf{p}}^{+}$, $\hat{a}_{\mathbf{p}}$ are the "creation" and "annihilation" operators for a free particle with momentum $\mathbf{p}$, i.e. in a state described (in volume V) by the wave function

$$\psi_{\mathbf{p}} = \frac{1}{\sqrt{V}}e^{i\mathbf{p}\cdot\mathbf{r}/\hbar}.$$

† See *Quantum Mechanics*, §125.

The cross-section for scattering into the solid-angle element do (in the centre-of-mass system) is $d\sigma = |f|^2\,do$ if the quantum identity of the particles is ignored. When this identity is taken into account, $d\sigma = 4|f|^2\,do$, and to obtain the total cross-section $d\sigma$ must be integrated over a hemisphere (not over the whole sphere).

‡ The quantity a is sometimes called the *scattering length*.

|| See *Quantum Mechanics*, §64.

The first term in (78.3) corresponds to the kinetic energy of the particle, and the second to its potential energy. In the latter term the summation is over all values of the momenta of a pair of particles such that the law of conservation of momentum in collisions is satisfied:

$$\mathbf{p}_1+\mathbf{p}_2 = \mathbf{p}_1'+\mathbf{p}_2';$$

only with this condition are the matrix elements

$$U^{\mathbf{p}_1'\mathbf{p}_2'}_{\mathbf{p}_1\mathbf{p}_2} = \frac{1}{V^2}\iint e^{i(\mathbf{p}_1-\mathbf{p}_1')\cdot\mathbf{r}_1/\hbar+i(\mathbf{p}_2-\mathbf{p}_2')\cdot\mathbf{r}_2/\hbar}\, U_{12}(\mathbf{r}_2-\mathbf{r}_1)\,\mathrm{d}V_1\,\mathrm{d}V_2$$
$$= \frac{1}{V}\int e^{-i\mathbf{p}\cdot\mathbf{r}/\hbar}U_{12}(r)\,\mathrm{d}V \tag{78.4}$$

different from zero ($\mathbf{p} = \mathbf{p}_2'-\mathbf{p}_2 = -(\mathbf{p}_1'-\mathbf{p}_1)$ is the particle momentum exchange in the collision). Since in our case the particle momenta are assumed small in accordance with (78.1), the matrix elements in all the terms of importance in the sum can be replaced by their values for $\mathbf{p} = 0$, putting

$$\hat{H} = \sum_{\mathbf{p}}\frac{p^2}{2m}\hat{a}_{\mathbf{p}}^+\hat{a}_{\mathbf{p}}+\frac{U_0}{2V}\sum \hat{a}_{\mathbf{p}_1'}^+\hat{a}_{\mathbf{p}_2'}^+\hat{a}_{\mathbf{p}_2}\hat{a}_{\mathbf{p}_1}. \tag{78.5}$$

The application of perturbation theory to the Hamiltonian (78.5) is based on the following remark. In the ground state of an ideal Bose gas, all the particles are in the "condensate", i.e. in a state of zero energy: $N_0 = N$, $N_{\mathbf{p}} = 0$ for $\mathbf{p} \neq 0$. In an almost ideal gas in low excited states (and in the ground state) the occupation numbers $N_{\mathbf{p}}$ are not zero but are very small compared with N_0. Since $\hat{a}_0{}^+\hat{a}_0 = N_0 \cong N$ is very large compared with unity, it follows that the expression $\hat{a}_0\hat{a}_0{}^+-\hat{a}_0{}^+\hat{a}_0 = 1$ is small compared with $\hat{a}_0$, $\hat{a}_0{}^+$ themselves, and so the latter may be regarded as ordinary numbers (equal to $\sqrt{N_0}$), ignoring the fact that they do not commute.

The application of perturbation theory now requires a formal expansion of the quadruple sum in (78.5) in powers of the small quantities $\hat{a}_{\mathbf{p}}$, $\hat{a}_{\mathbf{p}}{}^+$ ($\mathbf{p} \neq 0$). The zero-order term of the expansion is

$$a_0{}^+a_0{}^+a_0a_0 = a_0{}^4. \tag{78.6}$$

There are no first-order terms, since they cannot be such as to satisfy the law of conservation of momentum. The second-order terms are

$$a_0{}^2\sum_{\mathbf{p}\neq 0}(\hat{a}_{\mathbf{p}}\hat{a}_{-\mathbf{p}}+\hat{a}_{\mathbf{p}}{}^+\hat{a}_{-\mathbf{p}}{}^++4\hat{a}_{\mathbf{p}}{}^+\hat{a}_{\mathbf{p}}). \tag{78.7}$$

Restricting ourselves to accuracy as far as second-order quantities, in (78.7) we can replace $a_0{}^2 = N_0$ by the total number N of particles. In the term (78.6) we must use the more precise relation

$$a_0{}^2+\sum_{\mathbf{p}\neq 0}\hat{a}_{\mathbf{p}}{}^+\hat{a}_{\mathbf{p}} = N.$$

The sum of the terms (78.6) and (78.7) thus becomes

$$N^2+N\sum_{\mathbf{p}\neq 0}(\hat{a}_{\mathbf{p}}\hat{a}_{-\mathbf{p}}+\hat{a}_{\mathbf{p}}^{+}\hat{a}_{-\mathbf{p}}^{+}+2\hat{a}_{\mathbf{p}}^{+}\hat{a}_{\mathbf{p}}),$$

and after substitution in (78.5) we obtain for the Hamiltonian the expression

$$\hat{H}=\frac{N^2}{2V}U_0+\frac{N}{2V}U_0\sum_{\mathbf{p}\neq 0}(\hat{a}_{\mathbf{p}}\hat{a}_{-\mathbf{p}}+\hat{a}_{\mathbf{p}}^{+}\hat{a}_{-\mathbf{p}}^{+}+2\hat{a}_{\mathbf{p}}^{+}\hat{a}_{\mathbf{p}})+\sum_{\mathbf{p}}\frac{p^2}{2m}\hat{a}_{\mathbf{p}}^{+}\hat{a}_{\mathbf{p}}. \quad (78.8)$$

The integral U_0 which appears here has still to be expressed in terms of an actual physical quantity, the scattering amplitude. In the second-order terms in (78.8) (which are needed only to determine the energy spectrum; see below), this can be done directly from (78.2). In the first term (which is important in determining the energy of the ground state of the system) this formula, which corresponds only to the first approximation of perturbation theory, is insufficiently exact.

To obtain a more precise relation, we recall that, if the probability of a given quantum transition of a system under the action of a constant perturbation $\hat{V}$ is determined in the first approximation by the matrix element V_0^0, then in the second approximation V_0^0 is replaced by

$$V_0^0+\sum_n{}'\frac{V_n^0V_0^n}{E_0-E_n},$$

where the summation is over all states of the unperturbed system.† In the present case of a collision in a two-particle system, V_0^0 becomes $U_{00}^{00}=U_0/V$.

Using also the other matrix elements (78.4), we find that to go from the first to the second approximation we must replace U_0 by

$$U_0+\frac{1}{V}\sum_{\mathbf{p}\neq 0}\frac{|\int U_{12}e^{-i\mathbf{p}\cdot\mathbf{r}/\hbar}\,dV|^2}{-p^2/m}$$

or, again replacing‡ all the integrals by U_0 as in (78.5), by

$$U_0\left(1-\frac{U_0}{V}\sum_{\mathbf{p}\neq 0}\frac{m}{p^2}\right).$$

Instead of (78.2), we therefore have

$$a=\frac{m}{4\pi\hbar^2}U_0\left(1-\frac{U_0}{V}\sum_{\mathbf{p}\neq 0}\frac{m}{p^2}\right), \quad (78.9)$$

or, to the same accuracy,

$$U_0=\frac{4\pi\hbar^2}{m}a\left(1+\frac{4\pi\hbar^2 a}{V}\sum_{\mathbf{p}\neq 0}\frac{1}{p^2}\right).$$

† See *Quantum Mechanics*, §43.

‡ This replacement leads to a sum which diverges for large **p**, but this does not matter, since on subsequent substitution in the Hamiltonian a convergent expression is obtained, in which large values of **p** are unimportant.

Substituting this in (78.8), we find the Hamiltonian

$$\hat{H} = \frac{2\pi\hbar^2}{m} a \frac{N^2}{V}\left(1+\frac{4\pi\hbar^2 a}{V}\sum_{\mathbf{p}\neq 0}\frac{1}{p^2}\right)+$$

$$+\frac{2\pi\hbar^2}{m} a \frac{N}{V}\sum_{\mathbf{p}\neq 0}(\hat{a}_{\mathbf{p}}\hat{a}_{-\mathbf{p}}+\hat{a}_{\mathbf{p}}^{+}\hat{a}_{-\mathbf{p}}^{+}+2\hat{a}_{\mathbf{p}}^{+}\hat{a}_{\mathbf{p}})+$$

$$+\sum_{\mathbf{p}}\frac{p^2}{2m}\hat{a}_{\mathbf{p}}^{+}\hat{a}_{\mathbf{p}}. \tag{78.10}$$

To determine the energy levels, we must bring the Hamiltonian to diagonal form; this is achieved by an appropriate linear transformation of the operators $\hat{a}_{\mathbf{p}}$, $\hat{a}_{\mathbf{p}}^{+}$. We define new operators $\hat{b}_{\mathbf{p}}$, $\hat{b}_{\mathbf{p}}^{+}$ by

$$\hat{a}_{\mathbf{p}} = u_{\mathbf{p}}\hat{b}_{\mathbf{p}}+v_{\mathbf{p}}\hat{b}_{-\mathbf{p}}^{+},$$

$$\hat{a}_{\mathbf{p}}^{+} = u_{\mathbf{p}}\hat{b}_{\mathbf{p}}^{+}+v_{\mathbf{p}}\hat{b}_{-\mathbf{p}},$$

and require them to satisfy the commutation relations

$$\hat{b}_{\mathbf{p}}\hat{b}_{\mathbf{p}'}-\hat{b}_{\mathbf{p}'}\hat{b}_{\mathbf{p}} = 0, \qquad \hat{b}_{\mathbf{p}}\hat{b}_{\mathbf{p}'}^{+}-\hat{b}_{\mathbf{p}'}^{+}\hat{b}_{\mathbf{p}} = \delta_{\mathbf{pp}'},$$

which are similar to those for the operators $\hat{a}_{\mathbf{p}}$, $\hat{a}_{\mathbf{p}}^{+}$. It is easy to see that this imposes the condition $u_{\mathbf{p}}^2-v_{\mathbf{p}}^2 = 1$. Using this, we write the linear transformation in the form

$$\hat{a}_{\mathbf{p}} = \frac{\hat{b}_{\mathbf{p}}+L_{\mathbf{p}}\hat{b}_{-\mathbf{p}}^{+}}{\sqrt{(1-L_{\mathbf{p}}^2)}}, \quad \hat{a}_{\mathbf{p}}^{+} = \frac{\hat{b}_{\mathbf{p}}^{+}+L_{\mathbf{p}}\hat{b}_{-\mathbf{p}}}{\sqrt{(1-L_{\mathbf{p}}^2)}}. \tag{78.11}$$

The quantity $L_{\mathbf{p}}$ must be so defined that the non-diagonal terms $(\hat{b}_{\mathbf{p}}\hat{b}_{-\mathbf{p}}, \hat{b}_{\mathbf{p}}^{+}\hat{b}_{-\mathbf{p}}^{+})$ disappear from the Hamiltonian. A simple calculation gives

$$L_{\mathbf{p}} = \frac{mV}{4\pi a\hbar^2 N}\left\{\varepsilon(p)-\frac{p^2}{2m}-mu^2\right\}, \tag{78.12}$$

where

$$\varepsilon(p) = \sqrt{[u^2p^2+(p^2/2m)^2]}, \tag{78.13}$$

$$u = \sqrt{(4\pi\hbar^2 aN/m^2V)}. \tag{78.14}$$

The Hamiltonian then becomes

$$H = E_0+\sum_{\mathbf{p}\neq 0}\varepsilon(p)\hat{b}_{\mathbf{p}}^{+}\hat{b}_{\mathbf{p}}, \tag{78.15}$$

where

$$E_0 = \tfrac{1}{2}Nmu^2+\tfrac{1}{2}\sum_{\mathbf{p}\neq 0}\left\{\varepsilon(p)-\frac{p^2}{2m}-mu^2+\frac{m^3u^4}{p^2}\right\}. \tag{78.16}$$

The Hamiltonian (78.15) and the Bose commutation relations for the operators $\hat{b}_{\mathbf{p}}^{+}$, $\hat{b}_{\mathbf{p}}$ show that $\hat{b}_{\mathbf{p}}^{+}$ and $\hat{b}_{\mathbf{p}}$ are the "creation" and "annihilation" operators for quasi-particles (elementary excitations) with energy $\varepsilon(p)$, obeying Bose statistics. The quantity $\hat{b}_{\mathbf{p}}^{+}\hat{b}_{\mathbf{p}} = n_{\mathbf{p}}$ is the number of quasi-particles with momentum **p**, and formula (78.13) gives the relation between their energy and momentum; the occupation numbers for the quasi-particles are denoted by $n_{\mathbf{p}}$ to distinguish them from the occupation numbers $N_{\mathbf{p}}$ for the actual particles of the gas. This completely determines the energy spectrum of the low excited states of the gas considered; it is, of course, a Bose-type spectrum (§66).

The quantity E_0 is the energy of the ground state of the gas. Replacing the summation over discrete values of **p** (in the volume V) by integration over **p** after multiplication by $V/(2\pi\hbar)^3$, we obtain after some calculation

$$E_0 = \frac{2\pi\hbar^2 a}{m}\frac{N^2}{V}\left[1+\frac{128}{15\sqrt{\pi}}\sqrt{\frac{a^3N}{V}}\right] \tag{78.17}$$

(T. D. Lee and C. N. Yang 1957). This gives the first two terms of an expansion of this quantity in powers of $\sqrt{(a^3N/V)}$. However, even the next term can not be calculated by the above method; it must contain $1/V^2$, and a quantity of this order depends on ternary as well as binary collisions.

For large momenta ($p \gg mu$) the energy ε (78.13) of the quasi-particles tends to $p^2/2m$, i.e. to the kinetic energy of a single particle in the gas. For small momenta ($p \ll mu$) we have $\varepsilon = up$. It is easy to see that u is the velocity of sound in the gas, so that this expression is in accordance with the general statements in §66. At absolute zero the free energy is equal to E, and from the leading term in the expression for the latter we find the pressure

$$P = -\frac{\partial E}{\partial V} = \frac{2\pi\hbar^2 a}{m}\frac{N^2}{V^2}.$$

The velocity of sound is $u = \sqrt{(\partial P/\partial \varrho)}$, where $\varrho = mN/V$ is the gas density, and this agrees with (78.14)†.

It may be noted that, in the gas model considered, the scattering amplitude a must necessarily be positive (corresponding to repulsion between the particles). This is formally evident from the fact that imaginary terms appear in the above formulae for the energy if $a < 0$. The significance of the condition $a > 0$ is that it is necessary in order to satisfy the thermodynamic inequality $(\partial P/\partial V)_T < 0$ in this model of a Bose gas.

The statistical distribution of elementary excitations at a non-zero temperature is given simply by the Bose distribution formula (with zero chemical potential):

$$\overline{n_{\mathbf{p}}} = 1/(e^{\varepsilon/T}-1).$$

The momentum distribution of the actual particles of the gas is easily calculated to be

$$\overline{N_{\mathbf{p}}} = \overline{\hat{a}_{\mathbf{p}}^{+}\hat{a}_{\mathbf{p}}}.$$

† The two limiting forms of $\varepsilon(p)$ can legitimately be considered in the approximation used, since the change from the phonon region ($\varepsilon \cong up$) to the free-particle region ($\varepsilon \cong p^2/2m$) occurs at momenta $p/\hbar \sim mu/\hbar \sim \sqrt{(aN/V)}$, which satisfy the condition (78.1).

Substituting (78.11) and using the fact that the products $\hat{b}_{-\mathbf{p}}\hat{b}_{\mathbf{p}}$ and $\hat{b}_{\mathbf{p}}^{+}\hat{b}_{-\mathbf{p}}^{+}$ have no diagonal matrix elements, we find

$$\overline{N_{\mathbf{p}}} = \frac{\overline{n_{\mathbf{p}}}+L_{\mathbf{p}}^{2}(\overline{n_{\mathbf{p}}}+1)}{1-L_{\mathbf{p}}^{2}}. \tag{78.18}$$

This expression is, of course, valid only for $\mathbf{p} \neq 0$. The number of particles with zero momentum is

$$\overline{N_0} = 1-\sum_{\mathbf{p}\neq 0}\overline{N_{\mathbf{p}}} = 1-\frac{V}{(2\pi\hbar)^3}\int \overline{N_{\mathbf{p}}}\, d^3p. \tag{78.19}$$

In particular, at absolute zero $\overline{n_{\mathbf{p}}} = 0$ for $\mathbf{p} \neq 0$, and from (78.12) and (78.18) we obtain the distribution function in the form†

$$\overline{N_{\mathbf{p}}} = \frac{m^2u^4}{2\varepsilon(p)\{\varepsilon(p)+p^2/2m+mu^2\}}. \tag{78.20}$$

The departure of the Bose gas from the ideal state naturally causes the appearance of particles with non-zero momentum even at absolute zero; the integration in (78.19), with $\overline{N_{\mathbf{p}}}$ given by (78.20), is elementary, and the result is

$$\frac{\overline{N_0}}{N} = 1-\frac{8}{3\sqrt{\pi}}\left(\frac{Na^3}{V}\right)^{1/2}. \tag{78.21}$$

A further remark should be made concerning the spectrum considered here. For small p the derivative $d\varepsilon/dp > u$ (the curve $\varepsilon(p)$ turning upwards away from the initial tangent $\varepsilon = up$). It is easy to see that when the function $\varepsilon(p)$ is of this type the laws of conservation of energy and momentum allow a spontaneous decay of the quasi-particle (phonon) into two others. This means that the spectrum found is actually unstable throughout its extent (from small p onwards); the elementary excitations in it have a finite lifetime. The spontaneous decay time is large, however, so that the corresponding level width is small (being proportional to p^5 when p is small) and does not affect the results obtained in the approximations considered here.‡

† It may be noted that the maximum number of particles having a given magnitude of the momentum ($\sim p^2\overline{N_p}$) lies at $p/\hbar \sim \sqrt{(aN/V)}$, where the change occurs from one limiting form of $\varepsilon(p)$ to the other. This has already been mentioned in the first footnote to this section.

‡ The spectrum of an actual Bose-type quantum liquid (liquid He^4) does not possess such an instability. Here the curve $\varepsilon(p)$ turns downwards from the initial tangent $\varepsilon = up$, and so the spontaneous decay of the phonon is not possible. The finite lifetime of the phonon then depends only on its interaction (by collisions) with other quasi-particles, and when their concentration is small the lifetime is very long.

§79. A degenerate "almost ideal" Fermi gas with repulsion between the particles

For a degenerate "almost ideal" Fermi gas, models are in principle allowable with either a repulsive (scattering amplitude $a > 0$) or an attractive ($a < 0$) type of interaction between particles, but the properties of the gas are entirely different in these two cases. We shall first consider the case of repulsive interaction.

The state of a free particle with non-zero spin (which we shall assume to be $\frac{1}{2}$) is determined by the z-component σ of the spin as well as by the momentum $\mathbf{p}$. Accordingly, the second quantisation operators will be written with double suffixes, and (78.3) becomes

$$\hat{H} = \sum_{\mathbf{p},\sigma} \frac{p^2}{2m} \hat{a}_{\mathbf{p}\sigma}{}^{+} \hat{a}_{\mathbf{p}\sigma} + $$
$$+ \tfrac{1}{2} \sum U_{\mathbf{p}_1\sigma_1, \mathbf{p}_2\sigma_2}^{\mathbf{p}_1'\sigma_1', \mathbf{p}_2'\sigma_2'} \hat{a}_{\mathbf{p}_1'\sigma_1'}{}^{+} \hat{a}_{\mathbf{p}_2'\sigma_2'}{}^{+} \hat{a}_{\mathbf{p}_2\sigma_2} \hat{a}_{\mathbf{p}_1\sigma_1}. \tag{79.1}$$

As in (78.3), we replace all the matrix elements in the second term by the value

$$U_{0\sigma_1, 0\sigma_2}^{0\sigma_1', 0\sigma_2'}$$

which they have when the particle momenta are zero. Next, we note that, since the operators $\hat{a}_{\mathbf{p}_1\sigma_1}$, $\hat{a}_{\mathbf{p}_2\sigma_2}$ anticommute in Fermi statistics, their product is antisymmetric with respect to interchange of the suffixes, and the same is true of the products $\hat{a}_{\mathbf{p}_1'\sigma_1'}{}^{+}\hat{a}_{\mathbf{p}_2'\sigma_2'}{}^{+}$. Thus all terms in the second sum in (79.1) which contain the same pairs of suffixes σ_1, σ_2 or σ_1', σ_2' are zero. Physically this occurs because, in the limiting case of slow collisions of like particles, only particles with opposite spins can scatter each other.†

Using the notation‡

$$U_0/V = U_{0+, 0-}^{0+, 0-} - U_{0-, 0+}^{0+, 0-} \tag{79.2}$$

(the suffixes + and − henceforward denoting $\sigma = +\frac{1}{2}$ and $\sigma = -\frac{1}{2}$), we obtain the Hamiltonian in the form

$$\hat{H} = \sum_{\mathbf{p},\sigma} \frac{p^2}{2m} \hat{a}_{\mathbf{p}\sigma}{}^{+} \hat{a}_{\mathbf{p}\sigma} + \frac{U_0}{V} \sum \hat{a}_{\mathbf{p}_1'+}{}^{+} \hat{a}_{\mathbf{p}_2'-}{}^{+} \hat{a}_{\mathbf{p}_2-} \hat{a}_{\mathbf{p}_1+}, \tag{79.3}$$

the summation in the second term being over all values of the momenta subject to the conservation law $\mathbf{p}_1 + \mathbf{p}_2 = \mathbf{p}_1' + \mathbf{p}_2'$.

The eigenvalues of this Hamiltonian are calculated by quantum perturbation theory in its usual form, regarding the second term in (79.3) (the energy of interaction of the particles) as a small correction to the first term (the

† See *Quantum Mechanics*, §135. When $f \to$ constant, the amplitude in formula (135.3) tends to zero.

‡ If the interaction of the particles is independent of spin, the second term is zero (the collision does not change the spin of either particle separately).

kinetic energy). The latter is already in diagonal form, and its eigenvalues are

$$E^{(0)} = \sum_{\mathbf{p},\sigma} \frac{p^2}{2m} n_{\mathbf{p}\sigma}. \tag{79.4}$$

The first-order correction is given by the diagonal matrix elements of the interaction energy:

$$E^{(1)} = \frac{U_0}{V} \sum_{\mathbf{p}_1,\mathbf{p}_2} n_{\mathbf{p}_1+} n_{\mathbf{p}_2-}. \tag{79.5}$$

To find the second-order correction, we use the well-known formula of perturbation theory

$$E_n^{(2)} = \sum_m{}' \frac{|V_{nm}|^2}{E_n - E_m},$$

where the suffixes n and m label the states of the unperturbed system as a whole. A simple calculation gives†

$$\frac{2}{V^2} U_0^2 \sum_{\mathbf{p}_1,\mathbf{p}_2,\mathbf{p}_1'} \frac{n_{\mathbf{p}_1+} n_{\mathbf{p}_2-}(1-n_{\mathbf{p}_1'+})(1-n_{\mathbf{p}_2'-})}{(p_1^2+p_2^2-p_1'^2-p_2'^2)/2m} \tag{79.6}$$

(with $\mathbf{p}_1+\mathbf{p}_2 = \mathbf{p}_1'+\mathbf{p}_2'$). The structure of this expression is quite clear: the square of the matrix element for the transition $\mathbf{p}_1, \mathbf{p}_2 \to \mathbf{p}_1', \mathbf{p}_2'$ is proportional to the occupation numbers of the states $\mathbf{p}_1, \mathbf{p}_2$ and to the number of "vacancies" in the states $\mathbf{p}_1', \mathbf{p}_2'$.

The second-order terms in the energy, however, are not all given by this expression. A contribution of the same order arises from (79.5) when U_0 is expressed in terms of the scattering amplitude. By the same method as that used to derive (78.9), we now find‡

$$a = \frac{mU_0}{4\pi\hbar^2}\left[1 + \frac{2U_0}{V}\sum_{\mathbf{p}_1'} \frac{2m}{p_1^2+p_2^2-p_1'^2-p_2'^2}\right].$$

Hence, expressing U_0 in terms of a and substituting in (79.5), we obtain, as well as the first-order quantity

$$E^{(1)} = \frac{g}{V} \sum_{\mathbf{p}_1,\mathbf{p}_2} n_{\mathbf{p}_1+} n_{\mathbf{p}_2-}, \tag{79.7}$$

† This sum in the form (79.6) is divergent owing to the replacement of all the matrix elements in (79.3) by a constant; its divergence does not affect the subsequent discussion (cf. the sixth footnote to §78).

‡ By a we mean the scattering amplitude for slow particles, which is independent of their energy. The formula written here seems at first sight to involve a dependence on the momenta $\mathbf{p}_1, \mathbf{p}_2$. In reality, this dependence occurs only in the imaginary part of the amplitude (which appears when the correct method of summation is used), and this can be ignored, since we know that the final result will be real.

second-order terms which together with (79.6) give

$$E^{(2)} = \frac{2g^2}{V^2} \sum_{\mathbf{p}_1, \mathbf{p}_2, \mathbf{p}_1'} \frac{n_{\mathbf{p}_1+} n_{\mathbf{p}_2-} [(1-n_{\mathbf{p}_1'+})(1-n_{\mathbf{p}_2'-})-1]}{(p_1^2+p_2^2-p_1'^2-p_2'^2)/2m};$$

for brevity the notation $g = 4\pi\hbar^2 a/m$ is used in the intermediate formulae. Expanding the numerator, we see that the term containing the product of four n is zero, since its numerator is symmetric and its denominator antisymmetric with respect to the interchange of $\mathbf{p}_1, \mathbf{p}_2$ and $\mathbf{p}_1', \mathbf{p}_2'$; the summation over these variables is symmetrical. Thus we have finally

$$E^{(2)} = -\frac{2g^2}{V^2} \sum_{\mathbf{p}_1, \mathbf{p}_2, \mathbf{p}_1'} \frac{n_{\mathbf{p}_1+} n_{\mathbf{p}_2-} (n_{\mathbf{p}_1'+} + n_{\mathbf{p}_2'-})}{(p_1^2+p_2^2-p_1'^2-p_2'^2)/2m}. \tag{79.8}$$

By means of these formulae we can, first of all, calculate the energy of the ground state of the gas. To do so it is necessary to put all the $n_{\mathbf{p}\sigma}$ equal to unity within the Fermi sphere $p < p_0$ and equal to zero outside it. In this connection it should be noted that, although in the original Hamiltonian the quantities $\hat{a}_{\mathbf{p}\sigma}^+ \hat{a}_{\mathbf{p}\sigma}$ give the occupation numbers of the states of the gas particles themselves, when it is diagonalised by means of perturbation theory we have a quasi-particle distribution function (which, as in §78, we denote by $n_{\mathbf{p}\sigma}$); in the zero-order approximation this function has the values stated.

Noting that

$$\sum_{\mathbf{p}} n_{\mathbf{p}+} = \sum_{\mathbf{p}} n_{\mathbf{p}-} = \tfrac{1}{2}N,$$

we obtain from (79.7) the first-order correction $E_0^{(1)} = \frac{1}{4}gN^2/V$. In (79.8) we replace the summation over the four momenta subject to the condition $\mathbf{p}_1+\mathbf{p}_2 = \mathbf{p}_1'+\mathbf{p}_2'$ by integration over $\mathbf{p}_1$, $\mathbf{p}_2$, $\mathbf{p}_1'$, $\mathbf{p}_2'$ after multiplication by

$$\frac{V^3}{(2\pi\hbar)^9} \delta(\mathbf{p}_1+\mathbf{p}_2-\mathbf{p}_1'-\mathbf{p}_2'),$$

so that

$$E_0^{(2)} = -\frac{8mVg^2}{(2\pi\hbar)^9} \iiiint \frac{\delta(\mathbf{p}_1+\mathbf{p}_2-\mathbf{p}_1'-\mathbf{p}_2')}{p_1^2+p_2^2-p_1'^2-p_2'^2} \, d^3p_1 \, d^3p_2 \, d^3p_1' \, d^3p_2',$$

the integration being over the region $p_1, p_2, p_1' < p_0 = \hbar(3\pi^2 N/V)^{1/3}$, the limiting momentum. The calculation of the integral† leads to the following final result for the energy of the ground state (K. Huang and C. N. Yang 1957):

$$E_0 = \frac{3}{10}\left(\frac{3\pi^2 N}{V}\right)^{2/3} \frac{\hbar^2}{m} N + \frac{\pi a \hbar^2}{m} \frac{N}{V} N \left[1 + \frac{6a}{35}\left(\frac{3N}{\pi V}\right)^{1/3} (11 - 2\log 2)\right]. \tag{79.9}$$

According to the general results given in §68 the spectrum of elementary excitations (i.e. the function $\varepsilon(\mathbf{p})$) and the function $f(\mathbf{p}, \hat{\mathbf{s}}; \mathbf{p}', \hat{\mathbf{s}}')$ which is of

† In practice it is simpler to take the calculation in a different order, first calculating the function f (see below).

importance in the theory of Fermi-type spectra are determined by the first and second variations of the total energy with respect to the quasi-particle distribution function. If we write E as a discrete sum over $\mathbf{p}$ and σ, we have by definition

$$\delta E = \sum_{\mathbf{p},\sigma} \varepsilon(\mathbf{p}, \sigma)\delta n_{\mathbf{p}\sigma} + \frac{1}{2V}\sum_{\mathbf{p},\sigma,\mathbf{p}',\sigma'} f(\mathbf{p}, \sigma; \mathbf{p}', \sigma')\, \delta n_{\mathbf{p}\sigma}\, \delta n_{\mathbf{p}'\sigma'}; \tag{79.10}$$

after differentiating the energy, we must replace $n_{\mathbf{p}\sigma}$ by unity inside the Fermi sphere and zero outside it.

There is no need to calculate the energy of the quasi-particles in this way, however, since the function $\varepsilon(\mathbf{p})$ is actually meaningful only near $p = p_0$ (see §68), and there it is determined by the single parameter m^*, which can also be found in a simpler manner (see below).

To calculate the function $f(\mathbf{p}, \sigma; \mathbf{p}', \sigma')$ we twice differentiate the sum of the expressions (79.7) and (79.8) and then put $p = p' = p_0$. Effecting this simple calculation and changing from summation to integration, we obtain

$$f(\mathbf{p}, \tfrac{1}{2}; \mathbf{p}', -\tfrac{1}{2}) = g - \frac{8mg^2}{(2\pi\hbar)^3}\iint\left\{\frac{\delta(\mathbf{p}+\mathbf{p}'-\mathbf{p}_1-\mathbf{p}_2)}{2p_0^2-p_1^2-p_2^2} + \right.$$
$$\left. + \frac{\delta(\mathbf{p}+\mathbf{p}_1-\mathbf{p}'-\mathbf{p}_2)+\delta(\mathbf{p}'+\mathbf{p}_1-\mathbf{p}-\mathbf{p}_2)}{2(p_1^2-p_2^2)}\right\} d^3p_1\, d^3p_2,$$

$$f(\mathbf{p}, \tfrac{1}{2}; \mathbf{p}', \tfrac{1}{2}) = f(\mathbf{p}, -\tfrac{1}{2}; \mathbf{p}', -\tfrac{1}{2})$$
$$= \frac{4mg^2}{(2\pi\hbar)^3}\iint \frac{\delta(\mathbf{p}+\mathbf{p}_1-\mathbf{p}'-\mathbf{p}_2')+\delta(\mathbf{p}'+\mathbf{p}_1-\mathbf{p}-\mathbf{p}_2)}{p_1^2-p_2^2}\, d^3p_1\, d^3p_2.$$

The integration in these formulae is comparatively simple, because the multiplicity of the integrals is less.

The final result must be put in a form independent of the choice of the z-axis along which the spin components are taken. This is achieved by using the operator $\hat{\mathbf{s}}_1\cdot\hat{\mathbf{s}}_2$ of the product of the spins, whose eigenvalues for parallel and antiparallel spins are respectively $\tfrac{1}{4}$ and $-\tfrac{3}{4}$. The result is

$$f = \frac{2\pi a\hbar^2}{m}\left[1+2a\left(\frac{3N}{\pi V}\right)^{1/3}\left(2+\frac{\cos\theta}{2\sin\frac{1}{2}\theta}\log\frac{1+\sin\frac{1}{2}\theta}{1-\sin\frac{1}{2}\theta}\right)\right] -$$
$$-\frac{8\pi a\hbar^2}{m}\hat{\mathbf{s}}_1\cdot\hat{\mathbf{s}}_2\left[1+2a\left(\frac{3N}{\pi V}\right)^{1/3}\left(1-\tfrac{1}{2}\sin\tfrac{1}{2}\theta\log\frac{1+\sin\frac{1}{2}\theta}{1-\sin\frac{1}{2}\theta}\right)\right], \tag{79.11}$$

where θ is the angle between the vectors $\mathbf{p}$ and $\mathbf{p}'$ (A. Abrikosov and I. Khalatnikov 1957).†

† The function (79.11) tends logarithmically to infinity when $\theta = \pi$. This is a consequence of the approximations made. A more precise investigation shows that, although $\theta = \pi$ is in fact a singular point of the function, the latter is zero there, not infinite. The inapplicability of formula (79.11) near $\theta = \pi$ is unimportant in subsequent applications, which contain integrals convergent at that point.

The effective mass of the quasi-particles is obtained by integrating f, using formula (68.11), and is given by

$$\frac{m^*}{m} = 1+\frac{8}{15}a^2\left(\frac{3N}{\pi V}\right)^{2/3}(7\log 2-1). \tag{79.12}$$

From formula (68.14) we can find the velocity of sound in the gas under consideration:

$$u^2 = \frac{\pi^{4/3}\hbar^2}{3^{1/3}m^2}\left(\frac{N}{V}\right)^{2/3}+\frac{2\pi a\hbar^2}{m^2}\frac{N}{V}\left[1+\frac{4}{15}a\left(\frac{3N}{\pi V}\right)^{1/3}(11-2\log 2)\right]. \tag{79.13}$$

Integrating u^2m/N with respect to N, we then obtain, in accordance with formula (68.12), the chemical potential μ of the gas (at absolute zero), and a further integration with respect to N gives the expression (79.9) for the energy of the ground state: $E_0 = \int\mu\, dN$.

Formula (79.9) represents the leading terms in an expansion of the energy of the gas in powers of $a(N/V)^{1/3}$. By similar though considerably more laborious calculations we could obtain some further terms in the expansion, since in a Fermi gas the ternary collisions contribute to the energy only in a comparatively high approximation. Among three colliding particles, at least two have the same spin projection; the co-ordinate wave function of the system must be antisymmetric with respect to these two particles. This means that the orbital angular momentum of the relative motion of these particles is equal to at least 1 (p state). The corresponding wave function† contains an extra power of the wave number k (as compared with the wave function of the s state), and therefore the probability of such a collision contains an extra factor k^2, i.e. is reduced by a factor $\sim (ka)^2 \sim a^2(N/V)^{2/3}$ in comparison with the probability of a "head-on" collision of particles not obeying PAULI's principle. Thus ternary collisions give a contribution to the energy only in terms which contain the volume as $V^{-2}.V^{-2/3}$. In other words, the characteristics of the binary collisions alone determine all terms in the expansion of the energy as far as those of order $(a\hbar^2N^2/mV)[a(N/V)^{1/3}]^4$ inclusive, i.e. three more terms after those written in (79.9). However, the characteristics of binary collisions will include not only the s-wave scattering amplitude for slow collisions (as in (79.9)) but also its energy derivatives and the p-wave scattering amplitude.

In conclusion we may make one further remark with a view to the comparison (in §80) with the properties of another type of Fermi gas. We have spoken here of quasi-particles whose number is equal to that of the gas particles; at $T = 0$ these quasi-particles occupy the Fermi sphere. This corresponds to the general treatment of a Fermi liquid given in §68 (where the

† See *Quantum Mechanics*, §33.

relation between the magnitude of the limiting momentum p_0 and the density of the liquid was formulated in this way). An equally reasonable view, however, is one according to which the elementary excitations appear only when $T \neq 0$ and the fully occupied Fermi sphere is unobservable. In this picture the elementary excitations correspond to quasi-particles outside the Fermi sphere and "holes" within it; the energy $\varepsilon = v(p-p_0)$ must be ascribed to the former, and $\varepsilon = v(p_0-p)$ to the latter. The statistical distribution of each is given by the Fermi distribution formula with zero chemical potential, in accordance with the fact that the number of quasi-particles is not constant but depends on the temperature (cf. (60.1)):

$$n_{\mathbf{p}} = 1/(e^{\varepsilon/T}+1). \tag{79.14}$$

§80. A degenerate "almost ideal" Fermi gas with attraction between the particles

At first sight the calculations given in §79 appear equally applicable for both repulsion and attraction between gas particles. In reality, however, in the case of attraction, the ground state of the system thus found is unstable with respect to a certain rearrangement which alters its nature and decreases the energy.

The origin of this rearrangement is indicated by the fact already mentioned in §79 that the expression (79.11) for the function $f(\theta)$ obtained by means of perturbation theory has a singularity at $\theta = \pi$, i.e. when the momenta of the two quasi-particles are in opposite directions. Near this singularity,

$$f \sim (1-4\hat{\mathbf{s}}_1\cdot\hat{\mathbf{s}}_2) \log (1-\sin \tfrac{1}{2}\theta),$$

i.e. the singularity exists only when the spins of the particles are antiparallel (since $1-4\mathbf{s}_1\cdot\mathbf{s}_2 = 0$ when the spins are parallel). The appearance of this singularity indicates the invalidity of perturbation theory (in the form used in §79) when applied to the interaction of pairs of particles which are (in **p**-space) near the Fermi surface and have opposite momenta and spins. As will be seen from the results derived below, in the case of attraction it is just this interaction which leads to novel effects.†

It is clear from the foregoing that the system of operators $\hat{a}_{\mathbf{p}\sigma}$, $\hat{a}_{\mathbf{p}\sigma}^+$ corresponding to free states of individual particles of the gas can not serve as a correct initial approximation of perturbation theory. Instead of these, we must immediately define new operators, which will be written as the linear

† The problem considered here is the basis of the theory of superconductivity due to Bardeen, Cooper and Schrieffer (1957). In the solution given below we mainly follow the method developed by N. N. Bogolyubov (1958).

combinations

$$\begin{aligned} \hat{b}_{\mathbf{p}-} &= u_p \hat{a}_{\mathbf{p}-} + v_p \hat{a}_{-\mathbf{p}+}^{+}, \\ \hat{b}_{\mathbf{p}+} &= u_p \hat{a}_{\mathbf{p}+} - v_p \hat{a}_{-\mathbf{p}-}^{+}, \end{aligned} \tag{80.1}$$

of the operators of particles with opposite momenta and spins; when the gas is isotropic, the coefficients u_p, v_p can depend only on the absolute magnitude of the momentum **p**. In order that these new operators should correspond to the creation and annihilation of quasi-particles, they must satisfy the same Fermi commutation rules as the previous operators:

$$\hat{b}_{\mathbf{p}\sigma}\hat{b}_{\mathbf{p}\sigma}^{+} + \hat{b}_{\mathbf{p}\sigma}^{+}\hat{b}_{\mathbf{p}\sigma} = 1 \tag{80.2}$$

and all other pairs of operators anticommute. For this to be so, the transformation coefficients must satisfy the condition

$$u_p^2 + v_p^2 = 1 \tag{80.3}$$

(u_p and v_p are assumed real, in order to make the quasi-particle occupation numbers real). The inverse transformation to (80.1) is

$$\begin{aligned} \hat{a}_{\mathbf{p}+} &= u_p \hat{b}_{\mathbf{p}+} + v_p \hat{b}_{-\mathbf{p}-}^{+}, \\ \hat{a}_{\mathbf{p}-} &= u_p \hat{b}_{\mathbf{p}-} - v_p \hat{b}_{-\mathbf{p}+}^{+}. \end{aligned} \tag{80.4}$$

For the same reasons (the importance of the interaction between particles with opposite momenta and spins), we retain in the second sum in the Hamiltonian (79.3) only the terms in which $\mathbf{p}_1 = -\mathbf{p}_2 \equiv \mathbf{p}$, $\mathbf{p}_1' = -\mathbf{p}_2' \equiv \mathbf{p}'$:

$$\hat{H} = \sum_{\mathbf{p},\sigma} \frac{p^2}{2m} \hat{a}_{\mathbf{p}\sigma}^{+}\hat{a}_{\mathbf{p}\sigma} - \frac{g}{V} \sum_{\mathbf{p},\mathbf{p}'} \hat{a}_{\mathbf{p}'+}^{+}\hat{a}_{-\mathbf{p}'-}^{+}\hat{a}_{-\mathbf{p}-}\hat{a}_{\mathbf{p}+}, \tag{80.5}$$

where $g = 4\pi\hbar^2 |a|/m$ (the scattering amplitude a is now negative).

In the subsequent calculations it will be convenient to avoid the necessity of explicitly taking account of the constancy of the number of particles (atoms) in the system. In accordance with the general rules of statistical physics (cf. §35), this must be done by replacing the Hamiltonian function H by the difference $H - \mu N$, where the number N of particles is itself regarded as a variable; the chemical potential is then determined, in principle, by the condition that the mean value $\bar{N}$ is equal to the given number of particles in the system. In the second quantisation method this means that the Hamiltonian $\hat{H}$ is replaced by the difference $\hat{H} - \mu\hat{N}$, where the operator

$$\hat{N} = \sum_{\mathbf{p},\sigma} \hat{a}_{\mathbf{p}\sigma}^{+}\hat{a}_{\mathbf{p}\sigma}.$$

We shall henceforward refer to this difference as the Hamiltonian and denote it by $\hat{H}$ simply.

We also use the notation

$$\xi_p = p^2/2m - \mu.$$

Since $\mu \cong p_0^2/2m$, near the Fermi surface we have

$$\xi_p = v(p-p_0), \tag{80.6}$$

where $v = p_0/m$. Subtracting $\mu\hat{N}$ from the expression (80.5), we can therefore write the initial Hamiltonian in the form

$$\hat{H} = \sum_{\mathbf{p},\sigma} \xi_p \hat{a}_{\mathbf{p}\sigma}{}^+ \hat{a}_{\mathbf{p}\sigma} - \frac{g}{V} \sum_{\mathbf{p},\mathbf{p}'} \hat{a}_{\mathbf{p}'+}{}^+ \hat{a}_{-\mathbf{p}'-}{}^+ \hat{a}_{-\mathbf{p}-} \hat{a}_{\mathbf{p}+}. \tag{80.7}$$

In this Hamiltonian we make the transformation (80.4), use the relations (80.2) and (80.3) and replace the summation suffix **p** by $-\mathbf{p}$, obtaining

$$\hat{H} = 2\sum_{\mathbf{p}} \xi_p v_p^2 + \sum_{\mathbf{p}} \xi_p (u_p^2 - v_p^2)(\hat{b}_{\mathbf{p}+}{}^+ \hat{b}_{\mathbf{p}+} + \hat{b}_{\mathbf{p}-}{}^+ \hat{b}_{\mathbf{p}-}) +$$

$$+ 2\sum_{\mathbf{p}} \xi_p u_p v_p (\hat{b}_{\mathbf{p}+}{}^+ \hat{b}_{-\mathbf{p}-}{}^+ + \hat{b}_{-\mathbf{p}-} \hat{b}_{\mathbf{p}+}) - \frac{g}{V} \sum_{\mathbf{p},\mathbf{p}'} B_{\mathbf{p}'}{}^+ B_{\mathbf{p}}, \tag{80.8}$$

$$B_{\mathbf{p}} = u_p^2 \hat{b}_{-\mathbf{p}-} \hat{b}_{\mathbf{p}+} - v_p^2 \hat{b}_{\mathbf{p}+}{}^+ \hat{b}_{-\mathbf{p}-}{}^+ + v_p u_p (\hat{b}_{-\mathbf{p}-} \hat{b}_{-\mathbf{p}-}{}^+ - \hat{b}_{\mathbf{p}+}{}^+ \hat{b}_{\mathbf{p}+}).$$

The coefficients u_p, v_p are now chosen from the condition of minimum energy E of the system for given entropy. The entropy is essentially defined by the combinatorial expression

$$S = -\sum_{\mathbf{p},\sigma} [n_{\mathbf{p}\sigma} \log n_{\mathbf{p}\sigma} + (1-n_{\mathbf{p}\sigma}) \log (1-n_{\mathbf{p}\sigma})]. \tag{80.9}$$

The condition mentioned is therefore equivalent to that of minimum energy for given occupation numbers $n_{\mathbf{p}\sigma}$ of the quasi-particles.

In the Hamiltonian (80.8) the diagonal matrix elements contain only terms which include the products

$$\hat{b}_{\mathbf{p}\sigma}{}^+ \hat{b}_{\mathbf{p}\sigma} = n_{\mathbf{p}\sigma}, \qquad \hat{b}_{\mathbf{p}\sigma} \hat{b}_{\mathbf{p}\sigma}{}^+ = 1 - n_{\mathbf{p}\sigma}.$$

We therefore find

$$E = 2\sum_{\mathbf{p}} \xi_p v_p^2 + \sum_{\mathbf{p}} \xi_p (u_p^2 - v_p^2)(n_{\mathbf{p}+} + n_{\mathbf{p}-}) - \frac{g}{V} \Big[\sum_{\mathbf{p},\mathbf{p}'} u_p v_p (1 - n_{\mathbf{p}+} - n_{\mathbf{p}-})\Big]^2. \tag{80.10}$$

Varying this expression with respect to the parameters u_p (using the relation $u_p^2 + v_p^2 = 1$), we obtain as the condition for a minimum

$$\frac{\delta E}{\delta u_p} = -\frac{2}{v_p}(1 - n_{\mathbf{p}+} - n_{\mathbf{p}-}) \times$$

$$\times \Big[2\xi_p u_p v_p - \frac{g}{V}(u_p^2 - v_p^2) \sum_{\mathbf{p}'} u_{p'} v_{p'} (1 - n_{\mathbf{p}'+} - n_{\mathbf{p}'-})\Big] = 0.$$

Hence we have the equation†

$$2\xi_p u_p v_p = \Delta(u_p^2 - v_p^2), \tag{80.11}$$

with the notation

$$\Delta = \frac{g}{V}\sum_{\mathbf{p}'} u_{p'}v_{p'}(1 - n_{\mathbf{p}'+} - n_{\mathbf{p}'-}). \tag{80.12}$$

From (80.11) and (80.3) we can express u_p and v_p in terms of ξ_p and Δ

$$u_p^2 = \frac{1}{2}\left(1 + \frac{\xi_p}{\sqrt{(\Delta^2+\xi_p^2)}}\right), \qquad v_p^2 = \frac{1}{2}\left(1 - \frac{\xi_p}{\sqrt{(\Delta^2+\xi_p^2)}}\right). \tag{80.13}$$

Substituting these values in (80.12), we obtain an equation which determines Δ:

$$\frac{g}{2V}\sum_{\mathbf{p}} \frac{1 - n_{\mathbf{p}+} - n_{\mathbf{p}-}}{\sqrt{(\Delta^2+\xi_p^2)}} = 1. \tag{80.14}$$

Let us now examine the relations just derived. We shall see that the quantity Δ plays a fundamental part in the theory of spectra of the type under consideration. Let us first calculate its value for $T = 0$ (denoted by Δ_0).

For $T = 0$ there are no quasi-particles: $n_{\mathbf{p}+} = n_{\mathbf{p}-} = 0$. It may be noted at once that the equation then obtained for Δ certainly could not have a solution for $g < 0$, i.e. in the case of repulsion (the signs of the two sides of the equation then being necessarily different).

Changing from summation to integration in (80.14), we obtain the equation

$$\frac{g}{2(2\pi\hbar)^3}\int \frac{4\pi p^2\,dp}{\sqrt{(\Delta_0^2+\xi_p^2)}} = 1. \tag{80.15}$$

The main contribution to this integral comes from the range of momenta where $\Delta_0 \ll v|p_0 - p| \ll vp_0 \sim \mu$ and the integral is logarithmic; the smallness of Δ_0 in comparison with μ is confirmed by the result.‡ Then

$$\int \frac{p^2\,dp}{\sqrt{[\Delta_0^2 + v^2(p_0-p)^2]}} \cong \frac{p_0^2}{v}\int \frac{d\xi}{\sqrt{(\Delta_0^2+\xi^2)}} \cong \frac{p_0^2}{v}\cdot 2\log\frac{\mu\beta}{\Delta_0},$$

† It may be noted that by virtue of this relation all the terms cancel in the Hamiltonian (80.8) which contain one pair of the operators $\hat{b}_{\mathbf{p}+}{}^{+}\hat{b}_{-\mathbf{p}-}{}^{+}$, i.e. the terms which correspond to the creation of one pair of quasi-particles with opposite momenta. These are just the terms which in the first order of perturbation theory might lead to divergent integrals.

‡ For $p \gg p_0$, the quantity $\xi_p \sim p^2$ and the integral as written here diverges as p. In reality, however, this divergence is spurious, and is removed by "renormalising" the relation between the constant g (i.e. the scattering amplitude a) and the interaction potential, in the same way as in §§78 and 79. A consistent carrying out of the fairly complicated calculation makes it possible to determine also the coefficient β in the logarithm in (80.16).

where β is a numerical coefficient. Thus we find

$$\frac{gp_0 m}{2\pi^2\hbar^3}\log\frac{\beta\mu}{\Delta_0} = 1, \tag{80.16}$$

whence

$$\Delta_0 \sim \mu e^{-2\pi^2\hbar^3/gmp_0} = \mu e^{-\pi\hbar/2p_0|a|}. \tag{80.17}$$

Since $p_0|a|/\hbar \ll 1$, Δ_0 is exponentially small compared with μ.

The most interesting result is the form of the energy spectrum of the system, i.e. the energy $\varepsilon_{\mathbf{p}+} = \varepsilon_{\mathbf{p}-} \equiv \varepsilon(\mathbf{p})$ of the elementary excitations. We shall find this from the change in the energy E of the whole system when the occupation numbers of the quasi-particles vary, i.e. by varying E with respect to $n_{\mathbf{p}\sigma}$. Since the values of u_p and v_p have already been chosen by equating to zero the derivatives of E with respect to them, the variation of E with respect to $n_{\mathbf{p}\sigma}$ can be effected with u_p and v_p constant. We then have from (80.10)

$$\varepsilon = \left(\frac{\delta E}{\delta n_{\mathbf{p}\sigma}}\right)_{u_p,\, v_p} = \xi_p(u_p{}^2 - v_p{}^2) + \frac{2g}{V}u_p v_p \sum_{\mathbf{p}'} u_{p'}v_{p'}(1 - n_{\mathbf{p}'+} - n_{\mathbf{p}'-}),$$

and substitution of (80.12) and (80.13) gives

$$\varepsilon(p) = \sqrt{(\Delta + \xi_p{}^2)}. \tag{80.18}$$

This demonstrates a remarkable property of the energy spectrum of the system under consideration: the energy of a quasi-particle cannot be less than Δ, and this value is reached when $p = p_0$. In other words, the excited states of the system are separated from the ground state by an *energy gap*. The quasi-particles must appear in pairs, since they have a half-integral spin. In this sense we can say that the magnitude of the gap is 2Δ.

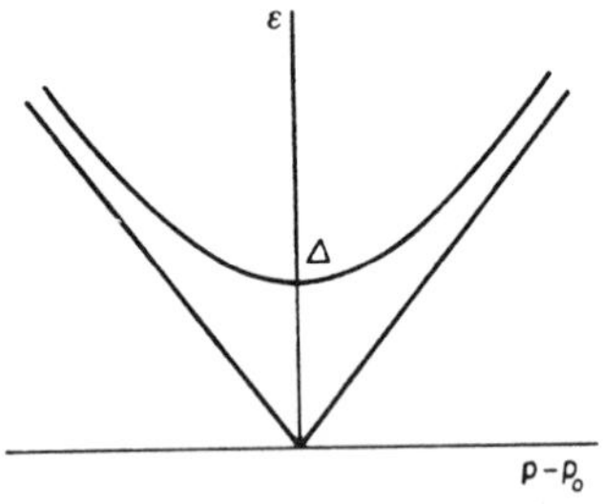

FIG. 11

The spectrum (80.18) satisfies the superfluidity condition derived in §67: the minimum value of ε/p is not zero. A Fermi gas with attraction between the particles must therefore be a superfluid.

Fig. 11 gives a comparison of the dispersion relations for quasi-particles in a superfluid (upper curve) and a normal system (these names referring to systems with spectra as considered in §§80 and 79 respectively). In the normal system the relation is given by the two straight lines $\varepsilon = v|p - p_0|$ in accordance with the discussion at the end of §79.

The gap width Δ depends on the temperature, i.e. the shape of the spectrum itself depends on the statistical distribution of quasi-particles—a situation similar to that which occurs for a normal Fermi liquid (§68). Since, as the temperature increases, the occupation numbers of the quasi-particles increase (tending to unity), we see from (80.14) that Δ decreases and becomes zero at some finite temperature T_c: the system passes from the superfluid to the normal state. This point is a phase transition of the second kind, like the transition in liquid helium (see §67).

The presence of the energy gap in the spectrum under consideration can be intuitively regarded as resulting from the formation of bound states by pairs of attracting particles; then 2Δ is the binding energy of such a pair, which must be expended in order to disrupt it. It is worthy of note that this effect occurs in a Fermi gas for any attraction, however weak. Having zero spin, the pairs behave as Bose objects, and a finite number can reach the level of lowest energy, i.e. the level with zero total momentum. In this intuitive interpretation this effect is entirely analogous to the steady increase of particles in a state of zero energy in a Bose gas (Bose–Einstein condensation, see §59).

The idea of bound pairs must not, of course, be taken too literally. It would be more precise to speak of a correlation between the states of a pair of particles in **p**-space, leading to a finite probability of the particles' having zero total momentum. The spread δp of momentum values in the region of correlation corresponds to an energy of the order of Δ, i.e. $\delta p \sim \Delta/v$. The corresponding length $l \sim \hbar/\delta p \sim \hbar v/\Delta$ gives the order of magnitude of the distances between particles with correlated momenta. This quantity for $T = 0$ is

$$l_0 \sim (\hbar/p_0)e^{\pi\hbar/2p_0|a|}, \tag{80.19}$$

and, since in a degenerate Fermi gas $\hbar/p_0$ is equal in order of magnitude to the distances between atoms, we see that l_0 is very large in comparison with the latter. This shows very clearly the arbitrariness of the idea of bound pairs.

Let us now determine explicitly the temperature dependence of the gap $\Delta(T)$. Writing equation (80.14) as

$$-1 + \frac{g}{2V}\sum_{\mathbf{p}} \frac{1}{\varepsilon} = \frac{g}{V}\sum_{\mathbf{p}} \frac{n_{\mathbf{p}}}{\varepsilon}$$

($n_{\mathbf{p}+} = n_{\mathbf{p}-} \equiv n_{\mathbf{p}}$), we note that the sum on the left-hand side differs from that for $T = 0$ only in that Δ_0 is replaced by Δ. Thus, using (80.16), we see that the left-hand side of the equation is equal to $(gp_0 m/2\pi^2\hbar^3)\log(\Delta_0/\Delta)$. On the right-hand side we substitute for $n_{\mathbf{p}}$ the Fermi distribution function (with zero chemical potential; cf. the end of §79) and change from summation to integration over p ($dp = d\xi/v$):

$$\log\frac{\Delta_0}{\Delta} = \int_{-\infty}^{\infty}\frac{d\xi}{\varepsilon(e^{\varepsilon/T}+1)} \equiv 2I(\Delta/T), \tag{80.20}$$

where

$$I(u) = \int_0^{\infty}\frac{dx}{\sqrt{(x^2+u^2)}(e^{\sqrt{(x^2+u^2)}}+1)};$$

the limits of integration can be taken as $\pm\infty$ because of the rapid convergence of the integral.

At low temperatures ($T \ll \Delta$) the calculation of the integral is simple† and

$$\Delta = \Delta_0[1-\sqrt{(2\pi T/\Delta_0)}e^{-\Delta_0/T}]. \tag{80.21}$$

In the region near the transition point, Δ is small, and the leading terms of the expansion of the integral $I(\Delta/T)$ give‡

$$\log\frac{\Delta_0}{\Delta} = \log\frac{\pi T}{\gamma\Delta}+\frac{7\zeta(3)}{8\pi^2}\frac{\Delta^2}{T^2}. \tag{80.22}$$

† For large u the first term in the expansion of $I(u)$ in powers of $1/u$ is

$$I(u) \cong \int_0^{\infty}\frac{dx}{u}e^{-u(1+x^2/2u^2)} = \sqrt{(\pi/2u)}e^{-u}.$$

‡ To expand the integral $I(u)$ for $u \to 0$ we add and subtract the integral

$$I_1 = \frac{1}{2}\int_0^{\infty}\left[\frac{1}{\sqrt{(x^2+u^2)}}-\frac{1}{x}\tanh\frac{1}{2}x\right]dx.$$

Then $I = I_1+I_2$, where

$$I_2 = \frac{1}{2}\int_0^{\infty}\left[\frac{1}{x}\tanh\frac{1}{2}x-\frac{1}{\sqrt{(x^2+u^2)}}\tanh\frac{1}{2}\sqrt{(x^2+u^2)}\right]dx.$$

In I_1 the integration of the first term in the integrand is elementary, and the second can be integrated by parts:

$$I_1 = -\log\frac{1}{2}u+\frac{1}{2}\int_0^{\infty}\frac{\log x}{\cosh^2\frac{1}{2}x}dx.$$

The integral in this formula is equal to $2\log(\pi/2\gamma)$ (where $\log\gamma = C = 0.577$ is EULER's constant), and so $I_1 = \log(\pi/\gamma u)$.

The integral I_2 vanishes when $u = 0$. The first term in its expansion in powers of u^2 is

$$I_2 = -\tfrac{1}{2}u^2\int_0^{\infty}\frac{dx}{x}\left(\frac{1}{x}\tanh\tfrac{1}{2}x\right)'.$$

Substituting the well-known expansion

$$\tanh\tfrac{1}{2}x = 4x\sum_{n=0}^{\infty}\frac{1}{\pi^2(2n+1)^2+x^2},$$

Hence, first of all, we see that Δ is zero at a temperature given by

$$T_c = \gamma\Delta_0/\pi = 0.57\Delta_0, \tag{80.23}$$

which is small compared with the degeneracy temperature $T_0 \sim \mu$. Then we have to the first order in $T_c - T$

$$\begin{aligned}\Delta &= \sqrt{\frac{8\pi^2}{7\zeta(3)}}T_c\sqrt{\left(1-\frac{T}{T_c}\right)} \\ &= 3.06T_c\sqrt{\left(1-\frac{T}{T_c}\right)}.\end{aligned} \tag{80.24}$$

It remains for us to calculate the thermodynamic quantities for the gas. Let us first consider the region of low temperatures, $T \ll \Delta$. To calculate the specific heat in this region, it is simplest to start from the formula

$$\delta E = \sum_{\mathbf{p}} \varepsilon_p(\delta n_{\mathbf{p}+} + \delta n_{\mathbf{p}-}) = 2\sum_{\mathbf{p}} \varepsilon_p \delta n_{\mathbf{p}}$$

for the change in the total energy when the quasi-particle occupation numbers are varied. Dividing by δT and changing from summation to integration, we obtain the specific heat:

$$C = V\frac{mp_0}{\pi^2\hbar^3}\int_{-\infty}^{\infty} \varepsilon \frac{\partial n}{\partial T}\, d\xi.$$

For $T \ll \Delta$, the quasi-particle distribution function $n \cong e^{-\varepsilon/T}$, and so we have

$$\begin{aligned}C &= V\frac{mp_0}{\pi^2\hbar^3 T^2} 2\int_0^{\infty} \varepsilon^2 e^{-\varepsilon/T}\, d\xi \\ &= V\frac{2mp_0\Delta^2}{\pi^2\hbar^3 T^2} e^{-\Delta/T}\int_0^{\infty} e^{-\xi^2/2T\Delta}\, d\xi,\end{aligned}$$

or, finally,

$$C = V\frac{mp_0\Delta_0}{\pi^2\hbar^3}\left(\frac{\Delta_0}{T}\right)^{3/2} e^{-\Delta_0/T}. \tag{80.25}$$

---

we obtain

$$\begin{aligned}I_2 &= 4u^2\sum_0^{\infty}\int_0^{\infty}\frac{dx}{[(2n+1)^2\pi^2+x^2]^2} \\ &= \frac{u^2}{\pi^2}\sum_0^{\infty}\frac{1}{(2n+1)^3} \\ &= u^2\cdot 7\zeta(3)/8\pi^2.\end{aligned}$$

Thus when $T \to 0$ the specific heat decreases exponentially—a direct consequence of the gap in the energy spectrum.

In the subsequent calculations it is convenient to start from the thermodynamic potential Ω, since the whole discussion takes place for a constant chemical potential of the system, not a constant number of particles in it.† We use the formula

$$(\partial\Omega/\partial g)_{T,V,\mu} = \overline{\partial\hat{H}/\partial g} \tag{80.26}$$

(cf. (11.4) and (15.11)), where the parameter λ is taken to be the coefficient g in the second term in the Hamiltonian (80.7), which describes the interaction of the gas particles. The mean value of this term is given by the last term in (80.10), which from (80.12) is equal to $-V\Delta^2/g$. We therefore have $\partial\Omega/\partial g = -V\Delta^2/g$. When $g \to 0$, the quantity Δ_0 tends to zero, and therefore so does Δ. Hence, integrating this equation with respect to g over the range 0 to g, we find the difference between the thermodynamic potential Ω in the superfluid state and the value which it would have in the normal state ($\Delta = 0$) at the same temperature:‡

$$\Omega_s - \Omega_n = -V\int_0^g \frac{\Delta^2}{g^2}\,dg. \tag{80.27}$$

According to the general rules (see (24.16)), this small correction, when expressed in terms of the appropriate variables, is the same for all the thermodynamic potentials.

At absolute zero $\Delta = \Delta_0$, and from (80.17) we have $d\Delta_0/dg = 2\pi^2\hbar^3\Delta_0/mp_0g^2$. Changing the integration over g in (80.27) into one over Δ_0, we find the following expression for the energy difference of the ground levels of the superfluid and normal systems:

$$E_s - E_n = -V(mp_c/4\pi^2\hbar^3)\,\Delta_0^2. \tag{80.28}$$

The negative sign of this difference indicates that, as already mentioned at the beginning of this section, the "normal" ground state of the system is unstable for the case of attraction between the gas particles.

† The chemical potential of the gas itself should not be confused with that of the quasi-particle gas (which is equal to zero).

‡ Here the following remark is needed, on account of the approximations which we have made from the start. When $g = 0$, no interaction between the particles remains in the Hamiltonian (80.7), and it might be thought that the result is an ideal Fermi gas, not a "normal" non-ideal gas. In reality, however, the Hamiltonian (80.7) already involves approximations such that there can be no question of calculating the absolute value of the energy. Interaction terms have been omitted which give a contribution to the energy (although they do not affect the form of the spectrum or the difference $\Omega_s - \Omega_n$), and this contribution is large compared with the exponentially small quantity (80.27); it is in fact the contribution proportional to Ng which has been calculated in (79.9).

Let us now consider the opposite case, $T \to T_c$. Differentiating equation (80.22) with respect to g, we find

$$\frac{7\zeta(3)}{4\pi^2 T^2}\Delta\, \mathrm{d}\Delta = \frac{\mathrm{d}\Delta_0}{\Delta_0} = \frac{2\pi^2\hbar^3}{mp_0}\frac{\mathrm{d}g}{g^2}.$$

Substituting $\mathrm{d}g/g^2$ from here in formula (80.27), and taking this formula as the difference of free energies, we have

$$F_s - F_n = -V\frac{7\zeta(3)mp_0}{8\pi^4\hbar^3 T^2}\int_0^{\Delta}\Delta^3\,\mathrm{d}\Delta$$

and finally, using (80.24),

$$F_s - F_n = -V\frac{2mp_0 T_c^2}{7\zeta(3)\hbar^3}\left(1-\frac{T}{T_c}\right)^2. \tag{80.29}$$

The entropy difference is therefore

$$S_s - S_n = -V\frac{4mp_0 T_c}{7\zeta(3)\hbar^3}\left(1-\frac{T}{T_c}\right).$$

The difference of specific heats tends to a finite value as $T \to T_c$:

$$C_s - C_n = \frac{4mp_0 T_c}{7\zeta(3)\hbar^3}V,$$

i.e. the specific heat has a discontinuity at the transition point, with $C_s > C_n$. The specific heat of the normal state is given (in the first approximation) by the ideal-gas formula (57.6), and, expressed in terms of p_0, it is $C_n = mp_0 TV/3$. The ratio of specific heats at the transition point is therefore

$$\frac{C_s(T_c)}{C_n(T_c)} = \frac{12}{7\zeta(3)}+1 = 2.43. \tag{80.30}$$

As regards its superfluidity, the gas is characterised by the division of its density ϱ into normal and superfluid "parts". According to (67.3) the normal part of the density is

$$\varrho_n = -\frac{4\pi}{3(2\pi\hbar)^3}\int p^4\frac{\mathrm{d}n}{\mathrm{d}\varepsilon}\,\mathrm{d}p$$

$$\cong -\frac{p_0^4}{3\pi^2\hbar^3 v}\int_{-\infty}^{\infty}\frac{\mathrm{d}n}{\mathrm{d}\varepsilon}\,\mathrm{d}\xi.$$

The total density of the gas is related to p_0 by $\varrho = mN/V = 8\pi p_0^3 m/3(2\pi\hbar)^3$, and hence

$$\frac{\varrho_n}{\varrho} = -2\int_0^{\infty}\frac{\mathrm{d}n}{\mathrm{d}\varepsilon}\,\mathrm{d}\xi. \tag{80.31}$$

This integral need not be specially calculated, since it can be reduced to the known function $\Delta(T)$. Differentiating equation (80.20) with respect to T and comparing the resulting integral with (80.31), we see that

$$\frac{\varrho}{\varrho_n} = 1 - \frac{\Delta}{T\Delta'}. \tag{80.32}$$

Substituting the limiting expressions (80.21), (80.24), we obtain

$$\begin{aligned} \varrho_n/\varrho &= \sqrt{(2\pi\Delta_0/T)}e^{-\Delta_0/T} && \text{as } T \to 0, \\ \varrho_s/\varrho &= 2(1-T/T_c) && \text{as } T \to T_c. \end{aligned} \tag{80.33}$$

Finally, two further comments are needed concerning the validity of the above formulae throughout the temperature range from 0 to T_c. Although the formulae given for small $T_c - T$ have a range of applicability, they actually become invalid sufficiently close to the transition point. Processes of mutual scattering of quasi-particles (ignored by us) must here cause the appearance of a singularity in the thermodynamic quantities, the nature of which is as yet unknown. This problem is related to the still unsolved general problem of the singularity of the thermodynamic quantities at a phase transition point of the second kind (see §138). Owing to the presence of a small parameter (the coupling constant g) in the model considered, the range of influence of this singularity will extend only to the immediate neighbourhood of the transition point.†

As in an "ordinary" superfluid (§67), so also in the Fermi gas considered here (unlike the Fermi gas with repulsion; cf. the penultimate footnote to §68) sound can be propagated, with a velocity $u \sim p_0/m$ determined in the usual way by the compressibility of the medium. This means that, together with the Fermi-type excitation spectrum discussed here, the spectrum of such a gas contains also a phonon (Bose) branch of excitations. The specific heat due to the phonons is proportional to T^3 with a very small coefficient, but as $T \to 0$ it must ultimately predominate over the exponentially decreasing specific heat (80.25).‡

† The size of the neighbourhood can be estimated from the magnitude of fluctuations in the model considered: the theory becomes inapplicable when the fluctuations are no longer small. This estimate gives the condition $(T-T_c)/T_c \sim (T_c/\mu)^4$; see V. L. Ginzburg, *Soviet Physics Solid State* **2**, 1824, 1961.

‡ This effect does not occur in the charged "electron liquid" in superconductors.

CHAPTER VIII

PHASE EQUILIBRIUM

§81. Conditions of phase equilibrium

THE (equilibrium) state of a homogeneous body is determined by specifying any two thermodynamic quantities, for example the volume V and the energy E. There is, however, no reason to suppose that for every given pair of values of V and E the state of the body corresponding to thermal equilibrium will be homogeneous. It may be that for a given volume and energy in thermal equilibrium the body is not homogeneous, but separates into two homogeneous parts in contact which are in different states.

Such states of matter which can exist simultaneously in equilibrium with one another and in contact are described as different *phases*.

Let us write down the conditions for equilibrium between two phases. First of all, as for any bodies in equilibrium, the temperatures T_1 and T_2 of the two phases must be equal:

$$T_1 = T_2.$$

The pressures in the two phases must also be equal:

$$P_1 = P_2,$$

since the forces exerted by the two phases on each other at their surface of contact must be equal and opposite. Finally, the chemical potentials of the two phases must be equal:

$$\mu_1 = \mu_2;$$

this condition is derived for the two phases in exactly the same way as in §25 for any two adjoining regions of a body. If the potentials are expressed as functions of pressure and temperature, and the common temperature and pressure are denoted by T and P, we have

$$\mu_1(P, T) = \mu_2(P, T), \tag{81.1}$$

whence the pressure and temperature of phases in equilibrium can be expressed as functions of each other. Thus two phases can not be in equilibrium with each other at all pressures and temperatures; when one of these is given, the other is completely determined.

If the pressure and temperature are plotted as co-ordinates, the points a which phase equilibrium is possible will lie on a curve (the phase equilibriur curve), and the points lying on either side of the curve will represent homo geneous states of the body. When the state of the body varies along a lin which intersects the equilibrium curve, the phases separate at the point o intersection and the body then changes to the other phase. It may be note that, when the state of the body changes slowly, it may sometimes remai homogeneous even when the phases should separate in complete equilibrium Examples are supercooled vapours and superheated liquids, but such state are only metastable.

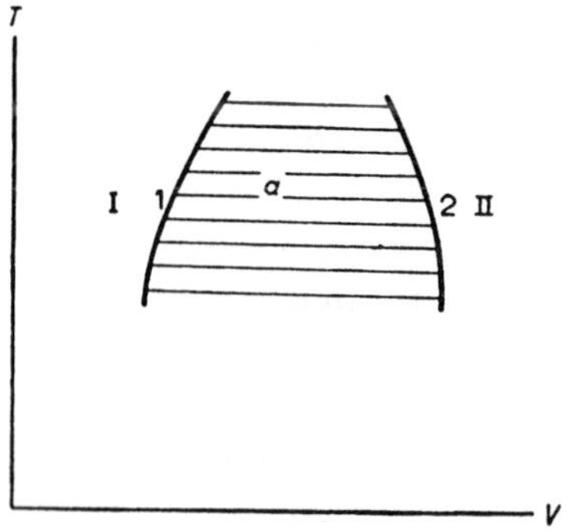

FIG. 12

If the equilibrium of phases is plotted in a diagram with temperature an volume (of a fixed quantity of matter) as co-ordinates, then the states i which two phases exist simultaneously will occupy a whole region of th plane, and not simply a curve. This difference from the (P, T) diagram arise because the volume V, unlike the pressure, is not the same for the two phases The resulting diagram is of the kind shown in Fig. 12. Points in the regions and II on either side of the hatched area correspond to homogeneous firs and second phases. The hatched area represents states in which the two phase are in equilibrium: at any point a the phases I and II are in equilibrium, wit specific volumes given by the abscissae of the points 1 and 2 which lie on horizontal line through a. It is easily deduced directly from the mass balanc that the quantities of phases I and II are inversely proportional to the length of the segments $a1$ and $a2$; this is called the *lever rule*.

In a similar way to the conditions for equilibrium of two phases, th equilibrium of three phases of the same substance is governed by the equa tions

$$P_1 = P_2 = P_3, \qquad T_1 = T_2 = T_3, \qquad \mu_1 = \mu_2 = \mu_3. \qquad (81.2)$$

If the common values of the pressure and temperature of the three phases are again denoted by P and T, we have the conditions

$$\mu_1(P, T) = \mu_2(P, T) = \mu_3(P, T). \tag{81.3}$$

These give two equations in the two unknowns P and T, and their solutions are specific pairs of values of P and T. The states in which three phases are simultaneously present (called ***triple points***) in the (P, T) diagram are represented by isolated points which are the points of intersection of the equilibrium curves of each pair of phases (Fig. 13, where regions I, II, III are those of the three homogeneous phases). The equilibrium of more than three phases of the same substance is clearly impossible.

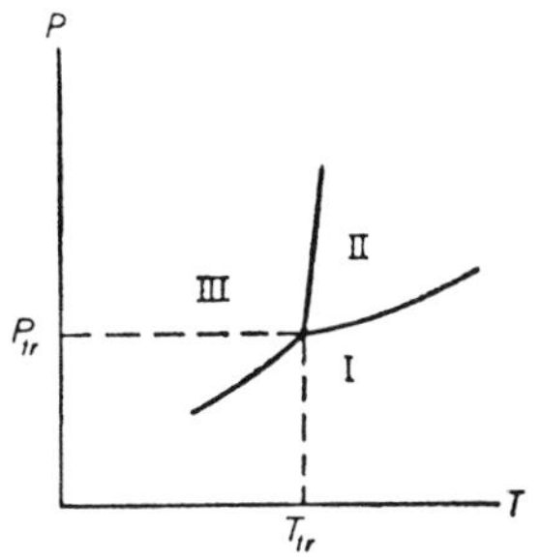

FIG 13

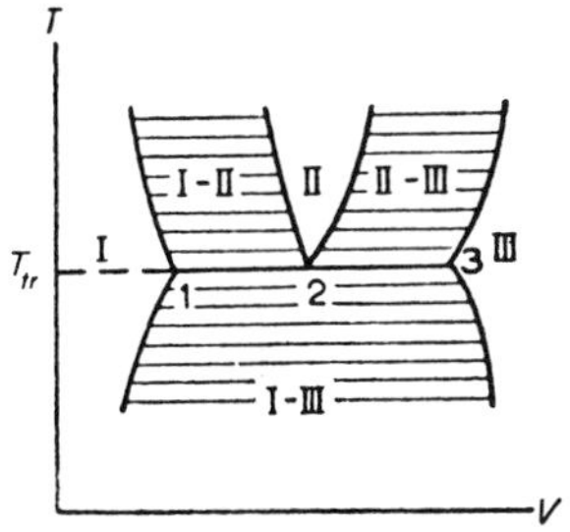

FIG. 14

In the (T, V) diagram the neighbourhood of the triple point has the appearance shown in Fig. 14, where the hatched areas are those of equilibrium of two phases; the specific volumes of the three phases in equilibrium at the triple point (at the temperature T_{tr}) are given by the abscissae of the points 1, 2, 3.

The change from one phase to another is accompanied by the evolution or absorption of a certain quantity of heat, called the *latent heat of transition* or simply the *heat of transition*. According to the conditions of equilibrium such a transition occurs at constant pressure and temperature. But in a process occurring at constant pressure the quantity of heat absorbed by the body is equal to the change in its heat function. The heat of transition q per molecule is therefore

$$q = w_2 - w_1, \tag{81.4}$$

where w_1 and w_2 are the heat functions per molecule of the two phases. The quantity q is positive if heat is absorbed by the body in changing from the first to the second phase, and negative if heat is evolved.

Since, for bodies consisting of a single substance, μ is the thermodynamic potential per molecule, we can write $\mu = \varepsilon - Ts + Pv$ (where ε, s, v are the molecular energy, entropy and volume). The condition $\mu_1 = \mu_2$ therefore gives

$$(\varepsilon_2-\varepsilon_1)-T(s_2-s_1)+P(v_2-v_1) = (w_2-w_1)-T(s_2-s_1) = 0,$$

where T and P are the temperature and pressure of both phases; hence

$$q = T(s_2-s_1). \tag{81.5}$$

We may note that this formula also follows directly from $q = \int T\,ds$ with T constant; the latter formula is applicable here, since the transition is reversible: the two phases remain in equilibrium during the transition process.

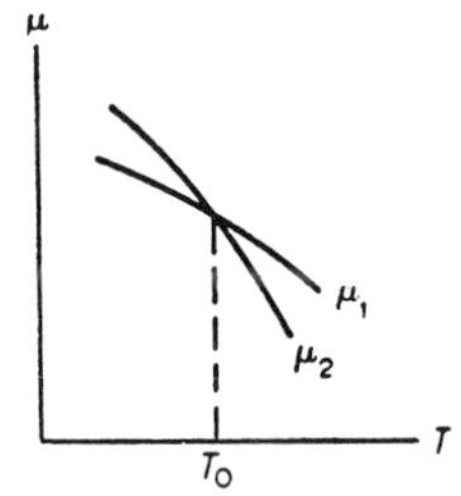

FIG. 15

Let the two curves in Fig. 15 represent the chemical potentials of the two phases as functions of temperature (at a given pressure). The point of intersection of the two curves gives the temperature T_0 for which (at the given pressure) the two phases can exist in equilibrium. At any other temperature only one or the other phase can exist. It is easy to see that at temperatures below T_0 the first phase exists, i.e. is stable, and at temperatures above T_0 the second phase. This follows because the stable state is the one where μ is smaller, since the thermodynamic potential tends to a minimum for given P and T. At the point of intersection of the two curves, the derivative $\partial\mu_1/\partial T$ is greater than $\partial\mu_2/\partial T$, i.e. the entropy of the first phase, $s_1 = -\partial\mu_1/\partial T$, is less than that of the second phase, $s_2 = -\partial\mu_2/\partial T$. The heat of transition $q = T(s_2-s_1)$ is therefore positive. Thus we conclude that, if the body goes from one phase to another when the temperature is raised, heat is absorbed in the process. This result could also be derived from LE CHATELIER's principle.

PROBLEMS

PROBLEM 1. Determine the temperature dependence of the saturated vapour pressure above a solid. The vapour is regarded as an ideal gas, and both the gas and the solid have constant specific heats.

SOLUTION. The chemical potential of the vapour is given by formula (43.3) and that of the solid by (62.6); since the saturated vapour pressure is relatively small, the quantity PV may be neglected for the solid, and Φ taken as equal to F. Equating the two expressions, we find

$$P = \text{constant}\times T^{(c_{p2}-c_1)}e^{(\varepsilon_{01}-\varepsilon_{02})/T},$$

where the suffix 1 refers to the solid and 2 to the vapour.

In the same approximation, the heat function of the solid may be taken as equal to its energy; the heat of transition (heat of sublimation) $q = w_2 - w_1$ is

$$q = (c_{p2} - c_1)T + (\varepsilon_{02} - \varepsilon_{01}).$$

In particular, for $T = 0$ the heat of transition is $q_0 = \varepsilon_{02} - \varepsilon_{01}$, so that we can write

$$P = \text{constant} \times T^{(c_{p2}-c_1)} e^{-q_0/T}.$$

PROBLEM 2. Determine the rate of evaporation from a condensed state into a vacuum.

SOLUTION. The rate of evaporation into a vacuum is determined by the number of particles which leave unit surface area of the body per unit time. Let us consider a body in equilibrium with its saturated vapour. Then the number of particles leaving the surface is equal to the number which strike and "adhere to" this surface in the same time, i.e. $P_0(1-R)/\sqrt{(2\pi mT)}$, where $P_0 = P_0(T)$ is the saturated vapour pressure, and R a mean "reflection coefficient" for gas particles colliding with the surface (see (39.2)). If P_0 is not too large, the number of particles leaving the surface of the body is independent of whether there is vapour in the surrounding space, so that the above expression gives the required rate of evaporation into a vacuum.

§82. The Clapeyron–Clausius formula

Let us differentiate both sides of the equilibrium condition $\mu_1(P, T) = \mu_2(P, T)$ with respect to temperature, bearing in mind, of course, that the pressure P is not an independent variable but a function of temperature determined by this same equation. We therefore write

$$\frac{\partial \mu_1}{\partial T} + \frac{\partial \mu_1}{\partial P}\frac{dP}{dT} = \frac{\partial \mu_2}{\partial T} + \frac{\partial \mu_2}{\partial P}\frac{dP}{dT};$$

since $(\partial\mu/\partial T)_P = -s$, $(\partial\mu/\partial P)_T = v$ (see (24.12)), this gives

$$\frac{dP}{dT} = \frac{s_1 - s_2}{v_1 - v_2}, \tag{82.1}$$

where s_1, v_1, s_2, v_2 are the molecular entropies and volumes of the two phases.

In this formula the difference $s_1 - s_2$ may conveniently be expressed in terms of the heat of transition from one phase to the other. Substituting $q = T(s_2 - s_1)$, we obtain the *Clapeyron–Clausius formula*:

$$\frac{dP}{dT} = \frac{q}{T(v_2 - v_1)}. \tag{82.2}$$

This gives the change in the pressure of phases in equilibrium when the temperature changes or, in other words, the change in pressure with temperature along the phase equilibrium curve. The same formula written as

$$\frac{dT}{dP} = \frac{T(v_2 - v_1)}{q}$$

gives the change in the temperature of the transition between phases (e.g. freezing point or boiling point) when the pressure changes. Since the molecular volume of the gas is always greater than that of the liquid, and heat is absorbed in the passage from liquid to vapour, it follows that the boiling point always rises when the pressure increases ($\mathrm{d}T/\mathrm{d}P > 0$). The freezing point may rise or fall with increasing pressure, according as the volume increases or decreases on melting.

All these consequences of formula (82.2) are in full agreement with LE CHATELIER's principle. Let us consider, for example, a liquid in equilibrium with its saturated vapour. If the pressure is increased, the boiling point must rise, and so some of the vapour will become liquid; this in turn will cause a decrease in pressure, so that the system acts as if to oppose the interaction which disturbs its equilibrium.

Let us consider the particular case of formula (82.2) which relates to equilibrium between a solid or liquid and its vapour. Then formula (82.2) determines the change in the saturated vapour pressure with temperature.

The volume of a gas is usually much greater than that of a liquid or solid containing the same number of particles. We can therefore neglect the volume v_1 in (82.2) in comparison with v_2 (the second phase being taken to be a gas), i.e. write $\mathrm{d}P/\mathrm{d}T = q/Tv_2$. Regarding the vapour as an ideal gas, we can express its volume in terms of the pressure and temperature by $v_2 = T/P$; then $\mathrm{d}P/\mathrm{d}T = qP/T^2$, or

$$\mathrm{d}\log P/\mathrm{d}T = q/T^2. \tag{82.3}$$

We may note that, in temperature intervals over which the heat of transition may be regarded as constant, the saturated vapour pressure varies exponentially with the temperature ($\sim e^{-q/T}$).

PROBLEMS

PROBLEM 1. Determine the specific heat of a vapour along the equilibrium curve of the liquid and its saturated vapour (i.e. the specific heat for a process in which the liquid is always in equilibrium with its saturated vapour). The vapour is regarded as an ideal gas.

SOLUTION. The required specific heat $h = T\,\mathrm{d}s/\mathrm{d}T$, where $\mathrm{d}s/\mathrm{d}T$ is the derivative along the equilibrium curve:

$$h = T\frac{\mathrm{d}s}{\mathrm{d}T} = T\left(\frac{\partial s}{\partial T}\right)_P + T\left(\frac{\partial s}{\partial P}\right)_T\frac{\mathrm{d}P}{\mathrm{d}T} = c_p - T\left(\frac{\partial v}{\partial T}\right)_P\frac{\mathrm{d}P}{\mathrm{d}T}.$$

Substituting the expression given by (82.3) for $\mathrm{d}P/\mathrm{d}T$, and $v = T/P$, we find

$$h = c_p - q/T.$$

At low temperatures, h is negative, i.e. if heat is removed in such a way that the vapour is always in equilibrium with the liquid, its temperature can increase.

PROBLEM 2. Determine the change in the volume of a vapour with temperature in a process where the vapour is always in equilibrium with the liquid (i.e. along the equilibrium curve of the liquid and its vapour).

SOLUTION. We have to determine the derivative $\mathrm{d}v/\mathrm{d}T$ along the equilibrium curve:

$$\frac{\mathrm{d}v}{\mathrm{d}T} = \left(\frac{\partial v}{\partial T}\right)_P + \left(\frac{\partial v}{\partial P}\right)_T \frac{\mathrm{d}P}{\mathrm{d}T}.$$

Substituting from (82.3), and $v = T/P$, we find

$$\frac{\mathrm{d}v}{\mathrm{d}T} = \frac{1}{P}\left(1 - \frac{q}{T}\right).$$

At low temperatures $\mathrm{d}v/\mathrm{d}T < 0$, i.e. the vapour volume decreases with increasing temperature in the process considered.

§83. The critical point

The phase equilibrium curve (in the *PT*-plane) may terminate at a certain point (Fig. 16), called the *critical point*; the corresponding temperature and pressure are the *critical temperature* and the *critical pressure*. At temperatures above T_c and pressures higher than P_c, no difference of phases exists, the substance is always homogeneous, and we can say that at the critical point the two phases become identical. The concept of the critical point was first used by D. I. MENDELEEV (1860).

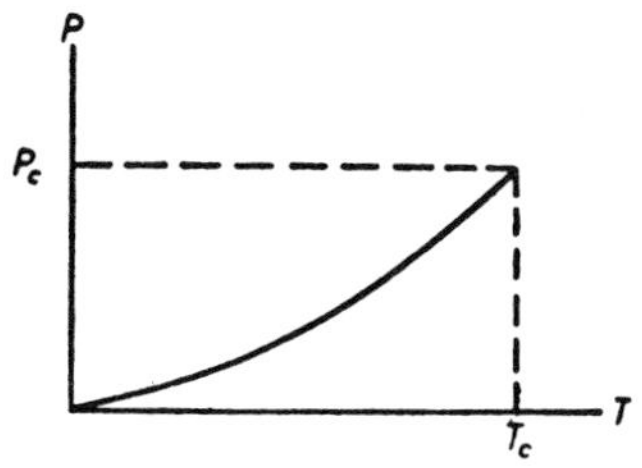

FIG. 16

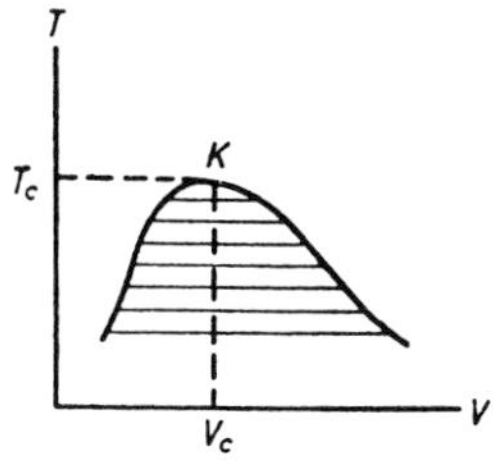

FIG. 17

In the co-ordinates T, V, when there is a critical point, the equilibrium diagram appears as in Fig. 17. As the temperature approaches its critical value, the specific volumes of the phases in equilibrium become closer, and at the critical point (K in Fig. 17) they coincide. The diagram in the co-ordinates P, V has a similar form.

When there is a critical point, a continuous transition can be effected between any two states of the substance without its ever separating into two phases. To achieve this, the state must be varied along a curve which passes round the critical point and nowhere intersects the equilibrium curve. In this sense, when there is a critical point, the concept of different phases is itself arbitrary, and it is not possible to say in every case which states have one phase and which have two. Strictly speaking, there can be said to be two phases only when they exist simultaneously and in contact—that is, at points lying on the equilibrium curve.

It is clear that the critical point can exist only for phases such that the difference between them is purely quantitative, for example a liquid and a gas differing only in the degree of interaction between the molecules.

On the other hand, such phases as a liquid and a solid (crystal), or different crystal modifications of a substance, are qualitatively different, since they have different internal symmetry (this is discussed further in Chapter XIII). It is clear that we can say only that a particular symmetry property (symmetry element) exists or does not exist; it can appear or disappear only as a whole, not gradually. In each state the body will have one symmetry or the other, and so we can always say to which of the two phases it belongs. The critical point therefore cannot exist for such phases, and the equilibrium curve must either go to infinity or terminate by intersecting the equilibrium curves of other phases.

An ordinary phase transition point is not a mathematical singularity of the thermodynamic quantities of the substance. For each of the phases can exist (though in a metastable state) beyond the transition point; the thermodynamic inequalities are not violated at that point. At the transition point the chemical potentials of the two phases are equal: $\mu_1(P, T) = \mu_2(P, T)$; but this point has no special property with respect to either one of the functions $\mu_1(P, T)$ and $\mu_2(P, T)$.†

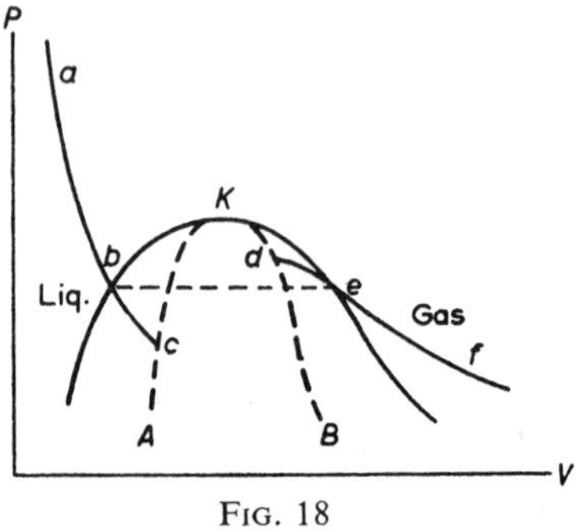

FIG. 18

Let us plot in the PV-plane an *isotherm* of the liquid and gas, i.e. the curve of P as a function of V in an isothermal expansion of a homogeneous body (*abc* and *def* in Fig. 18). According to the thermodynamic inequality $(\partial P/\partial V)_T < 0$, P is a decreasing function of V. This slope of the isotherms must continue

† It must be noted, however, that there is some degree of arbitrariness in these statements, due to an indeterminateness of $\mu(P, T)$ in the region of metastability. The metastable state is one of partial equilibrium, having a certain relaxation time, in this case for the process of formation of nuclei of a new phase (see §150). The thermodynamic functions in such a state can therefore be defined only without taking account of these processes, and they can not be regarded as the analytic continuation of the functions from the region of stability corresponding to the complete equilibrium states of the substance.

for some distance beyond their intersections (b and e) with the liquid-gas equilibrium curve; the segments bc and ed of the isotherms correspond to metastable superheated liquid and supercooled vapour, in which the thermodynamic inequalities are still satisfied.† If we use the fact that the points b and e, which correspond to liquid and gas in equilibrium with each other, have the same ordinate P, it is clear that the two isotherms cannot pass continuously into each other; there must be a discontinuity between them. The isotherms terminate at points (c and d) where the thermodynamic inequality ceases to hold, i.e. $(\partial P/\partial V)_T$ becomes equal to zero. By constructing the locus of the points of termination of the isotherms of the liquid and gas, we obtain a curve (AKB in Fig. 18) on which the thermodynamic inequalities are violated (for a homogeneous body), and which is the boundary of a region in which the body can never exist in a homogeneous state. The regions between this curve and the phase equilibrium curve correspond to superheated liquid and supercooled vapour. It is evident that at the critical point the two curves must touch, as shown in Fig. 18.

Of the points lying on the curve AKB itself, only one, namely the critical point K, corresponds to an actually existing state of the homogeneous body; this is the only point where the curve reaches the region of stable homogeneous states. According to the above discussion, we have in the critical state

$$(\partial P/\partial V)_T = 0. \tag{83.1}$$

It will be shown in §84 that, for such a state to be stable, the second derivative must also be zero:

$$(\partial^2 P/\partial V^2)_T = 0. \tag{83.2}$$

The conditions (83.1) and (83.2) are two equations in two unknowns, and can be satisfied only at an isolated point, the critical point of the substance.

It is worth mentioning that the condition (83.1) at the critical point can also be derived from the following simple considerations. Near the critical point, the specific volumes of the liquid and the vapour are almost the same; denoting them by V and $V+\delta V$, we can write the condition for equal pressures of the two phases as

$$P(V, T) = P(V+\delta V, T). \tag{83.3}$$

Expanding the right-hand side in powers of δV and dividing by the small but finite quantity δV, we have

$$\left(\frac{\partial P}{\partial V}\right)_T + \tfrac{1}{2}\delta V \left(\frac{\partial^2 P}{\partial V^2}\right)_T + \ldots = 0. \tag{83.4}$$

† A complete-equilibrium isothermal change of state between the points b and e corresponds, of course, to the horizontal straight line be, on which separation into two phases occurs.

Hence we see that, when δV tends to zero, i.e. at the critical point, $(\partial P/\partial V)_T$ must tend to zero.

In connection with the discussion of metastable states of a liquid, the following interesting point may be noted. The segment of the isotherm which corresponds to superheated liquid (*bc* in Fig. 18) may lie partly below the axis of abscissae. That is, a superheated liquid may have a negative pressure. Such a liquid exerts an inward force on its boundary surface. Thus the pressure is not necessarily positive, and there can exist in Nature states (though only metastable ones) of a body with negative pressures, as already mentioned in §12.

§84. Properties of matter near the critical point

There is very good reason to suppose that the boundary of the region where a homogeneous body cannot exist (the line *AKB* in Fig. 18) is a line of singularities of the thermodynamic quantities. However, no theoretical analysis of this problem has yet been made, and the nature of the singularity is unknown; the theory given in this section is essentially based on the hypothesis that on the line in question, and in particular at the critical point itself, the thermodynamic quantities of the substance (as functions of the variables V and T) have no mathematical singularity, so that this curve is characterised only by the vanishing of $(\partial P/\partial V)_T$.† In this situation it is impossible to say which of the results of the discussion will be retained in a correct theory and which will undergo substantial changes.

With this restriction in mind, let us begin by deriving the conditions for stability of the state of the substance at the critical point itself. In the derivation of the thermodynamic inequalities in §21 we started from the condition (21.1), which led to the inequality (21.2), which is satisfied if the conditions (21.3),(21.4) hold. The case $(\partial P/\partial V)_T = 0$ of interest here corresponds to the particular case of the extremum conditions with the equality sign in (21.4):

$$\frac{\partial^2 E}{\partial S^2}\frac{\partial^2 E}{\partial V^2}-\left(\frac{\partial^2 E}{\partial V \partial S}\right)^2 = 0. \tag{84.1}$$

The quadratic form in (21.2) may now be either positive or zero, depending on the values of δS and δV, and so the question whether $E - T_0 S + P_0 V$ has a minimum requires further investigation.

We must obviously examine the case where in fact the equality sign occurs in (21.2):

$$\frac{\partial^2 E}{\partial S^2}(\delta S)^2 + 2\frac{\partial^2 E}{\partial S \partial V}\delta S\,\delta V + \frac{\partial^2 E}{\partial V^2}(\delta V)^2 = 0. \tag{84.2}$$

† As functions of the variables P, T, the thermodynamic quantities have a singularity due to the vanishing of the Jacobian $\partial(P, T)/\partial(V, T)$ of the transformation of variables.

Using (84.1), this equation may be written

$$\frac{1}{\partial^2 E/\partial S^2}\left(\frac{\partial^2 E}{\partial S^2}\delta S+\frac{\partial^2 E}{\partial S\partial V}\delta V\right)^2 = \frac{1}{\partial^2 E/\partial S^2}\left[\delta\frac{\partial E}{\partial S}\right]^2 = (C_v/T)(\delta T)^2 = 0.$$

Thus the equation (84.2) implies that we must consider deviations from equilibrium at constant temperature ($\delta T = 0$).

At constant temperature the original inequality (21.1) becomes $\delta F+P\delta V > 0$. Expanding δF in powers of δV and making use of the assumption that $\partial^2 F/\partial V^2 = -(\partial P/\partial V)_T = 0$, we find

$$\frac{1}{3!}\left(\frac{\partial^2 P}{\partial V^2}\right)_T\delta V^3+\frac{1}{4!}\left(\frac{\partial^3 P}{\partial V^3}\right)_T\delta V^4+\ldots < 0.$$

If this inequality holds for all δV, we must have

$$(\partial^2 P/\partial V^2)_T = 0, \qquad (\partial^3 P/\partial V^3)_T < 0. \tag{84.3}$$

It may be noted that the case of equality in (21.3) ($\partial^2 E/\partial S^2 = 0$, or what is the same thing, $C_v = \infty$) is impossible in this discussion, since the condition (21.4) would then be violated. The simultaneous vanishing of the two expressions (21.3) and (21.4) is also impossible: if we add a further condition to the vanishing of $(\partial P/\partial V)_T$ and $(\partial^2 P/\partial V^2)_T$, there result three equations in two unknowns, which in general have no common solution.†

Let us now consider the equation of state of a substance near the critical point. We shall use the notation‡

$$T-T_c = t, \qquad V-V_c = v,$$

and consider the properties of the substance when v and t are small. Taking only the leading terms in the expansion, we write

$$-(\partial P/\partial V)_T = At+Bv^2. \tag{84.4}$$

There is no term proportional to v, since the coefficient of v is the second derivative of the pressure with respect to the volume, which is zero at the critical point. The term containing the product tv is always less than At, and the same is true of the term proportional to t^2. The term Bv^2 must be retained, on the other hand, since, although both t and v are assumed small, nothing is assumed regarding their relative magnitude, so that the term Bv^2 is not necessarily less than At.

† The vanishing of the derivatives $(\partial P/\partial V)_T$ and $(\partial^2 P/\partial V^2)_T$ at the critical point may apparently be regarded as confirmed by experimental data. It may not be so certain that the third derivative $(\partial^3 P/\partial V^3)_T$ is finite. There are also results which indicate that the specific heat C_v becomes infinite at the critical point, which the theory described here does not give reason to assume.

‡ The v used in this section should not be confused with that in other sections, which denotes the molecular volume.

Since $(\partial^3 P/\partial V^3)_T < 0$ at the critical point, the coefficient B must be positive. Moreover, we must have $-(\partial P/\partial V)_T > 0$ at all points near the critical point which represent a stable state of the substance. This applies, in particular, to all points with $t > 0$ (where separation into phases never occurs). Hence it follows that the coefficient A is also positive. From (84.4) we obtain for the pressure the expression

$$P = -Atv - \tfrac{1}{3}Bv^3 + f(t), \tag{84.5}$$

where $f(t)$ is a function of t only, which is unimportant for our purposes.

Formula (84.5) determines the form of the isotherms of a homogeneous substance near the critical point. For $t > 0$, the isotherm $P(v)$ is a monotonically decreasing function (curve 1 in Fig. 19). The isotherm (curve 2) which corresponds to the critical temperature ($t = 0$) has a point of inflection at the critical point ($v = 0$).

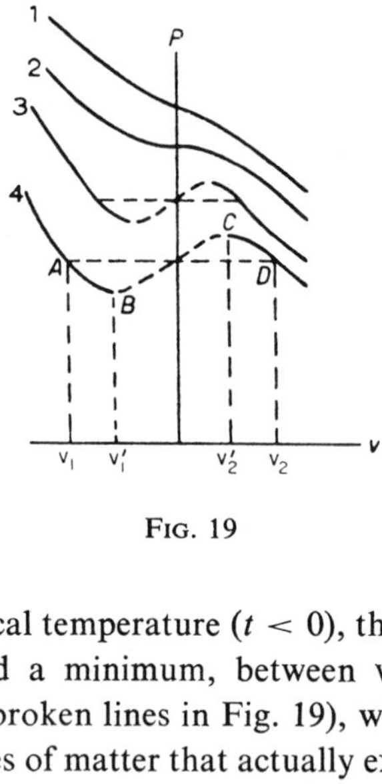

FIG. 19

Finally, below the critical temperature ($t < 0$), the isotherms (curves 3 and 4) have a maximum and a minimum, between which is a segment with $(\partial P/\partial v)_t > 0$ (shown by broken lines in Fig. 19), which does not correspond to any homogeneous states of matter that actually exist in Nature.†

As has been explained in §83, the equilibrium passage from liquid to gas corresponds to a straight horizontal segment (AD on isotherm 4), AB is the isotherm of the superheated liquid, and DC that of the supercooled vapour.

Let us determine the abscissae of the points A and D, i.e. the volumes v_1 and v_2 of liquid and gas in equilibrium.‡ We can write down the condition

† In reality we may expect that there will be no curve BC in a theory which correctly takes account of the singularity of the thermodynamic quantities at the boundary of the metastable states.

‡ By the volume we everywhere mean, of course, the volume of a given quantity of the substance.

of phase equilibrium $\mu_1 = \mu_2$ in the form

$$\int_1^2 \mathrm{d}\mu = 0,$$

where the integral is taken along the path of transition from a state of one phase to a state of the other phase. Let us take the integration along the isotherm *ABCD*. Since for $T =$ constant we have $\mathrm{d}\mu = V\,\mathrm{d}P = (V_c+v)\,\mathrm{d}P$, it follows that

$$\int_1^2 \mathrm{d}\mu = \int_1^2 v\,\mathrm{d}P + V_c(P_2-P_1) = 0;$$

but the pressures of the two phases are equal in equilibrium, $P_1 = P_2$, so that we have finally

$$\int_1^2 v\,\mathrm{d}P = \int_{v_1}^{v_2} v(\partial P/\partial v)_t\,\mathrm{d}v = 0. \tag{84.6}$$

Substituting the expression (84.4) near the critical point, we find that the integrand is an odd function of v, and so we must clearly have $v_1 = -v_2$. Using now the condition $P_1 = P_2$ and formula (84.5), we find

$$Atv_1+\tfrac{1}{3}Bv_1^3 = -Atv_1-\tfrac{1}{3}Bv_1^3,$$

i.e. $At+\frac{1}{3}Bv_1^2 = 0$. Hence

$$v_1 = -v_2 = -\sqrt{(-3At/B)}. \tag{84.7}$$

Thus v_1 and v_2 are equal in absolute magnitude and proportional to the square root of T_c-T. In other words, the phase equilibrium curve in the (T, v) diagram has a simple maximum at the critical point.

It is easy to determine also the volumes v_1' and v_2' which correspond to the boundaries of the metastable regions (the points *B* and *C* on isotherm 4 in Fig. 19). At these points

$$-(\partial P/\partial v)_t = At+Bv^2 = 0,$$

whence

$$v_1' = -v_2' = -\sqrt{(-At/B)}. \tag{84.8}$$

These volumes are also proportional to the square root of T_c-T, and smaller than the volumes v_1 and v_2 for the same temperature by a factor of $\sqrt{3}$.

The heat of transition (latent heat of evaporation) is zero at the critical point. Since the temperature of the two phases in equilibrium is the same, and the volume difference near the critical point is small, we can write the heat of transition as

$$q = T(S_2 - S_1) \cong T_c(\partial S/\partial v)_t(v_2 - v_1). \tag{84.9}$$

Since the difference $v_2 - v_1$ is proportional to $\sqrt{(T_c - T)}$, the heat of transition is proportional to the same square root.

From the formula

$$C_p - C_v = -T\frac{[(\partial P/\partial t)_v]^2}{(\partial P/\partial v)_t}$$

it follows that the specific heat C_p becomes infinite at the critical point as $(\partial P/\partial v)_t$ vanishes. Substituting (84.4) in this formula, we find

$$C_p \sim 1/(At + Bv^2). \tag{84.10}$$

In particular, for states on the equilibrium curve, v is proportional to $\sqrt{t}$, and so $C_p \sim 1/t$.

§85. The law of corresponding states

VAN DER WAALS' interpolation formula for the equation of state,

$$\left(P + \frac{N^2 a}{V^2}\right)(V - Nb) = NT,$$

is in qualitative agreement with the properties of the liquid-vapour transition which have been described in the preceding sections. The isotherms determined by this equation are shown in Fig. 20. They are easily seen to be similar to

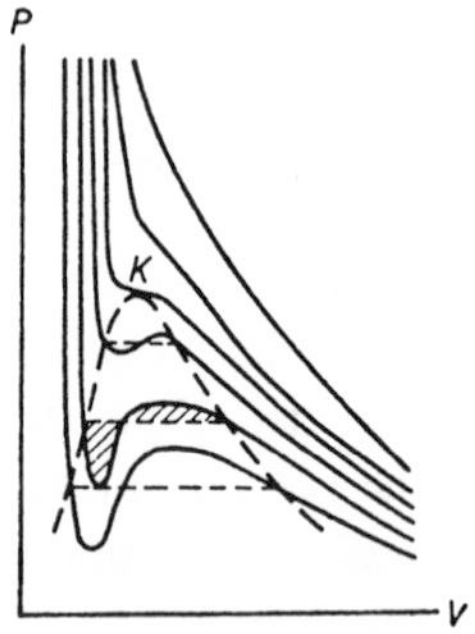

FIG. 20

those shown in Fig. 19. Here also horizontal straight segments correspond to the equilibrium transition from liquid to vapour; the position of these segments is given by the conditions of equilibrium:

$$\int_1^2 V\,\mathrm{d}P = 0, \tag{85.1}$$

where the integral is taken in this case along the van der Waals isotherm from the beginning to the end of the horizontal segment. Geometrically, this condition signifies that the areas shown hatched in Fig. 20 for one isotherm are equal.

The critical temperature, critical pressure and critical volume can be expressed in terms of the parameters a and b which appear in VAN DER WAALS' equation. To do this, we write the equation as

$$P = \frac{NT}{V-Nb} - \frac{N^2a}{V^2}$$

and equate to zero the derivatives

$$\left(\frac{\partial P}{\partial V}\right)_T = -\frac{NT}{(V-Nb)^2} + \frac{2N^2a}{V^3} = 0,$$

$$\left(\frac{\partial^2 P}{\partial V^2}\right)_T = \frac{2NT}{(V-Nb)^3} - \frac{6N^2a}{V^4} = 0.$$

These three equations give

$$T_c = \frac{8}{27}\frac{a}{b}, \qquad V_c = 3Nb, \qquad P_c = \frac{1}{27}\frac{a}{b^2}. \tag{85.2}$$

We now use instead of T, P, V the quantities

$$T' = T/T_c, \qquad P' = P/P_c, \qquad V' = V/V_c. \tag{85.3}$$

These are called the *reduced* temperature, pressure and volume, and are all equal to unity at the critical point.

If we express T, P, V in VAN DER WAALS' equation in terms of T', P', V', we obtain

$$\left(P' + \frac{3}{V'^2}\right)(3V'-1) = 8T'. \tag{85.4}$$

This is the *reduced van der Waals' equation.* It contains only V', P' and T', and not quantities pertaining to a given substance. Equation (85.4) is therefore the equation of state for all bodies to which VAN DER WAALS' equation is applicable. The states of two bodies for which their values of T', P', V' are equal are called *corresponding states* (clearly the critical states of all bodies are corresponding states). It follows from (85.4) that, if two bodies have equal values of two of the three quantities T', P', V', then the values of the third quantity are also equal, i.e. they are in corresponding states (the *law of corresponding states*).

The "reduced" isotherms $P' = P'(V')$ given by equation (85.4) are the same for all substances. The positions of the straight segments which give the liquid-gas transition points are therefore also the same. We can therefore conclude that for equal reduced temperatures the following quantities must be the same for all substances: (1) the reduced saturated vapour pressure, (2) the reduced specific volume of the saturated vapour, (3) the reduced specific volume of the liquid in equilibrium with the saturated vapour.

The law of corresponding states can also be applied to the heat of transition from the liquid to the gaseous state. Here the "reduced heat of evaporation" must be represented by a dimensionless quantity, q/T_c. We can therefore write†

$$q/T_c = f(T/T_c). \tag{85.5}$$

In conclusion, we may note that the law of corresponding states does not apply only to VAN DER WAALS' equation. The parameters of a specific substance disappear when reduced quantities are used in any equation of state containing only two such parameters. The law of corresponding states, taken as a general theorem not pertaining to any specific form of the equation of state, is somewhat more accurate than VAN DER WAALS' equation, but its applicability is in general very restricted.

† At temperatures considerably below the critical temperature, the ratio q/T_c is approximately 10 (where q is the molecular heat of evaporation).

CHAPTER IX

SOLUTIONS

§86. Systems containing different particles

HITHERTO we have considered only bodies consisting of identical particles. Let us now go on to discuss systems which contain different particles. These include all kinds of mixtures of more than one substance; if the mixture contains much more of one substance than of the others, it is called a *solution* of the other substances in the predominant substance (the *solvent*).

The number of *independent components* of the system customarily signifies the number of substances whose quantities in a state of complete equilibrium can be specified arbitrarily. All the thermodynamic quantities for a system in complete equilibrium are entirely determined, for example, by the temperature, the pressure and the numbers of particles of the independent components. The number of independent components may not be the same as the total number of different substances in the system if a chemical reaction can occur between the latter; if such a system is in partial equilibrium only, the determination of its thermodynamic quantities requires, in general, a knowledge of the amounts of all the substances present in it.

It is easy to generalise the results of §24 to bodies consisting of different substances. Firstly, all the thermodynamic quantities must be homogeneous functions of the first order in all the additive variables—the numbers of the different particles and the volume.

Next, in formulae (24.5), (24.7)–(24.9) the term $\mu \, \mathrm{d}N$ must now be replaced by the sum $\sum \mu_i \, \mathrm{d}N_i$, where N_i is the number of particles of the ith kind and the quantities μ_i are called the chemical potentials of the corresponding substances. Accordingly, in formulae (24.6) and (24.10) the chemical potential and the number of particles must now have the suffix i. In order to find the chemical potential of any substance in the mixture we must differentiate E, F, Φ or W with respect to the corresponding number of particles. In particular

$$\mu_i = (\partial \Phi / \partial N_i)_{P,\,T}. \tag{86.1}$$

The chemical potentials are then expressed as functions of the pressure, the temperature and the *concentrations*, i.e. the ratios of the numbers of particles of the different substances. These numbers of particles can appear in μ_i only as ratios, since Φ is a homogeneous function of the first order in the N_i, and the chemical potentials must therefore be homogeneous functions of zero order in these variables.

From the fact that Φ is a homogeneous function of the first order in the N_i, we have, using EULER's theorem,

$$\Phi = \sum_i N_i \partial\Phi/\partial N_i = \sum_i \mu_i N_i, \tag{86.2}$$

which is a generalisation of the formula $\Phi = N\mu$.

For the potential Ω we now have

$$\Omega = F - \sum \mu_i N_i$$

and hence again $\Omega = -PV$. The last formula ceases to be valid only for bodies in an external field, when the pressure in different parts of the bodies is different.

The results of §25 can also be generalised immediately: the conditions of equilibrium for a system in an external field require the temperature and also the chemical potential of each component to be constant throughout the system:

$$\mu_i = \text{constant}. \tag{86.3}$$

Finally, the Gibbs distribution for systems consisting of different particles becomes

$$w_{nN_1N_2\ldots} = \exp\left\{\frac{\Omega + \sum \mu_i N_i - E_{nN_1N_2\ldots}}{T}\right\}, \tag{86.4}$$

an obvious generalisation of formula (35.2).

§87. The phase rule

Let us now consider a system consisting of different substances and comprising r phases in contact (each phase containing, in general, all the substances).

Let the number of independent components in the system be n. Then each phase is described by its pressure, temperature and n chemical potentials. We have seen in §81 that the condition for equilibrium of phases consisting of identical particles is that temperature, pressure and chemical potential should be equal. It is evident that, in the general case of more than one component, the phase equilibrium condition will be that the temperature, pressure and each chemical potential are equal. Let T and P be the common temperature and pressure of the phases. In order to distinguish the chemical potentials belonging to different phases and components we shall write them with a roman index for the phase and an arabic suffix for the component.

Then the phase equilibrium conditions may be written

$$\left.\begin{array}{l}\mu_1^{\mathrm{I}} = \mu_1^{\mathrm{II}} = \ldots = \mu_1^r,\\ \mu_2^{\mathrm{I}} = \mu_2^{\mathrm{II}} = \ldots = \mu_2^r,\\ \ldots \quad \ldots \quad \ldots \quad \ldots\\ \mu_n^{\mathrm{I}} = \mu_n^{\mathrm{II}} = \ldots = \mu_n^r.\end{array}\right\} \tag{87.1}$$

Each of these potentials is a function of $n+1$ independent variables: P, T, and $n-1$ concentrations of different components in the phase concerned (each phase contains n independent numbers of particles of different kinds, giving $n-1$ independent ratios).

The conditions (87.1) form a set of $n(r-1)$ equations. The number of unknowns is $2+r(n-1)$. If these equations have solutions, the number of equations must certainly not be greater than the number of unknowns, i.e. $n(r-1) \leqslant 2+r(n-1)$, or

$$r \leqslant n+2. \tag{87.2}$$

In other words, in a system consisting of n independent components, not more than $n+2$ phases can be in equilibrium simultaneously. This is called *Gibbs' phase rule*. We have seen a particular case of it in §81: when there is one component, the number of phases that exist in contact at one time cannot exceed three.

If the number r of coexisting phases is less than $n+2$, $n+2-r$ of the variables in equations (87.1) can obviously take arbitrary values. That is, we can arbitrarily vary any $n+2-r$ variables without destroying the equilibrium; the other variables must, of course, be varied in a definite manner. The number of variables which can be arbitrarily varied without destroying the equilibrium is called the number of *thermodynamic degrees of freedom* of the system. If this is denoted by f, the phase rule may be written

$$f = n+2-r, \tag{87.3}$$

where f can not, of course, be less than zero. If the number of phases has its maximum possible value $n+2$, then $f = 0$, i.e. all the variables in equations (87.1) have definite values, and none of them can be varied without destroying the equilibrium and causing one of the phases to disappear.

§88. Weak solutions

We shall now consider (in §§88–93) the thermodynamic properties of weak solutions, i.e. those in which the number of molecules of the dissolved substances (the *solutes*) is much less than the number of solvent molecules. Let us first take the case of a solution with only one solute; the generalisation to a solution with more than one solute is immediate.

Let N be the number of solvent molecules in the solution, and n the number of solute molecules. The ratio $c = n/N$ is the *concentration* of the solution, and from the above hypothesis $c \ll 1$.

Let us derive an expression for the thermodynamic potential of the solution. Let $\Phi_0(P, T, N)$ be the thermodynamic potential of the pure solvent (containing no solute). According to the formula $\Phi = N\mu$ (which is valid for pure substances) it can be written $\Phi_0 = N\mu_0(P, T)$, where $\mu_0(P, T)$ is the chemical potential of the pure solvent. Let $\alpha = \alpha(P, T, N)$ denote the small change which would occur in the thermodynamic potential if one molecule of solute were added to the solvent. Since the solution is assumed weak, the solute molecules in it are comparatively far apart, and their interaction is therefore weak. Neglecting this interaction, we can then say that the change in the thermodynamic potential when n molecules are added to the solvent is $n\alpha$. But the expression $\Phi_0+n\alpha$ thus obtained fails to take account of the fact that all the molecules of the solute are identical. This is the expression which would be obtained from formula (31.5) if all the solute particles were regarded as different in calculating the partition function. As we know (cf. (31.7)), the partition function thus calculated must in fact be divided by $n!$.†

This leads to an additional term $T \log n!$ in the free energy, and therefore in the potential Φ. Thus

$$\Phi = N\mu_0(P, T)+n\alpha(P, T, N)+T \log n!.$$

Next, since n is itself a very large number, though small in comparison with N, we can write $\log n! = n \log (n/e)$ in the last term. Then

$$\begin{aligned}\Phi &= N\mu_0+n[\alpha+T \log (n/e)] \\ &= N\mu_0+nT \log [(n/e)e^{\alpha/T}].\end{aligned}$$

We now take into consideration the fact that Φ must be a homogeneous function of the first order in n and N. For this to be so it is clearly necessary that the argument of the logarithm should be of order zero in n and N, and hence $e^{\alpha/T}$ must be inversely proportional to N, i.e. must be of the form $f(P, T)/N$. Thus

$$\Phi = N\mu_0+nT \log [(n/eN)f(P, T)].$$

Defining a new function $\psi(P, T) = T \log f(P, T)$, we finally have for the thermodynamic potential of the solution the expression

$$\Phi = N\mu_0(P, T)+nT \log (n/eN)+n\psi(P, T). \tag{88.1}$$

The assumption, made at the beginning of this section, about the addition of a term of the form $n\alpha$ to the potential of the pure solvent amounts

† Here we neglect quantum effects, which is always permissible for a weak solution, as it is for a sufficiently rarefied gas.

essentially to an expansion in powers of n, retaining only the leading terms. The term in Φ which is of the next order in n is $n^2 f_1(P, T, N)$; but, since Φ must be a homogeneous function of N and n, $f_1(P, T, N)$ must be inversely proportional to N, i.e. $f_1(P, T, N) = \beta(P, T)/2N$, where β is a function of P and T only. Thus the thermodynamic potential of a weak solution as far as the second-order terms is

$$\Phi = N\mu_0(P, T) + nT\log(n/eN) + n\psi(P, T) + (n^2/2N)\beta(P, T). \tag{88.2}$$

For a weak solution of more than one substance, the thermodynamic potential will obviously be, instead of (88.1),

$$\Phi = N\mu_0 + \sum_i n_i T\log(n_i/eN) + \sum_i n_i\psi_i, \tag{88.3}$$

where the n_i are the numbers of molecules of the various solutes, and the $\psi_i(P, T)$ are various functions. The expression (88.2) is similarly generalised to

$$\Phi = N\mu_0 + \sum_i n_i T\log(n_i/eN) + \sum_i n_i\psi_i + \sum_{i,k}(n_i n_k/2N)\beta_{ik}. \tag{88.4}$$

From (88.1) we can easily find the chemical potentials for the solvent (μ) and the solute (μ'): the former is

$$\mu = \partial\Phi/\partial N = \mu_0 - Tn/N = \mu_0 - Tc, \tag{88.5}$$

and the latter is

$$\mu' = \partial\Phi/\partial n = T\log(n/N) + \psi = T\log c + \psi. \tag{88.6}$$

§89. Osmotic pressure

In this and the following sections we shall discuss some properties of solutions, again assuming them weak and therefore using the results of §88.

Let us suppose that two solutions of the same substance in the same solvent but with different concentrations c_1 and c_2 are separated by a partition through which solvent molecules can pass but solute molecules cannot (a semi-permeable membrane). The pressures on the two sides of the membrane will then be different; the argument in §12 to prove the equality of pressures is invalid here, because of the presence of the semi-permeable membrane. The difference between the pressures is called the *osmotic pressure*.

The condition of equilibrium between the two solutions is (apart from the equality of their temperatures) that the chemical potentials of the solvent in them should be equal. The chemical potentials of the solute need not be the same, since the semi-permeability of the membrane means that there is equilibrium only with respect to the solvent.

Denoting the pressures in the two solutions by P_1 and P_2, and using the expression (88.5), we obtain the equilibrium condition in the form

$$\mu_0(P_1, T) - c_1 T = \mu_0(P_2, T) - c_2 T. \tag{89.1}$$

The pressure difference $P_2 - P_1 = \Delta P$ (i.e. the osmotic pressure) is relatively small for weak solutions, and so we can expand $\mu_0(P_2, T)$ in powers of ΔP, retaining only the first two terms:

$$\mu_0(P_2, T) = \mu_0(P_1, T) + \Delta P \cdot \partial\mu_0/\partial P.$$

Substitution in (89.1) gives

$$\Delta P \cdot \partial\mu_0/\partial P = (c_2 - c_1)T.$$

But $\partial\mu_0/\partial P$ is just the molecular volume v of the pure solvent. Thus

$$\Delta P = (c_2 - c_1)T/v. \tag{89.2}$$

In particular, if there is pure solvent on one side of the membrane ($c_1 = 0$, $c_2 = c$), the osmotic pressure is

$$\Delta P = cT/v = nT/V, \tag{89.3}$$

where n is the number of solute molecules in a volume V of solvent; since the solution is weak, V is almost exactly equal to the total volume of the solution. Formula (89.3) is called *van 't Hoff's formula.* It should be pointed out that this formula is applicable to weak solutions independently of the particular solvent and solute concerned, and that it resembles the equation of state of an ideal gas. The gas pressure is replaced by the osmotic pressure, the gas volume by the solution volume, and the number of particles in the gas by the number of molecules of solute.

The generalisation of formulae (89.2) and (89.3) to the case of solutions of more than one substance is obvious: in this case the osmotic pressure is the sum of the osmotic pressures of the various solutes, i.e. the pressures which would exist if each solute were dissolved alone.

§90. Solvent phases in contact

In this section we shall consider the equilibrium of two solvent phases in contact, with a certain amount of the same substance dissolved in each. The equilibrium conditions are (apart from the equality of pressures and temperatures) the equality of the chemical potentials of the solvent and those of the solute in the two phases. Here we shall use the first condition, writing it in the form

$$\mu_0^{\mathrm{I}}(P, T) - c_{\mathrm{I}} T = \mu_0^{\mathrm{II}}(P, T) - c_{\mathrm{II}} T, \tag{90.1}$$

where c_{I}, c_{II} are the concentrations and μ_0^{I}, μ_0^{II} the chemical potentials of the two phases of the pure solvent.

It must be noted that the system considered here, consisting of two components and two phases, has two thermodynamic degrees of freedom. Of the four quantities P, T, c_{I}, c_{II}, therefore, only two may be chosen arbitrarily; if we choose P or T and one of the concentrations, for example, then the other concentration has a definite value.

If the two solvent phases contained no solute, the condition for their equilibrium would be

$$\mu_0^{\mathrm{I}}(P_0, T_0) = \mu_0^{\mathrm{II}}(P_0, T_0), \tag{90.2}$$

the temperature and pressure of both phases being denoted by T_0 and P_0.

Thus, whereas in the equilibrium of pure solvent phases the relation between pressure and temperature is given by equation (90.2), when any substance is dissolved in these phases the relation is given by equation (90.1). For weak solutions the two equations are not greatly different.

Let us now expand $\mu_0^{\mathrm{I}}(P, T)$ and $\mu_0^{\mathrm{II}}(P, T)$ in equation (90.1) in powers of $P-P_0 = \Delta P$ and $T-T_0 = \Delta T$, where P_0 and T_0 are the pressure and temperature at some point on the equilibrium curve of the pure solvent phases close to a given point P, T on the equilibrium curve of the solution phases. Retaining in the expansion only the first-order terms in ΔP and ΔT, and using (90.2), we have from (90.1)

$$\frac{\partial \mu_0^{\mathrm{I}}}{\partial T}\Delta T + \frac{\partial \mu_0^{\mathrm{I}}}{\partial P}\Delta P - c_{\mathrm{I}}T = \frac{\partial \mu_0^{\mathrm{II}}}{\partial T}\Delta T + \frac{\partial \mu_0^{\mathrm{II}}}{\partial P}\Delta P - c_{\mathrm{II}}T.$$

But $-\partial\mu_0/\partial T$ and $\partial\mu_0/\partial P$ are just the entropy s and the volume v of the pure solvent (per molecule). Adding the suffix denoting the phase, we have

$$-(s_{\mathrm{I}}-s_{\mathrm{II}})\,\Delta T + (v_{\mathrm{I}}-v_{\mathrm{II}})\,\Delta P = (c_{\mathrm{I}}-c_{\mathrm{II}})T. \tag{90.3}$$

According to formula (81.5), we have $(s_{\mathrm{II}}-s_{\mathrm{I}})T = q$, where q is the latent heat of transition of the solvent from phase I to phase II. Thus (90.3) may be written

$$(q/T)\Delta T + (v_{\mathrm{I}}-v_{\mathrm{II}})\Delta P = (c_{\mathrm{I}}-c_{\mathrm{II}})T. \tag{90.4}$$

Let us examine two particular cases of this formula. We first choose the point P_0, T_0 such that $P_0 = P$. Then ΔT will be the horizontal distance between the two curves, i.e. the change in the temperature of transition between the two phases when the solute is added, or the difference between the transition temperature T (at pressure P) when both phases are solutions and the transition temperature T_0 (at the same pressure) for the pure solvent. Since $\Delta P = 0$ here, (90.4) gives

$$\Delta T = T^2(c_{\mathrm{I}}-c_{\mathrm{II}})/q. \tag{90.5}$$

If one of the phases (I, say) is the pure solvent ($c_{II} = 0, c_I = c$), then

$$\Delta T = T^2 c/q. \tag{90.6}$$

This formula determines, in particular, the change in the freezing point when the solute is added, if the solute is insoluble in the solid phase; the two phases are then the liquid solution and the solid solvent, and ΔT is the difference between the temperature at which the solvent freezes out of the solution and that at which the pure solvent freezes. On freezing, heat is liberated, i.e. q is negative. Hence $\Delta T < 0$ also; i.e. if the pure solvent freezes out, the addition of solute lowers the freezing point.

The relation (90.6) also determines the change in the boiling point when the solute is added, if the solute is not volatile; the two phases are then the liquid solution and the solvent vapour, and ΔT is the difference between the temperature at which the solvent boils off from the solution and that at which the pure solvent boils. Since heat is absorbed in boiling, $q > 0$ and therefore $\Delta T > 0$, i.e. the boiling point is raised by the addition of the solute.

All these consequences of formula (90.6) are fully in accordance with LE CHATELIER's principle. For example, let a liquid solution be in equilibrium with the solid solvent. If the concentration of the solution is increased, then by LE CHATELIER's principle the freezing point must be lowered so that part of the solid solvent is added to the solution and the concentration is thereby lowered. The system as it were counteracts its disturbance from the equilibrium state. Similarly, if the concentration of the liquid solution in equilibrium with the solvent vapour is increased, the boiling point must be raised so that part of the vapour condenses into the solution and the concentration is lowered.

Let us now consider another particular case of formula (90.4), choosing the point P_0, T_0 so that $T = T_0$. Then ΔP is the vertical distance between the two curves, i.e. the difference between the pressure of the two solution phases in equilibrium and that of the two pure solvent phases in equilibrium (at the same temperature). Here $\Delta T = 0$, and from (90.4) we have

$$\Delta P = T(c_I - c_{II})/(v_I - v_{II}). \tag{90.7}$$

The ratio

$$\Delta P/\Delta T = q/T(v_I - v_{II})$$

is in accordance with the Clapeyron–Clausius formula (applied to the pure solvent), as it should be, since ΔP and ΔT are relatively small.

Let us apply formula (90.7) to an equilibrium between liquid and gaseous phases. Then the volume of one phase (the liquid) may be neglected in comparison with that of the other, and (90.7) becomes

$$\Delta P = T(c_I - c_{II})/v, \tag{90.8}$$

where v is the molecular volume of the gas phase (I). Noting that $Pv = T$, and substituting to the same accuracy $P \approx P_0$ (where P_0 is the saturated vapour pressure over the pure solvent), we can write this formula as

$$\Delta P = P_0(c_{\mathrm{I}} - c_{\mathrm{II}}). \tag{90.9}$$

If the gas phase is the pure solvent vapour ($c_{\mathrm{I}} = 0$, $c_{\mathrm{II}} = c$), then (90.9) becomes

$$\Delta P/P_0 = -c, \tag{90.10}$$

where c is the concentration of the solution. This formula gives the difference between the saturated vapour pressure of the solvent over the solution (P) and over the pure solvent (P_0). The relative decrease in the saturated vapour pressure when the solute is added is equal to the concentration of the solution (*Raoult's law*).†

§91. Equilibrium with respect to the solute

Let us now consider a system consisting of two solutions in contact, the solutions being of the same substance in different solvents (for instance, in two immiscible liquids), and their concentrations being denoted by c_1 and c_2.

The equilibrium condition for this system is that the chemical potentials of the solute in the two solutions should be equal. Using (88.6), we can write this condition in the form

$$T \log c_1 + \psi_1(p, T) = T \log c_2 + \psi_2(p, T).$$

The functions ψ_1 and ψ_2 are, of course, different for the different solvents. Hence we find

$$c_1/c_2 = e^{(\psi_2 - \psi_1)/T}. \tag{91.1}$$

The right-hand side of this equation is a function of P and T only. Thus the solute is distributed between the solvents in such a way that the ratio of concentrations is always the same (for given pressure and temperature), independently of the total quantities of the solute and solvents (the *distribution law*). The same law obviously applies to a solution of one substance in two adjacent phases of the same solvent.

Now let us consider the equilibrium between a gas (assumed ideal) and a solution of it in a liquid or solid solvent. The equilibrium condition, i.e. the equality of the chemical potentials of the pure gas and the dissolved gas, can be written (using (42.6) and (88.6)) in the form

$$T \log c + \psi(P, T) = T \log P + \chi(T),$$

† It will be remembered that c denotes the molecular concentration (ratio of numbers of molecules, n/N).

whence

$$c = Pe^{(\chi-\psi)/T}. \tag{91.2}$$

The function $\psi(P, T)$ describes the properties of the liquid (or solid) solution. At low pressures, the properties of a liquid depend only very slightly on the pressure. Hence the dependence of $\psi(P, T)$ on the pressure is unimportant, and we can suppose that the coefficient of P in (91.2) is a constant independent of the pressure:

$$c = P \times \text{constant}. \tag{91.3}$$

Thus, when a gas dissolves, the concentration of the (weak) solution is proportional to the gas pressure (*Henry's law*).†

PROBLEM

Find the variation of concentration with height for a solution in a gravitational field.

SOLUTION. We apply the equilibrium condition (86.3) in an external field, writing it fo- the solute: $T \log c + \psi(P, T) + mgz$ = constant, since the potential energy of a solute moler cule in the gravitational field is mgz (z being the height, and m the mass of the molecule). We differentiate this equation with respect to z, noting that the temperature is constant by one of the conditions of equilibrium:

$$\frac{T}{c}\frac{\mathrm{d}c}{\mathrm{d}z} + mg + \frac{\partial\psi}{\partial P}\frac{\mathrm{d}P}{\mathrm{d}z} = 0.$$

Since the volume of the solution is

$$\frac{\partial\Phi}{\partial P} = N\frac{\partial\mu_0}{\partial P} + n\frac{\partial\psi}{\partial P}$$

(substituting for Φ the expression (88.1)), the quantity $\partial\psi/\partial P$ may be called the volume v' per molecule of solute. Hence

$$\frac{T}{c}\frac{\mathrm{d}c}{\mathrm{d}z} + mg + v'\frac{\mathrm{d}P}{\mathrm{d}z} = 0.$$

In order to find P as a function of z, we use the equilibrium condition for the solvent:‡

$$v\,\mathrm{d}P/\mathrm{d}z + Mg = 0,$$

where $v = \partial\mu_0/\partial P$ is the molecular volume and M the mass of a solvent molecule. Substituting $\mathrm{d}P/\mathrm{d}z$ in the previous condition, we find

$$\frac{T}{c}\frac{\mathrm{d}c}{\mathrm{d}z} + mg - Mg\frac{v'}{v} = 0.$$

If the solution may be regarded as incompressible, i.e. v and v' are constants, this gives

$$c = c_0 e^{-(gz/T)(m - v'M/v)},$$

where c_0 is the concentration of the solution when $z = 0$, i.e. the usual barometric formula corrected in accordance with ARCHIMEDES' principle.

† It is assumed that the gas molecules dissolve unchanged. If they dissociate (as in the dissolution of hydrogen H_2 in certain metals), the dependence of the concentration on the pressure is different; see §104, Problem 2.

‡ In this condition the term involving the concentration ($-T\,\mathrm{d}c/\mathrm{d}z$) is small and may be omitted; in the condition for the solute, it contained c in the denominator and was therefore not small.

§92. Evolution of heat and change of volume on dissolution

The process of dissolution is accompanied by the evolution or absorption of heat. Let us now calculate the quantity of heat involved, and first determine the maximum work which can be done as a result of the dissolution process.

Let us suppose that the dissolution occurs at constant pressure and temperature. In this case the maximum work is determined by the change in the thermodynamic potential. Let us calculate it for a process in which a small number δn of solute molecules are dissolved in a solution already of concentration c. The change $\delta\Phi$ in the total thermodynamic potential of the system is equal to the sum of the changes in the potentials of the solution and the pure solute. Since δn molecules of solute are added to the solution, the change in its thermodynamic potential is

$$\delta\Phi_{\text{sol}} = \frac{\partial\Phi_{\text{sol}}}{\partial n}\delta n = \mu'\delta n,$$

where μ' is the chemical potential of the solute in the solution. The change in the potential Φ_0' of the pure solute is

$$\delta\Phi_0' = -\frac{\partial\Phi_0'}{\partial n}\delta n = -\mu_0'\delta n,$$

since the number of molecules of it decreases by δn, μ_0' being the chemical potential of the pure solute. The total change in the thermodynamic potential in this process is therefore

$$\delta\Phi = \delta n(\mu' - \mu_0'). \tag{92.1}$$

We have now only to substitute μ' from (88.6):

$$\delta\Phi = -\delta n(\mu_0' - \psi - T\log c)$$

or

$$\delta\Phi = -T\delta n \log\frac{c_0(P, T)}{c}, \tag{92.2}$$

where

$$c_0(P, T) = e^{(\mu_0' - \psi)/T} \tag{92.3}$$

is the solubility, i.e. the concentration of a saturated solution (that is, one which is in equilibrium with the pure solute). This follows immediately from the fact that in equilibrium Φ must have a minimum, i.e. we must have $\delta\Phi = 0$. Formula (92.3) can also be derived directly from the condition for equilibrium between the solution and the pure solute, i. e. from the equality of the chemical potentials of the pure solute and that in the solution.

It should be noted that c_0 may be identified with the concentration of the saturated solution only if c_0 is small, since all the formulae in the last few sections are applicable only to small concentrations.

The expression obtained gives the required quantity of work: $|\delta\Phi|$ is the maximum work which can be done by the dissolution of δn molecules, and is also the minimum work which is needed to separate δn molecules of solute from a solution of concentration c.

There is now no difficulty in calculating the heat δQ_P absorbed in dissolution at constant pressure (if $\delta Q_P < 0$, this means that heat is evolved). The quantity of heat absorbed in a process which occurs at constant pressure is equal to the change in the heat function (§14). Since, on the other hand,

$$W = -T^2\left(\frac{\partial}{\partial T}\frac{\Phi}{T}\right)_P,$$

we have†

$$\delta Q_P = -T^2\left(\frac{\partial}{\partial T}\frac{\delta\Phi}{T}\right)_P. \tag{92.4}$$

Substituting the expression (92.2) in this formula, we find the required quantity of heat:

$$\delta Q_P = T^2\delta n\, \partial \log c_0/\partial T. \tag{92.5}$$

Thus the quantity of heat involved in the dissolution process is related to the temperature dependence of the solubility. We see that δQ_P is simply proportional to δn; this formula is therefore applicable also to the dissolution of any finite quantity of substance (so long as the solution remains weak, of course). The quantity of heat absorbed in the dissolution of n molecules is

$$Q_P = T^2 n\, \partial \log c_0/\partial T. \tag{92.6}$$

We may also determine the change in volume on dissolution, i.e. the difference between the volume of the solution and the sum of the volumes of the pure solute and the solvent in which it is dissolved. Let us calculate this change δV in the dissolution of δn molecules. The volume is the derivative of the thermodynamic potential with respect to the pressure. The change in volume is therefore equal to the derivative, with respect to pressure, of the change in the thermodynamic potential for a given process, i.e.

$$\delta V = \frac{\partial}{\partial P}\delta\Phi. \tag{92.7}$$

† The corresponding formula for the quantity of heat in a process which occurs at constant volume is

$$\delta Q_V = -T^2\left(\frac{\partial}{\partial T}\frac{\delta F}{T}\right)_V. \tag{92.4a}$$

Substituting $\delta\Phi$ from (92.2), we find

$$\delta V = -T\,\delta n \frac{\partial}{\partial P}\log c_0. \tag{92.8}$$

In conclusion, it may be noted that formula (92.6) is in accordance with LE CHATELIER's principle. Let us suppose, for example, that Q_P is negative, i.e. that heat is evolved on dissolution, and let us consider a saturated solution. If this is cooled, then by LE CHATELIER's principle the solubility must increase so that more dissolution occurs. Heat is then evolved, i.e. the system as it were counteracts the cooling which disturbs its equilibrium. The same follows from (92.6), since in this case $\partial c_0/\partial T$ is negative. Similar arguments show that formula (92.8) is also in accordance with LE CHATELIER's principle.

PROBLEMS

PROBLEM 1. Find the maximum work that can be done in the formation of a saturated solution.

SOLUTION. Before dissolution, the thermodynamic potential of the pure solvent was $N\mu_0$, and that of the pure solute $n\mu_0'$. The potential of the whole system was $\Phi_1 = N\mu_0 + n\mu_0'$. After dissolution, the thermodynamic potential $\Phi_2 = N\mu_0 + nT\log(n/eN) + n\psi$. The maximum work is

$$\begin{aligned} R_{\max} &= \Phi_1 - \Phi_2 \\ &= -nT\log(n/eN) + n(\mu_0' - \psi) \\ &= nT\log(ec_0/c); \end{aligned}$$

this may also be derived by integration of (92.2). If a saturated solution is formed, i.e. $c = c_0$ and $n = Nc = Nc_0$, then

$$R_{\max} = nT = Nc_0T.$$

PROBLEM 2. Find the minimum work which must be done to raise the concentration of a solution from c_1 to c_2 by removing some of the solvent.

SOLUTION. Before the removal, the thermodynamic potential of the solution was $\Phi_1 = N\mu_0 + Nc_1T\log(c_1/e) + Nc_1\psi$ (the number of solute molecules was Nc_1, where N was the original number of solvent molecules). In order to raise the concentration of the solution to c_2, we must remove from it $N(1 - c_1/c_2)$ solvent molecules. The sum of the thermodynamic potentials of the remaining solution and the solvent removed gives $\Phi_2 = N\mu_0 + Nc_1T\log(c_2/e) + Nc_1\psi$. The minimum work is

$$R_{\min} = \Phi_2 - \Phi_1 = Nc_1T\log(c_2/c_1).$$

§93. The mutual interaction of solutes

Let us consider a weak solution of two different substances in the same solvent. If each substance were dissolved separately, their solubilities (the concentrations of their saturated solutions) would be c_{01} and c_{02}†; let the

† It is assumed, of course, that the saturated solution is still so weak that all the formulae used remain valid.

solubilities of the two substances when both are present be $c_{01}' = c_{01}+\delta c_{01}$ and $c_{02}' = c_{02}+\delta c_{02}$. We shall determine the relation between δc_{01} and δc_{02}.

To solve this problem we must obviously take account of those terms in the thermodynamic potential which contain the concentrations of both solutes. The second-order terms include one of this type. The thermodynamic potential of the solution of the two substances is, from (88.4),

$$\Phi = N\mu_0+n_1T\log(n_1/eN)+n_2T\log(n_2/eN)+n_1\psi_1+n_2\psi_2+ \\ +\tfrac{1}{2}n_1^2\beta_{11}/N+\tfrac{1}{2}n_2^2\beta_{22}/N+n_1n_2\beta_{12}/N$$

as far as second-order terms. The chemical potentials of the two solutes are

$$\mu_1' = \partial\Phi/\partial n_1 = T\log c_1+\psi_1+c_1\beta_{11}+c_2\beta_{12}, \\ \mu_2' = \partial\Phi/\partial n_2 = T\log c_2+\psi_2+c_1\beta_{12}+c_2\beta_{22}, \tag{93.1}$$

where $c_1 = n_1/N$, $c_2 = n_2/N$. Let μ_{01}' and μ_{02}' be the chemical potentials of the pure solutes. The solubilities c_{01} and c_{02} are determined from the condition of equilibrium for each of the pure solutes with that solute in solution, i.e.

$$\mu_{01}' = T\log c_{01}+\psi_1+c_1\beta_{11}, \\ \mu_{02}' = T\log c_{02}+\psi_2+c_2\beta_{22}. \tag{93.2}$$

The solubilities c_{01}' and c_{02}' are determined from the equilibrium conditions

$$\mu_{01}' = T\log c_{01}'+\psi_1+c_1\beta_{11}+c_2\beta_{12}, \\ \mu_{02}' = T\log c_{02}'+\psi_2+c_2\beta_{22}+c_1\beta_{12}. \tag{93.3}$$

Subtracting (93.2) term by term from (93.3) and assuming that the changes in solubility are small ($\delta c_1 \ll c_{01}$, $\delta c_{02} \ll c_{02}$), we have approximately

$$T\delta c_{01}/c_{01} = -c_{02}\beta_{12}, \qquad T\delta c_{02}/c_{02} = -c_{01}\beta_{12},$$

since $\log c_0' - \log c_0 \approx \delta c_0/c_0$. Hence

$$\delta c_{01} = \delta c_{02}, \tag{93.4}$$

i.e. the changes in solubility of the two substances are equal.

Similarly we can determine the change in the saturated vapour pressures of two substances in the same solution. Let P_1 and P_2 be the saturated vapour pressures of the two substances above solutions of concentrations c_1 and c_2; let $P_1' = P_1+\delta P_1$, $P_2' = P_2+\delta P_2$ be the vapour pressures of the same substances above the solution of both together (with the same concentrations). The chemical potentials of the two substances in the vapour are $T\log P_1+\chi_1(T)$ and $T\log P_2+\chi_2(T)$. The pressures P_1 and P_2 are therefore given by the relations

$$T\log P_1+\chi_1(T) = T\log c_1+\psi_1+c_1\beta_{11}, \\ T\log P_2+\chi_2(T) = T\log c_2+\psi_2+c_2\beta_{22}, \tag{93.5}$$

and P_1' and P_2' by

$$T \log P_1' + \chi_1 = T \log c_1 + \psi_1 + c_1\beta_{11} + c_2\beta_{12},$$
$$T \log P_2' + \chi_2 = T \log c_2 + \psi_2 + c_2\beta_{22} + c_1\beta_{12}. \tag{93.6}$$

Subtracting (93.5) from (93.6) and assuming the changes δP_1 and δP_2 small, we find

$$T\delta P_1/P_1 = c_2\beta_{12}, \qquad T\delta P_2/P_2 = c_1\beta_{12},$$

whence

$$\delta P_1/\delta P_2 = P_1 c_2/P_2 c_1. \tag{93.7}$$

Thus the relative changes in pressure of saturated solutions, $\delta P_1/P_1$ and $\delta P_2/P_2$, are inversely proportional to the corresponding concentrations c_1 and c_2.

§94. Solutions of strong electrolytes

The method of expanding the thermodynamic quantities in powers of the concentration used in the preceding sections is completely inapplicable in the important case of solutions of *strong electrolytes*, that is, substances which dissociate almost completely into ions when dissolved. The slow decrease of the Coulomb interaction forces between ions with increasing distance leads to terms proportional to a power of the concentration lower than the second (namely, the 3/2 power).

It is easy to see that the problem of determining the thermodynamic quantities of a weak solution of a strong electrolyte reduces to the problem of a completely ionised gas discussed in §75. This result may be derived by starting from the fundamental statistical formula (31.5) for the free energy. The integration in the partition function will be carried out in two stages, first integrating over the co-ordinates and momenta of the solvent molecules. Then the partition function becomes

$$\int e^{-F(p,q)/T}\, d\Gamma,$$

where the integration is now taken only over the phase space of the electrolyte particles, and $F(p, q)$ is the free energy of the solvent with the ions "fixed" in it, the ion co-ordinates and momenta being regarded as parameters. We know from electrodynamics that the free energy of a system of charges in a medium (of given volume and temperature) can be deduced from the energy of the charges in empty space by dividing the products of each pair of charges by the dielectric constant ε of the medium.† The second step in

† This assumes that the distances between ions are large compared with molecular dimensions, but we know from §75 that in the approximation considered the main contribution to the thermodynamic quantities comes in fact from these distances.

calculating the free energy of the solution is therefore identical with the calculations given in §75.

Thus the required contribution of the strong electrolyte to the free energy of the solution is given, according to (75.12), by

$$-\frac{2e^3}{3\varepsilon^{3/2}}\left(\frac{\pi}{TV}\right)^{1/2}\left(\sum_a n_a z_a^2\right)^{3/2},$$

where the summation is over all the kinds of ion in the solution; in accordance with the notation used in this chapter, n_a denotes the total number of ions of the ath kind (in the whole volume of the solution). The same expression gives the contribution to the thermodynamic potential for given temperature and pressure. Putting $V \cong Nv$, where $v(P, T)$ is the molecular volume of the solvent, we can write the thermodynamic potential of the solution in the form

$$\Phi = N\mu_0 + \sum_a n_a T \log(n_a/eN) + \sum_a n_a \psi_a - \frac{2e^3}{3\varepsilon^{3/2}}\left(\frac{\pi}{vT}\right)^{3/2}\left(\frac{\sum n_a z_a^2}{N}\right)^{3/2}. \qquad (94.1)$$

From this we can find, by the usual rules, any of the thermodynamic properties of the electrolyte solution. For example, to calculate the osmotic pressure we write chemical potential of the solvent as

$$\mu = \mu_0 - \frac{T}{N}\sum_a n_a + \frac{e^3}{3\varepsilon^{3/2}}\left(\frac{\pi}{vT}\right)^{1/2}\left(\frac{\sum n_a z_a^2}{N}\right)^{3/2}. \qquad (94.2)$$

As in §89, we find from this the osmotic pressure (at a boundary with the pure solvent)

$$\Delta P = \frac{T}{V}\sum_a n_a - \frac{e^3}{3\varepsilon^{3/2}}\left(\frac{\pi}{T}\right)^{1/2}\left(\frac{\sum n_a z_a^2}{V}\right)^{3/2}. \qquad (94.3)$$

The heat function of the solution is

$$W = -T^2\left(\frac{\partial}{\partial T}\frac{\Phi}{T}\right)_P = Nw_0 - T^2\sum_a n_a \frac{\partial}{\partial T}\frac{\psi_a}{T} + \frac{2e^3}{3}\left(\frac{\pi}{N}\right)^{1/2}\left(\sum n_a z_a^2\right)^{3/2} T^2 \frac{\partial}{\partial T}\left(\frac{1}{\varepsilon^{3/2}T^{3/2}v^{1/2}}\right). \qquad (94.4)$$

From this we can find the "heat of solution" Q which is liberated when the solution is diluted (at constant P and T) with a very large amount of solvent (so that the concentration tends to zero). This quantity of heat is given by the change in the heat function during the process. The terms linear in the number of particles obviously give zero difference, and we find from (94.4)

$$Q = \frac{2e^3\pi^{1/2}}{3} N\left(\frac{\sum n_a z_a^2}{N}\right)^{3/2} T^2 \frac{\partial}{\partial T}\left(\frac{1}{\varepsilon^{3/2}T^{3/2}v^{1/2}}\right). \qquad (94.5)$$

The only condition for the above formulae to be valid is that the concentration should be sufficiently small. For the fact that the electrolyte is strong means that the energy of attraction between ions of different kinds is always less than T. Hence it follows that the interaction energy is certainly small compared with T at distances large compared with molecular distances. But the condition $n \ll N$ for the solution to be weak means precisely that the mean distance between ions is large in comparison with molecular dimensions. Thus this condition necessarily implies that the condition of weak interaction

$$n/V \ll (\varepsilon T/z^2e^2)^3$$

(cf. (75.2)) is satisfied, and this is the basis of the approximations used in §75.

PROBLEM

Find the change in the solubility (assumed small) of a strong electrolyte when a certain quantity of another electrolyte is added to the solution (all the ions of the second electrolyte being different from those of the first).

SOLUTION. The solubility (i.e. the concentration of a saturated solution) of the strong electrolyte is given by the equation

$$\mu_s(P, T) = \sum_a \nu_a\mu_a = T\sum_a \nu_a \log (n_a/N) + \sum_a \nu_a\psi_a -$$
$$-\frac{e^3}{N\varepsilon^{3/2}}\left(\frac{\pi}{vT}\right)^{1/2}\Big(\sum_a \nu_a z_a^2\Big)\Big(\sum_b n_b z_b^2\Big)^{1/2}. \qquad (1)$$

Here μ_s is the chemical potential of the pure solid electrolyte, and ν_a the number of ions of the ath kind per molecule of the electrolyte. When other ions are added to the solution, the chemical potentials of the original ions are changed because of the change in the sum $\sum n_b z_b^2$, which must include all ions present in the solution. Having defined the solubility c_0 by $n_a/N = \nu_a c_0$, we find the change in it by varying the expression (1) for given P and T:

$$\delta c_0 = \frac{\pi^{1/2}e^3\delta\big(\sum n_b z_b^2\big)}{2\varepsilon^{3/2}v^{1/2}T^{3/2}N^2\sum \nu_a}.$$

The sum following δ includes only the added kinds of ion. It should be noted that the solubility is raised under the conditions assumed.

§95. Mixtures of ideal gases

The additivity of the thermodynamic quantities (such as energy and entropy) holds good only so long as the interaction between the various parts of a body is negligible. For a mixture of several substances, e.g. a mixture of several liquids, the thermodynamic quantities are therefore not equal to the sums of the thermodynamic quantities for the individual components of the mixture.

An exception is formed by mixtures of ideal gases, since the interaction between their molecules is by definition negligible. For example, the entropy of such a mixture is equal to the sum of the entropies which each of the gases forming the mixture would have if the other gases were absent and the volume of the one gas were equal to that of the mixture, its pressure therefore being equal to its partial pressure in the mixture. The partial pressure P_i of the ith gas is expressed in terms of the pressure P of the whole mixture by

$$P_i = N_i T/V = N_i P/N, \tag{95.1}$$

where N is the total number of molecules in the mixture, and N_i the number of molecules of the ith gas. Hence, by (42.7), the entropy of a mixture of two gases is

$$S = N_1 \log (eV/N_1) + N_2 \log (eV/N_2) - N_1 f_1'(T) - N_2 f_2'(T), \tag{95.2}$$

or, from (42.8),

$$\begin{aligned} S &= -N_1 \log P_1 - N_2 \log P_2 - N_1 \chi_1'(T) - N_2 \chi_2'(T) \\ &= -(N_1+N_2) \log P - N_1 \log (N_1/N) - N_2 \log (N_2/N) - \\ &\qquad - N_1 \chi_1'(T) - N_2 \chi_2'(T). \end{aligned} \tag{95.3}$$

The free energy of the mixture is, by (42.4),

$$\begin{aligned} F = & -N_1 T \log (eV/N_1) - N_2 T \log (eV/N_2) + \\ & + N_1 f_1(T) + N_2 f_2(T), \end{aligned} \tag{95.4}$$

and similarly (42.6) gives for the potential Φ

$$\begin{aligned} \Phi &= N_1 T \log P_1 + N_2 T \log P_2 + N_1 \chi_1(T) + N_2 \chi_2(T) \\ &= N_1 (T \log P + \chi_1) + N_2 (T \log P + \chi_2) + \\ &\qquad + N_1 T \log (N_1/N) + N_2 T \log (N_2/N). \end{aligned} \tag{95.5}$$

This expression shows that the chemical potentials of the two gases in the mixture are

$$\begin{aligned} \mu_1 &= T \log P_1 + \chi_1 = T \log P + \chi_1 + T \log (N_1/N), \\ \mu_2 &= T \log P_2 + \chi_2 = T \log P + \chi_2 + T \log (N_2/N), \end{aligned} \tag{95.6}$$

i.e. each has the same form as the chemical potential of a pure gas with pressure P_1 or P_2.

It may be noted that the free energy (95.4) of a mixture of gases has the form

$$F = F_1(N_1, V, T) + F_2(N_2, V, T),$$

where F_1 and F_2 are the free energies of the two gases as functions of the number of particles, volume and temperature. No similar formula is valid for the thermodynamic potential, however: the potential Φ of the mixture has the form

$$\Phi = \Phi_1(N_1, P, T) + \Phi_2(N_2, P, T) + N_1 T \log (N_1/N) + N_2 T \log (N_2/N).$$

Let us suppose that we have two different gases with numbers of particles N_1 and N_2 in vessels of volumes V_1 and V_2 at the same temperature and pressure, the two vessels then being connected and the gases mixed. The volume of the mixture becomes V_1+V_2, and the pressure and temperature obviously remain the same. The entropy, however, changes: before mixing, the entropy of the two gases is equal to the sum of their entropies,

$$S_0 = N_1 \log (eV_1/N_1)+N_2 \log (eV_2/N_2)- -N_1 f_1'(T)-N_2 f_2'(T),$$

while after mixing the entropy is, by (95.2),

$$S = N_1 \log [e(V_1+V_2)/N_1]+N_2 \log [e(V_1+V_2)/N_2]- -N_1 f_1'-N_2 f_2'.$$

The change in entropy is

$$\Delta S = S-S_0 = N_1 \log [(V_1+V_2)/V_1]+N_2 \log [(V_1+V_2)/V_2],$$

or, since the volume is proportional to the number of particles for given pressure and temperature,

$$\Delta S = N_1 \log (N/N_1)+N_2 \log (N/N_2). \tag{95.7}$$

This quantity is positive, i.e. the entropy increases on mixing, as it should, because the process is clearly irreversible. The quantity ΔS is called the *entropy of mixing*.

If the two gases were identical, the entropy after connecting the vessels would be

$$S = (N_1+N_2) \log [(V_1+V_2)/(N_1+N_2)]-(N_1+N_2)f,$$

and, since $(V_1+V_2)/(N_1+N_2) = V_1/N_1 = V_2/N_2$ (the pressures and temperatures being equal), the change in entropy would be zero.

Thus the change in entropy on mixing is due to the difference in the molecules of the gases that are mixed. This is in accordance with the fact that some work must be done in order to separate again the molecules of the two gases

§96. Mixtures of isotopes

A mixture of different isotopes (in any aggregate state) is a kind of "solution". For simplicity and definiteness we shall speak of a mixture of two isotopes of any element, but the same results apply to a mixture of any number of isotopes and also to chemical compounds in which different molecules contain different isotopes.

In classical mechanics, the difference between isotopes is simply a difference in mass, the laws of interaction between atoms of different isotopes being identical. This enables us to express the thermodynamic quantities for the mixture very simply in terms of those for the pure isotopes. In calculating the partition function for the mixture, the essential difference is that the phase volume element should be divided not by $N!$ as for a pure substance but by the product $N_1!\, N_2!$ of the factorials of the numbers of atoms of the two components of the system. This gives in the free energy the further terms

$$N_1 T \log (N_1/N) + N_2 T \log (N_2/N)$$

(where $N = N_1 + N_2$), which correspond to the "entropy of mixing" discussed in §95 for the case of a mixture of gases.

Similar terms appear in the thermodynamic potential of the mixture, which may be written

$$\Phi = N_1 T \log (N_1/N) + N_2 T \log (N_2/N) + \\ + N_1 \mu_{01} + N_2 \mu_{02}. \tag{96.1}$$

Here μ_{01} and μ_{02} are the chemical potentials of the pure isotopes, which differ only by a constant times the temperature:

$$\mu_{01} - \mu_{02} = -\tfrac{3}{2} T \log (m_1/m_2), \tag{96.2}$$

where m_1 and m_2 are the atomic masses of the two isotopes. This difference arises from the integration over the atomic momenta in the partition function; for gases, (96.2) is simply the difference between the chemical constants multiplied by T.

The difference (96.2) is the same for all phases of a given substance. The equation of phase equilibrium (the condition that the chemical potentials of the phases are equal) is therefore the same for every isotope. In particular, we can say that in the classical approximation the saturated vapour pressures of the various isotopes are equal.

The situation is no longer so simple when the substance cannot be described by means of classical statistics. In quantum theory, the difference between isotopes becomes considerably more profound, because of the differences in the vibrational and rotational levels, nuclear spins, etc.

It is important to note, however, that, even when the first correction terms (of order $\hbar^2$; see §33) in the thermodynamic quantities are taken into account, the thermodynamic potential of the mixture may be written in the form (96.1), since the terms in question form a sum, with each term containing the mass of only one atom (see formula (33.15) for the free energy). These terms may therefore be grouped so as to include them in the chemical potentials μ_{01} and μ_{02}, and hence formula (96.1) (but not, of course, (96.2)) remains valid.

It should be pointed out that the thermodynamic potential (96.1) is formally identical with that of a mixture of any two gases (§95). Mixtures having this property are called *ideal mixtures*. Thus mixtures of isotopes are ideal mixtures up to and including terms of order $\hbar^2$. In this sense, mixtures of isotopes form an exceptional case, since condensed (solid or liquid) mixtures of different substances which are not isotopes can be ideal mixtures only to a very rough approximation.

Within the limits of validity of formula (96.1) we can draw certain conclusions about the vapour pressure of the isotopes over the condensed mixture. The chemical potentials of the two components of the mixture are

$$\mu_1 = T \log c_1 + \mu_{01},$$
$$\mu_2 = T \log c_2 + \mu_{02}$$

(where $c_1 = N_1/N$, $c_2 = N_2/N$ are the concentrations of the isotopes). Equating these to the chemical potentials in the gas phase (which have the forms $T \log P_1 + \chi_1(T)$ and $T \log P_2 + \chi_2(T)$), we find for the partial vapour pressures

$$P_1 = P_{01} c_1, \qquad P_2 = P_{02} c_2, \tag{96.3}$$

where P_{01} and P_{02} denote the vapour pressures of the two pure isotopes (at a given temperature). Thus the partial vapour pressures of the two isotopes are proportional to their concentrations in the condensed mixture.

In the classical approximation we have for the saturated vapour pressures of the pure isotopes $P_{01} = P_{02}$, as already mentioned. When quantum effects are taken into account, however, the two vapour pressures are no longer equal. The difference cannot be calculated in a general form applicable to all substances. Such a calculation can be made only for monatomic elements (the inert gases) as far as the terms of order $\hbar^2$ (K. F. HERZFELD and E. TELLER 1938).

The correction to the thermodynamic potential of a liquid phase is given by formula (33.15)†; taking the value per atom, we find the chemical potential

$$\mu = \mu_{\text{cl}} + (\hbar^2/24mT)\overline{F^2},$$

where

$$\overline{F^2} = \overline{\left(\frac{\partial U}{\partial x}\right)^2} + \overline{\left(\frac{\partial U}{\partial y}\right)^2} + \overline{\left(\frac{\partial U}{\partial z}\right)^2}$$

is the mean square of the force exerted on one atom by the other atoms in the liquid. The chemical potential of the gas remains equal to its classical value, since the interaction between atoms in the gas is negligible. Equating the chemical potentials of the liquid and the gas, we find the correction to the classical value of the vapour pressure, and the required difference of vapour

† We again make use of the fact that small corrections to the various thermodynamic potentials, when expressed in terms of the corresponding variables, are equal (§15).

pressures between the two isotopes is

$$P_{01}-P_{02} = P_0 \frac{\hbar^2\overline{F^2}}{24T^2}\left(\frac{1}{m_1}-\frac{1}{m_2}\right), \tag{96.4}$$

where P_0 is the common classical value of P_{01} and P_{02}. We see that the sign of the difference is determined by that of the difference of the reciprocal masses of the isotopes, the vapour pressure of the lighter isotope being the greater.

§97. Vapour pressure over concentrated solutions

Let us now consider the equilibrium of a solution with the vapour over it, which in general also contains both substances. The solution may be either weak or strong, i.e. the quantities of the two substances in it are arbitrary. It will be remembered that the results derived in §90 apply only to weak solutions.

Since the solution and the vapour are in equilibrium, the chemical potentials μ_1 and μ_2 in the solution and in the vapour are equal. If the numbers of particles of the two substances in the solution are N_{1s} and N_{2s}, we can write the expression (24.14) for the solution in the form

$$d\Omega = -N_{1s}\,d\mu_1 - N_{2s}\,d\mu_2 - S_s\,dT - P\,dV_s, \tag{97.1}$$

where S_s and V_s are the entropy and volume of the solution; the temperature T and pressure P are the same for the solution and the vapour.

We shall assume that the vapour over the solution is so rarefied that it may be regarded as an ideal gas; its pressure is small. Then we can neglect in (97.1) the terms proportional to P, viz. $P\,dV$ and $d\Omega$. Let us first consider all derivatives to be taken at constant temperature. Then (97.1) gives

$$N_{1s}\,d\mu_1 + N_{2s}\,d\mu_2 = 0. \tag{97.2}$$

For the gas phase we have

$$\mu_{1g} = T\log P_1 + \chi_1(T),$$
$$\mu_{2g} = T\log P_2 + \chi_2(T),$$

where P_1 and P_2 are the partial pressures of the two components of the vapour. Differentiating these expressions (with T constant), we find

$$d\mu_{1g} = T\,d\log P_1, \quad d\mu_{2g} = T\,d\log P_2.$$

Substitution in (97.2) gives

$$N_{1s}\,d\log P_1 + N_{2s}\,d\log P_2 = 0. \tag{97.3}$$

The concentration ξ of the solution can be defined as the ratio of the number of particles of the first component to the total number of particles:

$$\xi = N_{1s}/(N_{1s}+N_{2s}),$$

and we can similarly define the concentration x of the vapour. The partial pressures P_1 and P_2 are equal to the total pressure P of the vapour multiplied by the concentrations of the corresponding components, i.e. $P_1 = xP$, $P_2 = (1-x)P$. Substituting these values in (97.3) and dividing this equation by the total number of particles in the solution, $N = N_{1s}+N_{2s}$, we find

$$\xi \,\mathrm{d} \log Px+(1-\xi)\,\mathrm{d} \log P(1-x) = 0,$$

whence

$$\mathrm{d} \log P = (x-\xi)\,\mathrm{d}x/x(1-x),$$

or

$$\xi = x-x(1-x)\partial \log P/\partial x. \tag{97.4}$$

This equation relates the solution and vapour concentrations to the dependence of the vapour pressure on the vapour concentration.

One further general relation can be obtained by considering the dependence of quantities on temperature. The condition for equality of the chemical potentials of one component, say the first, in the vapour and in the solution is $\mu_{1g} = \partial\Phi_s/\partial N_{1s}$. Dividing both sides by T and using the fact that the derivative with respect to the number of particles is taken at constant temperature, we write

$$\frac{\mu_{1g}}{T} = \frac{\partial}{\partial N_{1s}}\frac{\Phi_s}{T},$$

and then take the total derivative of each side with respect to temperature. In doing so we may assume with sufficient accuracy that the thermodynamic potential of the condensed phase (the solution) is independent of pressure. Noting also that the partial derivative with respect to temperature is

$$\frac{\partial}{\partial T}\frac{\Phi}{T} = -\frac{1}{T^2}\left(\Phi-T\frac{\partial\Phi}{\partial T}\right) = -\frac{W}{T^2},$$

we obtain the relation

$$T^2\frac{\partial \log P_1}{\partial T} = w_{1g}-\frac{\partial W_s}{\partial N_{1s}}. \tag{97.5}$$

Here w_{1g} is the molecular heat function of the first substance as a gas; the derivative $\partial W_s/\partial N_{1s}$ gives the change in the heat function of the solution when one molecule of that substance is added to it. The quantity on the right of (97.5) is therefore the heat absorbed when one particle of the first substance goes from the solution to the vapour.

For the first substance in the pure state, the relation (97.5) becomes the ordinary Clapeyron–Clausius equation,

$$T^2\frac{\partial \log P_{10}}{\partial T} = w_{1g}-w_{1l},$$

where P_{10} is the vapour pressure of the first substance in the pure state, and w_{1l} its molecular heat function when liquid. Subtracting this equation term by term from (97.5), we have finally

$$T^2 \frac{\partial}{\partial T} \log \frac{P_1}{P_{10}} = -q_1, \tag{97.6}$$

where $q_1 = \partial W_s/\partial N_{1s} - w_{1l}$ denotes the molecular "heat of dilution", i.e. the quantity of heat absorbed when one particle from the liquid first substance goes into the solution. A similar relation can, of course, be written down for the second substance also.

§98. Thermodynamic inequalities for solutions

It has been shown in §21 that a body can exist only in states for which certain conditions called *thermodynamic inequalities* are satisfied. These conditions were derived, however, for bodies consisting of identical particles. We shall now give a corresponding analysis for solutions, taking only the case of a mixture of two substances.

In §21 the condition of equilibrium used was not the condition of maximum entropy of a closed system as a whole but the equivalent condition which requires that the minimum work needed to bring any small part of the system from the equilibrium state to any neighbouring state should be positive.

We now use a similar procedure, considering some small part of the solution, which contains N solvent and n solute particles, say. In the equilibrium state the temperature, pressure and concentration in this small part are equal to their values in the rest of the solution (which acts as an "external medium"). Let us determine the minimum work needed to bring the temperature, pressure and number of solute particles in the small part considered (containing a fixed number N of solvent particles) to values which differ by small but finite amounts δT, δP and δn from their equilibrium values.

The minimum work will be done if the process occurs irreversibly. The work done by an external source is then equal to the change in the energy of the system, i.e.

$$\delta R_{\text{min}} = \delta E + \delta E_0;$$

quantities without suffix refer to the small part considered, and those with suffix zero refer to the remainder of the system. We express δE_0 in terms of the changes in the independent variables:

$$\delta R_{\text{min}} = \delta E + T_0 \delta S_0 - P_0 \delta V_0 + \mu_0' \delta n_0,$$

where μ_0' is the chemical potential of the solute in the medium; the number of solvent particles is unchanged in the process considered, and so the corresponding term for the solvent may be omitted.† From the reversibility of the process it follows that $\delta S_0 = -\delta S$, and from the conservation of the total volume and quantity of solute in the whole solution we have $\delta V = -\delta V_0$, $\delta n = -\delta n_0$. Substituting these, we obtain the final expression for the work:

$$\delta R_{\text{min}} = \delta E - T_0 \delta S + P_0 \delta V - \mu_0' \delta n. \tag{98.1}$$

Thus the condition of equilibrium can be taken to be that for any small part of the solution the inequality

$$\delta E - T_0 \delta S + P_0 \delta V - \mu_0' \delta n > 0 \tag{98.2}$$

holds. Henceforward, as in §21, we shall omit the suffix zero in expressions which are coefficients of the deviations of quantities from their equilibrium values; the values of these expressions in the equilibrium state will always be meant.

We expand δE in powers of δV, δS and δn (regarding E as a function of V, S and n). As far as the second-order terms this gives

$$\delta E = \frac{\partial E}{\partial S}\delta S + \frac{\partial E}{\partial V}\delta V + \frac{\partial E}{\partial n}\delta n + $$
$$+\tfrac{1}{2}\left[\frac{\partial^2 E}{\partial S^2}(\delta S)^2 + \frac{\partial^2 E}{\partial V^2}(\delta V)^2 + \frac{\partial^2 E}{\partial n^2}(\delta n)^2 + \right.$$
$$\left. +2\frac{\partial^2 E}{\partial S \partial V}\delta S \delta V + 2\frac{\partial^2 E}{\partial S \partial n}\delta S \delta n + 2\frac{\partial^2 E}{\partial V \partial n}\delta V \delta n\right].$$

But $\partial E/\partial V = -P$, $\partial E/\partial S = T$, $\partial E/\partial n = \mu'$. Thus the first-order terms cancel on substitution in (98.2), leaving

$$2\delta R_{\text{min}} = \frac{\partial^2 E}{\partial S^2}(\delta S)^2 + \frac{\partial^2 E}{\partial V^2}(\delta V)^2 + \frac{\partial^2 E}{\partial n^2}(\delta n)^2 + $$
$$+2\frac{\partial^2 E}{\partial S \partial V}\delta S \delta V + 2\frac{\partial^2 E}{\partial S \partial n}\delta S \delta n + 2\frac{\partial^2 E}{\partial V \partial n}\delta V \delta n > 0. \tag{98.3}$$

It is known from the theory of quadratic forms that, for a form in three variables (here δS, δV, δn) to be everywhere positive, its coefficients must

† The differential of the energy of the medium (at constant N) is

$$dE_0 = T_0 dS_0 - P_0 dV_0 + \mu_0' \, dn_0.$$

Since the quantities T_0, P_0, μ_0' may be regarded as constant, integration of this relation leads to a similar relation between the finite variations of the quantities E_0, S_0, V_0, N_0.

The quantity μ_0' should not be confused with the chemical potential of the pure solute.

satisfy three conditions, which for the form (98.3) are

$$\begin{vmatrix} \partial^2E/\partial V^2 & \partial^2E/\partial V\partial S & \partial^2E/\partial V\partial n \\ \partial^2E/\partial S\partial V & \partial^2E/\partial S^2 & \partial^2E/\partial S\partial n \\ \partial^2E/\partial n\partial V & \partial^2E/\partial n\partial S & \partial^2E/\partial n^2 \end{vmatrix} > 0, \tag{98.4}$$

$$\begin{vmatrix} \partial^2E/\partial V^2 & \partial^2E/\partial V\partial S \\ \partial^2E/\partial S\partial V & \partial^2E/\partial S^2 \end{vmatrix} > 0, \qquad \partial^2E/\partial S^2 > 0.$$

Substituting the values of the derivatives of E with respect to V, S and n, we can write these conditions as

$$\begin{vmatrix} \partial P/\partial V & \partial P/\partial S & \partial P/\partial n \\ \partial T/\partial V & \partial T/\partial S & \partial T/\partial n \\ \partial \mu'/\partial V & \partial \mu'/\partial S & \partial \mu'/\partial n \end{vmatrix} < 0,$$

$$\begin{vmatrix} \partial P/\partial V & \partial P/\partial S \\ \partial T/\partial V & \partial T/\partial S \end{vmatrix} < 0, \qquad \partial T/\partial S > 0.$$

Each derivative is taken with the other two of the three variables V, S and n constant. These determinants are Jacobians:

$$\frac{\partial(P, T, \mu')}{\partial(V, S, n)} < 0, \quad \left(\frac{\partial(P, T)}{\partial(V, S)}\right)_n < 0, \quad \left(\frac{\partial T}{\partial S}\right)_{V,n} > 0. \tag{98.5}$$

The second and third conditions give the already known inequalities $(\partial P/\partial V)_{T,n} < 0$ and $C_v > 0$. The first condition may be transformed as follows:

$$\frac{\partial(P, T, \mu')}{\partial(V, S, n)} = \frac{\partial(P, T, \mu')/\partial(P, T, n)}{\partial(V, S, n)/\partial(P, T, n)}$$

$$= \frac{(\partial\mu'/\partial n)_{P,T}}{(\partial(V, S)/\partial(P, T))_n} < 0.$$

Since the denominator is negative by the second condition (98.5), we must have

$$(\partial\mu'/\partial n)_{P,T} > 0. \tag{98.6}$$

Using instead of n the concentration $c = n/N$, we find (since N is constant)

$$(\partial\mu'/\partial c)_{P,T} > 0. \tag{98.7}$$

Thus, as well as the inequalities $(\partial P/\partial V)_{T,c} < 0$, $C_v > 0$, the inequality (98.7) must also be satisfied in solutions.

It may be noted that for weak solutions $\partial\mu'/\partial c = T/c$, so that the inequality (98.7) is always satisfied.

The case where

$$(\partial\mu'/\partial c)_{P,T} = 0 \tag{98.8}$$

needs special consideration. Such a state is called a *critical point* of the solution; other aspects of this concept are discussed in §99.

The equality (98.8) corresponds to the vanishing of the first determinant in (98.4) (the third-order determinant). In this case the quadratic form (98.3) may vanish for certain values of δS, δV and δn, and higher-order terms in its expansion must be examined in order to ascertain the conditions for the inequality (98.2) to be satisfied (cf. §84).

The quadratic form (98.3) may be written in the identical form

$$\begin{aligned}2\delta R_{\min} &= \delta S\delta(\partial E/\partial S)_{V,n}+\delta V\delta(\partial E/\partial V)_{S,n}+\\&\qquad\qquad+\delta n\delta(\partial E/\partial n)_{S,V}\\&= \delta S\delta T-\delta V\delta P+\delta n\delta\mu'.\end{aligned} \tag{98.9}$$

When $(\partial\mu'/\partial n)_{P,T} = 0$, we have

$$\delta\mu' = (\partial\mu'/\partial T)\delta T+(\partial\mu'/\partial P)\delta P;$$

thus, if δT and δP are zero, $\delta\mu'$ is also zero, and so is the whole expression (98.9).† The case where the quadratic form vanishes can therefore be treated by simply considering deviations from equilibrium at constant T and P. For such deviations, the inequality (98.2) may be written $\delta\Phi-\mu'\delta n > 0$. Expanding $\delta\Phi$ in powers of δn for constant P and T, and using the fact that $\partial\Phi/\partial n = \mu'$, we find

$$\frac{1}{2}\frac{\partial\mu'}{\partial n}(\delta n)^2+\frac{1}{6}\frac{\partial^2\mu'}{\partial n^2}(\delta n)^3+\frac{1}{24}\frac{\partial^3\mu'}{\partial n^3}(\delta n)^4+\ldots > 0,$$

where all the derivatives are taken with P and T constant. If $\partial\mu'/\partial n = 0$, this inequality can be satisfied for all δn only if the coefficient of $(\delta n)^3$ also vanishes and that of $(\delta n)^4$ is positive.

Thus, at a critical point we must have, together with (98.8),

$$(\partial^2\mu'/\partial c^2)_{P,T} = 0, \tag{98.10}$$

$$(\partial^3\mu'/\partial c^3)_{P,T} > 0. \tag{98.11}$$

The equations (98.8) and (98.10) define a line (the *critical line*) in the co-ordinates P, T, c.

It should be emphasised, however, that the foregoing discussion of critical points in solutions is subject to the same reservation as that made in §84 regarding the theory of critical points for pure substances: it is based on the assumption that the thermodynamic quantities have no singularity (as functions of the variables c, V, T); since this assumption cannot be justified, we do not know to what extent the results derived are valid.

† The second and third expressions in (98.4) cannot vanish, since this would violate other conditions (cf. §84).

§99. Equilibrium curves

The state of a body consisting of identical particles is defined by the values of any two quantities, for instance P and T. To define the state of a system having two components (a *binary mixture*) it is necessary to specify three quantities, for instance P, T and the concentration. In this and subsequent sections, the concentration of the mixture will be defined as the ratio of the quantity of one of the substances to the total quantity of both, and will be denoted by x; clearly x takes values from 0 to 1. The state of a binary mixture may be represented by a point in a three-dimensional co-ordinate system, whose axes correspond to these three quantities (just as the state of a system of identical particles was represented by a point in the PT-plane).

According to the phase rule, a two-component system can consist of not more than four phases in contact. The number of degrees of freedom of such a system is two when there are two phases, one for three phases, and none for four phases. The states in which two phases are in equilibrium are therefore represented by points forming a surface in the three-dimensional co-ordinate system; states with three phases (triple points) by points forming a line (called the *line of triple points* or the *three-phase line*) and states with four phases by isolated points.

It has already been shown in §81 that, for systems with only one component, the states in which two phases are in equilibrium are represented by a curve in the PT-plane; each point on this curve determines the pressure and temperature (which are the same in both phases, from the conditions of equilibrium). Points not lying on the curve represent homogeneous states of the system. If the temperature and volume are taken as co-ordinates, the phase equilibrium is represented by a curve such that points within it represent states where there is separation into two phases represented by the points of intersection of a straight line T = constant with the equilibrium curve.

The situation is similar for mixtures. If we take as co-ordinates P, T and the chemical potential of one component (i.e. quantities which have equal values for phases in contact), equilibrium of two phases is represented by a surface, each point of which determines P, T and μ for the two phases in equilibrium. When three phases are present, the points representing their equilibrium (triple points) will lie on the curves of intersection of the equilibrium surfaces for each pair of them.

The use of the variables P, T, μ is inconvenient, however, and in what follows we shall use P, T, x as independent variables. In terms of these variables the equilibrium of two phases is represented by a surface whose points of intersection with a straight line P = constant, T = constant represent the states of the two phases in contact for the relevant values of P and T (i.e. determine the concentrations of the phases, which may of course be different). The points on this line between the two points of intersection represent states in which a homogeneous body is unstable and therefore separates into two phases (represented by the points of intersection). Since the surface represents the equilibrium between the two phases, it must clearly be such that the number of its intersections with any straight line parallel to the x-axis is even.

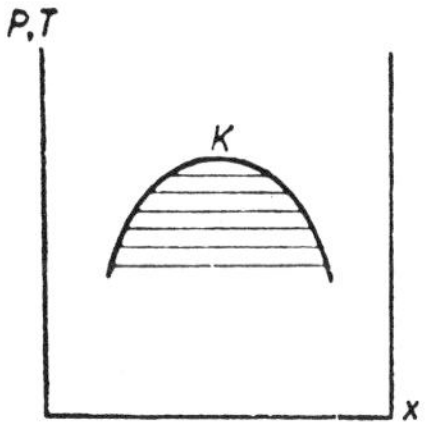

FIG. 21

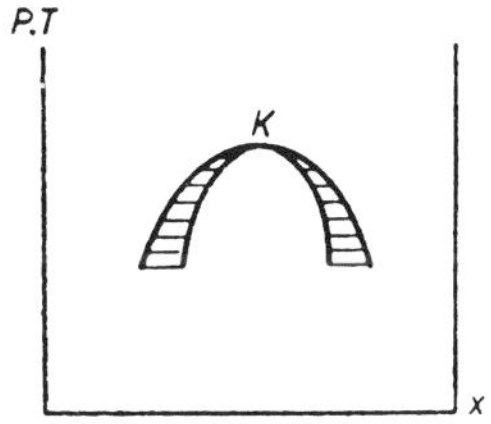

FIG. 22

We shall generally use two-dimensional diagrams with P and x, or T and x, as co-ordinates; the lines of intersection of the equilibrium surface with the planes of constant temperature or pressure can then be drawn. We shall call these lines *equilibrium curves*.

Let us consider the points on an equilibrium curve at which the concentration becomes equal in the two phases. Two cases are possible: (1) at such a point all other properties of the two phases also become equal, i.e. the phases become identical, (2) at such a point two distinct phases continue to exist. In case (1) the point is said to be a *critical point*, in case (2) it will be called a *point of equal concentration*.

Near a critical point the equilibrium curve has the form shown in Fig. 21, or a similar form with a minimum at the critical point K (the abscissa being x and the ordinate P or T; the curve is then the intersection of the equilibrium surface with a plane of constant temperature or constant pressure respectively). Points lying within this curve (in the hatched region) represent states in which there is separation into two phases; the concentrations in these phases are determined by the points of intersection of the curve with the appropriate horizontal line. At the point K the two phases coalesce; the fact that at this point they form a single phase is seen from the possibility of a continuous passage between the points coinciding at K along any path lying outside the hatched region, so that separation into two phases nowhere occurs.

Fig. 21 shows that near the critical point there exist states in which two phases are in equilibrium which have concentrations x and $x+\delta x$ differing by an arbitrarily small amount. For such phases the equilibrium condition is $\mu(P, T, x) = \mu(P, T, x+\delta x)$, where μ is the chemical potential of one of the substances in the mixture. Hence we see (cf. §83) that at the critical point the condition

$$(\partial\mu/\partial x)_{P,\,T} = 0 \tag{99.1}$$

must hold.

This condition is identical with (98.8), and hence the two definitions of the critical point (here and in §98) are equivalent. It may be noted that μ in (99.1) signifies the chemical potential of either of the two substances in the mixture; but the two conditions obtained by taking these two chemical potentials in (99.1) are actually equivalent. This is easily seen by noting that each of the chemical potentials is the derivative of Φ with respect to the corresponding number of particles, and Φ is a first-order homogeneous function in both numbers of particles.

The critical points clearly form a line on the equilibrium surface (as already mentioned in §98).

Near a point of equal concentration the equilibrium curves must have the form shown in Fig. 22, or a similar form with a minimum at the point K. The two curves touch at the maximum (or minimum). The region between the two curves is that where separation into phases occurs. At the point K the concentrations of the two phases in equilibrium become equal, but the different phases continue to exist, since any path between the points which coincide at K must pass through the region of separation into two phases. Like critical points, points of equal concentration lie on a curve on the equilibrium surface.

Let us now consider the properties of the equilibrium curves at low concentrations (i.e. when one of the substances is present in the mixture in a considerably smaller quantity than the other; x is close to zero or to unity).

It has been shown in §90 that at low concentrations (weak solutions) the difference between the phase equilibrium temperatures of solutions and of the pure substance (at a given pressure) is proportional to the difference of concentrations of the two phases. The same applies to the pressure difference at a given temperature. Moreover, it has been shown in §91 (also for low concentrations) that the ratio of concentrations in the two phases depends only on P and T, and so it may be regarded as constant in the neighbourhood of $x = 0$.

From the above it follows directly that at low concentrations the equilibrium curves have the form shown in Fig. 23, i.e. consist of two straight lines intersecting on the ordinate axis (or a similar form with the straight lines ascending). The region between the two lines is the region of separation into phases. The regions above and below the lines are the regions of the two different phases.

At the beginning of this section it has already been mentioned that a system with two components may consist of three phases in contact. Near a triple point the equilibrium curves appear as shown in Fig. 24. All three phases have equal pressure and temperature in equilibrium. The points A, B, C which determine their concentrations therefore lie on a straight line parallel to the axis of abscissae. The point A, which gives the concentration of the first phase at the triple point, is the point of intersection of the equilibrium curves 12 and 13 between the first and second, and first and third, phases. Similarly, the points B and C are the intersections of the equilibrium curves 12 and 23 between the first and second, and second and third, phases (B), and of the equilibrium curves 23 and 13 between the second and third, and

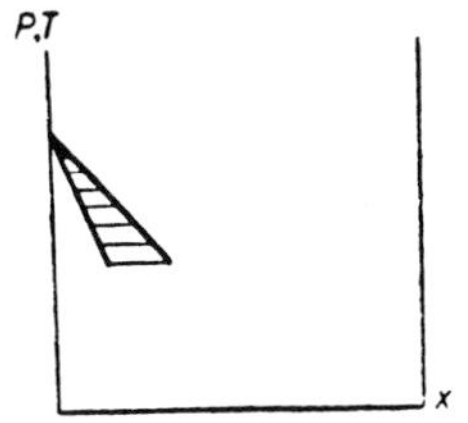

FIG. 23

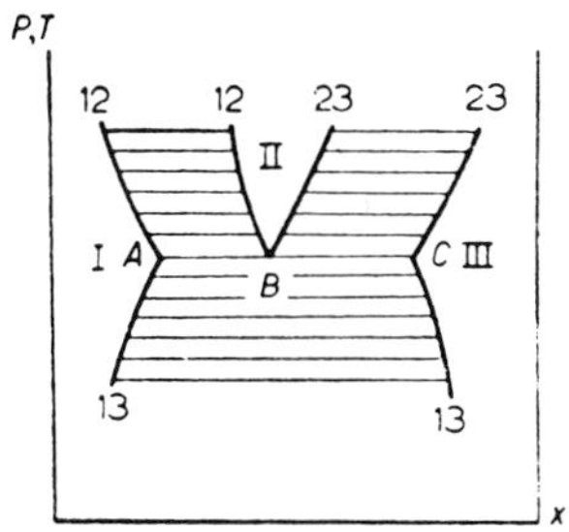

FIG. 24

first and third, phases (C). The points A, B, C are, of course, the points of intersection of the plane P = constant or T = constant with three lines on the equilibrium surface; we shall call the line corresponding to the point B a *line of triple points* or a *three-phase line*. The regions I, II, III represent states of the separate phases, first, second and third. The region between the two curves 13 below the line ABC is the region of separation into the first and third phases, and those between the two curves 12 and the two curves 23 (above the line ABC) are respectively the regions of separation into the first and second, and second and third, phases. Region II must obviously lie entirely above ABC (or entirely below ABC). At the points A, B and C the curves 12, 13 and 23 intersect, in general, at certain angles, and do not join smoothly. The directions of the curves 12, 13, 23 need not necessarily be as shown in Fig. 24, of course. The only essential feature is that the curves 12 and 23 and the curves 13 must lie on opposite sides of the straight line ABC.

If any of these singular lines on the equilibrium surface is projected on the PT-plane, the projection divides this plane into two parts. For a critical line, the points projected on one part are those corresponding to the two different phases and those corresponding to separation into these phases. The other part of the PT-plane contains the projections of points which represent homogeneous states, at none of which does separation into two phases occur. In Fig. 25 the dotted line represents the projection of a critical line on the

PT-plane. The letters *a* and *b* denote the two phases. The symbol *a-b* signifies that this part of the plane contains the projections of the two phases and those of states where these two phases are present in equilibrium. The symbol *ab* denotes the single phase into which the phases *a* and *b* merge above the critical points.

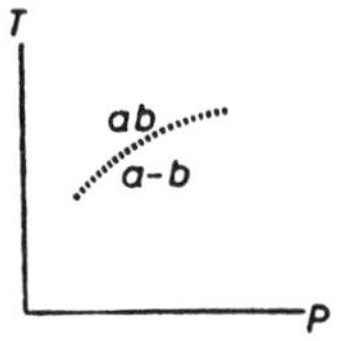

FIG. 25

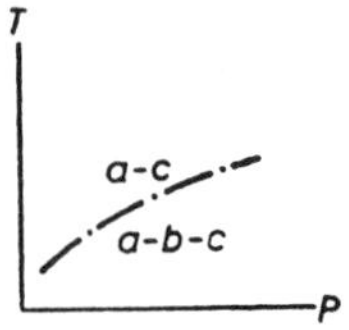

FIG. 26

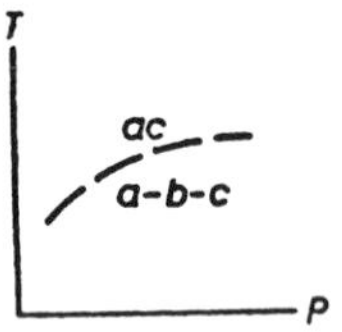

FIG. 27

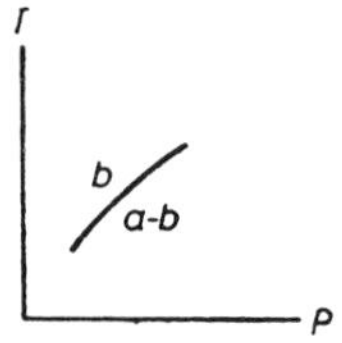

FIG. 28

The projection of a three-phase line similarly divides the *PT*-plane into two parts. Fig. 26 shows which points are projected on the two parts. The symbol *a-b-c* signifies that this region contains the projections of points which represent the phases *a*, *b*, *c* and states in which there is separation into phases *a* and *b* or *b* and *c*.

Fig. 27 shows a similar projection for a line of points of equal concentration, and Fig. 28 for a line of phase equilibrium for a pure substance (i.e. $x = 0$ or $x = 1$); the latter, of course, lies in the *PT*-plane. The letter *b* in Fig. 28 signifies that this part of the plane contains the projections of points corresponding to states of phase *b* only. In the sequence of letters in the symbols *a-b*, *a-b-c* the letter *b* will be understood to denote a phase whose concentration is higher than that of *a*, and *c* a phase whose concentration is higher than that of *b*.†

It may be noted that the four types of singular point on the equilibrium curves (triple point, point of equal concentration, critical point, and pure-substance point) correspond to the four possible types of maximum (or minimum) on these curves.

† To avoid misunderstanding, we should emphasise that the notation *a-b-c* for a line of equal concentration (unlike a three-phase line) is to some extent arbitrary: the letters *a* and *c* here denote states which are not two essentially different phases, since they never exist simultaneously in contact.

If any phase has the same fixed composition everywhere (i.e. independently of the values of P and T), the equilibrium curves become somewhat simpler near the points here considered. Such phases are a chemical compound of the two components or pure-substance phases, which always have concentration $x = 0$ (or $x = 1$).

Let us consider the form of the equilibrium curves when there are phases of constant composition, near points where the lines corresponding to these phases terminate. It is evident that such points must be maxima or minima of the equilibrium curves, and thus are among the types of point considered in this section.

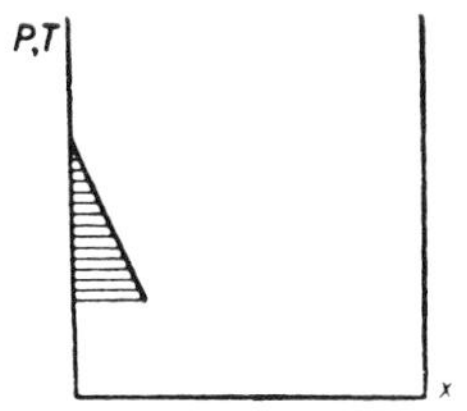

FIG. 29

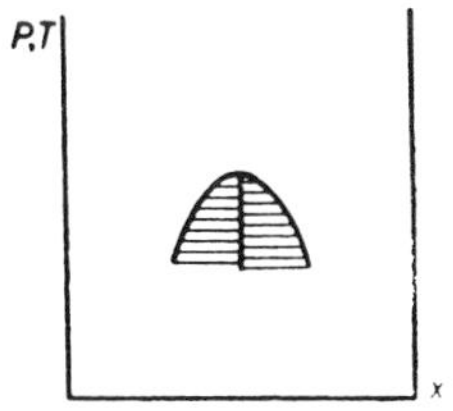

FIG. 30

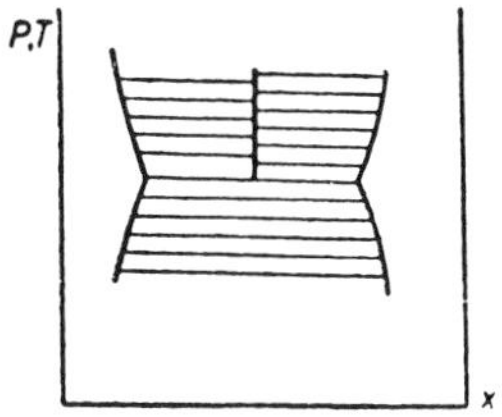

FIG. 31

If the phase of constant composition is a pure-substance phase with concentration $x = 0$, the corresponding line coincides with the P-axis or the T-axis and can terminate at a point of the kind shown in Fig. 29. This diagram gives the form of the equilibrium curve near such a point; one of the lines in Fig. 23 coincides with the axis of ordinates.

If one phase is a chemical compound of fixed composition, then near a point of equal concentration the equilibrium curve has the form shown in Fig. 30, i.e. the inner region in Fig. 22 becomes a vertical line. The hatched region on either side of this line is the region of separation into two phases, one a chemical compound whose composition is given by the vertical line. The curve has no break at the maximum, as in Fig. 22.

Similarly, near a triple point the equilibrium curves have the form shown in Fig. 31. The phase which is a chemical compound is represented by a vertical line, to which region II (Fig. 24) here reduces.

§100. Examples of phase diagrams

In this section we shall enumerate the principal types of equilibrium curve; in contrast to §99, their form will be considered in general, and not only near the singular points. These curves (also called *phase diagrams*) can have many forms, but in most cases they belong to one of the types given below, or are a combination of more than one of these. The hatched regions in all these diagrams are the regions of separation into phases, and the remaining regions are those of homogeneous states. The points of intersection of horizontal lines with the curves bounding the regions of separation into phases determine the composition of the phases into which separation occurs (for given

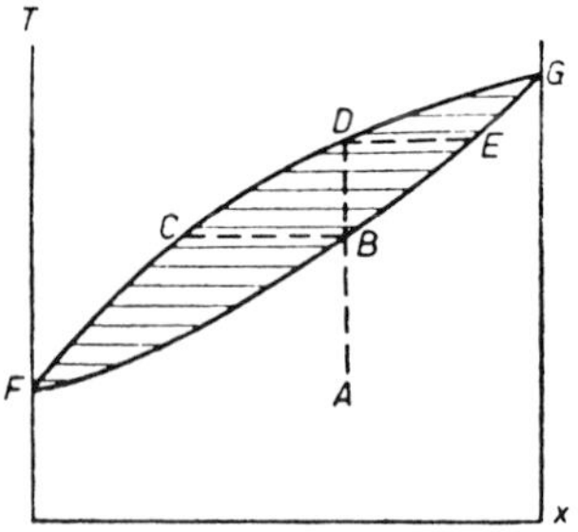

FIG. 32

P and T). The relative amounts of the two phases are determined by the "lever rule" already mentioned in §81.

In what follows we shall for definiteness discuss Tx-diagrams; similar types are possible in the co-ordinates P and x. The concentration x is taken as abscissa, and varies from 0 to 1.

1. There are two phases; each can have any concentration (i.e. the two components mix in any proportion in both phases). In the simplest case, where the curves have no maxima or minima (apart from the pure-substance points), the phase diagram has the "cigar" form shown in Fig. 32.

For example, let one of the phases be a liquid (the region below the cigar), and the other a vapour (the region above the cigar). Then the upper curve of the cigar is called the *condensation point curve*, and the lower curve the *boiling point curve*.†

If a liquid mixture of given composition is heated, then the liquid will begin to boil at a temperature determined by the intersection B of the vertical line AD (corresponding to the given concentration) with the lower curve of the cigar. Vapour boils off, whose composition is given by the point C, i.e. has

† The laws of boiling and condensation of liquid mixtures were established by D. P. KONOVALOV.

a lower concentration than the liquid. The concentration of the remaining liquid is obviously increased, and its boiling point accordingly rises. On further heating, the point which represents the state of the liquid phase will move upwards along the lower curve, and the point which represents the vapour leaving the liquid will move upwards along the upper curve. Boiling ceases at various temperatures, depending on the way in which the process takes place. If boiling occurs in a closed vessel, so that all the vapour generated remains permanently in contact with the liquid, the liquid will obviously boil away completely at a temperature where the concentration of the vapour is equal to the original concentration of the liquid (the point *D*). In this case, therefore, boiling begins and ends at temperatures given by the intersection

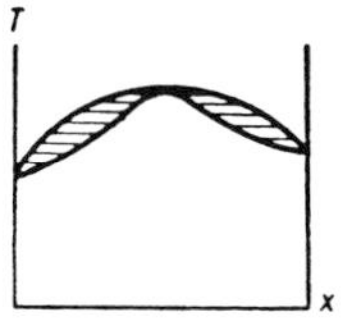

FIG. 33

of the vertical line *AD* with the lower and upper curves of the cigar. If the vapour boiling off is steadily removed (boiling in an open vessel), then only the vapour just evolved will be in equilibrium with the liquid at any given time. In this case, it is evident that boiling will cease at the boiling point *G* of the pure substance, where the liquid and vapour compositions are the same. The condensation of vapour into liquid occurs in a similar manner.

The situation is exactly analogous when the two phases are a liquid (above the cigar) and a solid (below the cigar).

2. The two components mix in any proportion in both phases (as in case 1), but there is a point of equal concentration. The phase diagram then has the form shown in Fig. 33 (or a similar form with a minimum). At the point of equal concentration, the two curves touch, and both have a maximum or a minimum.

The transition from one phase to another occurs in the same way as described for case 1, except that the process can terminate (if one phase is steadily removed, as for example by boiling a liquid in an open vessel) not only at the pure-substance point but also at the point of equal concentration. At the composition corresponding to this point, the entire process occurs at a single temperature.†

3. There are two phases, liquid and gas, in which the two components mix

† A mixture corresponding to the point of equal concentration is also said to be *azeotropic*.

in any proportion, and there is a critical point. The phase diagram is as shown in Fig. 34 (K being the critical point). The region to the right of the curve corresponds to liquid states, and that to the left to gaseous states. It should be remembered, however, that when there is a critical point the liquid and gaseous phases can, strictly speaking, be distinguished only when they are in equilibrium with each other.

A diagram of this type leads to the following curious effect. If a liquid whose composition is represented by the line AC (passing to the right of the point K) is heated in a closed vessel, then, after boiling begins (at the point B), the quantity of vapour will gradually increase as heating continues, but after a certain time it begins to decrease again, and the vapour disappears entirely at the point C. This is called *retrograde condensation.*

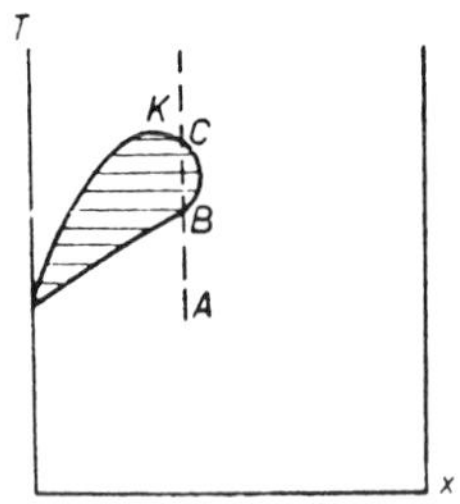

FIG. 34

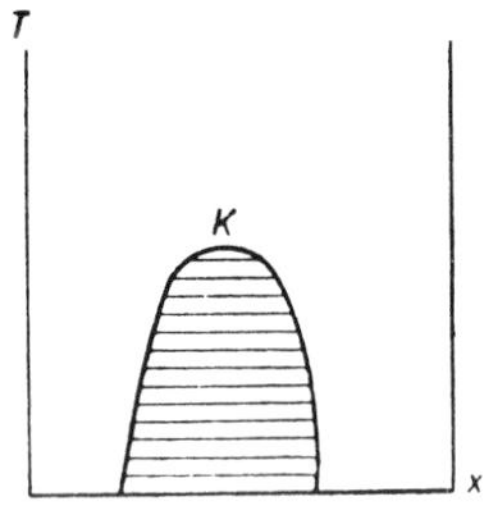

FIG. 35

4. Two liquids which mix, but not in all proportions. The phase diagram is as shown in Fig. 35. At temperatures above that of the critical point K, the components mix in any proportion. Below this temperature the components do not mix in the proportions represented by points within the hatched region. In this region there is separation into two liquid mixtures whose concentrations are given by the points of intersection of the corresponding horizontal line with the equilibrium curve. Similar diagrams are possible with K a minimum, or with two critical points, an upper and a lower, so that the region of separation into two phases (two solutions) is bounded by a closed curve.

5. In the liquid (or gaseous) state the two components mix in any proportion, but in the solid (or liquid) state they do not mix in all proportions (limited miscibility). In this case there is a triple point. According as the temperature of the triple point lies below the pure-component phase equilibrium temperatures (the points A and C) or between them (it obviously cannot lie above them on the assumption made here that the components mix in any proportion in the higher phase), the phase diagram appears as in Fig. 36 or Fig. 37 respectively. For example, let the phase of unlimited miscibility be a liquid, and that of limited miscibility be a solid. The region above the curve ABC (Fig. 36) or ADC (Fig. 37) is the region of liquid states; the

regions bounded by the curves *ADF* and *CEG* (Fig. 36) or *ABF* and *CEG* (Fig. 37) are the regions of homogeneous solid phases (solid solutions). At the triple point (whose temperature is given by the line *DBE*) the liquid and two solid solutions of different concentrations are in equilibrium. The point *B* in Fig. 36 is called the *eutectic point*. A liquid mixture whose concentration

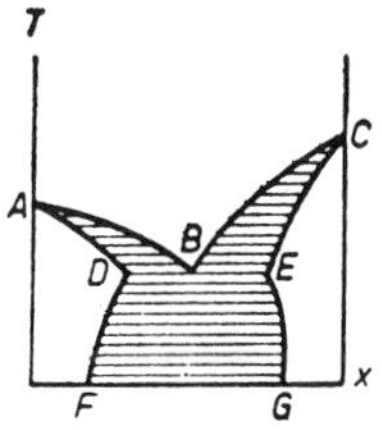

FIG. 36

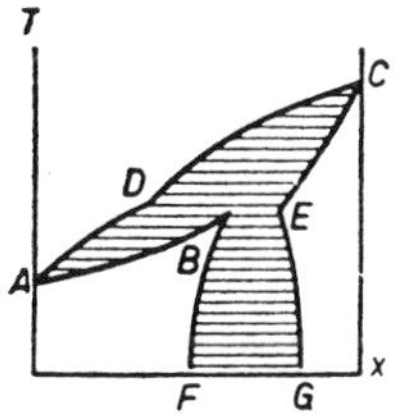

FIG. 37

corresponds to this point freezes completely, without change of concentration, whereas at other concentrations a solid mixture freezes out with a concentration different from that of the liquid. The regions *ADB* and *CBE* (Fig. 36) and *ADB* and *CDE* (Fig. 37) correspond to separation into a liquid phase and one of the solid phases; the regions *DEGF* (Fig. 36) and *BEGF* (Fig. 37) correspond to separation into two solid phases.

If, in a diagram such as Fig. 36, the components do not mix at all in the solid state, the phase diagram takes the form shown in Fig. 38. In the hatched

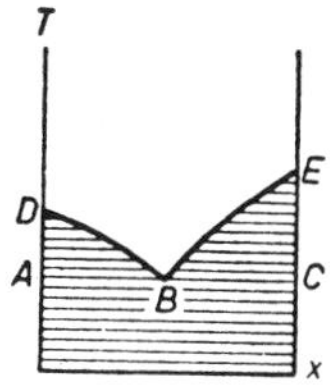

FIG. 38

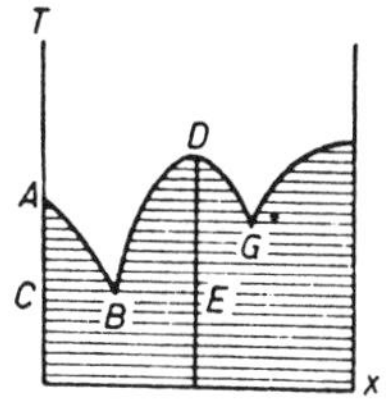

FIG. 39

regions above the line *ABC*, the mixed liquid phase is in equilibrium with the solid phase of one of the pure substances, and below *ABC* we have the two pure solid phases. When the temperature of the liquid mixture decreases, one or the other of the pure substances freezes out according as the concentration of the liquid lies to the right or the left of the eutectic point. As the temperature decreases further, the composition of the liquid varies along the curve *DB* or *EB*, and the liquid freezes completely at the eutectic point *B*.

6. In the liquid state the two components mix in any proportion, but in the solid state they do not mix at all, forming only a chemical compound of definite composition. The phase diagram is shown in Fig. 39. The straight line *DE* gives the composition of the chemical compound. There are two

triple points, *B* and *G*, at which there is equilibrium between the liquid phase, the solid chemical compound, and the solid phase of one of the pure components. Between the points *B* and *G* lies a point of equal concentration, *D* (cf. Fig. 30). It is easy to see where separation occurs, and which are the resulting phases: in the region *DBE* they are a liquid phase and the solid chemical compound; below the line *CE*, the chemical compound and one of the solid pure substances, and so on. The freezing of the liquid terminates at one of the eutectic points *G* and *B*, according as the concentration of the liquid lies to the right or to the left of the line *DE*.

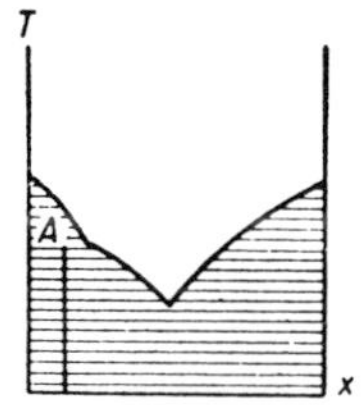

FIG. 40

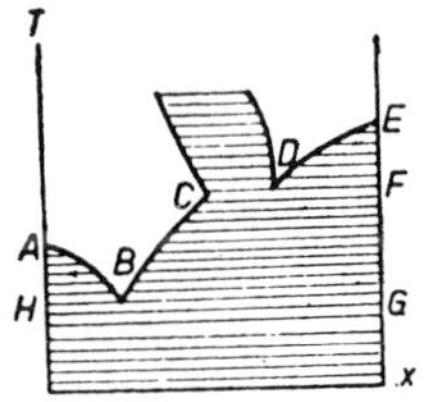

FIG. 41

7. In the liquid state the two components mix in any proportion, but in the solid state they do not mix at all, forming only a chemical compound; this compound, however, decomposes at a certain temperature, before it melts. The straight line defining the composition of this compound cannot terminate at a point of equal concentration as in case 6, since it does not reach the melting point. It can therefore terminate at a triple point of the type shown in Fig. 31, §99 (the point *A* in Fig. 40). In Fig. 40, which shows one possible form of the phase diagram for this case, it is easy to see which phases result from the separation at various points in the hatched region.

8. In the solid state the components do not mix at all, and in the liquid state they mix only in certain proportions. In this case there are two triple points, at which the liquid is in equilibrium with the two solid pure substances (the point *B* in Fig. 41) or one of the pure substances is in equilibrium with two mixed liquid phases of different concentrations (the point *D*). The regions not hatched in Fig. 41, above *ABC* and above *DE*, represent liquid states with various concentrations; the hatched region above *CD* is that of separation into two liquid phases, *DEF* is that of separation into a liquid and one of the solid pure substances, and so on.

§101. Intersection of singular curves on the equilibrium surface

The four kinds of line discussed in §99 (critical lines, three-phase lines, lines of equal concentration and pure-substance lines) all lie on the same surface, the equilibrium surface. They will therefore in general intersect.

Some properties of the points of intersection of such lines are described below.

It may be shown that no two critical lines can intersect, nor can two lines of equal concentration. We shall not pause to prove these statements here.

Let us now list (again without proof) the properties of the remaining points of intersection. All these properties follow almost immediately from the general properties of equilibrium curves given in §99. The diagrams will show the projections of the intersecting lines on the *PT*-plane (see §99); their form is, of course, chosen arbitrarily. A dotted line everywhere denotes a critical

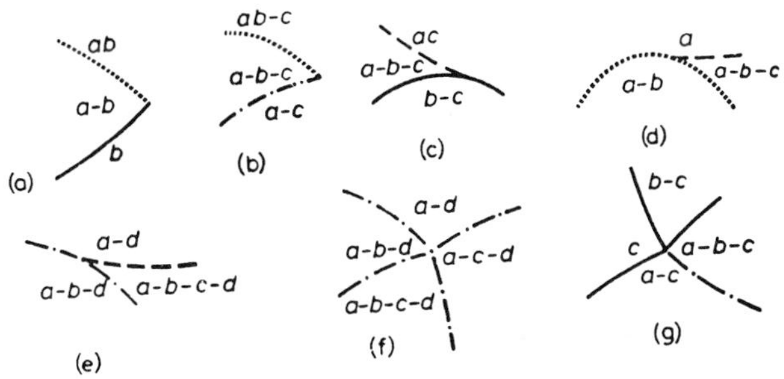

FIG. 42

line; a continuous line, a line of phase equilibrium for a pure substance; a broken line, a line of equal concentration; and a dot-and-dash line, a three-phase line. The letters have the same significance as in Figs. 25 to 28 (§99).

At a point of intersection between a critical line and a pure-substance line (Fig. 42a) both lines terminate, and similarly for a critical line and a three-phase line (Fig. 42b). When a pure-substance line intersects a line of equal concentration, only the latter terminates (Fig. 42c), the two curves touching at the point of intersection. The same occurs when a line of equal concentration meets a critical line (Fig. 42d) or a three-phase line (Fig. 42e). In each case the line of equal concentration terminates at the point of intersection, the two curves touching at this point.

The point of intersection of three-phase lines (Fig. 42f) is a quadruple point, i.e. a point where four phases are in equilibrium. Four three-phase lines meet at this point, corresponding to equilibrium between each three of the four phases.

Finally, the point where a pure-substance line intersects a three-phase line (Fig. 42g) must clearly be also a point of intersection between the three-phase line with all three pure-substance phase equilibrium lines (corresponding to equilibrium between each two of the three pure-substance phases).

§102. Gases and liquids

Let us now consider in more detail the equilibrium of liquid and gaseous phases consisting of two components.

When the temperature is sufficiently high (T large in comparison with the mean interaction energy of the molecules) all substances mix in any proportion. On the other hand, since a substance is a gas at such temperatures, we can say that all substances have unlimited miscibility in the gas phase (although when there are critical lines the difference between the liquid and the gas becomes to some extent arbitrary, and so likewise does the foregoing formulation).

In the liquid state, some substances mix in any proportion, others only in certain proportions (liquids of limited miscibility).

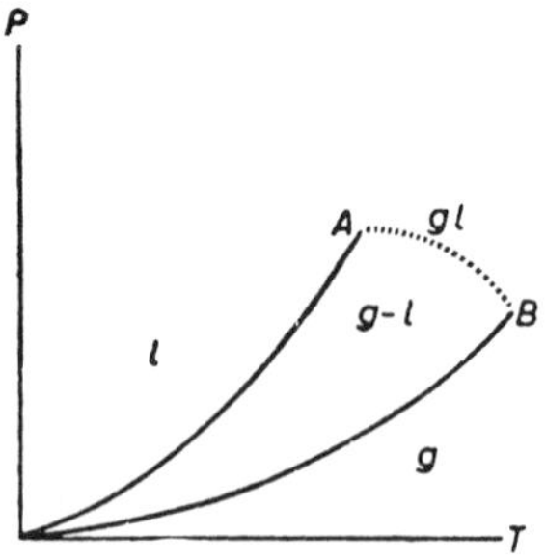

FIG. 43

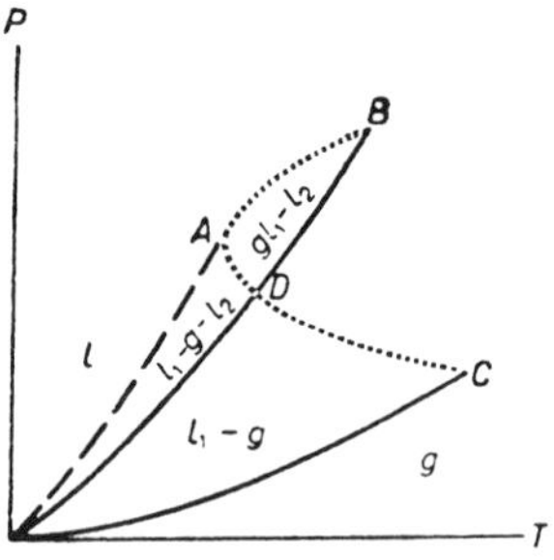

FIG. 44

In the former case, when the two components mix in any proportion in both phases, the phase diagrams contain no triple points, since the system cannot consist of more than two phases (all liquid states are one phase, and the same applies to gaseous states). Let us consider the projection of the singular lines of the equilibrium surface on the PT-plane. There are two lines of phase equilibrium for the pure substances (i.e. for concentrations $x = 0$ and $x = 1$ in both phases). One of these lines is itself in the PT-plane, and the other is in a plane parallel to it, so that its projection has the same form as the line itself. Each of these lines terminates at a point which is a critical point for phases of the corresponding pure substance. A critical line begins and ends at these points (at a point of intersection of a critical line and a pure-substance line, both terminate; see §101). Thus the projection of these various lines on the PT-plane has the form shown in Fig. 43; the notation is the same as in §§99 and 101. The letters g and l have a similar significance to a, b, c in the diagrams in §§99 and 101: g denotes gas, and l liquid. The regions g and l contain projections of gaseous and liquid states respectively; the region $g-l$ includes these and also states where separation into liquid and gas occurs; above the critical line, the difference between liquid and gas does not exist.

If there is also a line of equal concentration, the projection on the *PT*-plane is as shown in Fig. 44. The projection of the line of equal concentration lies either above the line from the origin O to B (as in Fig. 44), or below OC, but not between them. Only A, B, C are points of intersection of lines. The point D does not correspond to a true intersection of the pure-substance line with the critical line; these intersect only in projection. The letters l_1 and l_2 in the diagram denote liquid phases of different concentrations. Above the line of equal concentration there is only one liquid phase.†

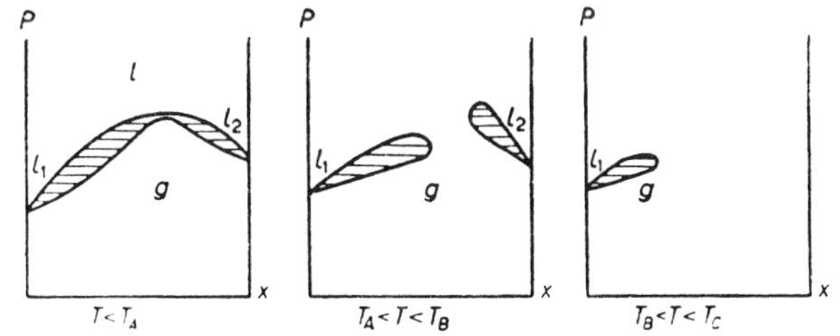

FIG. 45

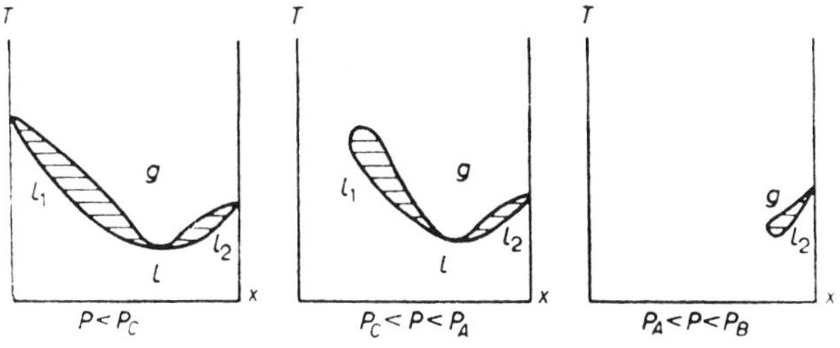

FIG. 46

All these properties of the projections of the singular lines on the *PT*-plane become obvious if we consider the phase diagrams corresponding to the cross-sections of the equilibrium surface by various planes of constant temperature (or pressure). For example, the cross-sections corresponding to pressures below that at B and to pressures between those of A and B in Fig. 43 give phase diagrams as shown in Figs. 32 and 34 respectively. Fig. 45 shows cross-sections for various successive temperature ranges in Fig. 44 (T_A, T_B, T_C being the temperatures corresponding to the points A, B, C): the region of separation into two phases "breaks up" at the point of equal concentration, and two critical points are formed; thereupon, first one and

† Not being concerned with solid phases, we shall conventionally show lines in all the (P, T) diagrams as starting from the origin, as if solidification did not occur.

then the other hatched region shrink to a point on the ordinate axis and disappear. Fig. 46 shows similar cross-sections for successive pressure ranges.

If the two components have limited miscibility in the liquid state, there is a three-phase line, which terminates at a point where it intersects a critical line

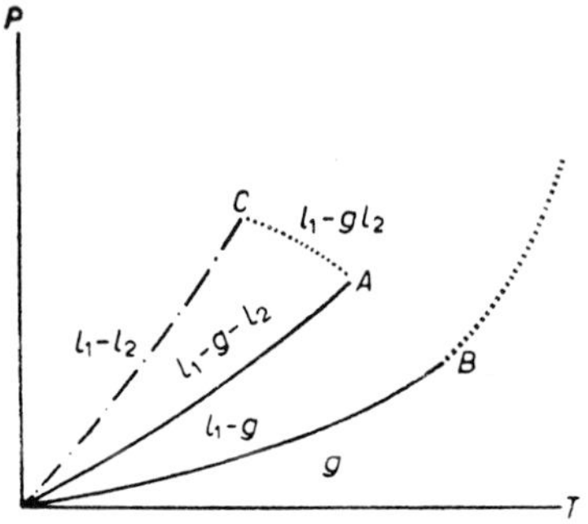

FIG. 47

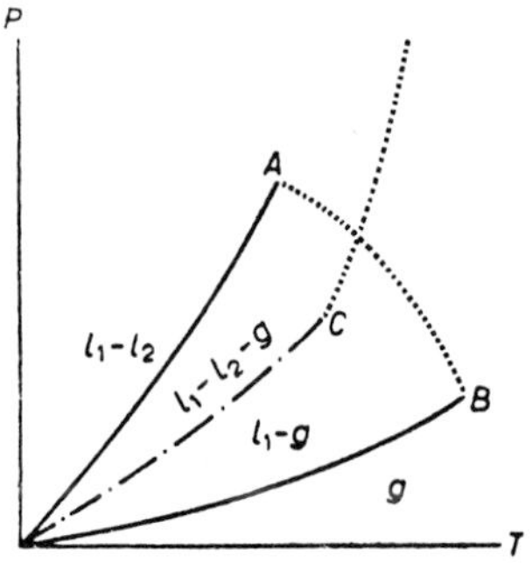

FIG. 48

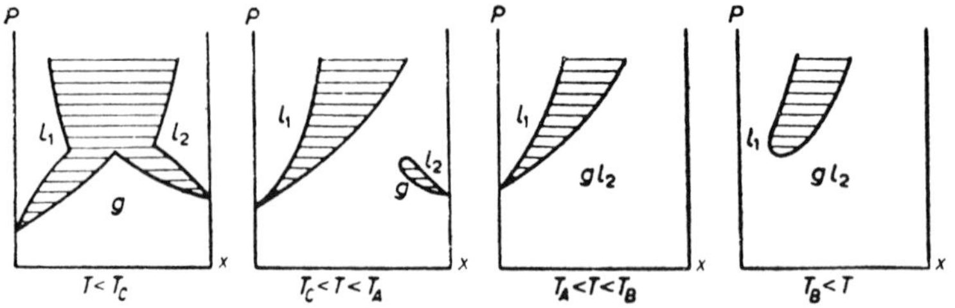

FIG. 49

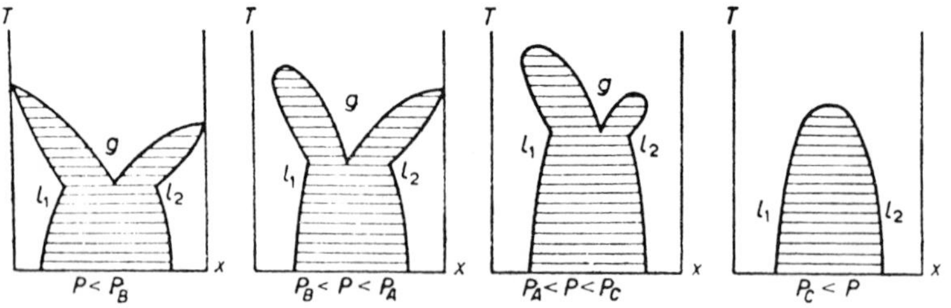

FIG. 50

starting from that point. Figs. 47 and 48 show the two essentially different types of (P, T) projection that can occur in this case. They differ in that in Fig. 47 the projection of the three-phase line lies above both the pure-substance lines, while in Fig. 48 it lies between them; the three-phase line cannot lie below both the pure-substance lines, since in the gaseous state the two components mix in any proportion. In both cases there are two critical lines, one of which runs out towards high pressures.

Figs. 49 and 50 show a number of successive cross-sections by Px and Tx planes for the case shown in Fig. 47.

In conclusion, it should be emphasised that the examples of (P, T) diagrams discussed in this section are only the most typical ones for equilibrium of liquid and gaseous phases; they do not exhaust all theoretically possible forms.

CHAPTER X

CHEMICAL REACTIONS

§103. The condition for chemical equilibrium

A CHEMICAL reaction occurring in a mixture of reacting substances ultimately leads to the establishment of an equilibrium state in which the quantity of each of the substances that take part in the reaction no longer changes. This case of thermodynamic equilibrium is called *chemical equilibrium.* Any chemical reaction can take place, in general, in either direction; until equilibrium is reached, one direction predominates, but in equilibrium the two opposite reactions occur at rates such that the total numbers of particles of each of the reacting substances remain constant. The object of thermodynamics as applied to chemical reactions is to study only the chemical equilibrium, not the course of the reaction leading to that equilibrium.

It is important to note that the state of chemical equilibrium is independent of how and under what conditions the reaction occurred†; it depends only on the conditions under which the mixture of reacting substances exists in equilibrium. In deriving the condition for chemical equilibrium, we can therefore make any desired assumptions concerning the course of the reaction.

First of all, we shall describe the method to be used for expressing the reaction. Chemical reactions are commonly written as symbolic equations, which, if all the terms are taken to one side, have the form

$$\sum_i \nu_i \mathrm{A}_i = 0, \qquad (103.1)$$

where the A_i are the chemical symbols of the reacting substances, and the coefficients ν_i are positive or negative integers. For example, in the reaction $2\,H_2+O_2 = 2\,H_2O$ or $2\,H_2+O_2-2\,H_2O = 0$ the coefficients are $\nu_{H_2} = 2$, $\nu_{O_2} = 1$, $\nu_{H_2O} = -2$.

Let us assume that the reaction occurs at constant temperature and pressure. In such processes the thermodynamic potential of the system tends to a minimum. In equilibrium, therefore, the potential Φ must have its least possible value (for given P and T). Let $N_1, N_2, \ldots$ be the numbers of particles of the various substances taking part in the reaction. Then the necessary condition for Φ to be a minimum can be written as the vanishing of the total

† In particular, it is independent of whether a catalyst took part in the reaction.

derivative of Φ (for given P and T) with respect to one of the N_i, say N_1:

$$\frac{\partial \Phi}{\partial N_1}+\frac{\partial \Phi}{\partial N_2}\frac{dN_2}{dN_1}+\frac{\partial \Phi_3}{\partial N_3}\frac{dN_3}{dN_1}+\ldots = 0.$$

The changes in the numbers N_i during the reaction are related by the reaction equation: it is clear that, if N_1 changes by ν_1, each of the other N_i will change by ν_i, so that $dN_i = (\nu_i/\nu_1)dN_1$, or $dN_i/dN_1 = \nu_i/\nu_1$. The foregoing equation may therefore be written

$$\sum_i \frac{\partial \Phi}{\partial N_i}\frac{\nu_i}{\nu_1} = 0.$$

Finally, putting $\partial\Phi/\partial N_i = \mu_i$ and multiplying by ν_1, we have

$$\sum_i \nu_i \mu_i = 0. \tag{103.2}$$

This is the required condition for chemical equilibrium. In order to obtain it, therefore, we must replace the symbols A_i by the corresponding chemical potentials μ_i in the equation of the chemical reaction. When several different reactions can occur in the mixture, the equilibrium condition will be a set of several equations such as (103.2), each obtained by the above method from the equation of the corresponding reaction.

It may be noted that the condition (103.2) retains its form even when the reacting substances are distributed in the form of solutes in two different phases in contact. This follows from the fact that in equilibrium the chemical potentials of each substance in either phase must be equal, in accordance with the conditions for phase equilibrium.

§104. The law of mass action

Let us apply the general condition for chemical equilibrium, derived in §103, to reactions taking place in a gas mixture, assuming that the gas may be regarded as an ideal one.

The chemical potential of each gas in the mixture is (see §95)

$$\mu_i = T\log P_i+\chi_i(T), \tag{104.1}$$

where P_i is the partial pressure of the ith gas in the mixture; $P_i = c_i P$ if P is the total pressure of the mixture and $c_i = N_i/N$ is the concentration of the gas in question, defined as the ratio of the number N_i of molecules of that gas to the total number $N = \sum N_i$ of molecules in the mixture.

It is now easy to write down the condition of chemical equilibrium for reactions in a mixture of gases. Substitution of (104.1) in (103.2) gives

$$\sum_i \nu_i\mu_i = T\sum_i \nu_i \log P_{0i}+\sum_i \nu_i\chi_i = 0,$$

where the P_{0i} are the partial pressures of the gases in a state of chemical equilibrium, or

$$\sum_i \nu_i \log P_{0i} = -\frac{1}{T}\sum_i \nu_i \chi_i.$$

Using the notation

$$K_p(T) = e^{-\Sigma \nu_i \chi_i / T}, \tag{104.2}$$

we thus have

$$\prod_i P_{0i}^{\nu_i} = K_p(T). \tag{104.3}$$

Instead of P_{0i} we can substitute Pc_{0i}, where the c_{0i} are the concentrations of the gases in chemical equilibrium. Then

$$\prod_i c_{0i}^{\nu_i} = P^{-\Sigma \nu_i} K_p(T) \equiv K_c(P, T). \tag{104.4}$$

The quantity on the right of (104.3) or (104.4) is a function only of temperature and pressure, and does not depend on the initial amounts of the reacting gases: this quantity is usually called the *chemical equilibrium constant*, and the law expressed by formula (104.3) or (104.4) is called the *law of mass action.*

The dependence of the gas reaction equilibrium constant on the pressure is entirely determined by the factor $P^{-\Sigma \nu_i}$ on the right-hand side of equation (104.4); if the quantities of reacting substances are expressed in terms of their partial pressures, the equilibrium constant is independent of pressure. The determination of its dependence on temperature, however, requires further assumptions concerning the properties of the gases.

For example, if the gases have constant specific heats, a comparison of the expression (104.1) with formula (43.3) for the thermodynamic potential of such a gas shows that the functions $\chi_i(T)$ are of the form

$$\chi_i(T) = \varepsilon_{0i} - c_{pi} T \log T - T\zeta_i, \tag{104.5}$$

where c_{pi} is the specific heat and ζ_i the chemical constant of the gas. Substituting this expression in (104.2), we obtain for the equilibrium constant the formula

$$K_p(T) = e^{\Sigma \nu_i \zeta_i} T^{\Sigma c_{pi} \nu_i} e^{-\Sigma \nu_i \varepsilon_{0i} / T}, \tag{104.6}$$

which is essentially an exponential function of temperature.

The law of mass action is valid also for reactions between solutes, provided that the solution may be regarded as weak. For the chemical potential of each solute has the form

$$\mu = T \log c_i + \psi_i(P, T), \tag{104.7}$$

obtained by differentiating the thermodynamic potential (88.3) with respect to n_i. The concentration c_i is here defined as the ratio of the number of particles of the solute in question to the number of solvent particles ($c_i = N_i/N$).

Substituting (104.7) in the equilibrium condition (103.2), we find in the same way

$$\prod_i c_{0i}^{\nu_i} = K(P, T), \tag{104.8}$$

with the equilibrium constant

$$K(P, T) = e^{-\Sigma \nu_i \psi_i / T}. \tag{104.9}$$

Unlike the case of gas reactions, the dependence of the equilibrium constant on the pressure here remains indeterminate.

If the reaction involves, as well as gases or solutes, substances in a pure condensed phase (i.e. not mixed with other substances), e.g. pure solids, then the equilibrium condition again leads to the law of mass action. Here, however, since the chemical potential of the pure phases depends only on the pressure and temperature, the left-hand side of the equation for the law of mass action will not involve the quantities of the pure phases, i.e. the product of the concentrations of the gases (or solutes) must be written as if the solids were absent. The latter affect only the dependence of the equilibrium constant on pressure and temperature.

If only gases and solids take part in the reaction, then, since the pressure of the gases is comparatively small, the chemical potential of the solids may be regarded as independent of the pressure, and the dependence of the equilibrium constant on the pressure remains the same as in (104.4). The sum $\Sigma \nu_i$ in the exponent must of course denote only the sum of the coefficients of the gaseous substances in the reaction equation.

Finally, the law of mass action is valid also for reactions in weak solutions where the solvent as well as the solutes takes part in the reactions. For, when the chemical potential is substituted in the condition for chemical equilibrium, the small terms which contain the concentration may be omitted, and the potential then reduces to a quantity which depends only on temperature and pressure. Thus we again obtain the equation of the law of mass action, and its left-hand side again involves only the concentrations of the reacting solutes, not that of the solvent.

PROBLEMS

PROBLEM 1. Find the equilibrium constant for the dissociation of a diatomic gas at high temperatures; the gas molecule consists of identical atoms and has zero spin and orbital angular momentum in the ground state.

SOLUTION. The reaction concerned is of the form $A_2 = 2A$. The specific heats of the gases A_2 and A are $c_{pA_2} = 9/2$, $c_{pA} = 5/2$, and the chemical constants are (see (45.4), (46.4), (49.8))

$$\zeta_A = \log [g_A(m/2\pi\hbar^2)^{3/2}], \qquad \zeta_{A_2} = \log [(I/\hbar^6\omega)(m/\pi)^{3/2}],$$

where m is the mass of the atom A (that of the molecule A_2 being $2m$), and g_A the statistical weight of the ground state of the atom A; at sufficiently high temperatures $g_A =$

$(2S+1)(2L+1)$, where S and L are the spin and orbital angular momentum of the atom. Substitution in (104.6) gives

$$K_p(T) = \frac{8I}{g_A{}^2\omega}\left(\frac{\pi}{m}\right)^{3/2}\frac{1}{\sqrt{T}}e^{\varepsilon_0/T}.$$

Here $\varepsilon_0 = 2\varepsilon_{0A}-\varepsilon_{0A_2}$ is the dissociation energy of the molecule.

PROBLEM 2. Determine the dependence of the concentration of hydrogen dissolved as H atoms in a metal on the pressure of H_2 gas over the metal.

SOLUTION. Regarding the process as a chemical reaction $H_2 = 2\,H$, we can write the equilibrium condition as $\mu_{H_2} = 2\mu_H$; μ_{H_2} is written as the chemical potential of an ideal gas, $\mu_{H_2} = T\log P+\chi(T)$, and μ_H as the chemical potential of the solute in a solution, $\mu_H = T\log c+\psi$. Since ψ depends only slightly on the pressure (cf. §91), we find c = constant$\times\sqrt{P}$.

§105. Heat of reaction

A chemical reaction is accompanied by the absorption or evolution of heat. In the former case the reaction is said to be *endothermic*, and in the latter case *exothermic*. It is evident that, if any particular reaction is exothermic, the reverse reaction will be endothermic, and *vice versa*.

The amount of heat involved in a reaction depends on the conditions under which the reaction occurs. Hence, for instance, we must distinguish the heats of reaction at constant volume and at constant pressure (although the difference is usually quite small).

As in calculating the heat of solution (§92), we first find the maximum work which can be obtained by means of the chemical reaction. We call a reaction between one group of molecules as shown by the reaction equation an "elementary reaction", and calculate the change in the thermodynamic potential of a mixture of reacting substances when a small number δn of elementary reactions take place, assuming that the reaction occurs at constant temperature and pressure. We have

$$\delta\Phi = \sum_i \frac{\partial\Phi}{\partial N_i}\delta N_i = \sum_i \mu_i\delta N_i.$$

The change in the number of molecules of the ith substance after δn elementary reactions is clearly $\delta N_i = -\nu_i\,\delta n$. Thus

$$\delta\Phi = -\delta n\sum_i \nu_i\mu_i. \qquad (105.1)$$

In equilibrium $\delta\Phi/\delta n$ is zero, as we should expect.

Formula (105.1) is the general expression for the minimum work which must be done in order to bring about δn elementary reactions. It is also the maximum work which can be obtained from that number of reactions occurring in the reverse direction.

Let us first suppose that the reaction is between gases. Using the expression (104.1) for μ_i, we find

$$\delta\Phi = -\delta n(T\sum_i \nu_i \log P_i + \sum_i \nu_i\chi_i),$$

or, in terms of the equilibrium constant,

$$\begin{aligned}\delta\Phi &= T\,\delta n[-\sum_i \nu_i \log P_i + \log K_p(T)]\\ &= T\,\delta n[-\sum_i \nu_i \log c_i + \log K_c(P, T)].\end{aligned} \tag{105.2}$$

For reactions in solution we similarly find, using the expression (104.7) for μ,

$$\delta\Phi = -\delta n\,(T\sum_i \nu_i \log c_i + \sum_i \nu_i\psi_i),$$

or, in terms of the equilibrium constant $K(P, T)$,

$$\delta\Phi = T\,\delta n\,[-\sum_i \nu_i \log c_i + \log K(P, T)]. \tag{105.3}$$

The sign of $\delta\Phi$ determines the direction in which the reaction takes place: since Φ tends to a minimum, for $\delta\Phi < 0$ the reaction occurs in the forward direction (i.e. "from left to right" in the equation of the chemical reaction), while if $\delta\Phi > 0$ the reaction will actually go in the opposite direction in the mixture concerned. It may be noted, however, that the direction of the reaction is also evident directly from the law of mass action: we form the product $\Pi P_i^{\nu_i}$ for the mixture in question and compare it with the value of the equilibrium constant for the reaction. If, for instance, we find that $\Pi P_i^{\nu_i} > K_p$, this means that the reaction will occur in the forward direction, so as to reduce the partial pressures of the original substances (which have positive ν_i in the reaction equation), and increase those of the reaction products (for which $\nu_i < 0$).

We can now determine also the heat absorbed (or evolved, according to sign), again for δn elementary reactions. Formula (92.4) shows that this heat δQ_p is, for a reaction at constant temperature and pressure,

$$\delta Q_p = -T^2\left(\frac{\partial}{\partial T}\frac{\delta\Phi}{T}\right)_P.$$

For reactions between gases we have, substituting (105.2),

$$\delta Q_p = -T^2\,\delta n\frac{\partial \log K_p(T)}{\partial T}. \tag{105.4}$$

Similarly, for solutions

$$\delta Q_p = -T^2\,\delta n\frac{\partial \log K(P, T)}{\partial T}. \tag{105.5}$$

We may note that δQ_p is simply proportional to δn and does not depend on the values of the concentrations at any instant. These formulae are therefore valid for any δn, whether small or not.

If $Q_p > 0$, i.e. the reaction is endothermic, $\partial \log K/\partial T < 0$, and the equilibrium constant decreases with increasing temperature. On the other hand, for an exothermic reaction ($Q_p < 0$) the equilibrium constant increases with temperature. An increase in the equilibrium constant signifies that the chemical equilibrium is shifted back towards the re-formation of the initial substances—the reaction goes "from right to left", so as to increase the product $\Pi c_{0i}^{\nu_i}$. Conversely, a decrease in the equilibrium constant signifies a shift of the equilibrium towards formation of the reaction products. In other words, we can formulate the following rule: heating shifts the equilibrium in the direction of the endothermic process, and cooling in the direction of the exothermic process. This rule is entirely in agreement with LE CHATELIER's principle.

For reactions between gases the heat of reaction at constant volume (and temperature) is also of interest. This quantity δQ_v is related in a simple manner to δQ_p. The quantity of heat absorbed in a process at constant volume is equal to the change in the energy of the system, whereas δQ_p is equal to the change in the heat function. Since $E = W - PV$, it is clear that $\delta Q_v = \delta Q_p - \delta(PV)$, or, substituting $PV = T\sum N_i$ and $\delta N_i = -\nu_i \,\delta n$,

$$\delta Q_v = \delta Q_p + T\,\delta n \sum_i \nu_i. \tag{105.6}$$

Finally, let us determine the change in volume of a mixture of reacting substances as a result of a reaction occurring at constant pressure (and temperature). For gases, the problem is trivial:

$$\delta V = (T/P)\delta N = -(T/P)\delta n \sum_i \nu_i. \tag{105.7}$$

In particular, reactions in which the total number of particles is unchanged ($\sum \nu_i = 0$) occur without change of volume.

For reactions in weak solutions we use the formula $\delta V = \partial\, \delta\Phi/\partial P$ and, substituting (105.3), we obtain

$$\delta V = T\,\delta n \frac{\partial \log K(P, T)}{\partial P}; \tag{105.8}$$

for gases this formula reduces to (105.7), of course, on substituting $K = K_p(T)P^{-\Sigma\nu_i}$.

Thus a change in volume in the reaction is due to a pressure dependence of the equilibrium constant. In a similar way to the previous discussion of the temperature dependence, we easily deduce that an increase in pressure favours reactions in which the volume decreases (i.e. shifts the equilibrium in the direction of such reactions), and a decrease in pressure favours reactions which lead to an increase in volume, again in complete agreement with LE CHATELIER's principle.

§106. Ionisation equilibrium

At sufficiently high temperatures, collisions between gas particles may cause their ionisation. The existence of such *thermal ionisation* leads to the establishment of a thermal equilibrium in which certain fractions of the total number of gas particles are in various stages of ionisation. Let us consider thermal ionisation of a monatomic gas; this is the most interesting case, since chemical compounds are usually completely dissociated before the onset of thermal ionisation.

Thermodynamically, ionisation equilibrium is a particular case of chemical equilibrium corresponding to a series of simultaneously occurring "ionisation reactions", which may be written

$$A_0 = A_1 + e^-, \qquad A_1 = A_2 + e^-, \ \ldots, \tag{106.1}$$

where the symbol A_0 denotes the neutral atom, A_1, A_2, ... the singly, doubly etc. ionised atoms and e^- the electron. For these reactions the application of the law of mass action gives the set of equations

$$c_{n-1}/c_n c = PK_p^{(n)}(T) \qquad (n = 1, 2, \ldots), \tag{106.2}$$

where c_0 is the concentration of neutral atoms, c_1, c_2, ... the concentrations of the various ions, and c the concentration of electrons (each defined as the ratio of the number of particles of the kind in question to the total number of particles, including electrons). To these equations we must add one which expresses the electrical neutrality of the gas as a whole:

$$c = c_1 + 2c_2 + 3c_3 + \ldots. \tag{106.3}$$

Equations (106.2) and (106.3) determine the concentrations of the various ions in ionisation equilibrium.

The equilibrium constants $K_p^{(n)}$ can be calculated without difficulty. All gases which take part in "reactions" (gases of neutral atoms, ions, or electrons) are monatomic and have constant specific heat $c_p = 5/2$, and their chemical constants are $\zeta = \log [g(m/2\pi\hbar^2)^{3/2}]$, where m is the mass of a particle of the gas considered, and g the statistical weight of its ground state; for electrons, $g = 2$, while for atoms and ions $g = (2L+1)(2S+1)$, where L and S are the orbital angular momentum and spin of the atom or ion.† Substituting these values in formula (104.6), we obtain the following expression for the required equilibrium constants:

$$K_p^{(n)}(T) = \frac{g_{n-1}}{2g_n}\left(\frac{2\pi}{m}\right)^{3/2}\frac{\hbar^3}{T^{5/2}}\, e^{I_n/T}, \tag{106.4}$$

† For reasons given below we may assume that all atoms and ions are in the ground state, even in a considerably ionised gas. If the atom (or ion) ground state has a fine structure, we assume that T is large compared with the intervals in this structure.

where m is the electron mass and $I_n = \varepsilon_{0n} - \varepsilon_{0,n-1}$ the energy of the nth ionisation (nth ionisation potential) of the atom.

The degree of ionisation (for n-fold ionisation) of the gas becomes of the order of unity as the temperature increases and the equilibrium constant $K_c^{(n)} = PK_p^{(n)}$ decreases to a value of the order of unity. It is very important to note that, despite the exponential dependence of the equilibrium constant on temperature, this stage is reached not when $T \sim I_n$ but at considerably lower temperatures. The reason is that the coefficient of the exponential $e^{I_n/T}$ is small: the quantity $(P/T)(\hbar^2/mT)^{3/2} = (N/V)(\hbar^2/mT)^{3/2}$ is in general very small, being for $T \sim I$ of the order of the ratio of the atomic volume to the volume V/N per atom in the gas.

Thus the gas will be considerably ionised even at temperatures which are small compared with the ionisation energy, but the number of excited atoms in the gas will still be small, since the excitation energy of the atom is in general of the same order as the ionisation energy.

When T becomes comparable with the ionisation energy, the gas is almost completely ionised. At temperatures of the order of the binding energy of the last electron in the atom, the gas may be regarded as consisting of electrons and bare nuclei only.

The binding energy I_1 of the first electron is usually much less than the subsequent ones I_n; there is therefore a range of temperatures in which the gas may be supposed to include only neutral atoms and singly charged ions. Defining the *degree of ionisation* α of the gas as the ratio of the number of ionised atoms to the total number of atoms, we have

$$c = c_1 = \alpha/(1+\alpha), \qquad c_0 = (1-\alpha)/(1+\alpha),$$

and equation (106.2) gives $(1-\alpha^2)/\alpha^2 = PK_p^{(1)}$, whence

$$\alpha = 1/\sqrt{(1+PK_p^{(1)})}. \tag{106.5}$$

This entirely determines the degree of ionisation as a function of pressure and temperature (in the temperature range considered).

§107. Equilibrium with respect to pair production

At extremely high temperatures, comparable with the rest energy† mc^2 of the electron, collisions of particles in matter may be accompanied by the formation of electron-positron pairs. The number of particles itself then ceases to be a given quantity, and depends on the conditions of thermal equilibrium.

Pair production (and the reverse process, annihilation) can be regarded thermodynamically as a "chemical reaction" $e^+ + e^- = \gamma$, where the symbols

† The energy $mc^2 = 0.51 \times 10^6$ eV, so that the temperature $mc^2/k = 6 \times 10^9$ degrees.

e^+ and e^- denote a positron and an electron, and γ denotes one or more photons. The chemical potential of the photon gas is zero (§60). The condition of equilibrium for pair production is therefore

$$\mu^- + \mu^+ = 0, \tag{107.1}$$

where μ^- and μ^+ are the chemical potentials of the electron and positron gases. It should be emphasised that μ here denotes the relativistic expression for the chemical potential, including the rest energy of the particles (cf. §27), which plays an important part in pair production.

Even at temperatures $T \sim mc^2$, the number of pairs formed per unit volume is very large in comparison with the atomic electron density (see the next footnote). We can therefore suppose with sufficient accuracy that the number of electrons is equal to the number of positrons. Then $\mu^- = \mu^+$, and the condition (107.1) gives $\mu^- = \mu^+ = 0$, i.e. in equilibrium the chemical potentials of the electrons and positrons must be zero.

Electrons and positrons obey Fermi statistics; their number is therefore obtained by integrating the distribution (55.3) with $\mu = 0$:

$$N^+ = N^- = \frac{V}{\pi^2\hbar^3}\int_0^\infty \frac{p^2\,dp}{e^{\varepsilon/T}+1}, \tag{107.2}$$

where ε is determined from the relativistic expression $\varepsilon = c\sqrt{(p^2+m^2c^2)}$.

For $T \ll mc^2$, this number is exponentially small ($\sim e^{-mc^2/T}$). In the opposite case ($T \gg mc^2$) we can put $\varepsilon = cp$, and formula (107.2) gives

$$N^+ = N^- = \frac{V}{\pi^2}\left(\frac{T}{\hbar c}\right)^3 \int_0^\infty \frac{x^2\,dx}{e^x+1}.$$

The integral in this formula can be expressed in terms of the ζ function (see the second footnote to §57), giving†

$$N^+ = N^- = \frac{3\zeta(3)}{2\pi^2}\left(\frac{T}{\hbar c}\right)^3 V = 0.183(T/\hbar c)^3 V. \tag{107.3}$$

The energy of the positron and electron gases is similarly

$$E^+ = E^- = \frac{VT}{\pi^2}\left(\frac{T}{\hbar c}\right)^3 \int_0^\infty \frac{x^3\,dx}{e^x+1} = 7\pi^2 VT^4/120(\hbar c)^3. \tag{107.4}$$

This quantity is $\frac{7}{8}$ of the energy of black-body radiation in the same volume.

† For $T \sim mc^2$ the volume per particle formed is of the order of $(\hbar/mc)^3$, i.e. the cube of the Compton wavelength. This volume is very small in comparison with the atomic dimensions (for example, in comparison with the cube of the Bohr radius, $(\hbar^2/me^2)^3$).

PROBLEM

Determine the equilibrium density of electrons and positrons for $T \ll mc^2$.

SOLUTION. Using the expression (46.1a) for the chemical potential (to which mc^2 must be added), we obtain

$$n^+ n^- = 4(mT/2\pi\hbar^2)^3 e^{-2mc^2/T},$$

where $n^- = N^-/V$ and $n^+ = N^+/V$ are the electron and positron densities. If n_0 is the initial electron density (in the absence of pair production), then $n^- = n^+ + n_0$, and we find

$$n^+ = n^- - n_0 = -\tfrac{1}{2}n_0 + [\tfrac{1}{4}n_0^2 + \tfrac{1}{2}(mT/2\pi\hbar^2)^3 e^{-2mc^2/T}]^{1/2}.$$

CHAPTER XI

PROPERTIES OF MATTER AT VERY HIGH DENSITY

§108. The equation of state of matter at high density

THE study of the properties of matter at extremely high density is of fundamental importance. Let us follow qualitatively the change in these properties as the density is gradually increased.

When the volume per atom becomes less than the usual size of the atom, the atoms lose their individuality, and so the substance is transformed into a highly compressed plasma of electrons and nuclei. If the temperature of the substance is not too high, the electron component of this plasma is a degenerate Fermi gas. An unusual property of such a gas has been mentioned at the end of §56: it becomes more nearly "ideal" as the density increases. Thus, when the substance is sufficiently compressed, the interaction of the electrons with the nuclei (and with one another) becomes unimportant, and the formulae for an ideal Fermi gas may be used. According to (56.9) this occurs when $n_e \gg (m_e e^2/\hbar^2)^3 Z^2$ holds, where n_e is the number density of electrons, m_e the electron mass, and Z some mean atomic number of the substance. We therefore find for the total mass density of the substance the inequality

$$\varrho \gg (m_e e^2/\hbar^2)^3 m' Z^2 \sim 20Z^2 \text{ g/cm}^3, \tag{108.1}$$

where m' is the mass per electron, so that $\varrho = n_e m'$.† The "gas of nuclei" may still be far from degeneracy, because of the large mass of the nucleus, but its contribution to the pressure of the substance, for example, is in any case entirely negligible in comparison with that of the electron gas.

Thus the thermodynamic quantities for a substance under the conditions in question are given by the formulae derived in §56, applied to the electron component. In particular, for the pressure we have‡

$$P = \frac{(3\pi^2)^{2/3}}{5} \frac{\hbar^2}{m_e} \left(\frac{\varrho}{m'}\right)^{5/3}. \tag{108.2}$$

† In all the numerical estimates given in this section it is assumed that the mean atomic weight of the substance is twice its mean atomic number, so that m' is twice the proton mass.

It may be mentioned that the degeneracy temperature of the electrons corresponding to a density $\varrho \sim 20Z^2$ g/cm³ is of the order of $10^6 Z^{4/3}$ degrees.

‡ This expression is the first term in an expansion in powers of the reciprocal density (the small parameter is $(1/n_e)(m_e e^2/\hbar^2)^3$, but a more precise criterion of smallness will

The condition (108.1) on the density gives for the pressure the numerical inequality $P \gg 5\times 10^8\, Z^{10/3}$ atm.

In the above formulae the electron gas is assumed non-relativistic. This implies that the Fermi limiting momentum p_0 is small compared with mc (see §58), giving the numerical inequalities

$$\varrho \ll 2\times 10^6 \text{ g/cm}^3, \qquad P \ll 10^{17} \text{ atm}.$$

When the density and pressure of the gas become comparable with these values, the electron gas becomes relativistic, and when the opposite inequalities hold we have the extreme relativistic case, where the equation of state is determined by formula (58.4):

$$P = \tfrac{1}{4}(3\pi^2)^{1/3}\hbar c(\varrho/m')^{4/3}. \qquad (108.3)$$

A further increase in density leads to states where nuclear reactions consisting in the capture of electrons by nuclei (with emission of neutrinos) are thermodynamically favoured. Such a reaction decreases the charge on the nucleus (leaving its atomic weight constant), and this in general causes a decrease in the binding energy of the nucleus, i.e. a decrease in its mass defect. The energy required to bring about such a process is more than counterbalanced at sufficiently high densities by the decrease in the energy of the degenerate electron gas because of the smaller number of electrons.

It is not difficult to write down the thermodynamic conditions which govern the "chemical equilibrium" of the nuclear reaction mentioned, which may be symbolically written as

$$\mathrm{A}_Z + e^- = \mathrm{A}_{Z-1} + \nu,$$

where A_Z denotes a nucleus of atomic weight Z, e^- an electron and ν a neutrino. The neutrinos are not retained by matter and leave the body; such a process must lead to a steady cooling of the body. Thus thermal equilibrium can be meaningfully considered in these conditions only if the temperature of the substance is taken as zero. The chemical potential of the neutrinos will not then appear in the equation of equilibrium. The chemical potential of the nuclei is mainly governed by their internal energy, which we denote by $-\varepsilon_{A,Z}$ (the term "binding energy" usually refers to the positive quantity $\varepsilon_{A,Z}$). Finally, let $\mu_e(n_e)$ denote the chemical potential of the electron gas as a function of the number density n_e of particles in it. Then the condition of chemical equilibrium takes the form $-\varepsilon_{A,Z}+\mu_e(n_e) = -\varepsilon_{A,Z-1}$, or, putting $\varepsilon_{A,Z}-\varepsilon_{A,Z-1} = \Delta$,

$$\mu_e(n_e) = \Delta.$$

---

depend on Z also). In the next approximation, which contains an extra power $\varrho^{-1/3}$, a contribution to the plasma energy is given by the Coulomb interaction of the electrons and nuclei. The minimum of this energy corresponds to an ordered arrangement of the nuclei in a "crystal lattice". A calculation of the relevant corrections to the equation of state is given by A. A. Abrikosov, *Soviet Physics JETP* **12**, 1254, 1961.

Using formula (58 2) for the chemical potential of an extreme relativistic degenerate gas, we thus find

$$n_e = \Delta^3/3\pi^2(c\hbar)^3. \qquad (108.4)$$

The equilibrium condition therefore gives a constant value of the electron density. This means that, as the density of the substance gradually increases, the nuclear reaction in question begins to occur when the electron density reaches the value (108.4). As the substance is further compressed, more and more nuclei will each capture an electron, so that the total number of electrons will decrease but their density will remain constant. Together with the electron density the pressure of the substance will also remain constant, being again determined mainly by the pressure of the electron gas: substitution of (108.4) in (108.3) gives

$$P = \Delta^4/12\pi^2(\hbar c)^3. \qquad (108.5)$$

This will continue until each nucleus has captured an electron.

At still higher densities and pressures the nuclei will capture further electrons, the nuclear charge being thus reduced further. Ultimately the nuclei will contain so many neutrons that they become unstable and break up. At a density $\varrho \sim 3\times 10^{11}$ g/cm^3 (and pressure $P \sim 10^{24}$ atm) the neutrons begin to be more numerous than the electrons, and when $\varrho \sim 10^{12}$ g/cm^3 the pressure due to the neutrons begins to predominate. This is the beginning of a density region in which matter may be regarded as essentially a degenerate neutron Fermi gas with a small number of electrons and various nuclei, whose concentrations are given by the equilibrium conditions for the corresponding nuclear reactions. The equation of state of matter in this range is

$$P = \frac{(3\pi^2)^{2/3}}{5}\frac{\hbar^2}{m_n^{8/3}}\varrho^{5/3}, \qquad (108.6)$$

where m_n is the neutron mass.

Finally, at densities $\varrho \gg 6\times 10^{15}$ g/cm^3, the degenerate neutron gas becomes extreme-relativistic, and the equation of state is

$$P = \frac{(3\pi^2)^{1/3}}{4}\hbar c\left(\frac{\varrho}{m_n}\right)^{4/3}. \qquad (108.7)$$

It should be remembered, however, that at densities of the order of that of nuclear matter the specifically nuclear forces (strong interaction of nucleons) become important. In this range of densities formula (108.7) can be only qualitative. In the present state of our knowledge concerning strong interactions we can draw no definite conclusions concerning the state of matter at densities considerably above the nuclear value. We shall merely mention that in this range other particles besides neutrons may be expected to appear. Since particles of each kind occupy a separate group of states, the conversion of neutrons into other particles may be thermodynamically favoured because of the decrease in the limiting energy of the Fermi distribution of neutrons.

§109. Equilibrium of bodies of large mass

Let us consider a body of very large mass, the parts of which are held together by gravitational attraction. Actual bodies of large mass that are known to us, namely the stars, continuously radiate energy and are certainly not in thermal equilibrium. It is, however, of fundamental interest to discuss an equilibrium body of large mass. We shall neglect the effect of temperature on the equation of state, i.e. consider a body at absolute zero—a "cold" body. Since in actual conditions the temperature of the outer surface is considerably lower than the internal temperature, a discussion of a body with a constant non-zero temperature is in any case devoid of physical meaning.

We shall further assume that the body is not rotating; then it will be spherical in equilibrium, and the density distribution will be symmetrical about the centre.

The equilibrium distribution of density (and of the other thermodynamic quantities) in the body will be determined by the following equations. The Newtonian gravitational potential ϕ satisfies the differential equation $\triangle\phi = 4\pi G\varrho$, where ϱ is the density of the substance and G the Newtonian constant of gravitation. In the case of spherical symmetry,

$$\frac{1}{r^2}\frac{\mathrm{d}}{\mathrm{d}r}\left(r^2\frac{\mathrm{d}\phi}{\mathrm{d}r}\right) = 4\pi G\varrho. \tag{109.1}$$

Moreover, in thermal equilibrium the condition (25.2) must be satisfied. In the gravitational field the potential energy of a particle of the body of mass m' is $m'\phi$, and so we have

$$\mu + m'\phi = \text{constant}, \tag{109.2}$$

where for brevity the suffix zero is omitted from the chemical potential of the substance in the absence of the field. Expressing ϕ in terms of μ by means of (109.2) and substituting in (109.1), we have

$$\frac{1}{r^2}\frac{\mathrm{d}}{\mathrm{d}r}\left(r^2\frac{\mathrm{d}\mu}{\mathrm{d}r}\right) = -4\pi m' G\varrho. \tag{109.3}$$

As the mass of the gravitating body increases, so of course does its mean density, as the following calculations will confirm. When the total mass M of the body is sufficiently large, therefore, we can, as shown in §108, regard the substance as a degenerate electron Fermi gas, initially non-relativistic and then at still greater masses relativistic.

The chemical potential of a non-relativistic degenerate electron gas is related to the density ϱ of the body by

$$\mu = \frac{(3\pi^2)^{2/3}}{2}\frac{\hbar^2}{m_e m'^{2/3}}\varrho^{2/3}; \tag{109.4}$$

see formula (56.3), with $\varrho = m'N/V$ (m' is the mass per electron and m_e the electron mass). Expressing ϱ in terms of μ and substituting in (109.3), we have†

$$\frac{1}{r^2}\frac{\mathrm{d}}{\mathrm{d}r}\left(r^2\frac{\mathrm{d}\mu}{\mathrm{d}r}\right) = -\lambda\mu^{3/2}, \qquad \lambda = \frac{2^{7/2}m_e^{3/2}m'^2G}{3\pi\hbar^3}. \tag{109.5}$$

The physically meaningful solutions of this equation must not have singularities at the origin: $\mu \to$ constant for $r \to 0$. This requirement necessarily imposes on the first derivative the condition

$$\mathrm{d}\mu/\mathrm{d}r = 0 \qquad \text{for} \qquad r = 0, \tag{109.6}$$

as follows immediately from equation (109.5) after integration over r:

$$\frac{\mathrm{d}\mu}{\mathrm{d}r} = -\frac{\lambda}{r^2}\int_0^r r^2\mu^{3/2}\,\mathrm{d}r.$$

A number of important results can be derived from equation (109.5) by simple dimensional considerations. The solution of (109.5) contains only two constants, λ and (for instance) the radius R of the body, a knowledge of which uniquely defines the solution. From these two quantities we can form only one quantity with the dimensions of length, the radius R itself, and one with the dimensions of energy, $1/\lambda^2R^4$ (the constant λ having dimensions $\mathrm{cm}^{-2}\ \mathrm{erg}^{-\frac{1}{2}}$). It is therefore clear that the function $\mu(r)$ must have the form

$$\mu(r) = \frac{1}{\lambda^2R^4}f\left(\frac{r}{R}\right), \tag{109.7}$$

where f is some function of the dimensionless ratio r/R only. Since the density ϱ is proportional to $\mu^{3/2}$, the density distribution must be of the form

$$\varrho(r) = \frac{\text{constant}}{R^6}F\left(\frac{r}{R}\right).$$

† It is easy to see that, for an electrically neutral gas consisting of electrons and atomic nuclei, the equilibrium condition can be written in the form (109.2) with the electron chemical potential as μ and the mass per electron as m'. For the derivation of this equilibrium condition (§25) involves considering the transport of an infinitesimal amount of substance from one place to another. In a gas consisting of both positively and negatively charged particles, such transport must be regarded as that of a certain quantity of neutral matter (i.e. electrons and nuclei together). The separation of the positive and negative charges is energetically very unfavourable, because of the resulting very large electric fields. We therefore obtain the equilibrium condition in the form

$$\mu_{\text{nuc}} + Z\mu_{\text{el}} + (m_{\text{nuc}} + Zm_{\text{el}})\phi = 0$$

(with Z electrons per nucleus). Owing to the large mass of the nuclei (compared with that of the electrons) their chemical potential is very small compared with μ_{el}. Neglecting μ_{nuc} and dividing the equation by Z, we obtain

$$\mu_{\text{el}} + m'\phi = 0.$$

If the atomic weight of the nuclei is assumed to be approximately twice their atomic number, m' can be taken as twice the proton mass ($m' = 2m_p$).

Thus, when the size of the sphere varies, the density distribution in it remains similar in form, the density at corresponding points being inversely proportional to R^6. In particular, the mean density of the sphere is inversely proportional to R^6:

$$\bar{\varrho} \sim 1/R^6.$$

The total mass M of the body is therefore inversely proportional to the cube of the radius:

$$M \sim 1/R^3.$$

These two relations may also be written

$$R \sim M^{-\frac{1}{3}}, \qquad \bar{\varrho} \sim M^2. \tag{109.8}$$

Thus the dimensions of an equilibrium sphere are inversely proportional to the cube root of its total mass, and the mean density is proportional to the square of the mass. The latter result confirms the assumption made above that the density of a gravitating body increases as its mass increases.

The fact that a gravitating sphere of non-relativistic degenerate Fermi gas can be in equilibrium for any total mass M can be seen *a priori* from the following qualitative argument. The total kinetic energy of the particles in such a gas is proportional to $N(N/V)^{2/3}$ (see (56.6)), or, what is the same thing, to $M^{5/3}/R^2$, and the gravitational energy of the gas as a whole is negative and proportional to M^2/R. The sum of two such expressions can have a minimum (as a function of R) for any M, and at the minimum $R \sim M^{-\frac{1}{3}}$.

Substituting (109.7) in (109.5) and using the dimensionless variable $\xi = r/R$, we find that the function $f(\xi)$ satisfies the equation

$$\frac{1}{\xi^2}\frac{\mathrm{d}}{\mathrm{d}\xi}\left(\xi^2\frac{\mathrm{d}f}{\mathrm{d}\xi}\right) = -f^{3/2} \tag{109.9}$$

with the boundary conditions $f'(0) = 0$, $f(1) = 0$. This equation cannot be solved analytically, and must be integrated numerically. It may be mentioned that as a result we find $f(0) = 178.2$, $f'(1) = -132.4$.

Using these numerical values it is easy to determine the value of the constant MR^3. Multiplying equation (109.1) by $r^2\,\mathrm{d}r$ and integrating from 0 to R, we obtain

$$GM = R^2[\mathrm{d}\phi/\mathrm{d}r]_{r=R} = -(R^2/m')[\mathrm{d}\mu/\mathrm{d}r]_{r=R} = -f'(1)/m'\lambda^2R^3,$$

whence†

$$MR^3 = 91.9\hbar^6/G^3m_e^{\,3}m'^5. \tag{109.10}$$

† In §108 we have seen that matter may be regarded as a non-relativistic degenerate electron gas at densities $\varrho \gg 20Z^2$ g/cm³. If this inequality is satisfied for the mean density of the sphere considered, its mass must satisfy the condition $M \gg 5\times10^{-3}Z\odot$, where $\odot = 2\times10^{33}$ g is the Sun's mass, and m' is taken equal to twice the mass of the proton. The corresponding radii are less than $5\times10^4\,Z^{-\frac{1}{3}}$ km.

For reference it may be noted that (with $m' = 2m$) $MR^3 = 1.40\times10^{60}$ g·cm³.

Finally, the ratio of the central density $\varrho(0)$ to the mean density $\bar{\varrho} = 3M/4\pi R^3$ is easily found to be

$$\varrho(0)/\bar{\varrho} = -f^{3/2}(0)/3f'(1) = 5.99. \tag{109.11}$$

Curve 1 in Fig. 51 shows the ratio $\varrho(r)/\varrho(0)$ as a function of r/R.

Let us now examine the equilibrium of a sphere consisting of a degenerate extreme-relativistic electron gas. The total kinetic energy of the particles of such a gas is proportional to $N(N/V)^{\frac{1}{3}}$ (see (58.3)), and hence to $M^{4/3}/R$; the gravitational energy is proportional to $-M^2/R$. Thus the two quantities depend on R in the same manner, and their sum will also be of the form constant$/R$. It follows that the body cannot be in equilibrium: if the constant is

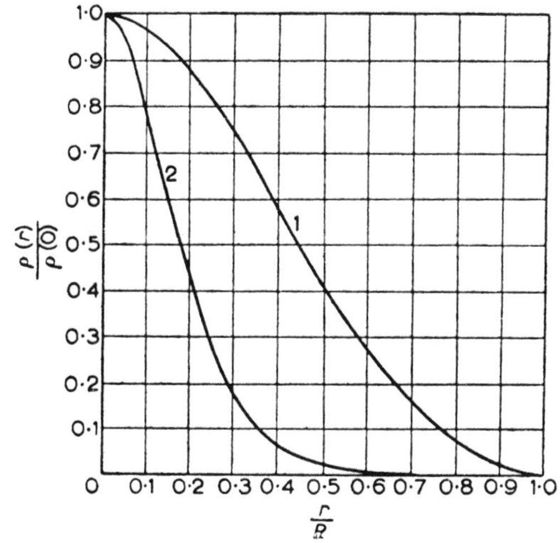

FIG. 51

positive, the body will tend to expand until the gas becomes non-relativistic; if the constant is negative, a decrease of R to zero corresponds to a decreased total energy, i.e. the body will contract without limit. The body can be in equilibrium only in the special case where the constant is zero, and the equilibrium is then neutral, the value of R being arbitrary.

This qualitative argument is, of course, entirely confirmed by exact quantitative analysis. The chemical potential of the relativistic gas considered is related to the density by

$$\mu = (3\pi^2)^{\frac{1}{3}}\hbar c(\varrho/m')^{\frac{1}{3}}. \tag{109.12}$$

(see (58.2)). Instead of (109.5) we now have

$$\frac{1}{r^2}\frac{\mathrm{d}}{\mathrm{d}r}\left(r^2\frac{\mathrm{d}\mu}{\mathrm{d}r}\right) = -\lambda\mu^3, \qquad \lambda = \frac{4Gm'^2}{3\pi c^3\hbar^3}. \tag{109.13}$$

Since λ now has dimensions erg $^{-2}$ cm $^{-2}$, we find that the chemical potential as a function of r must be of the form

$$\mu(r) = \frac{1}{R\sqrt{\lambda}} f\left(\frac{r}{R}\right), \tag{109.14}$$

and the density distribution

$$\varrho(r) = \frac{\text{constant}}{R^3} F\left(\frac{r}{R}\right).$$

Thus the mean density is inversely proportional to R^3, and the total mass $M \sim R^3\bar{\varrho}$ is independent of R:

$$\bar{\varrho} \sim 1/R^3, \qquad M = \text{constant} \equiv M_0. \tag{109.15}$$

M_0 is the only value of the mass for which equilibrium is possible: for $M > M_0$ the body will tend to contract indefinitely, and for $M < M_0$ it will expand.

For an exact calculation of the "critical mass" M_0, it is necessary to integrate numerically the equation

$$\frac{1}{\xi^2}\frac{\mathrm{d}}{\mathrm{d}\xi}\left(\xi^2 \frac{\mathrm{d}f}{\mathrm{d}\xi}\right) = -f^3, \qquad f'(0) = 0, \qquad f(1) = 0, \tag{109.16}$$

which is satisfied by the function $f(\xi)$ in (109.14). The result is $f(0) = 6.897$, $f'(1) = -2.018$. For the total mass we find

$$GM_0 = R^2[\mathrm{d}\phi/\mathrm{d}r]_{r=R} = -f'(1)/m'\sqrt{\lambda},$$

whence

$$M_0 = \frac{3.1}{m'^2}\left(\frac{\hbar c}{G}\right)^{3/2}. \tag{109.17}$$

Putting m' equal to twice the proton mass, we find $M_0 = 1.45\,\odot$. Finally, the ratio of the central density to the mean density is $\varrho(0)/\bar{\varrho} = -f^3(0)/3f'(1) = 54.2$. Curve 2 in Fig. 51 shows $\varrho(r)/\varrho(0)$ in the extreme relativistic case as a function of r/R.

The results obtained above concerning the relation between the mass and the radius of a "cold" spherical body in equilibrium can be represented by a single relation $M = M(R)$ for all radii R. For large R (and therefore for small densities), the electron gas may be regarded as non-relativistic, and the function $M(R)$ decreases as $1/R^3$. When R is sufficiently small, however, the density is so large that we have the extreme relativistic case, and the function $M(R)$ is almost a constant M_0; strictly $M(R) \to M_0$ when $R \to 0$. Fig. 52 shows the curve $M = M(R)$ calculated with $m' = 2m_p$.† It should be noted that the

† The intermediate part of the curve is constructed by numerical integration of equation (109.3) with the exact equation of state for a degenerate gas, i.e. with the chemical potential related to the density by

$$\varrho = m'p_0^3/3\pi^2\hbar^3 = \frac{m'}{3\pi^2\hbar^3}\left(\frac{\mu^2}{c^2} - m_e^2c^2\right)^{3/2},$$

where p_0 is the Fermi limiting momentum.

limiting value 1.45 ⊙ is reached only very gradually; this is because the density decreases rapidly away from the centre of the body, and so the extreme relativistic case may hold near the centre while the gas remains non-relativistic in a considerable part of the volume of the body. We may also mention that the initial part of the curve (R small) has no real physical significance: at sufficiently small radii the density becomes so large that nuclear reactions begin to occur. The pressure will then increase with density less rapidly than $\varrho^{4/3}$, and for such an equation of state no equilibrium is possible.†

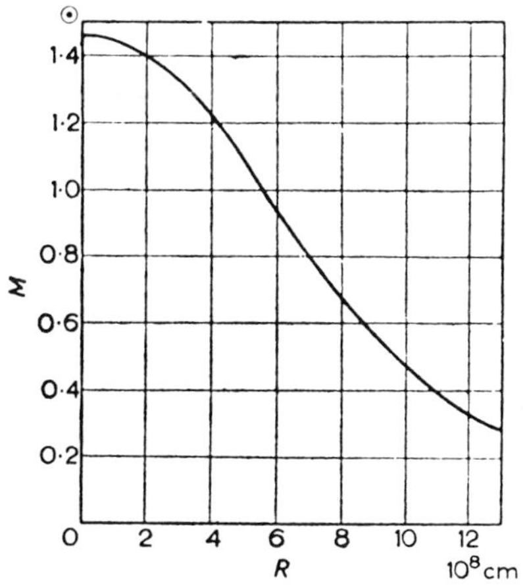

FIG. 52

Finally, this curve also has no meaning for large values of R (and small M): as has already been mentioned (see the second footnote to this section), in this range the equation of state used above becomes invalid. Here it should be pointed out that there is an upper limit to the possible size of a "cold" body, since on the curve in Fig. 52 large dimensions of the body correspond to small masses and small densities, but when the density is sufficiently small the substance will be in the ordinary "atomic" state and will be solid at the low temperatures here considered. The dimensions of a body consisting of such a substance will obviously decrease as its mass decreases further, and not increase as shown in Fig. 52. The true curve $R = R(M)$ must therefore have a maximum for some value of M.

The order of magnitude of the maximum radius can easily be determined by noting that it must correspond to the density at which the interaction

† If the chemical potential is proportional to a power of the density, $\mu \sim \varrho^n$ (and so $P \sim \varrho^{n+1}$), the internal energy of the body is proportional to $V\varrho^{n+1}$, i.e. to M^{n+1}/R^{3n}; the gravitational energy is again proportional to $-M^2/R$. It is easy to see that for $n < \frac{1}{3}$ the sum of two such expressions has an extremum as a function of R, but this extremum is a maximum, not a minimum.

between electrons and nuclei becomes important, i.e. for $\varrho \sim (m_e e^2/\hbar^2)^3 m' Z^2$ (see (108.1)). Combining this with equation (109.10), we obtain

$$R_{\max} \sim \hbar^2/G^{\frac{1}{2}} e m_e m' Z^{\frac{1}{3}} \sim 10^5 m_p/m' Z^{\frac{1}{3}} \text{ km}. \tag{109.18}$$

§110. **The energy of a gravitating body**

The gravitational potential energy E_{gr} of a body is given by the integral

$$E_{gr} = \tfrac{1}{2}\int \varrho\phi \, dV, \tag{110.1}$$

taken over the whole volume of the body. It will, however, be more convenient for us to start from a different expression for this quantity, which may be found as follows. Let us imagine the body to be gradually "built up" from material brought from infinity. Let $M(r)$ be the mass of substance within a sphere of radius r. Let us suppose that a mass $M(r)$ with a certain value of r has already been brought from infinity; then the work required to add a further mass $dM(r)$ is equal to the potential energy of that mass (in the form of a spherical shell of radius r and thickness dr) in the field of the mass $M(r)$, i.e. $-GM(r)\, dM(r)/r$. The total gravitational energy of a sphere of radius R is therefore

$$E_{gr} = -G\int \frac{M(r)\, dM(r)}{r}. \tag{110.2}$$

Differentiation of the equilibrium condition (109.2) gives

$$v\frac{dP}{dr} + m'\frac{d\phi}{dr} = 0;$$

the differentiation must be at constant temperature, and $(\partial\mu/\partial P)_T = v$ is the volume per particle. The derivative $-d\phi/dr$ is the force of attraction on unit mass at a distance r from the centre, and equals $-GM(r)/r^2$. Using also the density $\varrho = m'/v$, we have

$$\frac{1}{\varrho}\frac{dP}{dr} = -\frac{GM(r)}{r^2}. \tag{110.3}$$

From this equation we substitute in (110.2) $-GM(r)/r = (r/\varrho)\, dP/dr$, and write $dM(r)$ in the form $\varrho(r)\cdot 4\pi r^2\, dr$:

$$E_{gr} = 4\pi\int_0^R r^3 \frac{dP}{dr}\, dr,$$

and finally integrate by parts (bearing in mind that at the boundary of the body $P(R) = 0$ and that $r^3 P \to 0$ as $r \to 0$):

$$E_{gr} = -12\pi\int_0^R Pr^2\, dr,$$

or

$$E_{gr} = -3\int P\,dV. \tag{110.4}$$

Thus the gravitational energy of an equilibrium body can be expressed as an integral of the pressure over the volume.

Let us apply this formula to the degenerate Fermi gases considered in §109. We make the calculation for the general case, and take the chemical potential of the substance to be proportional to some power of its density:

$$\mu = K\varrho^n. \tag{110.5}$$

Since $d\mu = v\,dP = (m'/\varrho)\,dP$, we have

$$P = \frac{n}{n+1}\frac{K}{m'}\varrho^{n+1}. \tag{110.6}$$

In the equilibrium condition $\mu/m'+\phi =$ constant, the constant is just the potential at the boundary of the body, where μ vanishes; this potential is $-GM/R$ ($M = M(R)$ being the total mass of the body), and so we can write

$$\phi = -\frac{\mu}{m'}-\frac{GM}{R}.$$

We substitute this expression in the integral (110.1) which gives the gravitational energy, and use formulae (110.5), (110.6), obtaining

$$\begin{aligned} E_{gr} &= -\frac{1}{2m'}\int \mu\varrho\,dV-\frac{GM}{2R}\int \varrho\,dV = -\frac{K}{2m'}\int \varrho^{n+1}\,dV-\frac{GM^2}{2R} \\ &= -\frac{n+1}{2n}\int P\,dV-\frac{GM^2}{2R}. \end{aligned}$$

Finally, expressing the integral on the right in terms of E_{gr} by (110.4), we have

$$E_{gr} = \frac{n+1}{6n}E_{gr}-\frac{GM^2}{2R}$$

or

$$E_{gr} = -\frac{3n}{5n-1}\frac{GM^2}{R}. \tag{110.7}$$

Thus the gravitational energy of the body can be expressed by a simple formula in terms of its total mass and its radius.

A similar formula can also be obtained for the thermal internal energy E of the body. The internal energy per particle is $\mu - Pv$ (for zero temperature and entropy); the energy per unit volume is therefore

$$\frac{1}{v}(\mu-Pv) = \frac{\varrho\mu}{m'}-P,$$

or, substituting (110.5) and (110.6),

$$\frac{K}{m'}\frac{\varrho^{n+1}}{n+1} = \frac{P}{n}.$$

The internal thermal energy of the whole body is therefore

$$E = \frac{1}{n}\int P\,\mathrm{d}V = -\frac{1}{3n}E_{\mathrm{gr}} = \frac{1}{5n-1}\frac{GM^2}{R}. \tag{110.8}$$

Finally, the total energy of the body is

$$E_{\mathrm{tot}} = E+E_{\mathrm{gr}} = -\frac{3n-1}{5n-1}\frac{GM^2}{R}. \tag{110.9}$$

For a non-relativistic degenerate gas $n = \frac{2}{3}$, and so†

$$E_{\mathrm{gr}} = -\frac{6}{7}\frac{GM^2}{R}, \qquad E = \frac{3}{7}\frac{GM^2}{R}, \qquad E_{\mathrm{tot}} = -\frac{3}{7}\frac{GM^2}{R}. \tag{110.10}$$

In the extreme relativistic case, $n = \frac{1}{3}$, so that

$$E_{\mathrm{gr}} = -E = -\frac{3}{2}\frac{GM^2}{R}, \qquad E_{\mathrm{tot}} = 0. \tag{110.11}$$

The total energy is zero in this case, in accordance with the qualitative arguments given in §109 concerning the equilibrium of such a body.

§111. **Equilibrium of a "neutron" sphere**

For a body of large mass there are two possible equilibrium states. One corresponds to the state of matter consisting of electrons and nuclei, as assumed in the numerical estimates in §109. The other corresponds to the "neutron" state of matter, in which almost all the electrons have been captured by protons and the substance may be regarded as a neutron gas. When the body is sufficiently massive, the second possibility must always become thermodynamically more favourable than the first. Although the transformation of nuclei and electrons into free neutrons involves a considerable expenditure of energy, when the total mass of the body is sufficiently great this is more than counterbalanced by the release of gravitational energy owing to the decrease in size and increase in density of the body (see below).

First of all, let us examine the conditions in which the neutron state of a body can correspond to any thermodynamic equilibrium (which may be metastable). To do this, we start from the equilibrium condition $\mu+m_n\phi$ = constant, where μ is the chemical potential (the thermodynamic potential per neutron), m_n the neutron mass, and ϕ the gravitational potential.

† In this case $2E = -E_{\mathrm{gr}}$, in agreement with the virial theorem of mechanics, applied to a system of particles interacting according to Newton's law; see *Mechanics*, §10.

Since the pressure must be zero at the boundary of the body, it is clear that in an outer layer of the body the substance will be at low pressure and density and will therefore consist of electrons and nuclei. Although the thickness of this "shell" may be comparable with the radius of the dense inner neutron "core", the density of the outer layer is much lower, and so its total mass may be regarded as small compared with the mass of the core.†

Let us compare the values of $\mu+m_n\phi$ at two points: in the dense core near its boundary and near the outer boundary of the shell. The gravitational potential at these points may be taken as $-GM/R$ and $-GM/R'$, where R and R' are the core and shell radii, and M the mass of the core, which in the approximation used here is equal to the total mass of the body. The chemical potential at both points is determined mainly by the internal energy (binding energy) of the corresponding particles, which is large compared with their thermal energy. The difference between the two chemical potentials may therefore be taken as simply equal to the difference between the rest energy of a neutral atom (i.e. a nucleus and Z electrons) per unit atomic weight and the rest energy of the neutron; let this quantity be denoted by Δ.‡ Then, equating the values of $\mu+m_n\phi$ at the two points considered, we have

$$m_n MG\left(\frac{1}{R}-\frac{1}{R'}\right) = \Delta.$$

Hence, whatever the radius R', the mass and radius of the neutron core must certainly satisfy the inequality

$$m_n MG/R > \Delta. \tag{111.1}$$

Applying the results of §109 to a spherical body consisting of a degenerate (non-relativistic) neutron gas, we find that M and R are related by

$$MR^3 = 91.9\hbar^6/G^3 m_n{}^8 = 7.2\times10^{51} \text{ g}\cdot\text{cm}^3 \tag{111.2}$$

(formula (109.10) with m_e and m' replaced by m_n). Hence expressing M in terms of R and substituting in (111.1), we obtain an inequality for M, which in numerical form is $M > \sim 0.2\odot$. For example, with Δ for oxygen we get $M > 0.17\odot$, and for iron $M > 0.18\odot$. These masses correspond to radii $R < 26$ km.

This inequality gives a lower limit of mass, beyond which the "neutron" state of the body cannot be stable. It does not, however, ensure complete stability; the state may be metastable. To determine the limit of metastability, we must compare the total energies of the body in two states: the neutron state and the electron-nucleus state. The conversion of the whole mass M from the electron-nucleus state to the neutron state requires an expenditure

† There is, of course, no sharp boundary between the "core" and the "shell", and the transition between them is continuous.

‡ Δ/c^2 is just the difference of the nucleus and neutron "packing fractions", multiplied by the nuclear mass unit.

of energy $M\Delta/m_n$ to counterbalance the binding energy of the nuclei. In the process, energy is released because of the contraction of the body; according to formula (110.10), this gain of energy is

$$\frac{3GM^2}{7}\left(\frac{1}{R_n}-\frac{1}{R_e}\right),$$

where R_n is the radius of the body in the neutron state, given by formula (111.2), and R_e its radius in the electron-nucleus state, given by (109.10). Since $R_e \gg R_n$, the quantity $1/R_e$ may be neglected, and we obtain the following sufficient condition for complete stability of the neutron state of the body (omitting the suffix in R_n):

$$3GMm_n/7R > \Delta. \tag{111.3}$$

Comparing this condition with (111.1) and using (111.2), we see that the lower limit of mass determined by the inequality (111.3) is greater by a factor $(7/3)^{3/4} = 1.89$ than that given by (111.2). Numerically, the limit of metastability of the neutron state is therefore at a mass $M \cong \frac{1}{5}\odot$ (and radius $R \cong 22$ km).†

Let us now consider the upper limit of the range of mass values for which a "neutron" body can be in equilibrium. If we were to use the results of §109 (formula (109.17), with m_n in place of m'), the value obtained for this limit would be $6\odot$. In reality, however, these results are not applicable here, for the following reason. In a relativistic neutron gas, the kinetic energy of the particles is of the order of, or greater than, the rest energy.‡ In consequence it is no longer valid to use the Newtonian gravitational theory, and the calculations must be based on the general theory of relativity; and, as we shall see later, we find that the extreme relativistic case is no longer reached. The calculations must therefore make use of the exact equation of state of a degenerate Fermi gas—the parametric equation derived in §58, Problem 3.

The calculations are effected by numerical integration of the equations of a spherically symmetric static gravitational field, and the results are as follows.||

The limiting mass of a neutron sphere in equilibrium is found to be only $M_{max} = 0.76\odot$, and this value is reached at a finite radius $R_{min} = 9.42$ km. Fig. 53 shows a graph of the relation obtained between the mass M and the radius R. Stable neutron spheres of larger mass or smaller radius cannot,

† The mean density of the body is then 1.4×10^{13} g/cm³, and so the neutron gas may in fact still be regarded as non-relativistic, and the formulae used here are still valid.

‡ In the relativistic *electron* gas, the kinetic energy of the particles is comparable with the rest energy of the electrons, but is still small in comparison with the rest energy of the nuclei, which contribute most of the mass of the substance.

|| The details of the calculations are given in the original paper by J. R. Oppenheimer and G. M. Volkoff, *Physical Review* **55**, 374, 1939.

therefore, exist. It should be mentioned that the mass M here denotes the product $M = Nm_n$, where N is the total number of particles (i.e neutrons) in the sphere. This quantity is not equal to the gravitational mass M_{gr} of the body, which determines the gravitational field created by it in the surrounding space. Because of the "gravitational mass defect", in stable states we always have $M_{gr} < M$ (in particular, for $R = R_{min}$, $M_{gr} = 0.95M$).†

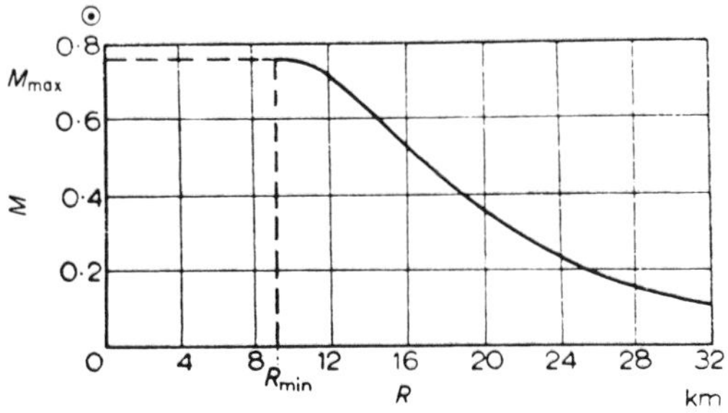

FIG. 53

The question arises of the behaviour of a spherical body of mass exceeding M_{max}. It is clear *a priori* that such a body must tend to contract indefinitely (*gravitational collapse*). The discovery of the nature and course of such a process requires an investigation of the non-stationary solutions of the gravitational equations. This can be carried out in a closed analytical form only for the simple case of matter in the form of "dust", with equation of state $P = 0$. Although such a description of matter is certainly inadequate for the later stages of contraction, the solution of such a problem still seems to give a correct idea of the nature of the process even for the general case of the exact equation of state. It is found that, as viewed by a distant observer (with a Galilean frame of reference at infinity), the sphere contracts in such a way that as $t \to \infty$ its radius tends asymptotically to the value $2M_{gr}G/c^2$ (called the *gravitational radius* of the body). The external observer's infinite time corresponds to a finite proper time in the local frame of reference; beyond this time, the matter continues to "fall" inward and reaches the centre, still after a finite interval of proper time.‡

† The point $R = R_{min}$ in Fig. 53 is in fact a maximum on the curve $M = M(R)$. This curve continues beyond the maximum as an inward spiral which asymptotically approaches a centre. The parameter which increases monotonically along the curve is the density at the centre of the sphere, which tends to infinity for a sphere corresponding to the limiting point of the spiral; see N. A. DMITRIEV and S. A. KHOLIN, *Voprosy kosmogonii* **9,** 254, 1963. However, no part of the curve for $R < R_{min}$ corresponds to a stable state of the sphere.

‡ The details are given in the original paper by J. R. OPPENHEIMER and H. SNYDER, *Physical Review* **56,** 455, 1939. See also *The Classical Theory of Fields*, §97, Problem 5; E. M. LIFSHITZ and I. M. KHALATNIKOV, *Soviet Physics JETP* **12,** 108, 1961.

It should be noted that the possibility in principle of gravitational collapse, which (for the model considered of a spherical body) is unavoidable for $M > M_{max}$, is not in fact restricted to large masses. A "collapsing" state exists for any mass, but for $M < M_{max}$ it is separated by a very high energy barrier from the static equilibrium state.†

† See YA. B. ZEL'DOVICH, *Soviet Physics JETP* **15**, 446, 1962.

CHAPTER XII

FLUCTUATIONS

§112. The Gaussian distribution

IT HAS already been stressed several times that the physical quantities which describe a macroscopic body in equilibrium are, almost always, very nearly equal to their mean values. Nevertheless, deviations from the mean values, though small, do occur (quantities are said to *fluctuate*), and the problem arises of finding the probability distribution of these deviations.

Let us consider some closed system, and let x be some physical quantity describing the system as a whole or some part of it (in the former case x must not, of course, be a quantity which is strictly constant for a closed system, such as its energy). In what follows it will be convenient to suppose that the mean value $\bar{x}$ has already been subtracted from x, and so we shall everywhere assume that $\bar{x} = 0$.

The discussion in §7 has shown that, if the entropy of a system is formally regarded as a function of the exact values of the energies of the subsystems, the function e^S will give the probability distribution for these energies (formula (7.17)). It is easy to see, however, that the discussion made no use of any specific properties of the energy. Similar arguments will therefore show that the probability for a quantity x to have a value in the interval from x to $x+dx$ is proportional to $e^{S(x)}$, where $S(x)$ is the entropy formally regarded as a function of the exact value of x. Denoting this probability by $w(x)\,dx$, we have†

$$w(x) = \text{constant} \times e^{S(x)}. \qquad (112.1)$$

Before proceeding to examine the consequences of this formula, let us consider its range of applicability. All the arguments leading to formula (112.1) tacitly assume that the quantity x behaves classically.‡ We must therefore find a condition which ensures that quantum effects are negligible.

As we know from quantum mechanics, the relation $\Delta E\, \Delta x \sim \hbar \dot{x}$ exists between the quantum uncertainties of energy and of some quantity x, $\dot{x}$ being the classical rate of change of x.||

† This formula was first applied to the study of fluctuations by A. EINSTEIN (1910).

‡ This does not mean, of course, that the whole system must be a classical one. Variables other than x pertaining to the system may have quantum behaviour.

|| See *Quantum Mechanics*, §16.

Let τ be a time† expressing the rate of change of the quantity x which is considered here and which has a non-equilibrium value; then $\dot{x} \sim x/\tau$, and so $\Delta E \Delta x \sim \hbar x/\tau$. It is clear that the quantity x may be said to have a definite value only if its quantum uncertainty is small: $\Delta x \ll x$, whence $\Delta E \gg \hbar/\tau$. Thus the quantum uncertainty of energy must be large in comparison with $\hbar/\tau$. The entropy of the system will then have uncertainty $\Delta S \gg \hbar/\tau T$.

If formula (112.1) is to be meaningful, it is clearly necessary for the uncertainty of entropy to be small compared with unity:

$$T \gg \hbar/\tau, \qquad \tau \gg \hbar/T. \tag{112.2}$$

This is the required condition. When the temperature is too low or when the quantity x varies too rapidly (τ is too small) the fluctuations cannot be treated thermodynamically, and the purely quantum fluctuations become of major importance.

Let us now return to formula (112.1). The entropy S has a maximum for $x = \bar{x} = 0$. Hence $\partial S/\partial x = 0$ and $\partial^2 S/\partial x^2 < 0$ for $x = 0$. In fluctuations the quantity x is very small. Expanding $S(x)$ in powers of x and retaining only terms of up to the second order, we obtain $S(x) = S(0) - \frac{1}{2}\beta x^2$, where β is a positive constant. Substitution in (112.1) gives the probability distribution in the form

$$w(x)\,\mathrm{d}x = Ae^{-\frac{1}{2}\beta x^2}\,\mathrm{d}x.$$

The normalisation constant A is given by the condition

$$A\int_{-\infty}^{\infty} e^{-\frac{1}{2}\beta x^2}\,\mathrm{d}x = 1.$$

Although we have used here the expression for $w(x)$ which is valid for small x- the integrand decreases so rapidly with increasing $|x|$ that the range of integration may be extended from $-\infty$ to ∞. The integration gives $A = \sqrt{(\beta/2\pi)}$.

Thus the probability distribution of the various values of the fluctuation x is given by the formula

$$w(x)\,\mathrm{d}x = \sqrt{(\beta/2\pi)}e^{-\frac{1}{2}\beta x^2}\,\mathrm{d}x. \tag{112.3}$$

Such a distribution is called a *Gaussian distribution*. It has a maximum when $x = 0$ and decreases rapidly and symmetrically as $|x|$ increases on either side of the maximum.

† The time τ need not be the same as the relaxation time for equilibrium to be reached with respect to x, and may be less than this time if x approaches $\bar{x}$ in an oscillatory manner. For example, if we consider the variation of pressure in a small region of the body (with linear dimensions $\sim a$), τ will be of the order of the period of acoustic vibrations with wavelength $\lambda \sim a$, i.e. $\tau \sim a/c$, where c is the velocity of sound.

The mean square fluctuation is

$$\overline{x^2} = \sqrt{\frac{\beta}{2\pi}} \int_{-\infty}^{\infty} x^2 e^{-\frac{1}{2}\beta x^2}\, dx = \frac{1}{\beta}. \tag{112.4}$$

Thus $\beta = 1/\overline{x^2}$, and we can write the Gaussian distribution in the form

$$w(x)\, dx = \frac{1}{\sqrt{(2\pi\overline{x^2})}} \exp\left(-\frac{x^2}{2\overline{x^2}}\right) dx. \tag{112.5}$$

As we should expect, the smaller $\overline{x^2}$, the sharper is the maximum of $w(x)$.

Knowing the mean square $\overline{x^2}$, we can find the corresponding quantity for any function $\phi(x)$. Since x is small, we have

$$\Delta\phi = [d\phi/dx]_{x=0} x,$$

and so

$$\overline{(\Delta\phi)^2} = [(d\phi/dx)^2]_{x=0} \overline{x^2}. \tag{112.6}$$

§113. **The Gaussian distribution for more than one variable**

In §112 we have discussed the probability of a deviation of any one thermodynamic quantity from its mean value, disregarding the values of other quantities.† In a similar manner we can determine the probability of a simultaneous deviation of several thermodynamic quantities from their mean values. Let these deviations be denoted by $x_1, x_2, \ldots, x_n$.

We define the entropy $S(x_1, \ldots, x_n)$ as a function of the quantities $x_1, x_2, \ldots, x_n$ and write the probability distribution in the form $w\, dx_1 \ldots dx_n$, with w given by (112.1). Let S be expanded in powers of the x_i; as far as the second-order terms, the difference $S - S_0$ is a negative-definite quadratic form:

$$S - S_0 = -\tfrac{1}{2} \sum_{i,k=1}^{n} \beta_{ik} x_i x_k$$

(clearly $\beta_{ik} = \beta_{ki}$). In the rest of this section we shall omit the summation sign, and summation from 1 to n over all repeated suffixes will be implied. Thus we write

$$S - S_0 = -\tfrac{1}{2}\beta_{ik} x_i x_k. \tag{113.1}$$

Substituting this expression in (112.1), we obtain for the required probability distribution

$$w = Ae^{-\frac{1}{2}\beta_{ik} x_i x_k}. \tag{113.2}$$

† This means that the function $S(x)$ used in §112 was the greatest possible value of the entropy for the given non-equilibrium value of x.

The constant A is determined from the normalisation condition

$$A \int_{-\infty}^{\infty} \dots \int_{-\infty}^{\infty} e^{-\frac{1}{2}\beta_{ik}x_i x_k} \, dx_1 \dots dx_n = 1.$$

Here, for the same reasons as in §112, the integration over each x_i can be taker from $-\infty$ to ∞. To calculate this integral we proceed as follows. The linean transformation

$$x_i = a_{ik}x'_k \tag{113.3}$$

of $x_1, x_2, \dots, x_n$ converts the quadratic form $\beta_{ik}x_i x_k$ into a sum of squares $x'_i x'_i = x'_i x'_k \delta_{ik}$ (where $\delta_{ik} = 0$ if $i \neq k$ and 1 if $i = k$) if the transformation coefficients a_{ik} satisfy the relations $\beta_{ik}a_{il}a_{km} = \delta_{lm}$, as is easily seen by substituting (113.3) in $\beta_{ik}x_i x_k$. The determinant of the sums $a_{il}b_{lk}$ is the product of the determinants $|a_{ik}|$ and $|b_{ik}|$, and similarly for a greater number of determinants. Since the determinant $|\delta_{ik}| = 1$, the above relation between the β_{ik} and the transformation coefficients in (113.3) shows that

$$\beta a^2 = 1, \tag{113.4}$$

where β and a denote the determinants $|\beta_{ik}|$ and $|a_{ik}|$.

When the transformation is applied to the normalisation integral, the result is

$$Aa \int_{-\infty}^{\infty} \dots \int_{-\infty}^{\infty} e^{-\frac{1}{2}x'_i x'_i} \, dx'_1 \dots dx'_n = 1,$$

since the Jacobian $\partial(x_1, \dots, x_n)/\partial(x'_1, \dots, x'_n) = a$. This integral now separates into a product of n integrals; calculating them and using the relation (113.4), we obtain

$$A = (2\pi)^{-\frac{1}{2}n} \sqrt{\beta}.$$

Thus we obtain finally the Gaussian distribution for more than one variable in the form

$$w = \frac{\sqrt{\beta}}{(2\pi)^{\frac{1}{2}n}} \exp\left(-\tfrac{1}{2}\beta_{ik}x_i x_k\right). \tag{113.5}$$

Let us define

$$X_i = -\partial S/\partial x_i = \beta_{ik}x_k, \tag{113.6}$$

and determine the mean values of the products $x_i X_k$:

$$\overline{x_i X_k} = \frac{\sqrt{\beta}}{(2\pi)^{\frac{1}{2}n}} \int \dots \int x_i \beta_{kl} x_l \cdot e^{-\frac{1}{2}\beta_{mp}x_m x_p} \, dx_1 \dots dx_n.$$

To calculate the integral let us assume for the moment that the mean values $\overline{x_i}$ are not zero but x_{i0}, say. Then in (113.5) x_i must be replaced by $x_i - x_{i0}$, and the definition of the mean gives

$$\bar{x_i} = \frac{\sqrt{\beta}}{(2\pi)^{\frac{1}{2}n}} \int \ldots \int x_i e^{-\frac{1}{2}\beta_{mp}(x_m - x_{m0})(x_p - x_{p0})} \, dx_1 \ldots dx_n = x_{i0}.$$

Differentiating this equation with respect to x_{k0} and again putting $x_{10}, \ldots, x_{n0}$ equal to zero, we have δ_{ik} on the right and the required integral on the left. Thus

$$\overline{x_i X_k} = \delta_{ik}. \tag{113.7}$$

Substituting (113.6) and, for convenience, renaming the suffixes, we find

$$\beta_{ml}\overline{x_l x_k} = \delta_{mk}.$$

Multiplying both sides by $\beta^{-1}{}_{im}$ (i.e. by an element of the matrix inverse to β_{im}) and summing over the suffix m gives

$$\beta^{-1}{}_{im}\beta_{ml}\overline{x_l x_k} = \beta^{-1}{}_{im}\delta_{mk} = \beta^{-1}{}_{ik}.$$

By the definition of the inverse matrix, $\beta^{-1}{}_{im}\beta_{ml} = \delta_{il}$, and so we have†

$$\overline{x_i x_k} = \beta^{-1}{}_{ik}. \tag{113.8}$$

Finally, let us determine $\overline{X_i X_k}$. According to (113.6) and (113.7), $\overline{X_i X_k} = \beta_{il}\overline{x_l X_k} = \beta_{il}\delta_{lk}$, and so

$$\overline{X_i X_k} = \beta_{ik}. \tag{113.9}$$

It is easy to determine also the mean square fluctuation of any function $f(x_1, \ldots, x_n)$ of the quantities $x_1, x_2, \ldots, x_n$. Since the deviations from the mean values are small, $\Delta f = (\partial f/\partial x_i)x_i$, where the $\partial f/\partial x_i$ denote the values of the derivatives for $x_1 = \ldots = x_n = 0$. Hence

$$\overline{(\Delta f)^2} = \frac{\partial f}{\partial x_i}\frac{\partial f}{\partial x_k}\overline{x_i x_k}$$

or, substituting (113.8),

$$\overline{(\Delta f)^2} = \frac{\partial f}{\partial x_i}\frac{\partial f}{\partial x_k}\beta^{-1}{}_{ik}. \tag{113.10}$$

If the fluctuations of any two of the x_i (x_1 and x_2, say) are statistically independent, the mean value $\overline{x_1 x_2}$ is equal to the product of the mean values $\bar{x}_1$ and $\bar{x}_2$; since each of these is zero, so is $\overline{x_1 x_2}$, and from (113.8) this implies that $\beta^{-1}{}_{12} = 0$. It is easy to see that for a Gaussian distribution the converse theorem is also valid: if $\overline{x_1 x_2} = 0$ (i.e. $\beta^{-1}{}_{12} = 0$), the fluctuations of x_1 and x_2 are statistically independent. For the probability distribution w_{12} of the quantities x_1 and x_2 is obtained by integrating the distribution (113.5) over all the

† The quantity

$$\overline{x_i x_k}/\sqrt{(\overline{x_i x_i}\;\overline{x_k x_k})}$$

is called the *correlation* of the quantities x_i and x_k.

other x_i; the result is an expression of the form

$$w_{12} = \text{constant}\times\exp\{-\tfrac{1}{2}\beta'_{11}x_1^2-\beta'_{12}x_1x_2-\tfrac{1}{2}\beta'_{22}x_2^2\}$$

(where the coefficients β'_{ik} are in general different from the corresponding β_{ik}). Applying formula (113.8) to this distribution, we find that $\overline{x_1x_2} = \beta'^{-1}_{12}$. If $\overline{x_1x_2} = 0$, then $\beta'^{-1}_{12} = 0$. But for a matrix of order two the vanishing of the inverse matrix element β'^{-1}_{12} implies that of the element β'_{12}.† Thus w_{12} separates into a product of two independent Gaussian distributions for the quantities x_1 and x_2, which are therefore statistically independent.

§114. Fluctuations of the fundamental thermodynamic quantities

We shall now calculate the mean square fluctuations of the fundamental thermodynamic quantities, pertaining to any small part of a body. This small part must still, of course, contain a sufficient number of particles. At very low temperatures, however, this condition may be weaker than (112.2), which ensures that quantum fluctuations are absent, as assumed; in this case the minimum permissible dimensions of the parts of the body will be determined by the latter condition.‡ To avoid misunderstanding, it should be emphasised that the degree of importance of quantum fluctuations has no bearing on the influence of quantum effects on the thermodynamic quantities (or equation of state) of the substance: the fluctuations may be purely thermodynamic while at the same time the equation of state is given by the formulae of quantum mechanics.

For quantities such as energy and volume, which have a purely mechanical significance as well as a thermodynamic one, the concept of fluctuations is self-explanatory, but it needs more precise treatment for quantities such as entropy and temperature, whose definition necessarily involves considering the body over finite intervals of time. For example, let $S(E, V)$ be the equilibrium entropy of the body as a function of its (mean) energy and (mean) volume. By the fluctuation of entropy we shall mean the change in the function $S(E, V)$, formally regarded as a function of the exact (fluctuating) values of the energy and volume.

As we have seen in the preceding sections, the probability w of a fluctuation is proportional to e^{S_t}, where S_t is the total entropy of a closed system, i.e. of the body as a whole. We can equally well say that w is proportional to $e^{\Delta S_t}$, where ΔS_t is the change in entropy in the fluctuation. According to (20.8) we

† For a matrix of order two we have

$$\beta^{-1}_{12} = \beta_{12}/(\beta_{12}^2-\beta_{11}\beta_{22}).$$

‡ For example, for pressure fluctuations the condition $\tau \gg \hbar/T$ with $\tau \sim a/c$ (see the last footnote to §112) gives $a \gg \hbar c/T$.

have $\Delta S_t = -R_{\min}/T_0$, where $R_{\min}$ is the minimum work needed to carry out reversibly the given change in the thermodynamic quantities in the small part considered (relative to which the remainder of the body acts as a "medium"). Thus

$$w \sim e^{-R_{\min}/T_0}. \tag{114.1}$$

Here we substitute for $R_{\min}$ the expression

$$R_{\min} = \Delta E - T_0 \Delta S + P_0 \Delta V,$$

where ΔE, ΔS, ΔV are the changes in the energy, entropy and volume of the small part of the body in the fluctuation, and T_0, P_0 the temperature and pressure of the "medium", i.e. the equilibrium (mean) values of the temperature and pressure of the body. In what follows we shall omit the suffix zero from all quantities which are the coefficients of fluctuations; the equilibrium values will always be meant. Thus we have

$$w \sim \exp\left(-\frac{\Delta E - T\Delta S + P\Delta V}{T}\right). \tag{114.2}$$

We may note that in this form the expression (114.2) is applicable to any fluctuations, small or large (a large fluctuation here meaning one in which, for example, ΔE is comparable with the energy of the small part of the body, though of course still small compared with the energy of the whole body). The application of formula (114.2) to small fluctuations (which are those generally occurring) leads to the following results.

Expanding ΔE in series, we obtain (cf. §21)

$$\Delta E - T\Delta S + P\Delta V = \tfrac{1}{2}\left[\frac{\partial^2 E}{\partial S^2}(\Delta S)^2 + 2\frac{\partial^2 E}{\partial S \partial V}\Delta S \Delta V + \frac{\partial^2 E}{\partial V^2}(\Delta V)^2\right].$$

It is easily seen that this expression may be written as

$$\tfrac{1}{2}\left[\Delta S \Delta\left(\frac{\partial E}{\partial S}\right)_V + \Delta V \Delta\left(\frac{\partial E}{\partial V}\right)_S\right] = \tfrac{1}{2}(\Delta S \Delta T - \Delta P \Delta V).$$

Thus we obtain the fluctuation probability (114.2) in the form

$$w \sim \exp\left(\frac{\Delta P \Delta V - \Delta T \Delta S}{2T}\right). \tag{114.3}$$

From this general formula we can find the fluctuations of various thermodynamic quantities. Let us first take V and T as independent variables. Then

$$\Delta S = \left(\frac{\partial S}{\partial T}\right)_V \Delta T + \left(\frac{\partial S}{\partial V}\right)_T \Delta V = \frac{C_v}{T}\Delta T + \left(\frac{\partial P}{\partial T}\right)_V \Delta V,$$

$$\Delta P = \left(\frac{\partial P}{\partial T}\right)_V \Delta T + \left(\frac{\partial P}{\partial V}\right)_T \Delta V;$$

see (16.3). Substituting these expressions in the exponent in formula (114.3) we find that the terms in $\Delta V \Delta T$ cancel, leaving

$$w \sim \exp\left\{-\frac{C_v}{2T^2}(\Delta T)^2+\frac{1}{2T}\left(\frac{\partial P}{\partial V}\right)_T(\Delta V)^2\right\}. \tag{114.4}$$

This expression separates into two factors, one depending only on ΔT and the other only on ΔV. In other words, the fluctuations of temperature and of volume are statistically independent, so that

$$\overline{\Delta T \Delta V} = 0. \tag{114.5}$$

Comparing successively each of the two factors of (114.4) with the general formula (112.5) for the Gaussian distribution, we find the following expressions for the mean square fluctuations of temperature† and volume:

$$\overline{(\Delta T)^2} = T^2/C_v, \tag{114.6}$$

$$\overline{(\Delta V)^2} = -T(\partial V/\partial P)_T. \tag{114.7}$$

These quantities are positive by virtue of the thermodynamic inequalities $C_v > 0$ and $(\partial P/\partial V)_T < 0$.

Let us now take P and S as the independent variables in (114.3). Then

$$\Delta V = \left(\frac{\partial V}{\partial P}\right)_S \Delta P+\left(\frac{\partial V}{\partial S}\right)_P \Delta S,$$

$$\Delta T = \left(\frac{\partial T}{\partial S}\right)_P \Delta S+\left(\frac{\partial T}{\partial P}\right)_S \Delta P = \frac{T}{C_p}\Delta S+\left(\frac{\partial T}{\partial P}\right)_S \Delta P.$$

From the formula $\mathrm{d}W = T\,\mathrm{d}S+V\,\mathrm{d}P$,

$$(\partial V/\partial S)_P = \partial^2 W/\partial P\partial S = (\partial T/\partial P)_S,$$

and so

$$\Delta V = \left(\frac{\partial V}{\partial P}\right)_S \Delta P+\left(\frac{\partial T}{\partial P}\right)_S \Delta S.$$

Substitution of ΔV and ΔT in (114.3) gives

$$w \sim \exp\left\{\frac{1}{2T}\left(\frac{\partial V}{\partial P}\right)_S(\Delta P)^2-\frac{1}{2C_p}(\Delta S)^2\right\}. \tag{114.8}$$

As in (114.4), this expression is a product of factors, one depending only on ΔP and the other only on ΔS. In other words, the fluctuations of entropy and of pressure are statistically independent,‡ so that

$$\overline{\Delta S\,\Delta P} = 0. \tag{114.9}$$

† If T is measured in degrees, $\overline{(\Delta T)^2} = kT^2/C_v$.

‡ The statistical independence of the pairs of quantities T, V and S, P is already obvious from the following argument. If we take $x_1 = \Delta S$ and $x_2 = \Delta V$ as the x_i in the formulae in §113, the corresponding X_i will be $X_1 = \Delta T/T$, $X_2 = -\Delta P/T$ (see §22). But $\overline{x_i X_k} = 0$ for $i \neq k$ according to the general formula (113.7), and this gives (114.5) and (114.9).

For the mean square fluctuations of entropy and pressure we find

$$\overline{(\Delta S)^2} = C_p, \tag{114.10}$$

$$\overline{(\Delta P)^2} = -T(\partial P/\partial V)_S. \tag{114.11}$$

These formulae show that the mean square fluctuations of additive thermodynamic quantities (volume and entropy) are proportional to the dimensions (the volume) of those parts of the body to which they relate. Accordingly the root-mean-square fluctuations of these quantities are proportional to the square root of the volume, and the relative fluctuations are inversely proportional to this square root, in complete agreement with the general results in §2 (formula (2.5)). But for quantities such as temperature and pressure the r.m.s. fluctuations themselves are inversely proportional to the square root of the volume.

Formula (114.7) determines the fluctuation of the volume of some part of the body, containing a certain number N of particles. Dividing both sides by N^2, we find the volume fluctuation per particle:

$$\overline{(\Delta(V/N))^2} = -\frac{T}{N^2}\left(\frac{\partial V}{\partial P}\right)_T. \tag{114.12}$$

This fluctuation must obviously be independent of whether we consider it for constant volume or for constant number of particles. From (114.12) we can therefore find the fluctuation of the number of particles in a fixed volume in the body. Since V is then constant, we must put $\Delta(V/N) = V\Delta(1/N) = -(V/N^2)\Delta N$. Substitution in (114.12) gives

$$\overline{(\Delta N)^2} = -(TN^2/V^2)(\partial V/\partial P)_T. \tag{114.13}$$

For certain calculations it is convenient to write this formula in a different form. Since the derivative $(\partial V/\partial P)_T$ is regarded as taken with N constant, we write

$$-\frac{N^2}{V^2}\left(\frac{\partial V}{\partial P}\right)_{T,N} = N\left(\frac{\partial}{\partial P}\frac{N}{V}\right)_{T,N}.$$

The number of particles N, as a function of P, T and V, must be of the form $N = Vf(P, T)$, as shown by considerations of homogeneity (cf. §24); that is, N/V is a function of P and T only, and it therefore does not matter whether N/V is differentiated at constant N or constant V. Hence we can write

$$N\left(\frac{\partial}{\partial P}\frac{N}{V}\right)_{T,N} = N\left(\frac{\partial}{\partial P}\frac{N}{V}\right)_{T,V} = \frac{N}{V}\left(\frac{\partial N}{\partial P}\right)_{T,V}$$
$$= \left(\frac{\partial N}{\partial P}\right)_{T,V}\left(\frac{\partial P}{\partial \mu}\right)_{T,V} = \left(\frac{\partial N}{\partial \mu}\right)_{T,V},$$

where we have used the equation $N/V = (\partial P/\partial\mu)_{T,V}$, which follows from formula (24.14): $\mathrm{d}\Omega = -V\,\mathrm{d}P = -S\,\mathrm{d}T - N\,\mathrm{d}\mu$. Thus we have for the fluctuation of the number of particles the formula†

$$\overline{(\Delta N)^2} = T(\partial N/\partial\mu)_{T,V}. \tag{114.14}$$

A body is characterised not only by the thermodynamic quantities considered above but also by the momentum $\mathbf{P}$ of its macroscopic motion relative to the medium. In a state of equilibrium there is no macroscopic motion, i.e. $\mathbf{P} = 0$. Motion may, however, result from fluctuation; let us determine the probability of such a fluctuation. The minimum work $R_{\min}$ in this case is simply equal to the kinetic energy of the body: $R_{\min} = P^2/2M = \frac{1}{2}Mv^2$, where M is its mass and $\mathbf{v} = \mathbf{P}/M$ the velocity of the macroscopic motion. Thus the required probability is

$$w \sim e^{-Mv^2/2T}. \tag{114.15}$$

It may be noted that the fluctuations of velocity are statistically independent of those of the other thermodynamic quantities. The mean square fluctuation of each Cartesian component of the velocity is equal to

$$\overline{(\Delta v_x)^2} = T/M, \tag{114.16}$$

and is inversely proportional to the mass of the body.

The foregoing formulae show that the mean square fluctuations of such quantities as energy, volume, pressure and velocity vanish at absolute zero (as the first power of the temperature). This is a general property of all thermodynamic quantities which also have a purely mechanical significance, but is not in general true of such purely thermodynamic quantities as entropy and temperature.

Formula (114.6) for the fluctuations of temperature can also be interpreted from a different point of view. As we know, the concept of temperature may

† This formula can also be easily derived directly from the Gibbs distribution. According to the definition of the mean value,

$$\overline{N} = e^{\Omega/T}\sum_N N e^{\mu N/T}\sum_n e^{-E_{nN}/T}.$$

Differentiation of this expression with respect to μ (for constant V and T) gives

$$\frac{\partial\overline{N}}{\partial\mu} = \frac{1}{T}e^{\Omega/T}\sum_N\left(N^2 + N\frac{\partial\Omega}{\partial\mu}\right)e^{\mu N/T}\sum_n e^{-E_{nN}/T} = \frac{1}{T}\left(\overline{N^2} + \overline{N}\frac{\partial\Omega}{\partial\mu}\right).$$

But $\partial\Omega/\partial\mu = -\overline{N}$, and so

$$\frac{\partial\overline{N}}{\partial\mu} = \frac{1}{T}(\overline{N^2} - \overline{N}^2) = \frac{1}{T}\overline{(\Delta N)^2},$$

which gives (114.14).

We could also use the Gibbs distribution to derive expressions for the fluctuations of the other thermodynamic quantities.

be introduced through the Gibbs distribution; it is then regarded as a parameter defining this distribution. As applied to an isolated body, the Gibbs distribution gives a complete description of the statistical properties, except for the inaccuracy that it leads to very small but non-zero fluctuations of the total energy of the body; these cannot in fact exist (see the end of §28). Conversely, if the energy is regarded as a given quantity, we cannot assign a definite temperature to the body, and we must suppose that the temperature undergoes fluctuations in accordance with (114.6), where C_v denotes the specific heat of the body as a whole. This quantity (114.6) obviously describes the accuracy with which the temperature of an isolated body can be defined.

PROBLEMS

PROBLEM 1. Find the mean square fluctuation of the energy (using V and T as independent variables).

SOLUTION. We have

$$\Delta E = \left(\frac{\partial E}{\partial V}\right)_T \Delta V + \left(\frac{\partial E}{\partial T}\right)_V \Delta T = \left[T\left(\frac{\partial P}{\partial T}\right)_V - P\right] \Delta V + C_v\, \Delta T.$$

Squaring and averaging, we obtain

$$\overline{(\Delta E)^2} = -\left[T\left(\frac{\partial P}{\partial T}\right)_V - P\right]^2 T\left(\frac{\partial V}{\partial P}\right)_T + C_v T^2.$$

PROBLEM 2. Find $\overline{(\Delta W)^2}$ (with variables P and S).

SOLUTION. $\overline{(\Delta W)^2} = -TV^2(\partial P/\partial V)_S + T^2 C_p$.

PROBLEM 3. Find $\overline{\Delta T\,\Delta P}$ (with variables V and T).

SOLUTION. $\overline{\Delta T\,\Delta P} = (T^2/C_v)(\partial P/\partial T)_V$.

PROBLEM 4. Find $\overline{\Delta V\,\Delta P}$ (with variables V and T).

SOLUTION. $\overline{\Delta V\,\Delta P} = -T$.

PROBLEM 5. Find $\overline{\Delta S\,\Delta V}$ (with variables V and T).

SOLUTION. $\overline{\Delta S\,\Delta V} = (\partial V/\partial T)_P T$.

PROBLEM 6. Find $\overline{\Delta S\,\Delta T}$ (with variables V and T).

SOLUTION. $\overline{\Delta S\,\Delta T} = T$.

PROBLEM 7. Find the mean square fluctuation deviation of a simple pendulum suspended vertically.

SOLUTION. Let l be the length of the pendulum, m its mass, and ϕ the angle of deviation from the vertical. The work $R_{\min}$ is here just the mechanical work done against gravity in the deviation of the pendulum; for small ϕ, $R_{\min} = \frac{1}{2}\, mg \cdot l\phi^2$. Hence

$$\overline{\phi^2} = T/mgl.$$

PROBLEM 8. Find the mean square fluctuation deviation of the points of a stretched string.

SOLUTION. Let l be the length of the string, and F its tension. Let us take a point at a distance x from one end of the string, and let y be its transverse displacement. To determine $\overline{y^2}$ we must consider the equilibrium form of the string when the displacement y of the point x is given; this consists of two straight segments from the fixed ends of the string to the point x, y. The work done in such a deformation of the string is

$$R_{\min} = F[\sqrt{(x^2+y^2)}-x]+F[\sqrt{\{(l-x)^2+y^2\}}-(l-x)] \cong \tfrac{1}{2}Fy^2\left(\frac{1}{x}+\frac{1}{l-x}\right).$$

Thus the mean square is

$$\overline{y^2} = (T/Fl)x(l-x).$$

PROBLEM 9. Determine the mean value of the product of the fluctuation displacements of two different points of the string.

SOLUTION. Let y_1, y_2 be the transverse movements of points at distances x_1, x_2 from one end of the string (with $x_2 > x_1$). The equilibrium form for given y_1, y_2 consists of three straight segments, and the work is

$$R_{\min} = \tfrac{1}{2}F\left(y_1^2\frac{x_2}{x_1(x_2-x_1)}+y_2^2\frac{l-x_1}{(l-x_2)(x_2-x_1)}-2y_1y_2\frac{1}{x_2-x_1}\right);$$

from formula (113.8) we then have

$$\overline{y_1y_2} = (T/Fl)x_1(l-x_2).$$

§115. Fluctuations in an ideal gas

The mean square fluctuation of the number of particles within some relatively small volume in an ordinary ideal gas is found by substituting $V = NT/P$ in formula (114.13). This gives the simple result

$$\overline{(\Delta N)^2} = N. \qquad (115.1)$$

The relative fluctuation of the number of particles is therefore just the reciprocal square root of the mean number of particles:

$$\frac{\sqrt{[\overline{(\Delta N)^2}]}}{N} = \frac{1}{\sqrt{N}}.$$

In order to calculate the fluctuation of the number of particles in an ideal Bose or Fermi gas, we must use formula (114.14), with the expression (55.5) substituted for N as a function of μ, T, V, obtained by integrating the corresponding distribution function. We shall not pause to write out here the fairly lengthy expressions which result, but simply note the following point. We have seen that, in a Bose gas at temperatures $T < T_0$ (see §59), the pressure is independent of the volume, i.e. the compressibility becomes infinite. According to formula (114.13) this would imply that the fluctuations of number of particles also become infinite. This means that, in calculating fluctuations in a gas obeying Bose statistics, the interaction between its particles cannot be neglected at low temperatures, however weak this interaction may be. When the interaction, which must exist in any actual gas, is taken into account, the resulting fluctuations are finite.

Next, let us examine fluctuations in the distribution of the gas particles over the various quantum states. We again consider the quantum states of the particles (including different states of their translational motion); let n_k be their occupation numbers.

Let us consider an assembly of n_k particles in the kth quantum state. Since this set of particles is statistically independent of the remaining particles in the gas (cf. §37), we can apply formula (114.14) to it:

$$\overline{(\Delta n_k)^2} = T\frac{\partial \bar{n}_k}{\partial \mu}. \tag{115.2}$$

For a Fermi gas we must substitute

$$\bar{n}_k = \frac{1}{e^{(\varepsilon_k-\mu)/T}+1}.$$

The differentiation gives

$$\overline{(\Delta n_k)^2} = \bar{n}_k(1-\bar{n}_k). \tag{115.3}$$

Similarly, for a Bose gas

$$\overline{(\Delta n_k)^2} = \bar{n}_k(1+\bar{n}_k). \tag{115.4}$$

For a Boltzmann gas the substitution $\bar{n}_k = e^{(\mu-\varepsilon_k)/T}$ naturally gives

$$\overline{(\Delta n_k)^2} = \bar{n}_k, \tag{115.5}$$

which is obtained from both (115.3) and (115.4) when $\bar{n}_k \ll 1$.

Summation of (115.3) or (115.4) over a group of G_j neighbouring states containing altogether $N_j = \sum n_k$ particles gives, by virtue of the statistical independence (already mentioned) of the fluctuations of the various n_k,

$$\overline{(\Delta N_j)^2} = G_j\bar{n}_j(1\mp\bar{n}_j) = \bar{N}_j(1\mp\bar{N}_j/G_j), \tag{115.6}$$

where $\bar{n}_j$ is the common value of the neighbouring $\bar{n}_k$, and $\bar{N}_j = \bar{n}_jG_j$.

These formulae can be applied, in particular, to black-body radiation (an equilibrium Bose gas of photons), for which we must put $\mu = 0$ in (115.4). Let us consider the set of quantum states for the photon (in a volume V) with neighbouring frequencies in a small interval $\Delta\omega_j$. The number of such states is $G_j = V\omega_j^2\Delta\omega_j/\pi^2c^3$; see (60.3). The total energy of the quanta in this frequency interval is $E_{\Delta\omega_j} = N_j\hbar\omega_j$. Multiplying (115.6) by $(\hbar\omega_j)^2$ and omitting the suffix j, we obtain the following expression for the fluctuation of the energy $E_{\Delta\omega}$ of black-body radiation in a given frequency interval $\Delta\omega$:

$$\overline{(\Delta E_{\Delta\omega})^2} = \hbar\omega\cdot E_{\Delta\omega}+\pi^2c^3(E_{\Delta\omega})^2/V\omega^2\Delta\omega, \tag{115.7}$$

a relation first derived by A. EINSTEIN (1924).

PROBLEM

Determine $\overline{(\Delta N)^2}$ for an electron gas at temperatures much lower than the degeneracy temperature.

SOLUTION. In calculating $(\partial N/\partial\mu)_{T,V}$ we can use the expression (56.3) for μ at absolute zero. A simple calculation gives

$$\overline{(\Delta N)^2} = \frac{3^{1/3}mT}{\pi^{4/3}\hbar^2}\left(\frac{N}{V}\right)^{1/3}V.$$

§116. Poisson's formula

Knowing the mean square fluctuation of the number of particles in a given volume of gas (115.1), we can write down the corresponding Gaussian probability distribution for fluctuations in this number of particles:

$$w(N)\,\mathrm{d}N = \frac{1}{\sqrt{(2\pi\bar{N})}} \exp\left\{-\frac{(N-\bar{N})^2}{2\bar{N}}\right\} \mathrm{d}N. \tag{116.1}$$

This formula is, however, valid only for small fluctuations: the deviation $N-\bar{N}$ must be small compared with the number $\bar{N}$ itself.

If the volume selected in the gas is sufficiently small, the number of particles in it is small, and we may also consider large fluctuations, where $N-\bar{N}$ becomes comparable with $\bar{N}$. This problem is meaningful only for a Boltzmann gas, since in a Fermi or Bose gas the probability of such fluctuations can become appreciable only in volumes so small that quantum fluctuations become important.

The solution of this problem is most simply found as follows. Let V_0 and N_0 be the total volume of the gas and the number of particles in it, and V a part of the volume, small compared with V_0. Since the gas is uniform, the probability that any given particle is in the volume V is obviously just the ratio V/V_0, and the probability that N given particles are simultaneously present in it is $(V/V_0)^N$. Similarly, the probability that a particle is not in the volume V is $(V_0-V)/V_0$, and the same probability for N_0-N given particles simultaneously is $[(V_0-V)/V_0]^{N_0-N}$. The probability w_N that the volume V contains N molecules in all is therefore given by

$$w_N = \frac{N_0!}{N!(N_0-N)!}\left(\frac{V}{V_0}\right)^N\left(1-\frac{V}{V_0}\right)^{N_0-N}, \tag{116.2}$$

where a factor has been included which gives the number of possible ways of choosing N out of N_0 particles.

In the case under consideration, $V \ll V_0$, and the number N, though it may differ considerably from its mean value $\bar{N}$, is of course assumed small compared with the total number N_0 of particles in the gas. Then we may put $N_0! \cong (N_0-N)!N_0^N$ and neglect N in the exponent N_0-N, obtaining

$$w_N = \frac{1}{N!}\left(\frac{N_0 V}{V_0}\right)^N\left(1-\frac{V}{V_0}\right)^{N_0}.$$

But N_0V/V_0 is just the mean number $\bar{N}$ of particles in the volume V. Hence

$$w_N = \frac{\bar{N}^N}{N!}\left(1-\frac{\bar{N}}{N_0}\right)^{N_0}.$$

Finally, using the well-known formula

$$\lim_{n\to\infty}\left(1-\frac{x}{n}\right)^n = e^{-x},$$

we replace $(1-\bar{N}/N_0)^{N_0}$ with N_0 large by $e^{-\bar{N}}$ and obtain the required probability distribution in the form†

$$w_N = \frac{\bar{N}^N e^{-\bar{N}}}{N!}. \tag{116.3}$$

This is called *Poisson's formula*. It is easily seen to satisfy the normalisation condition

$$\sum_{N=0}^{\infty} w_N = 1.$$

From this distribution we can calculate the mean square fluctuation of the number of particles:

$$\begin{aligned}\overline{N^2} &= \sum_{N=0}^{\infty} N^2 w_N = e^{-\bar{N}}\sum_{N=1}^{\infty}\frac{\bar{N}^N N}{(N-1)!} = e^{-\bar{N}}\sum_{N=2}^{\infty}\frac{\bar{N}^N}{(N-2)!}+e^{-\bar{N}}\sum_{N=1}^{\infty}\frac{\bar{N}^N}{(N-1)!}\\ &= \bar{N}^2+\bar{N}.\end{aligned}$$

Hence we find as before

$$\overline{(\Delta N)^2} = \overline{N^2}-\bar{N}^2 = \bar{N}, \tag{116.4}$$

and the mean square fluctuation of the number of particles is equal to $\bar{N}$ for any value of $\bar{N}$, and not only for large values.

We may note that formula (116.3) can also be derived directly from the Gibbs distribution. According to the latter, the distribution of N gas particles, considered simultaneously, among various quantum states is given by the expression $\exp\{(\Omega+\mu N-\sum\varepsilon_k)/T\}$, where $\sum\varepsilon_k$ is the sum of the energies of the individual particles. To derive the required probability w_N we must sum this expression over all particle states "belonging" to a given volume V. If we sum over the states of each particle independently, the result must be divided by $N!$ (cf. §41), giving

$$w_N = \frac{e^{\Omega/T}}{N!}\left(\sum_k e^{(\mu-\varepsilon_k)/T}\right)^N.$$

The sum is just the mean number of particles in the volume considered:

$$\sum_k e^{(\mu-\varepsilon_k)/T} = \bar{N}.$$

† For small fluctuations ($|N-\bar{N}| \ll \bar{N}$ with $\bar{N}$ large), this formula naturally becomes (116.1). This is easily seen by using STIRLING's asymptotic expression for the factorial of a large number N, $N! = \sqrt{(2\pi N)}\cdot N^N e^{-N}$, and expanding $\log w_n$ in powers of $N-\bar{N}$.

Thus we find

$$w_N = \text{constant} \times \bar{N}^N / N!,$$

and the normalisation condition then shows that the constant is $e^{-\bar{N}}$,† giving again formula (116.3).

§117. Fluctuations in solutions

The fluctuations of thermodynamic quantities in solutions can be calculated by the same method as that used in §114 for fluctuations in bodies consisting of identical particles. The calculations are considerably simplified by the following argument.

Let us take some small part of the solution, containing a given number N of solvent molecules, and try to calculate the mean fluctuation of the number n of solute molecules in that part of the solution or, what is the same thing, the fluctuation of the concentration $c = n/N$ in that part. To do so, we must consider the most complete equilibrium of the solution that is possible for a given non-equilibrium value of n; cf. the first footnote to §113. Taking a given value of the concentration does not affect the establishment of equilibrium between the small part considered and the remainder of the solution as regards exchange of energy between them or change in their volumes. The former means (see §9) that the temperature remains constant throughout the solution; the latter means that the pressure remains constant throughout the solution (§12). Thus to calculate the mean square $\overline{(\Delta c)^2}$, it is sufficient to consider the fluctuations of concentration occurring at constant temperature and pressure.

This fact in itself signifies that the fluctuations of concentration are statistically independent of those of temperature and pressure, i.e.‡

$$\overline{\Delta T \Delta c} = 0, \qquad \overline{\Delta c \Delta P} = 0. \tag{117.1}$$

The minimum work necessary to change the number n by Δn at constant pressure and temperature is, by (98.1), $R_{\min} = \Delta\Phi - \mu' \Delta n$, where μ' is the chemical potential of the solute. Expanding $\Delta\Phi$ in powers of Δn, we have

$$\Delta\Phi \cong \left(\frac{\partial \Phi}{\partial n}\right)_{P,T} \Delta n + \left(\frac{\partial^2 \Phi}{\partial n^2}\right)_{P,T} \cdot \tfrac{1}{2}(\Delta n)^2 = \mu' \Delta n + \left(\frac{\partial \mu'}{\partial n}\right)_{P,T} \cdot \tfrac{1}{2}(\Delta n)^2,$$

† That is, $\Omega = -PV = -\bar{N}T$, in accordance with the equation of state of an ideal gas.

‡ This may be more rigorously proved by the method indicated in the third footnote to §114. Using the formula $dE = T dS - P dV + \mu' dn$ (with N constant), we can rewrite (98.1) as

$$dR_{\min} = (T - T_0)\, dS - (P - P_0)\, dV + (\mu' - \mu'_0)\, dn.$$

Thus, if we take $x_1 = \Delta S$, $x_2 = \Delta V$, $x_3 = \Delta n$, the corresponding X_i are $X_1 = \Delta T/T$, $X_2 = -\Delta P/T$, $X_3 = \Delta\mu'/T$. The equations (117.1) then follow, since $\overline{x_3 X_1} = 0$, $\overline{x_3 X_2} = 0$.

and hence

$$R_{\min} = \tfrac{1}{2}\left(\frac{\partial \mu'}{\partial n}\right)_{P,\,T} (\Delta n)^2.$$

Substituting this expression in the general formula (114.1) and comparing with the Gaussian distribution formula (112.5), we obtain for the required square of the fluctuation of the number n

$$\overline{(\Delta n)^2} = \frac{T}{(\partial \mu'/\partial n)_{P,\,T}}, \qquad (117.2)$$

or, dividing by N^2, for the mean square fluctuation of concentration

$$\overline{(\Delta c)^2} = \frac{T}{N(\partial \mu'/\partial c)_{P,\,T}}. \qquad (117.3)$$

The latter expression is inversely proportional to the amount of matter (N) in the small part considered, as it should be (see the discussion following (114.11)).

For weak solutions, $\partial \mu'/\partial n = T/n$, and formula (117.2) gives

$$\overline{(\Delta n)^2} = n. \qquad (117.4)$$

It should be noticed that there is a complete analogy (as was to be expected) with formula (115.1) for the fluctuations of the number of particles in an ideal gas.

§118. Correlations of fluctuations

The statement that in a homogeneous isotropic substance (gas or liquid) all positions of the particles in space are equally probable applies to any given particle on condition that all other particles can occupy arbitrary positions, and is not, of course, in contradiction with the fact that the interaction between different particles must in fact cause some correlation in their positions. The latter means that, if we consider, say, two particles simultaneously, then for a given position of one particle the various positions of the other will not be equally probable.

To simplify the formulae below we shall consider only a monatomic substance, in which the position of each particle is fully determined by its three co-ordinates.

Let $n\,dV$ denote the number of particles present in a volume element dV at a given instant. Since dV is infinitesimal, it cannot contain more than one particle at a time; the probability of finding two particles in it at once is an infinitesimal quantity of a higher order. The mean number of particles $\bar{n}\,dV$ is therefore also the probability that a particle is present in dV.

Let us consider the mean value

$$\overline{(n_1-\bar{n}_1)(n_2-\bar{n}_2)} = \overline{n_1 n_2}-(\bar{n})^2, \tag{118.1}$$

where n_1 and n_2 are the particle number densities $n(\mathbf{r})$ at two different points in space, and $\bar{n}$ denotes the mean density, which is the same at every point ($\bar{n}_1 = \bar{n}_2 \equiv \bar{n}$) owing to the homogeneity of the body. If there were no correlation between the positions of different particles, we should have $\overline{n_1 n_2} = \bar{n}_1 \cdot \bar{n}_2 = (\bar{n})^2$, and the mean value (118.1) would be zero. Thus this quantity can serve as a measure of the correlation.

Let $n_{12}\,\mathrm{d}V_2$ denote the probability that there is a particle in the volume element $\mathrm{d}V_2$ when there is one in $\mathrm{d}V_1$; n_{12} is a function of the magnitude $r = |\mathbf{r}_2-\mathbf{r}_1|$ of the distance between the two elements.

Since, as already mentioned, the number $n\,\mathrm{d}V$ is 0 or 1, it is evident that the mean value

$$\overline{n_1\,\mathrm{d}V_1 \cdot n_2\,\mathrm{d}V_2} = \bar{n}_1\,\mathrm{d}V_1 \cdot n_{12}\,\mathrm{d}V_2,$$

or

$$\overline{n_1 n_2} = n_{12}\bar{n}.$$

In this relation, valid when $\mathbf{r}_1 \neq \mathbf{r}_2$, we cannot, however, go to the limit $\mathbf{r}_2 \to \mathbf{r}_1$, since the derivation ignores the fact that, if the points 1 and 2 coincide, a particle in $\mathrm{d}V_1$ is also in $\mathrm{d}V_2$. A relation which takes account of this is clearly

$$\overline{n_1 n_2} = \bar{n} n_{12}+\bar{n}\delta(\mathbf{r}_2-\mathbf{r}_1). \tag{118.2}$$

For let us take a small volume ΔV, multiply (118.2) by $\mathrm{d}V_1\,\mathrm{d}V_2$, and integrate over ΔV. The term $\bar{n}n_{12}$ then gives a second-order small quantity (proportional to $(\Delta V)^2$); the term containing the delta function gives ΔV, i.e. a first-order quantity. Thus we obtain

$$\overline{\left(\int_{\Delta V} n\,\mathrm{d}V\right)^2} = \bar{n}\Delta V,$$

as we should, since only 0 or 1 particle can be in the small volume, as far as first-order quantities. Substitution of (118.2) in (118.1) gives

$$\overline{(n_1-\bar{n}_1)(n_2-\bar{n}_2)} = \bar{n}\delta(\mathbf{r}_2-\mathbf{r}_1)+\bar{n}\nu(r), \tag{118.3}$$

where

$$\nu(r) = n_{12}-\bar{n} \tag{118.4}$$

is a function called the *correlation function*. It is evident that the correlation must tend to zero when the distance r increases to infinity:

$$\nu(\infty) = 0. \tag{118.5}$$

Let us now consider a finite volume V in the body, multiply equation (118.3) by $dV_1\,dV_2$ and integrate over V_1 and V_2. Since

$$\int (n_1-\overline{n_1})\,dV_1 = \int (n_2-\overline{n_2})\,dV_2 = N-\bar{N} \equiv \Delta N,$$

where N is the total number of particles in the volume V (so that $\bar{n}V = \bar{N}$), we find

$$\iint \nu(r)\,dV_1\,dV_2 = \frac{\overline{(\Delta N)^2}}{\bar{n}} - V.$$

Changing from the integration over V_1 and V_2 to one over V_1, say, and the relative co-ordinates $\mathbf{r} = \mathbf{r}_2-\mathbf{r}_1$ (the product of whose differentials we denote by dV), and bearing in mind that ν depends only on r, we have finally the following expression for the integral of the correlation function:

$$\int \nu\,dV = \frac{\overline{(\Delta N)^2}}{\bar{N}} - 1. \tag{118.6}$$

Thus the integral of the correlation function over a certain volume is related to the mean square fluctuation of the total number of particles in that volume. Using for the latter the thermodynamic formula (114.13), we can express the integral in terms of the thermodynamic quantities:

$$\int \nu\,dV = -\frac{TN}{V^2}\left(\frac{\partial V}{\partial P}\right)_T - 1. \tag{118.7}$$

In an ordinary (classical) ideal gas this gives $\int \nu\,dV = 0$, as it should: it is evident that in an ideal gas treated by classical mechanics there is no correlation between the positions of different particles, since the particles of an ideal gas are assumed not to interact with one another.

On the other hand, in a liquid (at temperatures not close to the critical point) the first term in (118.7) is small compared with unity, because the compressibility of a liquid is small. In this case we can write $\int \nu\,dV \cong -1$. This value of the integral of the correlation function corresponds in a sense to the mutual impenetrability of the liquid particles, regarded as closely packed solid spheres.

Next, let us multiply both sides of (118.3) by $e^{-i\mathbf{k}\cdot\mathbf{r}} = e^{-i\mathbf{k}\cdot(\mathbf{r}_2-\mathbf{r}_1)}$ and again integrate over V_1 and V_2. The result is

$$\overline{\iint (n_1-\bar{n})(n_2-\bar{n})e^{i\mathbf{k}\cdot(\mathbf{r}_1-\mathbf{r}_2)}\,dV_1\,dV_2} = \bar{N}+\bar{N}\int \nu e^{-i\mathbf{k}\cdot\mathbf{r}}\,dV,$$

or

$$\overline{\left|\int (n-\bar{n})e^{-i\mathbf{k}\cdot\mathbf{r}}\,dV\right|^2} = \bar{n}V\left(1+\int \nu e^{-i\mathbf{k}\cdot\mathbf{r}}\,dV\right). \tag{118.8}$$

This relation gives the Fourier components of the correlation function in terms of the mean squares of the Fourier components of the density n.

§119. Fluctuations at the critical point

At the critical point, the compressibility $(\partial V/\partial P)_T$ and the specific heat C_p of a substance become infinite (§84). The expressions (114.7) and (114.10) for the fluctuations of volume (i.e. of density) and of entropy likewise formally become infinite, but the fluctuations of temperature and of pressure remain finite. This means that at the critical point the fluctuations of density and entropy become anomalously large, and to calculate them $R_{\min}$ in formula (114.1) must be expanded as far as terms of a higher order of smallness than the second-order terms, which in this case vanish.† Here we shall discuss in detail the fluctuations of density near the critical point (ORNSTEIN and ZERNIKE 1917).

Since the fluctuations of density and temperature are statistically independent, the temperature may be regarded as constant in considering the fluctuations of density. The total volume of the body is also constant, by definition. Under these conditions the minimum work $R_{\min}$ is equal to the change ΔF_t in the total free energy of the body during the fluctuation, so that the probability of the fluctuation may be written as

$$w \sim e^{-\Delta F_t/T}. \tag{119.1}$$

The total free energy of the body can be written as the integral $F_t = \int F \, \mathrm{d}V$ over the whole volume of the body, F denoting the free energy per unit volume. Let $\bar{F}$ be the mean value of F, constant throughout the body. The fluctuation causes F to vary from point to point in the body, like the density, and

$$\Delta F_t = \int (F - \bar{F}) \, \mathrm{d}V. \tag{119.2}$$

Let the particle number density be n, its mean value $\bar{n}$, and let us expand $F - \bar{F}$ in powers of $n - \bar{n}$ at constant temperature.

The first term of the expansion is proportional to $n - \bar{n}$ and gives zero on integration over the volume, since the total number of particles in the body is constant: $\int n \, \mathrm{d}V = \int \bar{n} \, \mathrm{d}V$. The second-order term is of the form $\frac{1}{2}a(n-\bar{n})^2$, the positive coefficient a becoming zero at the critical point itself and being a small quantity near that point.‡ The coefficient in the third-order term is also small near the critical point (since both $\partial P/\partial n$ and $\partial^2 P/\partial n^2$ are zero at that point), and so the fourth-order term would have to be taken into account.

† The same is true of the fluctuations of concentration in solutions: at points on the critical line, $(\partial \mu'/\partial c)_{P,T} = 0$ (§98), and the expression (117.3) becomes infinite.

‡ The derivative $(\partial F/\partial n)_T$ is the chemical potential, and so the second derivative is

$$a = \left(\frac{\partial^2 F}{\partial n^2}\right)_T = \left(\frac{\partial \mu}{\partial n}\right)_T = \frac{1}{n}\left(\frac{\partial P}{\partial n}\right)_T.$$

In fact, however, the expansion of $F-\bar{F}$ contains terms of another type which are larger.

The reason is that so far we have always considered the thermodynamic quantities for homogeneous bodies. In an inhomogeneous body, the expansion of F may contain not only various powers of the density itself, but also those of its successive derivatives with respect to the co-ordinates. Because of the isotropy of the body, the first derivatives can appear in the expansion in terms of density only through the scalar combination $(\nabla n)^2$, and the second derivatives only in the Laplacian $\triangle n$. The integral over the volume of a term constant $\times \triangle n$ becomes an integral over the surface of the body, and represents a surface effect of no importance here. The integral of a term $n\triangle n$ is equivalent to that of $(\nabla n)^2$. Thus without loss of generality we can put

$$F-\bar{F} = \tfrac{1}{2}a(n-\bar{n})^2+\tfrac{1}{2}b(\nabla n)^2, \tag{119.3}$$

where b is a positive constant; if $b < 0$ the free energy could not have a minimum corresponding to n = constant. The quantity b need not be zero at the critical point and so is not small near that point.

The calculation of the mean fluctuations of density in particular small regions of the body is of relatively little interest; since (119.3) includes a term in the derivatives of the density, these fluctuations will depend on the shape of the region as well as on its size.† The problem of the fluctuations of the Fourier components of density near the critical point is of much greater interest.

Let us expand $n-\bar{n}$ as a Fourier series within the volume V of the body, writing it as

$$n-\bar{n} = \sum_{\mathbf{k}} n_{\mathbf{k}} e^{i\mathbf{k}\cdot\mathbf{r}}; \tag{119.4}$$

the components of the vector $\mathbf{k}$ take both positive and negative values, and the coefficients

$$n_{\mathbf{k}} = \frac{1}{V}\int (n-\bar{n})e^{-i\mathbf{k}\cdot\mathbf{r}}\,\mathrm{d}V$$

are related by $n_{-\mathbf{k}} = n_{\mathbf{k}}^*$, which follows from the fact that $n-\bar{n}$ is real. Substituting (119.4) in (119.3) and integrating over the volume, we have

$$\Delta F_t = \tfrac{1}{2}V\sum_{\mathbf{k}}(a+bk^2)\,|n_{\mathbf{k}}|^2. \tag{119.5}$$

Each term in this sum depends on only one of the $n_{\mathbf{k}}$; the fluctuations of the different $n_{\mathbf{k}}$ are therefore statistically independent. Each square $|n_{\mathbf{k}}|^2$

† At the critical point itself $a = 0$, and only the second term remains in (119.3). If the density undergoes a fluctuation in a region of linear dimensions $\sim l$, then $F-\bar{F} \sim b[(n-\bar{n})/l]^2$ and $\Delta F_t \sim b(n-\bar{n})^2 l$. The mean square fluctuation of density is therefore inversely proportional to l: $\overline{(\Delta n)^2} \sim 1/l$, i.e. inversely proportional to the cube root of the volume of the region, whereas away from the critical point the mean square fluctuation of density decreases in inverse proportion to the volume itself.

appears in the sum (119.5) twice (for $\pm\mathbf{k}$), so that the probability distribution for its fluctuations is

$$w \sim \exp\left\{-\frac{V}{T}(a+bk^2)\,|n_{\mathbf{k}}|^2\right\}.$$

Since $|n_{\mathbf{k}}|^2$ is the sum of the squares of two independent quantities ($n_{\mathbf{k}}$ being complex), we therefore find as the required mean square fluctuation

$$\overline{|n_{\mathbf{k}}|^2} = T/V(a+bk^2). \tag{119.6}$$

It should be emphasised that these formulae are applicable only when the magnitude of the wave vector k is not too large; otherwise, the expansion (119.3) can not be restricted to terms containing only the lowest derivatives of the density with respect to the co-ordinates.

The result derived above enables us to calculate the correlation function $\nu(r)$ (§118) near the critical point. According to the general formula (118.8) we have

$$\int \nu e^{-i\mathbf{k}\cdot\mathbf{r}}\,\mathrm{d}V = \frac{V}{\bar{n}}\overline{|n_{\mathbf{k}}|^2} - 1$$
$$= \frac{T}{\bar{n}(a+bk^2)} - 1.$$

The first term on the right is in general large compared with unity, since both a and k are assumed small. We can therefore write

$$\int \nu e^{-i\mathbf{k}\cdot\mathbf{r}}\,\mathrm{d}V = T/\bar{n}(a+bk^2), \tag{119.7}$$

and hence, by an inverse Fourier transformation, obtain†

$$\nu(r) = \frac{T}{4\pi\bar{n}b}\cdot\frac{1}{r}e^{-\sqrt{(a/b)}r}. \tag{119.8}$$

The coefficient of r in the exponent is small, since a is small. At the critical point, $a = 0$, and so the exponential factor becomes equal to unity:

$$\nu(r) = \frac{T}{4\pi\bar{n}b}\cdot\frac{1}{r}. \tag{119.9}$$

† If

$$\int \phi e^{-i\mathbf{k}\cdot\mathbf{r}}\,\mathrm{d}V = 4\pi(\varkappa^2+k^2), \tag{1}$$

the function ϕ is

$$\phi = e^{-\varkappa r}/r. \tag{2}$$

This is most simply seen by noting that the function (2) satisfies the differential equation

$$\triangle\phi - \varkappa^2\phi = -4\pi\delta(\mathbf{r}).$$

Multiplying both sides of this equation by $e^{-i\mathbf{k}\cdot\mathbf{r}}$ and integrating over all space (using a repeated integration by parts in the first term on the left), we obtain (1).

Thus near the critical point the correlation between the positions of different particles in the substance decreases very slowly with increasing distance, i.e. becomes much stronger than under ordinary conditions, where it is practically zero even at intermolecular distances.

Concerning the whole theory of fluctuations near the critical point as given in this section, the same reservation must be made as in §84: the proofs given here assume that there is no important singularity in the thermodynamic quantities at the critical point, and there can therefore be no certainty that the results obtained are correct.

§120. Correlations of fluctuations in an ideal gas

As already mentioned in §118, in a classical ideal gas there is no correlation between the positions of the various particles. In quantum mechanics, however, such a correlation exists because identical particles in an ideal gas "interact" indirectly owing to the principle of symmetry of wave functions. Correlations in a Fermi gas were first considered by V. Fursov (1937), and those in a Bose gas by A. Galanin (1940).

To simplify the formulae below we shall at first assume that the particles have no spin. Taking account of spin will not essentially affect the results.

The problem of calculating the correlation function can be most simply solved by the second quantisation method. In accordance with this method† we define the normalised wave functions

$$\psi_{\mathbf{k}} = \frac{1}{\sqrt{V}} e^{i\mathbf{k}\cdot\mathbf{r}}, \tag{120.1}$$

which describe states of a gas particle moving freely in a volume V with momentum $\mathbf{p} = \hbar\mathbf{k}$; in this section the wave vectors $\mathbf{k}$ will be used instead of the momenta $\mathbf{p}$. For a finite volume V the wave vector $\mathbf{k}$ takes an infinite set of discrete values, the intervals between which are very small, however, when V is sufficiently large.

Next, we define operators $\hat{a}_{\mathbf{k}}$ and $\hat{a}_{\mathbf{k}}^+$ which respectively decrease and increase by unity the numbers $n_{\mathbf{k}}$ of particles in the various quantum states $\psi_{\mathbf{k}}$, and the operators

$$\hat{\Psi}(\mathbf{r}) = \sum_{\mathbf{k}} \psi_{\mathbf{k}}(\mathbf{r})\hat{a}_{\mathbf{k}}, \qquad \hat{\Psi}^+(\mathbf{r}) = \sum_{\mathbf{k}} \psi_{\mathbf{k}}^*(\mathbf{r})\hat{a}_{\mathbf{k}}^+,$$

which respectively "remove" and "add" one particle at the point $\mathbf{r}$ in the system. The operator $\hat{\Psi}^+(\mathbf{r})\hat{\Psi}(\mathbf{r})\,dV$ is the operator of the number of particles with co-ordinates in $dx\,dy\,dz = dV$. Hence $\hat{\Psi}^+\hat{\Psi}$ can be regarded as an

† See *Quantum Mechanics*, §§64, 65.

operator $\hat{n}$, which in the second quantisation method represents the density of distribution of the gas particles in space:

$$\hat{n} = \hat{\Psi}^{+}(\mathbf{r})\hat{\Psi}(\mathbf{r})$$
$$= \sum_{\mathbf{k}}\sum_{\mathbf{k}'} \hat{a}_{\mathbf{k}}^{+}\hat{a}_{\mathbf{k}'}\psi_{\mathbf{k}}^{*}\psi_{\mathbf{k}'}. \qquad (120.2)$$

Here the summation with respect to $\mathbf{k}$ and $\mathbf{k}'$ is over all their possible values. It is easy to see that the "diagonal" terms of the sum ($\mathbf{k} = \mathbf{k}'$) give just the mean density $\bar{n}$. For, since the operator $\hat{a}_{\mathbf{k}}^{+}\hat{a}_{\mathbf{k}}$ is simply the number $n_{\mathbf{k}}$ of particles in the quantum state considered, and from (120.1) $|\psi_{\mathbf{k}}|^2 = 1/V$, these terms are equal to

$$\sum_{\mathbf{k}} \hat{a}_{\mathbf{k}}^{+}\hat{a}_{\mathbf{k}}|\psi_{\mathbf{k}}|^2 = \frac{1}{V}\sum_{\mathbf{k}} n_{\mathbf{k}} = N/V = n,$$

where N is the total number of particles in the volume V.

We can therefore write

$$\hat{n}-\bar{n} = \sum_{\mathbf{k}}\sum_{\mathbf{k}'}{}' \hat{a}_{\mathbf{k}}^{+}\hat{a}_{\mathbf{k}'}\psi_{\mathbf{k}}^{*}\psi_{\mathbf{k}'}, \qquad (120.3)$$

where the prime to the summation sign denotes that the term with $\mathbf{k}' = \mathbf{k}$ is to be omitted. Using this expression, we can easily calculate the mean value required, $\overline{(n_1-\bar{n})(n_2-\bar{n})}$.

This mean value is calculated in two stages. First of all, the quantum averaging, i.e. that with respect to the quantum states of the particles, is to be carried out. This amounts to taking the corresponding diagonal matrix element of the quantity concerned. Multiplying together the two operators (120.3) which belong to two different points $\mathbf{r}_1$ and $\mathbf{r}_2$, we obtain a sum of terms containing various products of the operators $\hat{a}_{\mathbf{k}}$ and $\hat{a}_{\mathbf{k}}^{+}$ taken four at a time. But among these products only those which contain two pairs of operators $\hat{a}_{\mathbf{k}}$, $\hat{a}_{\mathbf{k}}^{+}$ with the same suffix have diagonal matrix elements, i.e. the relevant terms are

$$\sum_{\mathbf{k}}\sum_{\mathbf{k}'}{}' \hat{a}_{\mathbf{k}}^{+}\hat{a}_{\mathbf{k}'}\hat{a}_{\mathbf{k}'}^{+}\hat{a}_{\mathbf{k}}\psi_{\mathbf{k}}^{*}(\mathbf{r}_1)\psi_{\mathbf{k}'}(\mathbf{r}_1)\psi_{\mathbf{k}'}^{*}(\mathbf{r}_2)\psi_{\mathbf{k}}(\mathbf{r}_2). \qquad (120.4)$$

These terms are diagonal matrices, and

$$\hat{a}_{\mathbf{k}'}\hat{a}_{\mathbf{k}'}^{+} = 1 \mp n_{\mathbf{k}'}, \qquad \hat{a}_{\mathbf{k}}^{+}\hat{a}_{\mathbf{k}} = n_{\mathbf{k}};$$

here and henceforward the upper sign refers to the case of Fermi statistics and the lower sign to that of Bose statistics. Substituting also the functions $\psi_{\mathbf{k}}$ from (120.1), we obtain

$$\frac{1}{V^2}\sum_{\mathbf{k}}\sum_{\mathbf{k}'}{}' (1 \mp n_{\mathbf{k}'})n_{\mathbf{k}}e^{i(\mathbf{k}-\mathbf{k}')\cdot(\mathbf{r}_2-\mathbf{r}_1)}.$$

This expression must now be averaged in the statistical sense, i.e. over the equilibrium distribution of the particles among the various quantum states. Since particles in different quantum states behave quite independently of one another, the numbers $n_{\mathbf{k}}$ and $n_{\mathbf{k}'}$ are averaged independently, i.e.

$$\overline{(1 \mp n_{\mathbf{k}'})n_{\mathbf{k}}} = (1 \mp \overline{n_{\mathbf{k}'}})\, \overline{n_{\mathbf{k}}}.$$

The mean values $\overline{n_{\mathbf{k}}}$ are determined by the Fermi or Bose distribution function.

Thus we obtain the following expression for the required mean value:

$$\overline{(n_1-\bar{n})(n_2-\bar{n})} = \frac{1}{V^2}\sum_{\mathbf{k}}\sum_{\mathbf{k}'}{}' (1 \mp \overline{n_{\mathbf{k}'}})\, \overline{n_{\mathbf{k}}} e^{i(\mathbf{k}-\mathbf{k}')\cdot(\mathbf{r}_2-\mathbf{r}_1)}. \tag{120.5}$$

Since, when the volume V is not too small, the wave vector $\mathbf{k}$ takes a practically continuous series of values, we can change from summation to integration, multiplying the expression (120.5) by

$$\frac{V\,\mathrm{d}^3k}{(2\pi)^3}\cdot\frac{V\,\mathrm{d}^3k'}{(2\pi)^3}.$$

The integral form of (120.5) separates into two parts: the first is†

$$\begin{aligned}\frac{1}{(2\pi)^6}\iint \overline{n_{\mathbf{k}}} e^{i(\mathbf{k}-\mathbf{k}')\cdot(\mathbf{r}_2-\mathbf{r}_1)}\,\mathrm{d}^3k'\,\mathrm{d}^3k &= \frac{1}{(2\pi)^3}\int e^{i\mathbf{k}\cdot(\mathbf{r}_2-\mathbf{r}_1)}\,\overline{n_{\mathbf{k}}}\,\delta(\mathbf{r}_2-\mathbf{r}_1)\,\mathrm{d}^3k \\ &= \frac{1}{(2\pi)^3}\delta(\mathbf{r}_2-\mathbf{r}_1)\int \overline{n_{\mathbf{k}}}\,\mathrm{d}^3k = \bar{n}\delta(\mathbf{r}_2-\mathbf{r}_1).\end{aligned}$$

This is just the first term in (118.3). The correlation function (the second term in (118.3)) is therefore

$$\begin{aligned}\nu(r) &= \mp\frac{1}{\bar{n}(2\pi)^6}\iint e^{i(\mathbf{k}-\mathbf{k}')\cdot\mathbf{r}}\overline{n_{\mathbf{k}}}\,\overline{n_{\mathbf{k}'}}\,\mathrm{d}^3k\,\mathrm{d}^3k' \\ &= \mp\frac{1}{\bar{n}(2\pi)^6}\left|\int e^{i\mathbf{k}\cdot\mathbf{r}}\overline{n_{\mathbf{k}}}\,\mathrm{d}^3k\right|^2.\end{aligned} \tag{120.6}$$

If the spin of the particles is taken into account throughout, formula (120.2) for the operator of the number density of particles in space must be written as

$$\hat{n} = \sum_{\sigma}\sum_{\mathbf{k}}\sum_{\mathbf{k}'} \hat{a}_{\mathbf{k}\sigma}{}^{+}\hat{a}_{\mathbf{k}'\sigma}\psi_{\mathbf{k}\sigma}{}^{*}\psi_{\mathbf{k}'\sigma},$$

where σ is the spin projection. Accordingly the expression (120.4) must also be summed over the spin variable σ, and the right-hand sides of formulae

† Here we use the formula

$$\int e^{i\mathbf{k}\cdot\mathbf{r}}\mathrm{d}^3k = (2\pi)^3\delta(\mathbf{r}).$$

(120.5) and (120.6) are therefore multiplied by $g = 2s+1$, with $\overline{n_{\mathbf{k}}}$ the mean number of particles in the quantum state with a given value of σ, i.e.

$$\overline{n_{\mathbf{k}}} = \frac{1}{e^{(\varepsilon-\mu)/T} \pm 1}. \qquad (120.7)$$

Thus we have finally the following formula for the correlation function:

$$\nu(r) = \mp \frac{g}{\bar{n}(2\pi)^6} \left| \int \frac{e^{i\mathbf{k}\cdot\mathbf{r}}\, d^3k}{e^{(\varepsilon-\mu)/T} \pm 1} \right|^2. \qquad (120.8)$$

Integration over the directions of $\mathbf{k}$ gives

$$\nu(r) = \mp \frac{g}{4\pi^4 \bar{n} r^2} \left| \int_0^\infty \frac{\sin kr \cdot k\, dk}{e^{(\varepsilon-\mu)/T} \pm 1} \right|^2. \qquad (120.9)$$

We may also state a formula for the mean squares of the Fourier components of the density fluctuations; this is easily obtained by substituting $\nu(r)$ from (120.8) in the general formula (118.8), and integrating over the co-ordinates:

$$\overline{\left| \int (n-\bar{n}) e^{-i\mathbf{f}\cdot\mathbf{r}} dV \right|^2} = \frac{gV}{(2\pi)^3} \int \overline{n_{\mathbf{k}}}(1 \mp \overline{n_{\mathbf{k}+\mathbf{f}}})\, d^3k. \qquad (120.10)$$

Formula (120.8) shows first of all that in a Fermi gas $\nu(r) < 0$, but in a Bose gas $\nu(r) > 0$. In other words, in a Bose gas the presence of a particle at some point increases the probability that another particle is near that point, i.e. the particles "attract" one another; in a Fermi gas they correspondingly "repel" one another (cf. the end of §55).

If we go to the limit of classical mechanics ($\hbar \to 0$), the correlation function tends to zero in accordance with the discussion at the beginning of this section; for, as $\hbar \to 0$, the frequency of the oscillating factor $e^{i\mathbf{k}\cdot\mathbf{r}} = e^{i\mathbf{p}\cdot\mathbf{r}/\hbar}$ in the integrand in (120.8) increases without limit and the integral tends to zero.†

As $r \to 0$ the function $\nu(r)$ tends to a finite limit: since

$$\frac{g}{(2\pi)^3} \int \overline{n_{\mathbf{k}}}\, d^3k = \bar{n},$$

(120.8) shows that

$$\nu(0) = \mp \bar{n}/g. \qquad (120.11)$$

Let us apply the formulae derived above to a degenerate Fermi gas at absolute zero. In this case the distribution function $\overline{n_{\mathbf{k}}} = 1$ for $k < k_0$ and 0 for $k > k_0$, where $k_0 = p_0/\hbar = (6\pi^2\bar{n}/g)^{1/3}$ is the limiting momentum of the

† To avoid misunderstanding (since it might seem that $\hbar$ does not appear in (120.8) and (120.9)) it must be remembered that $\varepsilon = p^2/2m = \hbar^2k^2/2m$.

Fermi distribution. We therefore have from (120.9)

$$\nu(r) = -\frac{g}{4\bar{n}\pi^4 r^2}\left|\int_0^{k_0} k \sin kr \cdot dk\right|^2.$$

We shall consider only distances which are greater than a certain value, assuming in fact that $k_0 r \gg 1$. Accordingly, we retain only the term in the lowest power of $1/r$ in the integral, obtaining

$$\nu(r) = -\frac{3}{2\pi^2 k_0 r^4}\cos^2 k_0 r. \tag{120.12}$$

If the rapidly varying squared cosine is averaged, this gives

$$\nu(r) = -\frac{3}{4\pi^2 k_0 r^4}. \tag{120.13}$$

Thus the correlation function decreases inversely as the fourth power of the distance.

PROBLEMS

PROBLEM 1. Determine the mean squares of the Fourier components (with small wave vectors) of the fluctuations of density in a Fermi gas at absolute zero.

SOLUTION. The integrand in (120.10) is non-zero (and equal to unity) only at points where $\overline{n_\mathbf{k}} = 1$, $\overline{n_{\mathbf{k}+\mathbf{f}}} = 0$, i.e. points in a sphere of radius k_0 centred at the origin which at the same time are not in a sphere of the same radius centred at **f**. A calculation of the volume of this region when $f \ll k_0$ gives

$$\overline{\left|\int (n-\bar{n})e^{i\mathbf{f}\cdot\mathbf{r}}\,dV\right|^2} = \frac{g\pi f k_0^2}{(2\pi)^3}V = 3fN/4k_0.$$

PROBLEM 2. Determine the correlation function for a Fermi gas at temperatures small compared with the degeneracy temperature.

SOLUTION. In the integral in (120.9) we put $\mu \cong \varepsilon_0 = \hbar^2 k_0^2/2m$ and transform it as follows:

$$I = \int_0^\infty \frac{k \sin kr \cdot dk}{e^{(\varepsilon-\varepsilon_0)/T}+1} = -\frac{\partial}{\partial r}\int_0^\infty \frac{\cos kr \cdot dk}{e^{(\varepsilon-\varepsilon_0)/T}+1}$$

$$= \frac{\partial}{\partial r}\int_{k=0}^\infty \frac{\sin kr}{r}\,d\left(\frac{1}{e^{(\varepsilon-\varepsilon_0)/T}+1}\right).$$

With the variable $x = \hbar^2 k_0(k-k_0)/mT$, taking into account the smallness of T and the rapid decrease of the integrand as $|x|$ increases, we can write

$$I = \frac{\partial}{\partial r}\int_{x=-\infty}^\infty \frac{1}{r}\sin(k_0 r + \lambda x r)\,d\left(\frac{1}{e^x+1}\right)$$

$$= \frac{\partial}{\partial r}\left\{\frac{\sin k_0 r}{r}\int_{x=-\infty}^\infty e^{i\lambda x r}\,d\left(\frac{1}{e^x+1}\right)\right\},$$

where $\lambda = mT/\hbar^2 k_0$. The resulting integral is transformed by the substitution $1/(e^x+1) = u$ into EULER's beta integral, giving

$$I = \frac{\partial}{\partial r}\left\{\frac{\pi\lambda}{\sinh(\pi\lambda r)} \sin k_0 r\right\}.$$

For distances $r \gg 1/k_0$, averaging the rapidly varying squared cosine gives finally

$$\nu(r) = -\frac{3(mT)^2}{4\hbar^4 k_0{}^3 r^2} \sinh^{-2}\left(\frac{\pi m T r}{\hbar^2 k_0}\right).$$

As $T \to 0$ this becomes formula (120.13).

§121. Correlations of fluctuations in time

Let us consider a physical quantity which describes a body in thermodynamic equilibrium, or part of such a body. This quantity will undergo small variations in time, fluctuating about its mean value. Let $x(t)$ again denote the difference between the quantity and its mean value (so that $\bar{x} = 0$).

There is some correlation between the values of $x(t)$ at different instants; this means that the value of x at a given instant t affects the probabilities of its various values at a later instant $t+\tau$. In the same way as for the spatial correlation discussed in the preceding sections, we can characterise the time correlation by the mean value of the product $x(t)x(t+\tau)$. The averaging here is, as usual, understood in a statistical sense, i.e. as an averaging over the probabilities of all possible values of the quantity x at the times t and $t+\tau$. As has been mentioned in §1, this statistical averaging is equivalent to a time averaging—in this case, over the time t for a given value of τ.

The quantity thus obtained is a function of τ only; let it be denoted by $\phi(\tau)$:

$$\phi(\tau) = \overline{x(t)x(t+\tau)}. \tag{121.1}$$

As τ increases indefinitely, the correlation clearly tends to zero, and accordingly the function $\phi(\tau)$ likewise tends to zero.

The Fourier component of the quantity $x(t)$ is defined by†

$$x_\omega = \frac{1}{2\pi}\int_{-\infty}^{\infty} x(t) e^{i\omega t}\, dt \tag{121.2}$$

and the inverse relation

$$x(t) = \int_{-\infty}^{\infty} x_\omega e^{-i\omega t}\, d\omega. \tag{121.3}$$

† The integral as written here is in fact divergent, since $x(t)$ does not tend to zero as $|t| \to \infty$. This is, however, unimportant as regards the formal results derived below, whose object is to calculate the mean squares, which are known to be finite.

The use of the Fourier components in this problem is due to S. M. RYTOV.

Substituting the latter relation in the definition $\phi(t'-t) = \overline{x(t)x(t')}$, we obtain

$$\phi(t'-t) = \int_{-\infty}^{\infty}\int_{-\infty}^{\infty} \overline{x_\omega x_{\omega'}}\, e^{-i(\omega t+\omega' t')}\, d\omega\, d\omega'.$$

If this interval is to be a function of the difference $\tau = t'-t$ only, the integrand must contain a delta function of $\omega+\omega'$, i.e. we must have

$$\overline{x_\omega x_{\omega'}} = (x^2)_\omega \delta(\omega+\omega'). \tag{121.4}$$

This relation is to be regarded as the definition of the quantity symbolically denoted by $(x^2)_\omega$. Although the quantities x_ω are complex, $(x^2)_\omega$ is clearly real. To show this, it is sufficient to note that the expression (121.4) is zero except when $\omega' = -\omega$, and taking the complex conjugate corresponds to changing the sign of ω, i.e. interchanging ω and ω'.

Substituting (121.4) in $\phi(\tau)$ and integrating over ω', we find

$$\phi(\tau) = \int_{-\infty}^{\infty} (x^2)_\omega e^{-i\omega\tau}\, d\omega. \tag{121.5}$$

In particular, $\phi(0)$ is just the mean square of the fluctuating quantity,

$$\overline{x^2} = \int_{-\infty}^{\infty} (x^2)_\omega\, d\omega. \tag{121.6}$$

We see that the "spectral density" of the mean square fluctuation is just $(x^2)_\omega$ (or $2(x^2)_\omega$ if the integral is taken only over positive values of ω). This quantity is also, by (121.5), the Fourier component of the correlation function. Conversely

$$(x^2)_\omega = \frac{1}{2\pi}\int_{-\infty}^{\infty} \phi(\tau) e^{i\omega\tau}\, d\tau. \tag{121.7}$$

By regarding the quantity $x(t)$ as a function of time we have implicitly assumed that it behaves classically. The above formulae can, however, easily be put in a form applicable to quantum variables also. To do this, we must replace x by the quantum operator $\hat{x}(t)$, with Fourier component

$$\hat{x}_\omega = \frac{1}{2\pi}\int_{-\infty}^{\infty} \hat{x}(t) e^{i\omega t}\, dt. \tag{121.8}$$

The operators $\hat{x}(t)$ and $\hat{x}(t')$ relating to different instants do not in general commute, and the correlation function must now be defined as

$$\phi(t'-t) = \tfrac{1}{2}\overline{[\hat{x}(t)\hat{x}(t')+\hat{x}(t')\hat{x}(t)]}, \tag{121.9}$$

the bar denoting averaging with respect to the exact probabilities given by quantum mechanics.† The quantity $(x^2)_\omega$ is defined by

$$\tfrac{1}{2}\overline{(\hat{x}_\omega \hat{x}_{\omega'} + \hat{x}_{\omega'} \hat{x}_\omega)} = (x^2)_\omega \delta(\omega + \omega'); \tag{121.10}$$

the relations (121.5)–(121.7) are then unchanged.

Let us assume that the quantity x is such that, if it has a definite value (considerably different from its mean fluctuation), a definite state of partial equilibrium can be described by it. In other words, the relaxation time for the establishment of partial equilibrium for a given value of x is assumed to be much less than the relaxation time required to reach the equilibrium value of x itself. This condition is satisfied by a wide class of quantities of physical interest. We shall call the fluctuations of such quantities *thermodynamic fluctuations*. In the rest of this section and in §§122–124 we shall consider fluctuations of this type, and moreover shall assume the quantities x to be classical.‡

We shall also assume in the rest of this section that, as complete equilibrium is approached, no other deviations from equilibrium arise in the system which would require the use of further quantities to describe them. In other words, at every instant the state of the body must be entirely defined by the value of x; a more general case will be discussed in §124.

Let the quantity $x(t)$ have at some instant t a value which is large compared with the mean fluctuation. Then we can say that at subsequent instants the body will tend to return to the equilibrium state, and accordingly the quantity x will decrease. Under the assumptions made above, its rate of change $\dot{x}$ will be at every instant entirely defined by the value of x at that instant: $\dot{x} = \dot{x}(x)$. If x is still small (compared with its range of possible values), then $\dot{x}(x)$ can be expanded in powers of x, keeping only the linear term:

$$\mathrm{d}x/\mathrm{d}t = -\lambda x, \tag{121.11}$$

where λ is a positive constant.

Let us also define a quantity $\xi_x(\tau)$ as the mean value of $x(t)$ at time $t+\tau$, given that at the preceding instant t it had the value x; this mean value is not in general zero. The correlation function $\phi(\tau)$ can obviously be written in terms of the function $\xi_x(\tau)$ as

$$\phi(\tau) = \overline{x\xi_x(\tau)}, \tag{121.12}$$

† It may again be mentioned that, according to the fundamental principles of statistical physics, the result of the averaging is independent of whether it is done mechanically over the exact wave function of the stationary state of the system or statistically by means of the Gibbs distribution. The only difference is that in the former case the result is expressed in terms of the energy of the body, and in the latter case as a function of its temperature.

‡ The final result for thermodynamic fluctuations of a quantum variable differs from that for a classical variable only by slight changes in the form of the expressions, discussed in §127 (see (127.22)).

where the averaging is only over the probabilities of the various values of x at the initial instant t.

For values of ξ_x large compared with the mean fluctuation, it follows from formula (121.11) that also

$$d\xi_x(\tau)/d\tau = -\lambda\xi_x(\tau), \tag{121.13}$$

and we must further expect this relation to be true for arbitrary (not necessarily large) $\xi_x(\tau)$. Integrating (121.13) we find, since by definition $\xi_x(0) = x$,

$$\xi_x(\tau) = xe^{-\lambda\tau},$$

and finally, substituting in (121.12), we obtain a formula for the time correlation function:

$$\phi(\tau) = \overline{x^2}e^{-\lambda\tau}.$$

It must be remembered, however, that this formula as it stands is valid only for $\tau > 0$, since in the foregoing derivation (equation (121.13)) it has been assumed that the instant $t+\tau$ is later than t. On the other hand, we have identically

$$\phi(\tau) = \overline{x(t)x(t+\tau)} = \overline{x(t-\tau)x(t)} = \phi(-\tau),$$

since this transformation amounts to a simple renaming of the variable ($t-\tau$ instead of t) over which the averaging is performed. Thus $\phi(\tau)$ is an even function of τ.

We can therefore write finally

$$\phi(\tau) = \overline{x^2}e^{-\lambda|\tau|}, \tag{121.14}$$

which is valid for both positive and negative τ. This function has two different derivatives at $\tau = 0$, the reason being that we have considered intervals of time long compared with the time for establishment of partial equilibrium (equilibrium with a given value of x). The consideration of short times, which is not possible within the thermodynamic theory, would of course show that $d\phi/d\tau = 0$ for $\tau = 0$, as must be true for any even function of τ.

An elementary integration leads to the following expression for the Fourier components of the function $\phi(\tau)$, as defined by (121.7):

$$(x^2)_\omega = \frac{\lambda}{\pi(\omega^2+\lambda^2)}\cdot\overline{x^2} = \lambda/\pi\beta(\omega^2+\lambda^2), \tag{121.15}$$

$\overline{x^2}$ being given by (112.4).

These results can also be written in another form which is often more convenient for practical applications.

The relation $\dot{x} = -\lambda x$ for the quantity x itself (rather than its mean value ξ_x) is valid, as already mentioned, only when x is large compared with the mean fluctuation of x. For arbitrary values of x we write

$$\dot{x} = -\lambda x + y, \tag{121.16}$$

thus defining a new quantity $y(t)$. Although the magnitude of the oscillations of y does not change with time, when x is large (in the sense already defined) y is relatively small and may be neglected in equation (121.16).

Multiplying this equation by $e^{i\omega t}$ and integrating over t from 0 to T (with integration by parts for the term $\dot{x}e^{i\omega t}$), we obtain $x_\omega = y_\omega/(\lambda - i\omega)$. Using now formulae (121.4) and (121.15), we have

$$(y^2)_\omega = \lambda\overline{x^2}/\pi. \tag{121.17}$$

It is worth noting that this quantity is independent of the frequency.

The quantity (121.17) is also the Fourier component of the mean value $\overline{y(t)y(t+\tau)}$ (just as $(x^2)_\omega$ is the Fourier component of the mean value $\overline{x(t)x(t+\tau)}$). A function whose Fourier components are independent of frequency is proportional to the delta function, and it is easy to see that

$$\overline{y(t)y(t+\tau)} = 2\lambda\overline{x^2}\delta(\tau). \tag{121.18}$$

The vanishing of this expression when $\tau \neq 0$ signifies that the values of $y(t)$ at different instants are entirely uncorrelated. In reality, of course, this statement is an approximation and signifies only that the values of $y(t)$ are correlated over time intervals of the order of the time for partial equilibrium to be established (equilibrium with a given value of x), which, as already mentioned, is regarded as negligibly small in the theory given here. In this connection it should be noted that all the formulae derived in this section for the Fourier components of various quantities are valid only for frequencies small compared with the reciprocal of the time for partial equilibrium to be established.

§122. The symmetry of the kinetic coefficients

Let us consider a closed system, not in a state of statistical equilibrium; let several thermodynamic quantities $x_1, x_2, \ldots, x_n$ which describe the whole system, or parts of it, all have non-equilibrium values (if they describe the whole system, they must not be quantities which remain constant for a closed system, such as energy or volume). As in §113, it will be convenient to suppose that the equilibrium values have been subtracted from these quantities, so that $x_1, x_2, \ldots$ themselves represent the degree of non-equilibrium of the system.

The quantities $x_1, \ldots, x_n$ will vary with time. We shall suppose that these quantities are such as to give a complete description of the approach to equilibrium, and that no other deviations from complete equilibrium occur during this process; cf. §121. Then the rate $\dot{x}_i$ of variation of the quantity x_i in any non-equilibrium state is a function of the values of $x_1, \ldots, x_n$ in that state:

$$\dot{x}_i = \dot{x}_i(x_1, \ldots, x_n). \tag{122.1}$$

Let us assume that the system is in a state fairly close to equilibrium, so that the quantities x_i can be regarded as small. Then, expanding the rates $\dot{x}_i$ in powers of $x_1, \ldots, x_n$, we need take only the first-order terms, i.e. write the $\dot{x}_i$ as linear sums of the form

$$\dot{x}_i = -\sum_{k=1}^{n} \lambda_{ik} x_k$$

with constant coefficients λ_{ik}. There can be no zero-order terms in this expansion, since in equilibrium (i.e. when $x_1 = 0$, $x_2 = 0, \ldots$) all the rates $\dot{x}_i$ must also be zero. In what follows, as in §113, we shall omit the summation signs; summation from 1 to n over all repeated suffixes will be implied. Thus†

$$\dot{x}_i = -\lambda_{ik} x_k. \tag{122.2}$$

We also define the derivatives

$$X_i = -\partial S/\partial x_i \tag{122.3}$$

of the entropy S of the system. In a state of equilibrium, the entropy is a maximum, so that

$$X_1 = 0, \qquad X_2 = 0, \ldots, \qquad X_n = 0. \tag{122.4}$$

and for small x_i, again taking only the first-order terms, we can write

$$X_i = \beta_{ik} x_k, \tag{122.5}$$

the β_{ik} being constant coefficients. These are the first derivatives of the X_i i.e. the second derivatives of S; hence

$$\beta_{ik} = \beta_{ki}, \tag{122.6}$$

i.e. the coefficients are symmetrical in the suffixes i and k; this is not true of the coefficients λ_{ik} in (122.2).

If the x_i are expressed in terms of the X_i by (122.5) and substituted in (122.2), the rates $\dot{x}_i$ are thereby also expressed as linear combinations of the X_i; the resulting equations are of the form

$$\dot{x}_i = -\gamma_{ik} X_k. \tag{122.7}$$

The quantities γ_{ik} are called *kinetic coefficients*. We shall now prove the *principle of the symmetry of the kinetic coefficients* (first discovered by

† In applications, cases occur where the complete equilibrium which is being approached depends on external parameters (such as volume or external field) which themselves vary slowly with time; the equilibrium (mean) values of the quantities considered therefore vary also. If this variation is sufficiently slow, we can again use the relations derived here, except that the mean values $\overline{x_i}$ can not be regarded as always equal to zero. If they are denoted by $x_i^{(0)}$, then (122.2), for example, must be replaced by

$$\dot{x}_i = -\lambda_{ik}(x_k - x_k^{(0)}). \tag{122.2a}$$

L. Onsager, 1931). This states that

$$\gamma_{ik} = \gamma_{ki}. \tag{122.8}$$

To prove this result, let us assume that the x_i are not equal to their mean values owing to a fluctuation of the system. We take the value of any one of the x_i at some instant t and that of another x_k at time $t+\tau$, and average the product $x_i(t)x_k(t+\tau)$ over time t (for a given positive value of τ). The equations of motion of the particles of the body (in the absence of an external magnetic field) are symmetrical under time reversal. It is therefore immaterial whether in the averaging we take x_k at the later time and x_i at the earlier time, or *vice versa*. The mean values of the products $x_i(t)x_k(t+\tau)$ and $x_i(t+\tau)x_k(t)$ must consequently be equal:

$$\overline{x_i(t)x_k(t+\tau)} = \overline{x_i(t+\tau)x_k(t)}. \tag{122.9}$$

We differentiate this equation with respect to τ and then put $\tau = 0$. The result is

$$\overline{x_i\dot{x}_k} = \overline{\dot{x}_i x_k}. \tag{122.10}$$

To avoid misunderstanding, the following comment should be made regarding the above derivation. By changing the variable t, over which the averaging is carried out, into $t-\tau$, we have identically

$$\overline{x_i(t)x_k(t+\tau)} = \overline{x_i(t-\tau)x_k(t)},$$

which may be written

$$\phi_{ki}(\tau) = \phi_{ik}(-\tau), \tag{122.11}$$

with the notation

$$\phi_{ik}(\tau) = \overline{x_i(t+\tau)x_k(t)}. \tag{122.12}$$

It might seem at first sight that, by differentiating (122.11) with respect to τ and then putting $\tau = 0$, we could show that $\dot{\phi}_{ik}(0) = 0$. In reality, however, as already mentioned in §121, in the approximation considered here the functions $\phi_{ik}(\tau)$ (like $\phi(\tau)$ in §121) have two different derivatives at $\tau = 0$ according as $\tau \to 0+$ or $0-$.

We now substitute in (122.10) the formula (122.7) for $\dot{x}_i$: $\overline{x_i\gamma_{kl}X_l} = \overline{\gamma_{il}X_l x_k}$. From (113.7), $\overline{x_iX_l} = \delta_{il}$, and so $\gamma_{kl}\delta_{il} = \gamma_{ki} = \gamma_{il}\delta_{ik} = \gamma_{ik}$. This proves (122.8).

The following two comments should, however, be made regarding this relation. The proof depends on the symmetry of the equations of motion with respect to time, and the formulation of this symmetry is somewhat altered for fluctuations in a uniformly rotating body and for bodies in an external magnetic field: in these cases the symmetry under time reversal holds only if the sign of the angular velocity of rotation $\boldsymbol{\Omega}$ or of the magnetic field $\mathbf{H}$ is simultaneously changed. Thus in these cases the kinetic coefficients depend on

Ω or $\mathbf{H}$ as parameter, and we have the relations

$$\begin{aligned}\gamma_{ik}(\Omega) &= \gamma_{ki}(-\Omega),\\ \gamma_{ik}(\mathbf{H}) &= \gamma_{ki}(-\mathbf{H}).\end{aligned} \tag{122.13}$$

Moreover, it has been tacitly assumed in the derivation that the quantities x_i are such as to remain unchanged under time reversal. But if these quantities are proportional to the velocities of some macroscopic motions, they will themselves change sign under time reversal. It is easy to see that, if any two quantities x_i and x_k both change sign, the relation (122.10) will still hold, and therefore $\gamma_{ik} = \gamma_{ki}$. But if one of x_i, x_k changes sign while the other does not, we have $\overline{x_i \dot{x}_k} = -\overline{\dot{x}_i x_k}$, and for the corresponding kinetic coefficients

$$\gamma_{ik} = -\gamma_{ki}. \tag{122.14}$$

From (122.7) and (122.8) it follows that the rates $\dot{x}_i$ can be written as the derivatives,

$$\dot{x}_i = -\partial f/\partial X_i, \tag{122.15}$$

of some "generating function" f, which is a quadratic form in the quantities X_i with coefficients $\frac{1}{2}\gamma_{ik}$:

$$f = \tfrac{1}{2}\gamma_{ik} X_i X_k. \tag{122.16}$$

This is an important function, since it determines the time derivative of the entropy: $\dot{S} = (\partial S/\partial x_i)\dot{x}_i = -X_i\dot{x}_i = X_i \partial f/\partial X_i$, and since f is a quadratic function of the X_i EULER's theorem gives

$$\dot{S} = 2f. \tag{122.17}$$

As the equilibrium state is approached, the entropy S must increase towards a maximum. The quadratic form f must therefore be positive-definite, and this imposes certain conditions on the coefficients γ_{ik}.

In an exactly similar way to the derivation of (122.8) we can show that, if the time derivatives of the X_i are expressed as linear functions of the x_i,

$$\dot{X}_i = -\zeta_{ik} x_k, \tag{122.18}$$

then the coefficients ζ_{ik} are symmetrical:

$$\zeta_{ik} = \zeta_{ki}. \tag{122.19}$$

We can therefore write the $\dot{X}_i$ as derivatives

$$\dot{X}_i = -\partial f/\partial x_i \tag{122.20}$$

of a quadratic function

$$f = \tfrac{1}{2}\zeta_{ik} x_i x_k. \tag{122.21}$$

Using formula (122.5) we find, since $\beta_{ik} = \beta_{ki}$,

$$dS = -X_k\, dx_k = -\beta_{ki} x_i\, dx_k = -x_i\, d(\beta_{ik} x_k) = -x_i\, dX_i,$$

and hence

$$-\partial S/\partial X_i = x_i, \tag{122.22}$$

where the entropy is now assumed to be expressed as a function of the X_i. The derivative $\dot{S}$ may therefore also be written

$$\dot{S} = (\partial S/\partial X_i)\dot{X}_i = -x_i\dot{X}_i = x_i\partial f/\partial x_i = 2f,$$

with f given by (122.21). Comparison with (122.17) shows that the two functions (122.16) and (122.21) are the same quantity expressed in terms of different variables.

For a system consisting of a body in an external medium we can transform (122.17) by using the fact that the change in entropy of a closed system in a deviation from equilibrium is $-R_{\min}/T_0$, where $R_{\min}$ is the minimum work needed to bring the system from the equilibrium state into the one considered (see (20.8)).† Putting also $R_{\min} = \Delta E - T_0\Delta S + P_0\Delta V$ (where E, S and V relate to the body, and T_0, P_0 are the temperature and pressure of the medium), we obtain

$$\dot{E} - T_0\dot{S} + P_0\dot{V} = -2fT_0. \tag{122.23}$$

In particular, if the deviation from equilibrium occurs with the temperature and pressure of the body constant and equal to T_0 and P_0,

$$\dot{\Phi} = -2fT, \tag{122.24}$$

and at constant temperature and volume

$$\dot{F} = -2fT. \tag{122.25}$$

§123. The dissipative function

The macroscopic motion of bodies surrounded by an external medium is in general accompanied by irreversible frictional processes, which ultimately bring the motion to a stop. The kinetic energy of the bodies is thereby converted into heat and is said to be *dissipated.*

A purely mechanical treatment of such a motion is clearly impossible: since the energy of macroscopic motion is converted into the energy of thermal motion of the molecules of the body and the medium, such a treatment would require the setting up of the equations of motion for all these molecules. The problem of setting up equations of motion in the medium which contain only the "macroscopic" co-ordinates of the bodies is therefore a problem of statistical physics.

This problem, however, cannot be solved in a general form. Since the internal motion of the atoms in a body depends not only on the motion of the body at a given instant but also on the previous history of the motion, the

†Owing to this relation between the value of the entropy and $R_{\min}$, the X_i can also be defined as

$$X_i = (1/T_0)\partial R_{\min}/\partial x_i, \tag{122.3a}$$

which is sometimes more convenient than (122.3); cf. (22.7).

equations of motion will in general contain not only the "macroscopic" co-ordinates $Q_1, Q_2, \ldots, Q_s$ of the bodies and their first and second time derivatives but also all the higher-order derivatives (more precisely, some integral operator of the co-ordinates). The Lagrangian for the macroscopic motion of the system does not then exist, of course, and the equations of motion will be entirely different in different cases.

The equations of motion can be derived in a general form if it may be assumed that the state of the system at a given instant is completely determined by the values of the co-ordinates Q_i and velocities $\dot{Q}_i$, and that the higher-order derivatives may be neglected; a more precise criterion of smallness has to be established in each particular case. We shall further suppose that the velocities $\dot{Q}_i$ are themselves so small that their higher powers may be neglected, and finally that the motion in question consists of small oscillations about certain equilibrium positions. The latter is the case usually met with in this connection. We shall assume the co-ordinates Q_i to be chosen so that $Q_i = 0$ in the equilibrium position. Then the kinetic energy $K(\dot{Q}_i)$ of the system will be a quadratic function of the velocities $\dot{Q}_i$ and independent of the co-ordinates Q_i themselves; the potential energy $U(Q_i)$ due to the external forces will be a quadratic function of the co-ordinates Q_i.

We define the generalised momenta P_i in the usual way:

$$P_i = \partial K(\dot{Q}_i)/\partial \dot{Q}_i. \tag{123.1}$$

These equations define the momenta as linear combinations of the velocities; using them to express the velocities in terms of the momenta and substituting in the kinetic energy, we obtain the latter as a quadratic function of the momenta, with

$$\dot{Q}_i = \partial K(P_i)/\partial P_i. \tag{123.2}$$

If the dissipative processes are entirely neglected, the equations of motion will be as in ordinary mechanics, according to which the time derivatives of the momenta are equal to the corresponding generalised forces:

$$\dot{P}_i = -\partial U/\partial Q_i. \tag{123.3}$$

First of all, let us note that equations (123.2), (123.3) are in formal agreement with the principle of the symmetry of the kinetic coefficients derived in §122, if the quantities $x_1, x_2, \ldots, x_{2s}$ used there are taken as the co-ordinates Q_i and momenta P_i. For the minimum work needed to bring the bodies from a state of rest in their equilibrium positions to the positions Q_i with momenta P_i is $R_{\min} = K(P_i) + U(Q_i)$. The quantities $X_1, X_2, \ldots, X_{2s}$ will therefore be the derivatives

$$X_{Q_i} = (1/T)\partial R_{\min}/\partial Q_i = (1/T)\partial U/\partial Q_i,$$
$$X_{P_i} = (1/T)\partial R_{\min}/\partial P_i = (1/T)\partial K/\partial P_i$$

(see the last footnote to §122), and equations (123.2), (123.3) will correspond to the relations (122.7) with $\gamma_{Q_iP_i} = -T = -\gamma_{P_iQ_i}$, in accordance with the rule (122.14); this is a case where one quantity (Q_i) remains invariant under time reversal but the other (P_i) changes sign.

In accordance with the general relations (122.7), we can now write the equations of motion allowing for dissipative processes, by adding to the right-hand sides of equations (123.2), (123.3) certain linear combinations of the quantities X_{Q_i}, X_{P_i}, such that the required symmetry of the kinetic coefficients is maintained. It is easy to see, however, that equations (123.2) must be left unchanged, since they are simply a consequence of the definition (123.1) of the momenta and do not depend on the presence or absence of dissipative processes. This shows that the terms added to equations (123.3) can only be linear combinations of the quantities X_{P_i} (i.e. of the derivatives $\partial K/\partial P_i$), since otherwise the symmetry of the kinetic coefficients would be violated.

Thus we have a set of equations of the form

$$\dot{P}_i = -\frac{\partial U}{\partial Q_i} - \sum_{k=1}^{s} \gamma_{ik} \frac{\partial K}{\partial P_k},$$

where the constant coefficients γ_{ik} are related by

$$\gamma_{ik} = \gamma_{ki}. \tag{123.4}$$

Putting $\partial K/\partial P_k = \dot{Q}_k$, we have finally

$$\dot{P}_i = -\frac{\partial U}{\partial Q_i} - \sum_{k=1}^{s} \gamma_{ik}\dot{Q}_k. \tag{123.5}$$

These are the required equations of motion. We see that the presence of dissipative processes leads, in this approximation, to the appearance of *frictional forces* which are linear functions of the velocities. Owing to the relations (123.4) these forces can be written as the derivatives, with respect to the corresponding velocities, of the quadratic function

$$f = \tfrac{1}{2}\sum_{i,k} \gamma_{ik}\dot{Q}_i\dot{Q}_k, \tag{123.6}$$

which is called the *dissipative function*. Then

$$\dot{P}_i = -\frac{\partial U}{\partial Q_i} - \frac{\partial f}{\partial \dot{Q}_i}. \tag{123.7}$$

Using the Lagrangian $L = K - U$, we can write these equations of motion as

$$\frac{\mathrm{d}}{\mathrm{d}t}\left(\frac{\partial L}{\partial \dot{Q}_i}\right) - \frac{\partial L}{\partial Q_i} = -\frac{\partial f}{\partial \dot{Q}_i}, \tag{123.8}$$

which differs from the usual form of LAGRANGE's equations by the presence of the derivative of the dissipative function on the right.

The existence of friction causes a decrease in the total mechanical energy $K+U$ of the moving bodies. In accordance with the general results of §122, the rate of decrease is determined by the dissipative function. This will be proved afresh here, since there is some difference in the notation used in the present section. We have

$$\frac{\mathrm{d}}{\mathrm{d}t}(K+U) = \sum_{i=1}^{s}\left(\frac{\partial K}{\partial P_i}\dot{P}_i+\frac{\partial U}{\partial Q_i}\dot{Q}_i\right) = \sum_i \dot{Q}_i\left(\dot{P}_i+\frac{\partial U}{\partial Q_i}\right),$$

or, substituting (123.7) and using the fact that the dissipative function is quadratic,

$$\frac{\mathrm{d}}{\mathrm{d}t}(K+U) = -\sum_i \dot{Q}_i\frac{\partial f}{\partial \dot{Q}_i} = -2f, \qquad (123.9)$$

as it should be.

Finally, we may mention that, when there is an external magnetic field, the equations of motion again take the form (123.5), but (123.4) is replaced by

$$\gamma_{ik}(\mathbf{H}) = \gamma_{ki}(-\mathbf{H}).$$

As a result, there is no dissipative function whose derivatives determine the friction forces. The equations of motion therefore cannot be written in the form (123.7).

§124. Time correlations of the fluctuations of more than one variable

The results obtained in §121 for the time correlation function of one fluctuating quantity can be generalised to fluctuations in which several thermodynamic quantities $x_1, x_2, \ldots, x_n$ simultaneously deviate from their equilibrium values.

By analogy with the definition (121.1), we define the correlation functions

$$\phi_{ik}(\tau) = \overline{x_i(t+\tau)x_k(t)}. \qquad (124.1)$$

These satisfy identically the relations

$$\phi_{ik}(\tau) = \phi_{ki}(-\tau) \qquad (124.2)$$

(see (122.11)).

Instead of (121.7) we now have

$$(x_i x_k)_\omega = \frac{1}{2\pi}\int_{-\infty}^{\infty}\phi_{ik}e^{i\omega\tau}\,\mathrm{d}\tau, \qquad (124.3)$$

where the quantities $(x_i x_k)_\omega$ are defined by

$$\overline{x_{i\omega}x_{k\omega'}} = (x_i x_k)_\omega\delta(\omega+\omega'). \qquad (124.4)$$

Let us consider the mean value $\overline{x_i(t)X_k(t+\tau)}$. Substituting $X_k = \beta_{kl}x_l$, we obtain

$$\overline{x_i(t)X_k(t+\tau)} = \beta_{kl}\phi_{li}(\tau).$$

On the other hand, using the mean values $\Xi_k(\tau)$ of the X_k and $\xi_k(\tau)$ of the x_k at $t+\tau$ for given values of all the $x_1, x_2, \ldots$ at time t, we can write (cf. (121.12))

$$\overline{x_i(t)X_k(t+\tau)} = \overline{x_i\Xi_k(\tau)}.$$

Differentiating this equation with respect to τ, substituting for the derivatives $d\Xi_k/d\tau$ the values $-\zeta_{kl}\xi_l$ with the same coefficients as in (122.18) and bearing in mind that $\overline{x_i\xi_l(\tau)} = \phi_{li}(\tau)$, we obtain the equations

$$\beta_{kl}\, d\phi_{li}/d\tau = -\zeta_{kl}\phi_{li} \qquad (\tau > 0), \tag{124.5}$$

which determine the ϕ_{ik} as functions of τ; it must be remembered that the equations in this form are valid only for $\tau > 0$ (cf. §121).

To calculate the Fourier components of the functions ϕ_{ik}, we multiply the equation (124.5) by $e^{i\omega\tau}$ and integrate over τ from 0 to ∞. Integrating by parts and using the fact that $\phi_{ik}(\infty) = 0$, we find

$$-\beta_{kl}\phi_{li}(0) - i\omega\beta_{kl}\int_0^\infty \phi_{li}(\tau)e^{i\omega\tau}\, d\tau = -\zeta_{kl}\int_0^\infty \phi_{li}(\tau)e^{i\omega\tau}\, d\tau.$$

But from (113.8)

$$\phi_{li}(0) = \overline{x_i x_l} = \beta^{-1}{}_{il},$$

and hence

$$(\zeta_{kl} - i\omega\beta_{kl})\int_0^\infty \phi_{li}(\tau)e^{i\omega\tau}\, d\tau = \delta_{ki}.$$

Thus we have

$$\int_0^\infty \phi_{li}e^{i\omega\tau}\, d\tau = (\zeta - i\omega\beta)^{-1}{}_{li},$$

where $(\zeta - i\omega\beta)^{-1}{}_{li}$ are the components of the matrix inverse to $(\zeta - i\omega\beta)_{li}$. Replacing τ by $-\tau$ and ω by $-\omega$, and using the relation (124.2), we obtain

$$\int_{-\infty}^0 \phi_{li}(\tau)e^{i\omega\tau}\, d\tau = (\zeta + i\omega\beta)^{-1}{}_{il}.$$

Finally, addition of these two equations gives

$$\int_{-\infty}^\infty \phi_{li}(\tau)e^{i\omega\tau}\, d\tau = (\zeta - i\omega\beta)^{-1}{}_{li} + (\zeta + i\omega\beta)^{-1}{}_{il}, \tag{124.6}$$

which determines the required Fourier components and thus generalises formula (121.15)†.

The matrix β_{li} is always symmetric. In the absence of a magnetic field the ζ_{li} are also symmetric, and therefore so is the matrix $\zeta_{li} \pm i\omega\beta_{li}$ and hence also its inverse.‡

As at the end of §121, we may write these results in a different form by using new variables Y_i defined by

$$\dot{X}_i = -\zeta_{ik}x_k + Y_i; \tag{124.7}$$

these quantities may be neglected when the x_k exceed their mean fluctuations. Exactly as in §121, a simple calculation leads to the formula

$$(Y_iY_k)_\omega = (1/2\pi)(\zeta_{ik} + \zeta_{ki}). \tag{124.8}$$

These quantities are again independent of ω.

For the quantities y_i defined by

$$\dot{x}_i = -\gamma_{ik}X_k + y_i, \tag{124.9}$$

there is the corresponding formula

$$(y_iy_k)_\omega = (1/2\pi)(\gamma_{ik} + \gamma_{ki}). \tag{124.10}$$

This formula is obvious without further calculation if we note that there is a reciprocal relation between the x_i and the X_i: the X_i are the derivatives of the entropy with respect to the x_i and *vice versa*.

We may also note the formula

$$y_i(t)y_k(t+\tau) = (\gamma_{ik} + \gamma_{ki})\delta(\tau), \tag{124.11}$$

which corresponds to (124.10) in the same way as (121.18) corresponds to (121.17).

For practical applications, formulae (124.8) and (124.10) have the advantage of containing the matrix elements ζ_{ik} and γ_{ik} themselves, not the matrix elements of the inverse matrix.

As an example of the use of the above formulae, let us consider fluctuations of a one-dimensional oscillator, i.e. a body which is at rest in the equilibrium position ($Q = 0$) but capable of executing small oscillations in some

† If there is only one quantity x, then we have

$$2\pi(x^2)_\omega = \frac{1}{\zeta - i\omega\beta} + \frac{1}{\zeta + i\omega\beta} = \frac{2\zeta}{\zeta^2 + \omega^2\beta^2}.$$

By the definition of λ, β, ζ we have in this case $X = \beta x$, $\dot{X} = -\zeta x$, $\dot{x} = -\lambda x$. Thus $\zeta = \lambda\beta$ and we return to formula (121.15).

‡ If one of the x_i (x_l, say) changes sign under time reversal, the corresponding matrix elements (124.6) must be antisymmetric. This is in fact so, since in that case $\zeta_{il} = -\zeta_{li}$ (cf. (122.14)) and the coefficients $\beta_{il} = 0$. The latter result follows because the β_{il} are coefficients of the products x_ix_l in the quadratic form which gives the change in entropy in a deviation from equilibrium. Since the entropy is invariant under time reversal while the product x_ix_l changes sign, the entropy cannot contain any such term, i.e. $\beta_{il} = 0$.

macroscopic co-ordinate Q. Because of fluctuations, the co-ordinate Q will in fact undergo deviations from the value $Q = 0$. The mean square of this deviation is determined directly from the coefficient in the quasi-elastic force which acts on the body during a deviation.

We write the potential energy of the oscillator in the form $U = \frac{1}{2}m\omega_0^2Q^2$, where m is the "mass" (i.e. the coefficient of proportionality between the generalised momentum P and the velocity $\dot{Q} : P = m\dot{Q}$), and ω_0 the natural frequency of the oscillator (in the absence of friction). Then the mean square fluctuation is (cf. §114, Problem 7) $\overline{Q^2} = T/m\omega_0^2$.

It is of more interest, however, to calculate the "Fourier components" of the fluctuations in the co-ordinate. We shall do this for the general case where the oscillations are accompanied by friction.

The equations of motion of an oscillator with friction are

$$\dot{Q} = P/m, \tag{124.12}$$

$$\dot{P} = -m\omega_0^2Q - \gamma P/m, \tag{124.13}$$

where $-\gamma P/m = -\gamma\dot{Q}$ is the "frictional force". As shown in §123, if Q and P are taken as x_1 and x_2, the corresponding X_1 and X_2 are $m\omega_0^2Q/T$ and P/mT. Equations (124.12) and (124.13) then represent the relations $\dot{x}_i = -\gamma_{ik}X_k$, so that

$$\gamma_{11} = 0, \qquad \gamma_{12} = -\gamma_{21} = -T, \qquad \gamma_{22} = \gamma T.$$

In order to apply these equations to the fluctuations, we write (124.13) as

$$\dot{P} = -m\omega_0^2Q - \gamma P/m + y; \tag{124.14}$$

equation (124.12), which is the definition of the momentum, must be left unchanged. According to (124.10) we have immediately $(y^2)_\omega = \gamma_{22}/\pi = \gamma T/\pi$.

Finally, in order to derive the required $(Q^2)_\omega$, we substitute $P = m\dot{Q}$ in (124.14), obtaining†

$$m\ddot{Q} + \gamma\dot{Q} + m\omega_0^2Q = y. \tag{124.15}$$

Multiplying by $e^{i\omega t}$ and integrating over time, we find

$$(-m\omega^2 - i\omega\gamma + m\omega_0^2)Q_\omega = y_\omega,$$

and hence finally

$$(Q^2)_\omega = \gamma T/\pi[m^2(\omega^2 - \omega_0^2)^2 + \omega^2\gamma^2]. \tag{124.16}$$

§125. The generalised susceptibility

It is not possible to derive a general formula for the spectral distribution of non-thermodynamic fluctuations analogous to formula (121.15) for the thermodynamic fluctuations, but in many cases it is possible to relate the

† If (124.15) is regarded as the "equation of motion" of a fluctuating oscillator, the quantity y is sometimes called the *random force* acting on the oscillator.

properties of non-thermodynamic fluctuations to quantities describing the behaviour of the body under certain external interactions. These may be fluctuations of either classical or quantum quantities.

Physical quantities of this type have the property that for each of them there exists an external interaction described by the presence, in the Hamiltonian of the body, of a perturbing operator of the type

$$\hat{V} = -\hat{x}f(t), \tag{125.1}$$

where $\hat{x}$ is the quantum operator of the physical quantity concerned, and the "perturbing force" f is a given function of time.†

The quantum mean value $\bar{x}$ is not zero when such a perturbation is present (whereas $\bar{x} = 0$ in the equilibrium state in the absence of the perturbation), and it can be written in the form $\hat{\alpha}f$, where $\hat{\alpha}$ is a linear integral operator whose effect on the function $f(t)$ is given by a formula of the type

$$\overline{x}(t) = \hat{\alpha}f = \int_0^\infty K(\tau)f(t-\tau)\,d\tau, \tag{125.2}$$

$K(\tau)$ being a function of time which depends on the properties of the body. The value of $\bar{x}$ at time t can, of course, depend only on the values of the "force" f at previous (not subsequent) times; the expression (125.2) satisfies this requirement.

Any perturbation depending on time can be reduced by means of a Fourier expansion to a set of monochromatic components with a time dependence $e^{-i\omega t}$. For such a perturbation, the relation between $\bar{x}$ and f is

$$\bar{x} = \alpha(\omega)f, \tag{125.3}$$

where the function $\alpha(\omega)$ is given by

$$\alpha(\omega) = \int_0^\infty K(\tau)e^{i\omega\tau}\,d\tau. \tag{125.4}$$

If this function is specified, the behaviour of the body under the perturbation in question is completely determined. We shall call α the *generalised susceptibility*. It plays a fundamental part in the theory described below, since, as we shall see, the fluctuations of the quantity x can be expressed in terms of it.‡

† For example, f may be an external electric field and x the electric dipole moment acquired by the body in that field.

‡ In the example given in the last footnote, α is the electric polarisability of the body. The quantity α thus defined is more convenient than the frequently used quantity $Z(\omega) = -1/i\omega\alpha$, called the *generalised impedance*, which is the coefficient in the relation $f = Z\dot{x}$.

The function $\alpha(\omega)$ is in general complex. Let its real and imaginary parts be denoted by α' and α'':

$$\alpha = \alpha' + i\alpha''. \qquad (125.5)$$

The definition (125.4) shows immediately that

$$\alpha(-\omega) = \alpha^*(\omega). \qquad (125.6)$$

Separating the real and imaginary parts, we find

$$\alpha'(-\omega) = \alpha'(\omega), \qquad \alpha''(-\omega) = -\alpha''(\omega), \qquad (125.7)$$

i.e. $\alpha'(\omega)$ is an even function of the frequency, and α'' an odd function. When $\omega = 0$ the function $\alpha''(\omega)$ changes sign, passing through zero (or in some cases through infinity).

It should be emphasised that the property (125.6) simply expresses the fact that the operator relation $\bar{x} = \hat{\alpha}f$ must lead to real values of $\bar{x}$ for every real f. If the function $f(t)$ is given by the real expression

$$f = \tfrac{1}{2}(f_0 e^{-i\omega t} + f_0{}^* e^{i\omega t}), \qquad (125.8)$$

then by applying the operator $\hat{\alpha}$ to each of the two terms we obtain

$$\bar{x} = \tfrac{1}{2}[\alpha(\omega) f_0 e^{-i\omega t} + \alpha(-\omega) f_0{}^* e^{i\omega t}]; \qquad (125.9)$$

the condition for this expression to be real is just (125.6).

As $\omega \to \infty$, the function $\alpha(\omega)$ tends to a real finite limit α_∞. For definiteness we shall suppose below that this limit is zero; a non-zero α_∞ requires only some obvious slight changes in some of the formulae.

The change in state of the body as a result of the "force" f is accompanied by absorption (dissipation) of energy; the source of this energy is the external interaction, and after absorption in the body it is converted into heat. This dissipation also can be expressed in terms of the quantity α. To do so, we use the equation $\mathrm{d}E/\mathrm{d}t = \overline{\partial H/\partial t}$, which states that the time derivative of the mean energy of the body is equal to the mean value of the partial derivative of the Hamiltonian of the body with respect to time (see §11). Since only the perturbation $\hat{V}$ in the Hamiltonian depends explicitly on the time, we have

$$\mathrm{d}E/\mathrm{d}t = -\bar{x}\,\mathrm{d}f/\mathrm{d}t. \qquad (125.10)$$

This relation is of importance in applications of the theory under discussion. If we know the expression for the change in energy in a particular process, a comparison with (125.10) will show which quantity is to be interpreted as the "force" f with respect to a given variable x.

The mean energy dissipation Q per unit time can be derived from (125.10) by substituting $\bar{x}$ from (125.9) and averaging over the period of the external interaction. The terms containing $e^{\pm 2i\omega t}$ vanish, and we obtain

$$Q = \tfrac{1}{4}i\omega(\alpha^* - \alpha)|f_0|^2 = \tfrac{1}{2}\omega\alpha''|f_0|^2. \qquad (125.11)$$

From this we see that the imaginary part of the susceptibility determines the dissipation of energy. Since any actual process is always accompanied by some dissipation ($Q > 0$), we reach the important conclusion that, for all positive values of the variable ω, the function α'' is positive and not zero.

It is possible to derive some very general relations concerning the function $\alpha(\omega)$ by using the methods of the theory of functions of a complex variable. We regard ω as a complex variable ($\omega = \omega' + i\omega''$) and consider the properties of the function $\alpha(\omega)$ in the upper half of the ω-plane. From the definition (125.4) and the fact that $K(\tau)$ is finite for all positive τ, it follows that $\alpha(\omega)$ is a one-valued regular function everywhere in the upper half-plane. For, when $\omega'' > 0$, the integrand in (125.4) includes the exponentially decreasing factor $e^{-\tau\omega''}$ and, since the function $K(\tau)$ is finite throughout the range of integration, the integral converges. The function $\alpha(\omega)$ has no singularity on the real axis ($\omega'' = 0$), except possibly at the origin.† It is useful to notice that the conclusion that $\alpha(\omega)$ is regular in the upper half-plane is, physically, a consequence of the causality principle. Owing to this principle, the integration in (125.2) is taken only over times previous to t, and the range of integration in (125.4) therefore extends from 0 to ∞ rather than from $-\infty$ to ∞.

It is evident also from the definition (125.4) that

$$\alpha(-\omega^*) = \alpha^*(\omega). \tag{125.12}$$

This generalises the relation (125.6) for real ω. In particular, for purely imaginary ω we have $\alpha(i\omega'') = \alpha^*(i\omega'')$, i.e. the function $\alpha(\omega)$ is real on the imaginary axis.

We shall prove the following theorem. The function $\alpha(\omega)$ does not take real values at any finite point in the upper half-plane except on the imaginary axis, where it decreases monotonically from a positive value $\alpha_0 > 0$ at $\omega = i0$ to zero at $\omega = i\infty$. Hence, in particular, it will follow that the function $\alpha(\omega)$ has no zeros in the upper half-plane.

To prove the theorem‡ we use a theorem from the theory of functions of a complex variable, according to which the integral

$$\frac{1}{2\pi i}\int \frac{d\alpha(\omega)}{d\omega}\frac{d\omega}{\alpha(\omega)-a}, \tag{125.13}$$

taken round some closed contour C, is equal to the difference between the number of zeros and the number of poles of the function $\alpha(\omega)-a$ in the region bounded by the contour. Let a be a real number and let C be taken as a contour consisting of the real axis and an infinite semicircle in the upper

† In the lower half-plane, the definition (125.4) is invalid, since the integral diverges. Hence the function $\alpha(\omega)$ can be defined in the lower half-plane only as the analytical continuation of the expression (125.4) from the upper half-plane, and in general has singularities in this region.

‡ The proof given here is due to N. N. Meĭman.

half-plane (Fig 54) Let us first suppose that α_0 is finite. Since in the upper half-plane the function $\alpha(\omega)$ has no pole, the same is true of $\alpha(\omega)-a$, and the integral in question gives simply the number of zeros of the difference $\alpha-a$, i.e. the number of points at which $\alpha(\omega)$ takes the real value a.

To calculate the integral, we write it as

$$\frac{1}{2\pi i}\int_{C'}\frac{d\alpha}{\alpha-a},$$

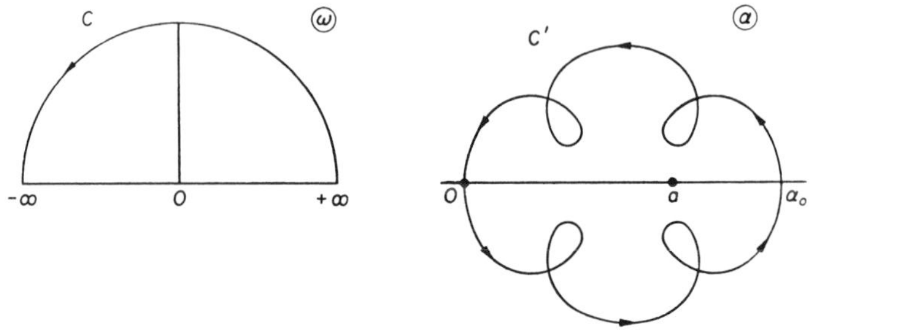

FIG. 54

the integration being round a contour C' in the plane of the complex variable α which is the map of the contour C in the ω-plane. The whole infinite semicircle is mapped on to the point $\alpha = 0$, and the origin ($\omega = 0$) is mapped on to another real point α_0. The right and left halves of the real axis of ω are mapped in the α-plane on to some very complicated (generally self-intersecting) curves which are entirely in the upper and lower half-planes respectively. It is important to note that these curves nowhere meet the real axis (except at $\alpha = 0$ and $\alpha = \alpha_0$), since α does not take real values for any real finite ω except $\omega = 0$. Because of this property of the contour C', the total change of the argument of the complex number $\alpha-a$ on passing round C' is 2π (if a lies between 0 and α_0 as shown in Fig. 54) or zero (if a lies outside that range), whatever the number of self-intersections of the contour. Hence it follows that the expression (125.13) is equal to 1 if $0 < a < \alpha_0$ and zero for any other value of a.

Thus we conclude that the function $\alpha(\omega)$ takes, in the upper half-plane of ω, each real value of a in this range once only, and values outside this range not at all. Hence we can deduce first of all that on the imaginary axis, where the function $\alpha(\omega)$ is real, it cannot have either a maximum or a minimum, since otherwise it would take some values at least twice. Consequently, $\alpha(\omega)$ varies monotonically on the imaginary axis, taking on that axis, and nowhere else, all real values from α_0 to zero once only.

If $\alpha_0 = \infty$ (i.e. $\alpha(\omega)$ has a pole at the point $\omega = 0$), the above proof is affected only in that on passing along the real axis (in the ω-plane) it is necessary to avoid the origin by means of an infinitesimal semicircle above it. The change in the contour C' in Fig. 54 can be regarded as the result of moving α_0 to infinity. The function $\alpha(\omega)$ then decreases monotonically from ∞ to 0 on the imaginary axis.

Let us now derive a formula relating the real and imaginary parts of the function $\alpha(\omega)$. To do so, we choose some real positive value ω_0 of ω and integrate the expression $\alpha/(\omega-\omega_0)$ round the contour shown in Fig. 55. This contour includes the whole of the real axis, indented upwards at the point $\omega = \omega_0 > 0$ (and also at the point $\omega = 0$ if the latter is a pole of the function $\alpha(\omega)$), and is completed by an infinite semicircle. At infinity, $\alpha \to 0$, and the

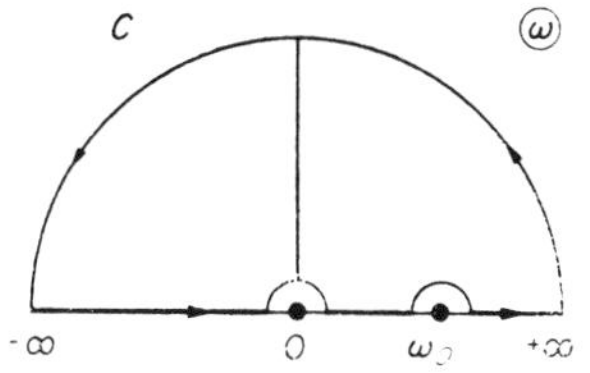

FIG. 55

function $\alpha/(\omega-\omega_0)$ therefore tends to zero more rapidly than $1/\omega$. The integral

$$\int\limits_C \frac{\alpha(\omega)}{\omega-\omega_0}\, d\omega$$

consequently converges; and since $\alpha(\omega)$ is regular in the upper half-plane, and the point $\omega = \omega_0$ has been excluded from the region of integration, the function $\alpha/(\omega-\omega_0)$ is analytic everywhere inside the contour C, and the integral is therefore zero.

The integral along the infinite semicircle is also zero. The point ω_0 is avoided by means of an infinitesimal semicircle whose radius ϱ tends to zero. The direction of integration is clockwise, and the contribution to the integral is $-i\pi\alpha(\omega_0)$. If α_0 is finite, the indentation at the origin is unnecessary, and the integration along the whole real axis therefore gives

$$\lim_{\varrho\to 0}\left\{\int\limits_{-\infty}^{\omega_0-\varrho} \frac{\alpha}{\omega-\omega_0}\, d\omega + \int\limits_{\omega_0+\varrho}^{\infty} \frac{\alpha}{\omega-\omega_0}\, d\omega\right\} - i\pi\alpha(\omega_0) = 0.$$

The first term is the principal value of the integral from $-\infty$ to ∞. Indicating this in the usual way, we have

$$i\pi\alpha(\omega_0) = P\int\limits_{-\infty}^{\infty} \frac{\alpha}{\omega-\omega_0}\, d\omega. \qquad (125.14)$$

Here the variable of integration ω takes only real values. We replace it by ξ, call the given real value ω instead of ω_0, and write the function $\alpha(\omega)$ of the real variable ω in the form $\alpha = \alpha' + i\alpha''$. Taking the real and imaginary parts of (125.14), we obtain the following two formulae:

$$\alpha'(\omega) = \frac{1}{\pi} P \int_{-\infty}^{\infty} \frac{\alpha''(\xi)}{\xi - \omega} d\xi, \tag{125.15}$$

$$\alpha''(\omega) = -\frac{1}{\pi} P \int_{-\infty}^{\infty} \frac{\alpha'(\xi)}{\xi - \omega} d\xi, \tag{125.16}$$

first derived by H. A. KRAMERS and R. DE L. KRONIG (1927). It should be emphasised that the only essential property of the function $\alpha(\omega)$ used in the proof is that it is regular in the upper half-plane.† Hence we can say that KRAMERS and KRONIG's formulae, like this property of $\alpha(\omega)$, are a direct consequence of the causality principle.

Using the fact that $\alpha''(\xi)$ is an odd function, we can rewrite (125.15) as

$$\alpha'(\omega) = \frac{1}{\pi} P \int_{0}^{\infty} \frac{\alpha''(\xi)}{\xi - \omega} d\xi + \frac{1}{\pi} P \int_{0}^{\infty} \frac{\alpha''(\xi)}{\xi + \omega} d\xi,$$

or

$$\alpha'(\omega) = \frac{2}{\pi} P \int_{0}^{\infty} \frac{\xi \alpha''(\xi)}{\xi^2 - \omega^2} d\xi. \tag{125.17}$$

If the function $\alpha(\omega)$ has a pole at the point $\omega = 0$, near which $\alpha = iA/\omega$, the semicircle avoiding this pole gives a further real term $-A/\omega_0$, which must be added to the left-hand side of equation (125.14). Thus formula (125.16) becomes

$$\alpha''(\omega) = -\frac{1}{\pi} P \int_{-\infty}^{\infty} \frac{\alpha'(\xi)}{\xi - \omega} d\xi + \frac{A}{\omega}, \tag{125.18}$$

but (125.15) and (125.17) remain unchanged.

We may also derive a formula which expresses the values of $\alpha(\omega)$ on the positive imaginary axis in terms of the values of $\alpha''(\omega)$ on the real axis. To do so, we consider the integral

$$\int \frac{\omega \alpha(\omega)}{\omega^2 + \omega_0^2} d\omega$$

† The property $\alpha \to 0$ as $\omega \to \infty$ is not essential: if the limit α_∞ were other than zero, we should simply take $\alpha - \alpha_\infty$ in place of α, with corresponding obvious changes in formulae (125.15), (125.16).

taken along a contour consisting of the real axis and an infinite semicircle in the upper half-plane (ω_0 being a real number). This integral can be expressed in terms of the residue of the integrand at the pole $\omega = i\omega_0$. The integral along the infinite semicircle is zero, and so we have

$$\int_{-\infty}^{\infty} \frac{\omega\alpha(\omega)}{\omega^2+\omega_0^2}\,d\omega = i\pi\alpha(i\omega_0).$$

On the left-hand side the real part of the integral is zero, since the integrand is an odd function. Replacing ω by ξ and ω_0 by ω, we have finally

$$\alpha(i\omega) = \frac{2}{\pi}\int_0^{\infty} \frac{\xi\alpha''(\xi)}{\omega^2+\xi^2}\,d\xi. \tag{125.19}$$

Integration of this with respect to ω gives

$$\int_0^{\infty} \alpha(i\omega)\,d\omega = \int_0^{\infty} \alpha''(\omega)\,d\omega. \tag{125.20}$$

§126. Non-thermodynamic fluctuations of a single variable

Let a body to which the quantity x refers be in a particular (nth) stationary state. The mean value (121.10) is calculated as the corresponding diagonal matrix element

$$\tfrac{1}{2}(\hat{x}_\omega\hat{x}_{\omega'}+\hat{x}_{\omega'}\hat{x}_\omega)_{nn} = \tfrac{1}{2}\sum_m [(x_\omega)_{nm}(x_{\omega'})_{mn}+(x_{\omega'})_{nm}(x_\omega)_{mn}], \tag{126.1}$$

where the summation is over the whole spectrum of energy levels; since the operator $\hat{x}_\omega$ is complex, the two terms in the brackets are not equal.

The time dependence of the operator $\hat{x}$ means that its matrix elements must be calculated by means of the time-dependent wave functions. We therefore have

$$(x_\omega)_{nm} = \frac{1}{2\pi}\int_{-\infty}^{\infty} x_{nm}e^{i(\omega_{nm}+\omega)t}\,dt = x_{nm}\,\delta(\omega_{nm}+\omega), \tag{126.2}$$

where x_{nm} is the ordinary time-independent matrix element of the operator $\hat{x}$, expressed in terms of the co-ordinates of the particles of the body, and $\omega_{nm} = (E_n-E_m)/\hbar$ is the frequency of the transition between the states n and m. Thus

$$\tfrac{1}{2}(\hat{x}_\omega\hat{x}_{\omega'}+\hat{x}_{\omega'}\hat{x}_\omega)_{nn} = \tfrac{1}{2}\sum_m |x_{nm}|^2\,[\delta(\omega_{nm}+\omega)\,\delta(\omega_{mn}+\omega')+$$
$$+\delta(\omega_{nm}+\omega')\,\delta(\omega_{mn}+\omega)],$$

where we have used the fact that $x_{nm} = x_{mn}^*$, since x is real. The products of delta functions in the brackets can clearly be written as

$$\delta(\omega_{nm}+\omega)\,\delta(\omega+\omega')+\delta(\omega_{mn}+\omega)\,\delta(\omega+\omega').$$

A comparison with (121.10) then gives

$$(x^2)_\omega = \tfrac{1}{2}\sum_m |x_{nm}|^2\,[\delta(\omega+\omega_{nm})+\delta(\omega+\omega_{mn})]. \tag{126.3}$$

The following comment may be made concerning the way in which this expression is written. Although the energy levels of a macroscopic body are, strictly speaking, discrete, they are so close together that in practice they form a continuous spectrum. Formula (126.3) may be written without the delta functions if it is averaged over small frequency intervals (which nevertheless contain many levels). If $\Gamma(E)$ is the number of energy levels less than E, then

$$(x^2)_\omega = \tfrac{1}{2}\hbar\,|x_{nm}|^2\left[\frac{d\Gamma}{dE_m}+\frac{d\Gamma}{dE_m'}\right], \tag{126.4}$$

where $E_m = E_n+\hbar\omega$, $E_m' = E_n-\hbar\omega$.

Let us now assume that the body is subject to a periodic perturbation (with frequency ω), described by the operator

$$\hat{V} = -f\hat{x} = -\tfrac{1}{2}(f_0e^{-i\omega t}+f_0^*e^{i\omega t})\hat{x}. \tag{126.5}$$

Under the effect of the perturbation the system makes transitions, and the probability (per unit time) of the transition $n \to m$ is given by†

$$w_{nm} = \frac{\pi|f_0|^2}{2\hbar^2}|x_{mn}|^2\{\delta(\omega+\omega_{mn})+\delta(\omega+\omega_{nm})\}. \tag{126.6}$$

The two terms in this formula correspond to those in (126.5). In each transition the system absorbs or emits a quantum $\hbar\omega_{mn}$. The sum

$$Q = \sum_m w_{nm}\hbar\omega_{mn}$$

is the mean energy absorbed by the body per unit time; this energy is supplied by the external perturbation, and after absorption in the body it is dissipated there. Substitution of (126.6) gives

$$Q = \frac{\pi}{2\hbar}|f_0|^2\sum_m |x_{nm}|^2\{\delta(\omega+\omega_{mn})+\delta(\omega+\omega_{nm})\}\omega_{mn}$$

or, since the delta functions are zero except when their argument is zero,

$$Q = \frac{\pi}{2\hbar}\omega\,|f_0|^2\sum_m |x_{nm}|^2\{\delta(\omega+\omega_{nm})-\delta(\omega+\omega_{mn})\}. \tag{126.7}$$

† See *Quantum Mechanics*, §42.

Comparison of (126.7) and (125.11) gives

$$\alpha''(\omega) = \frac{\pi}{\hbar} \sum |x_{nm}|^2 \{\delta(\omega+\omega_{nm}) - \delta(\omega+\omega_{mn})\}. \tag{126.8}$$

The quantities $(x^2)_\omega$ and α'' thus calculated are related in a simple manner, but the relation appears only when these quantities are expressed in terms of the temperature of the body. To do this, we average by means of the Gibbs distribution (cf. the second footnote to §121). For $(x^2)_\omega$ we have

$$(x^2)_\omega = \tfrac{1}{2} \sum_{n,\, m} \varrho_n |x_{nm}|^2 \{\delta(\omega+\omega_{nm}) + \delta(\omega+\omega_{mn})\},$$

where for brevity we have put $\varrho_n = e^{(F-E_n)/T}$, E_n denoting the energy levels of the body and F its free energy. Since the summation is now over both suffixes m and n, these can be interchanged. If this is done in the second term, we obtain

$$\begin{aligned}(x^2)_\omega &= \tfrac{1}{2} \sum_{m,\, n} (\varrho_n + \varrho_m) |x_{nm}|^2 \, \delta(\omega+\omega_{nm}) \\ &= \tfrac{1}{2} \sum_{m,\, n} \varrho_n (1 + e^{\hbar\omega_{nm}/T}) |x_{nm}|^2 \, \delta(\omega+\omega_{nm})\end{aligned}$$

or, because of the delta function in the summand,

$$(x^2)_\omega = \tfrac{1}{2}(1+e^{-\hbar\omega/T}) \sum_{m,\, n} \varrho_n |x_{nm}|^2 \, \delta(\omega+\omega_{nm}).$$

In an exactly similar manner we obtain

$$\alpha'' = \frac{\pi}{\hbar} (1-e^{-\hbar\omega/T}) \sum_{m,\, n} \varrho_n |x_{nm}|^2 \, \delta(\omega+\omega_{nm}).$$

A comparison of these two expressions gives

$$(x^2)_\omega = \frac{\hbar\alpha''}{2\pi} \coth\frac{\hbar\omega}{2T} = \frac{\hbar\alpha''}{\pi} \left\{ \tfrac{1}{2} + \frac{1}{e^{\hbar\omega/T}-1} \right\}. \tag{126.9}$$

The mean square of the fluctuating quantity itself is given by the integral

$$\overline{x^2} = \frac{\hbar}{\pi} \int_0^\infty \alpha''(\omega) \coth\frac{\hbar\omega}{2T} \, d\omega. \tag{126.10}$$

These important formulae (derived by H. B. Callen and T. A. Welton, 1951) relate the fluctuations of physical quantities to the dissipative properties of the system when it is subject to an external interaction. It should be noted that the factor in the braces in (126.9) is formally the mean energy (in units of $\hbar\omega$) of an oscillator of frequency ω at temperature T; the term $\frac{1}{2}$ corresponds to the zero-point oscillations.

The results obtained above can be written in a different form by regarding the spontaneous fluctuations of the quantity x purely formally as due to the action of some fictitious "random forces" f. It is convenient to write the formulae in terms of the "Fourier components" x_ω and f_ω as if x were an ordinary (not an operator) quantity. The relation between them is

$$x_\omega = \alpha(\omega) f_\omega, \tag{126.11}$$

and the mean square fluctuations can then be written in the form

$$\overline{x_\omega x_{\omega'}} = \alpha(\omega)\alpha(\omega')\overline{f_\omega f_{\omega'}} = (x^2)_\omega \delta(\omega+\omega') = |\alpha|^2 (f^2)_\omega\, \delta(\omega+\omega').$$

Hence we have from (126.9) for the spectral density of the mean square random force

$$(f^2)_\omega = \frac{\hbar\alpha''}{2\pi|\alpha|^2}\coth\frac{\hbar\omega}{2T}. \tag{126.12}$$

This treatment of the fluctuations may offer certain advantages in particular applications of the theory.

At temperatures $T \gg \hbar\omega$ we have $\coth(\hbar\omega/2T) \approx 2T/\hbar\omega$, and formula (126.9) becomes

$$(x^2)_\omega = (T/\pi\omega)\alpha''(\omega). \tag{126.13}$$

The constant $\hbar$ no longer appears, in accordance with the fact that under these conditions the fluctuations are classical.

If the inequality $T \gg \hbar\omega$ is valid for all frequencies of importance (those for which $\alpha''(\omega)$ is significantly different from zero), then we can take the classical limit in the integral formula (126.10) also:

$$\overline{x^2} = \frac{2T}{\pi}\int_0^\infty \frac{\alpha''(\omega)}{\omega}\, d\omega.$$

But from (125.17) this integral can be expressed in terms of the static value $\alpha'(0) = \alpha(0)$, and hence

$$\overline{x^2} = T\alpha(0).$$

But $\alpha(0) = 1/\beta T$ (see formula (127.20) below), and we return to the known result (112.4). This is not surprising, since this formula depends only on the fact that x is classical, not on the fluctuations' being thermodynamic.

PROBLEM

Derive formula (125.17) by a direct quantum-mechanical calculation of the mean value of x in the perturbed system.

SOLUTION. Let $\Psi_n^{(0)}$ be the wave functions of the unperturbed system. Following the general method†, we seek the wave functions of the perturbed system, in the first approximation, in the form

$$\Psi_n = \Psi_n^{(0)} + \sum_m a_{mn}\Psi_m^{(0)},$$

where the coefficients a_{mn} satisfy the equations

$$i\hbar\frac{da_{mn}}{dt} = V_{mn}e^{i\omega_{mn}t} = -\tfrac{1}{2}x_{mn}e^{i\omega_{mn}t}(f_0 e^{-i\omega t} + f_0{}^* e^{i\omega t}).$$

† See *Quantum Mechanics*, §40.

Hence

$$a_{mn} = \frac{1}{2\hbar} x_{mn} e^{i\omega_{mn}t} \left(\frac{f_0 e^{-i\omega t}}{\omega_{mn}-\omega} + \frac{f_0{}^* e^{i\omega t}}{\omega_{mn}+\omega} \right)$$

where we assume that $|\omega|$ is not equal to any of the frequencies ω_{mn}. Using the resulting function Ψ_n, we calculate the mean value of x as the corresponding diagonal matrix element of the operator $\hat{x}$; in the same approximation, we have

$$\begin{aligned}\bar{x} &= \int \Psi_n{}^* \hat{x} \Psi_n \, dq = \sum_m (a_{mn} x_{nm} e^{i\omega_{nm}t} + a_{mn}{}^* x_{mn} e^{i\omega_{mn}t}) \\ &= \frac{1}{2\hbar} \sum_m \left[\frac{x_{mn} x_{nm}}{\omega_{mn}-\omega} + \frac{x_{mn} x_{nm}}{\omega_{mn}+\omega} \right] (f_0 e^{-i\omega t} + f_0{}^* e^{i\omega t}) \\ &= \frac{1}{\hbar} \sum_m \frac{\omega_{mn} |x_{nm}|^2}{\omega_{mn}{}^2 - \omega^2} (f_0 e^{-i\omega t} + f_0{}^* e^{i\omega t}).\end{aligned}$$

Comparing this expression with the definition (125.3), we find

$$\alpha'(\omega) = \frac{2}{\hbar} \sum_m \frac{\omega_{mn} |x_{nm}|^2}{\omega_{mn}{}^2 - \omega^2}; \tag{1}$$

the imaginary part of α is absent, of course, since we have assumed that $|\omega| \neq \omega_{mn}$. If we substitute (1) and (126.8) in (125.17), it is easy to see that the latter is in fact satisfied identically, noting that in the integration over positive ξ only one of the delta functions in $\alpha''(\xi)$ can be non-zero.

§127. Non-thermodynamic fluctuations of more than one variable

The results given above can easily be generalised to the case where several fluctuating quantities x_i are considered simultaneously. The derivation will be given without repeating in detail calculations which are exactly analogous to those in §126.

Let x_i and x_k be any two of the physical quantities under consideration. We define the quantum-mechanical mean values of the symmetrised operator products:

$$\tfrac{1}{2}\overline{(\hat{x}_{i\omega}\hat{x}_{k\omega'} + \hat{x}_{k\omega'}\hat{x}_{i\omega})} = (x_i x_k)_\omega \, \delta(\omega+\omega'), \tag{127.1}$$

a generalisation of (121.10). A calculation similar to the derivation of (126.3) gives

$$\begin{aligned}(x_i x_k)_\omega = \tfrac{1}{2} \sum_m \{&(x_i)_{nm}(x_k)_{mn} \, \delta(\omega+\omega_{nm}) + \\ &+ (x_k)_{nm}(x_i)_{mn} \, \delta(\omega+\omega_{mn})\}.\end{aligned} \tag{127.2}$$

The perturbation acting on the system may be written

$$\hat{V} = -f_i \hat{x}_i = -\tfrac{1}{2}(f_{0i} e^{-i\omega t} + f_{0i}{}^* e^{i\omega t}) \hat{x}_i. \tag{127.3}$$

The amount of energy absorbed by the system per unit time is calculated in the same way as (126.7):

$$Q = \frac{\pi}{2\hbar} \omega \sum_n f_{0i} f_{0k}{}^* [(x_i)_{mn}(x_k)_{nm} \, \delta(\omega+\omega_{nm}) - (x_i)_{nm}(x_k)_{mn} \, \delta(\omega+\omega_{mn})]. \tag{127.4}$$

The definition (125.9) is generalised as follows:

$$\overline{x_i} = \tfrac{1}{2}(\alpha_{ik} f_{0k} e^{-i\omega t} + \alpha_{ik}^* f_{0k}^* e^{i\omega t}) \tag{127.5}$$

or

$$\overline{x_i} = \alpha_{ik} f_k, \tag{127.6}$$

if all quantities are expressed in complex form ($\sim e^{-i\omega t}$). The change in energy is given in terms of the external perturbation by

$$\dot{E} = -\dot{f}_i \overline{x_i}. \tag{127.7}$$

This formula, like (125.10), is generally used in specific applications of the theory in order to establish the actual correspondence between the quantities x_i and f_i.

Substituting (127.5) in (127.7) and averaging also over the period of the perturbation, we have instead of (125.11) the following expression for the energy dissipation:

$$Q = \tfrac{1}{4} i\omega(\alpha_{ik}^* - \alpha_{ki}) f_{0i} f_{0k}^*. \tag{127.8}$$

A comparison with (127.4) gives

$$\alpha_{ik}^* - \alpha_{ki} = -\frac{2\pi i}{\hbar} \sum_m [(x_i)_{mn}(x_k)_{nm}\, \delta(\omega+\omega_{nm}) - \\ -(x_i)_{nm}(x_k)_{mn}\, \delta(\omega+\omega_{mn})]. \tag{127.9}$$

Averaging this expression and (127.2) over the Gibbs distribution as in §126, we find the following generalisation of (126.9):

$$(x_i x_k)_\omega = \frac{i\hbar}{4\pi} (\alpha_{ki}^* - \alpha_{ik}) \coth\frac{\hbar\omega}{2T}. \tag{127.10}$$

As in formulae (126.11), (126.12), formula (127.10) can be expressed in terms of fictitious "random forces", the action of which produces results equivalent to the spontaneous fluctuations of the quantities x_i. To do so, we write

$$x_{i\omega} = \alpha_{ik} f_{k\omega}, \qquad f_{i\omega} = \alpha^{-1}{}_{ik} x_{k\omega} \tag{127.11}$$

and

$$(f_i f_k)_\omega = \alpha^{-1}{}_{il} \alpha^{-1}{}_{km} (x_l x_m)_\omega.$$

Substituting (127.10), we obtain

$$(f_i f_k)_\omega = \frac{i\hbar}{4\pi} (\alpha^{-1}{}_{ik} - \alpha^{-1}{}_{ki}{}^*) \coth\frac{\hbar\omega}{2T}. \tag{127.12}$$

From these formulae we can derive some conclusions concerning the symmetry properties of the quantities $\alpha_{ik}(\omega)$.† Let us first suppose that the quantities x_i, x_k are such that they are invariant under time reversal; then the corresponding operators $\hat{x}_i$, $\hat{x}_k$ are real. We shall further suppose that the body has

† The results given below are due to H. B. Callen, M. L. Barasch, J. L. Jackson and R. F. Greene (1952).

no "magnetic structure" (see the first footnote to §128) and is not in an external magnetic field; then the wave functions of its stationary states are also real,† and consequently so are the matrix elements of the quantities x. Since the matrices x_{nm} are Hermitian, we have $x_{nm} = x_{mn}^* = x_{mn}$. We also see that the right-hand side of (127.9), and consequently the left-hand side, are symmetrical in the suffixes i and k. Hence $\alpha_{ik}^* - \alpha_{ki} = \alpha_{ki}^* - \alpha_{ik}$ or $\alpha_{ik} + \alpha_{ik}^* = \alpha_{ki} + \alpha_{ki}^*$, i.e. we conclude that the real part of α_{ik} is symmetric.

But the real and imaginary parts (α_{ik}' and α_{ik}'') of each α_{ik} are related by linear integral equations, namely KRAMERS and KRONIG's formulae. Hence the symmetry of α_{ik}' implies that of α_{ik}'' and therefore that of α_{ik} itself. Our final result is therefore

$$\alpha_{ik} = \alpha_{ki}. \tag{127.13}$$

These relationships are somewhat modified if the body is in a constant external magnetic field $\mathbf{H}$. The wave functions of a system in a magnetic field are not real, but have the property $\psi^*(\mathbf{H}) = \psi(-\mathbf{H})$. Accordingly the matrix elements of the quantities x are such that $x_{nm}(\mathbf{H}) = x_{mn}(-\mathbf{H})$, and the expression on the right of (127.9) is unchanged, when the suffixes i and k are transposed, only if the sign of $\mathbf{H}$ is simultaneously changed. We therefore obtain the relation

$$\alpha_{ik}^*(\mathbf{H}) - \alpha_{ki}(\mathbf{H}) = \alpha_{ki}^*(-\mathbf{H}) - \alpha_{ik}(-\mathbf{H}).$$

Another relation is given by KRAMERS and KRONIG's formula (125.14), according to which

$$\alpha_{ki} = i\hat{J}(\alpha_{ki}),$$

where $\hat{J}$ is a real linear operator. Adding this to the Hermitian conjugate equation $\alpha_{ik}^* = -i\hat{J}(\alpha_{ik}^*)$, we obtain

$$\alpha_{ik}^* + \alpha_{ki} = -i\hat{J}(\alpha_{ik}^* - \alpha_{ki});$$

here all the α_{ik} are, of course, taken for a fixed value of $\mathbf{H}$. Hence we see that, if the difference $\alpha_{ik}^* - \alpha_{ki}$ has a particular symmetry property, then so has the sum $\alpha_{ik}^* + \alpha_{ki}$, and therefore α_{ik} itself. Thus

$$\alpha_{ik}(\mathbf{H}) = \alpha_{ki}(-\mathbf{H}). \tag{127.14}$$

Finally, let the quantities x include some which change sign under time reversal. The quantum-mechanical operator corresponding to such a quantity is purely imaginary, and so $x_{nm} = x_{mn}^* = -x_{mn}$. If the two quantities x_i, x_k are both of this kind, the derivation of (127.13) is unaffected, but if only one of them changes sign under time reversal, the right-hand side of equation

† The exact energy levels of a system of interacting particles can be degenerate only with respect to the directions of the total angular momentum of the system. This source of degeneracy can be eliminated by assuming the body to be enclosed in a vessel with immovable walls. The energy levels of the body will not then be degenerate, and so the corresponding exact wave functions can be taken as real.

(127.9) changes sign when the suffixes i, k are interchanged. Accordingly (127.13) becomes

$$\alpha_{ik} = -\alpha_{ki}. \tag{127.15}$$

Similarly in a magnetic field we have instead of (127.14)

$$\alpha_{ik}(\mathbf{H}) = -\alpha_{ki}(-\mathbf{H}). \tag{127.16}$$

All these relations can, of course, be derived also from formula (127.10) as a consequence of the symmetry of the fluctuations with respect to time. In the Fourier components, the effect of time reversal is to replace ω by $-\omega$ (if the quantity x itself is invariant under time reversal). In the expressions (127.1) (which are in fact different from zero only when $\omega' = -\omega$) this means interchanging ω and ω' or, equivalently, interchanging i and k. The time symmetry of the fluctuations therefore implies that $(x_i x_k)_\omega = (x_k x_i)_\omega$, i.e. the left-hand side of equation (127.10), and therefore the right-hand side, are symmetrical in the suffixes i, k, and we again obtain the relations (127.13). This derivation of the symmetry properties of the α_{ik} is analogous to the usual derivation of ONSAGER's principle of the symmetry of the kinetic coefficients, and we shall see below that formulae (127.13)—(127.16) may be regarded as a generalisation of that principle.

We shall now show the relationship between the foregoing general theory and the theory of thermodynamic fluctuations. The quantities whose fluctuations may be regarded as thermodynamic have the property that they satisfy equations of the form $\dot{x}_i = -\gamma_{ik}X_k$, which describe the behaviour of a closed system not in equilibrium. If the system is not closed but is subject to external forces, the right-hand sides of these equations must include additional forces which we denote by y_i:

$$\dot{x} = -\gamma_{ik}X_k + y_i. \tag{127.17}$$

It is easy to express the y_i in terms of the quantities f_i which describe the perturbation in question.†

To do so, we assume that static forces act on the body, i.e. the f_i are constant. This interaction causes a "displacement" of the equilibrium state, in which the mean values of the X_i are no longer zero. These new mean values can be expressed in terms of the f_i as follows. The energy of a body subject to a constant perturbation is $E = E_0 - f_i\overline{x_i}$, where E_0 is the energy of the body in the absence of the perturbation. The differential of E is

$$dE = T\,dS + (\partial E/\partial f_i)\,df_i.$$

† It should be emphasised that another interpretation of equation (127.17) is also possible: the quantities y_i (or f_i) may be regarded not as resulting from some external interaction on a system far from equilibrium but as "random forces", the inclusion of which in the equation makes it applicable to the fluctuating quantities x_i in a closed system. This interpretation corresponds to the form (127.12) of the fundamental formula.

But, from the general formula (11.4),

$$\partial E/\partial f_i = \overline{\partial \hat{H}/\partial f_i} = \overline{\partial \hat{V}/\partial f_i} = -\bar{x}_i,$$

and so $dE = d(E_0 - f_i\bar{x}_i) = T\,dS - \bar{x}_i\,df_i$, or

$$dE_0 = T\,dS + f_i\,d\bar{x}_i.$$

Thus we find the equilibrium values

$$X_i = -(\partial S/\partial \bar{x}_i)_{E_0} = f_i/T.$$

On the other hand, the right-hand sides of equations (127.17) must be zero in equilibrium. We see, therefore, that these equations can be written in terms of the f_i as

$$\dot{x}_i = -\gamma_{ik}(X_k - f_k/T). \tag{127.18}$$

We can now derive a relation between the generalised susceptibilities α_{ik} and the kinetic coefficients γ_{ik}. To do so, we substitute x_i from (127.5) in (127.18), and write the X_i as the linear combinations

$$X_i = \beta_{ik}x_k. \tag{127.19}$$

Separating the terms in $e^{-i\omega t}$ and $e^{i\omega t}$ in (127.18), we obtain

$$i\omega\alpha_{im}f_{0m} = \gamma_{ik}\beta_{kl}\alpha_{lm}f_{0m} - \frac{1}{T}\gamma_{im}f_{0m},$$

whence, since the f_{0m} are arbitrary, we have the relations

$$i\omega\alpha_{im} - \gamma_{ik}\beta_{kl}\alpha_{lm} = -\frac{1}{T}\gamma_{im}$$

or

$$\alpha_{ik} = \frac{1}{T}(\beta - i\omega\gamma^{-1})^{-1}{}_{ik}, \tag{127.20}$$

where the exponents -1 denote the inverse matrices. These are the required relations.

The quantities β_{ik} are by definition symmetric with respect to their suffixes (since $\beta_{ik} = -\partial^2 S/\partial x_i \partial x_k$). Hence the symmetry of the α_{ik} implies that of the γ_{ik}, i.e. the ordinary principle of the symmetry of the kinetic coefficients.

Substituting (127.20) in (127.12), we obtain

$$(f_i f_k)_\omega = \frac{\hbar\omega T}{4\pi}(\gamma^{-1}{}_{ik} + \gamma^{-1}{}_{ki})\coth\frac{\hbar\omega}{2T}$$

or, for the $y_i = -\gamma_{ik} f_k/T$,

$$(y_i y_k)_\omega = \frac{\hbar\omega}{4\pi T}(\gamma_{ik}+\gamma_{ki}) \coth \frac{\hbar\omega}{2T}. \tag{127.21}$$

This relation differs from formula (124.10) for the fluctuations of a classical quantity x by the factor

$$(\hbar\omega/2T) \coth (\hbar\omega/2T). \tag{127.22}$$

In the classical limit ($\hbar\omega \ll T$), this factor tends to unity, and so (127.21) becomes the same as (124.10).

CHAPTER XIII

THE SYMMETRY OF CRYSTALS

§128. Symmetry of particle configuration in a body

THE most usual properties of symmetry of macroscopic bodies relate to the symmetry of the configuration of particles in them.

Atoms and molecules in motion do not occupy precisely defined places in a body, and for an exact statistical description of their arrangement we must use a "density function" $\varrho(x, y, z)$, which gives the probability of various configurations of the particles: $\varrho\, dV$ is the probability that an individual particle is in the volume element dV. The symmetry properties of the configuration of the particles are determined by the co-ordinate transformations (translations, rotations and reflections) which leave the function $\varrho(x, y, z)$ invariant. The set of all such *symmetry transformations* for a given body forms what is called its *symmetry group*.

If the body consists of different kinds of atom, the function ϱ must be determined for each kind of atom separately; this, however, is unimportant here, since all these functions in an actual body will in practice possess the same symmetry. We could also use the function ϱ defined as the total electron density due to all the atoms at each point in the body.†

The highest symmetry is that of *isotropic* bodies (bodies whose properties are the same in all directions), which include gases, liquids and amorphous solids. It is evident that in such a body all positions in space of any given particle must be equally probable, i.e. we must have ϱ = constant. For if some positions of particles were more probable than others, the properties of the body would be different in different directions (e.g. in directions passing and not passing through any two maxima of the probability).

In *anisotropic* crystalline solids, on the other hand, the density function is not simply a constant. In this case it is a triply periodic function (with periods equal to those of the crystal lattice) and has sharp maxima at the lattice points. Besides translational symmetry, the lattice (i.e. the function $\varrho(x, y, z)$) also has, in general, symmetry under certain rotations and reflections. The lattice

† Moving electrons can cause not only a mean charge density $e\varrho$ but also a mean current density $\mathbf{j}(x, y, z)$. Bodies in which there are non-zero currents are those having a "magnetic structure", and the symmetry of the vector function $\mathbf{j}(x, y, z)$ determines the symmetry of that structure. This is discussed in *Electrodynamics of Continuous Media*, §28.

points which can be made to coincide by any symmetry transformation are said to be *equivalent*. The types of crystal symmetry will be discussed in detail in §§130–134.

A problem of fundamental interest is whether bodies can exist in Nature for which the density function depends not on three but only on one or two co-ordinates (R. E. PEIERLS 1934, and L. LANDAU 1937).

For example, a body with $\varrho = \varrho(x)$ could be regarded as consisting of parallel planes regularly arranged and lying perpendicular to the x-axis, with the atoms randomly distributed in each plane. When $\varrho = \varrho(x, y)$, the atoms would be randomly distributed along lines parallel to the z-axis, but these lines themselves would be regularly arranged.

To discuss this question, let us consider the displacements undergone by small parts of the body as a result of continually occurring fluctuations. It is clear that, if such displacements increase without limit as the size of the body increases, there will necessarily be a "smoothing-out" of the function ϱ, in contradiction with hypothesis. In other words, only those distributions ϱ can occur for which the mean displacement remains finite when the dimensions of the body become arbitrarily large.

Let us first confirm that this condition is satisfied in an ordinary crystal. Let $\mathbf{u}(x, y, z)$ denote the vector of the fluctuation displacement of a small region with co-ordinates x, y, z and let $\mathbf{u}$ be represented as a Fourier series:

$$\mathbf{u} = \sum_{\mathbf{k}} \mathbf{u}_{\mathbf{k}} e^{i\mathbf{k}\cdot\mathbf{r}}; \tag{128.1}$$

this series will include only terms with not too large wave numbers, $k \lesssim 1/d$, where d is the linear dimension of the region undergoing displacement. We shall consider the fluctuations $\mathbf{u}$ at constant temperature; then their probability is given by formulae (119.1), (119.2).

To calculate ΔF_t, we must expand $F - \bar{F}$ in powers of the displacement. The expansion will involve not the function $\mathbf{u}(x, y, z)$ itself but only its derivatives (cf. §119), since $F - \bar{F}$ must vanish when $\mathbf{u}$ = constant, corresponding to a simple displacement of the body as a whole. As regards the various derivatives of $\mathbf{u}$ with respect to the co-ordinates it is evident, first of all, that the terms in the expansion which are linear in these derivatives must be absent, since otherwise F could not have a minimum for $\mathbf{u} = 0$. Next, owing to the smallness of the wave numbers k, we need go only as far as the terms quadratic in the first derivatives of $\mathbf{u}$ in the expansion of the free energy, neglecting the terms containing the higher-order derivatives. Hence we find that ΔF_t has the form

$$\Delta F_t = V \sum_{\mathbf{k}} |\mathbf{u}_{\mathbf{k}}|^2 \phi_{\text{II}}(k_x, k_y, k_z),$$

where $\phi_{\text{II}}(k_x, k_y, k_z)$ is a quadratic function of the components of the vector $\mathbf{k}$.

Hence it follows that the mean square of the fluctuation $\mathbf{u_k}$ is

$$\overline{|\mathbf{u_k}|^2} \sim \frac{T}{V}\frac{1}{\phi_{\mathrm{II}}(k_x,k_y,k_z)},$$

and for the mean square of the total displacement $\mathbf{u}$ we obtain

$$\overline{u^2} = \sum_{\mathbf{k}} \overline{|\mathbf{u_k}|^2} \sim T\iiint \frac{\mathrm{d}k_x\,\mathrm{d}k_y\,\mathrm{d}k_z}{\phi_{\mathrm{II}}(k_x,k_y,k_z)}; \tag{128.2}$$

the summation over $\mathbf{k}$ is approximately replaced, in the usual manner, by multiplication by $V\,\mathrm{d}k_x\,\mathrm{d}k_y\,\mathrm{d}k_z$ and integration. This integral converges proportionally to k at the lower limit ($\mathbf{k} \to 0$). Thus the mean square of the fluctuation displacement is a finite quantity independent of the size of the body, as it should be.

Next, let us consider a body with density function $\varrho = \varrho(x)$. Since $\varrho =$ constant along the y and z axes in such a body, no displacement along these axes can "smooth out" the density function, and such displacements are consequently of no interest here. We need therefore consider only a displacement u_x. Moreover, it is easy to see that the first derivatives $\partial u_x/\partial y$, $\partial u_x/\partial z$ cannot appear in the expansion of the free energy, since, if the body is rigidly rotated about the y or z axis, these derivatives change, whereas the free energy must obviously remain constant. Thus in the expansion of $F-\bar{F}$ we have to consider the following terms quadratic in the displacement:

$$\left(\frac{\partial u_x}{\partial x}\right)^2,\quad \frac{\partial u_x}{\partial x}\left(\frac{\partial^2 u_x}{\partial y^2}+\frac{\partial^2 u_x}{\partial z^2}\right),\quad \left(\frac{\partial^2 u_x}{\partial y^2}+\frac{\partial^2 u_x}{\partial z^2}\right)^2;$$

the derivatives with respect to y and z appear symmetrically, owing to the complete symmetry in the yz-plane. Substitution of (128.1) leads to terms of the types

$$|u_{x\mathbf{k}}|^2 k_x^2,\quad |u_{x\mathbf{k}}|^2(k_y^2+k_z^2)k_x,\quad |u_{x\mathbf{k}}|^2(k_y^2+k_z^2)^2.$$

Although the two latter expressions include powers of the wave vector components higher than the first expression, they may be of the same order of magnitude, since nothing is known *a priori* concerning the relative magnitude of k_x and k_y, k_z.

Thus the change in the free energy will be of the form

$$\Delta F_t = V\sum_{\mathbf{k}} |u_{x\mathbf{k}}|^2 \phi_{\mathrm{II}}(k_x, k_y^2+k_z^2), \tag{128.3}$$

where ϕ_{II} is a quadratic function of two variables, k_x and $k_y^2+k_z^2$. Instead of (128.2) we now have

$$\overline{u_x^2} \sim T\iiint \frac{\mathrm{d}k_x\,\mathrm{d}k_y\,\mathrm{d}k_z}{\phi_{\mathrm{II}}(k_x, k_y^2+k_z^2)}. \tag{128.4}$$

This integral is easily seen to diverge logarithmically as $\mathbf{k} \to 0$.

The divergence of the mean fluctuation of the displacement u_x implies that a point to which a particular value of $\varrho(x)$ corresponds may be displaced through very large distances; in other words, the density $\varrho(x)$ is "smoothed out" through the whole body, so that no function $\varrho = \varrho(x)$ is possible except the trivial case ϱ = constant.

Similar arguments for a body with $\varrho = \varrho(x, y)$ give the following expression for the mean squares of the displacements u_x, u_y:

$$\overline{u_x^2}, \overline{u_y^2} \sim T \iiint \frac{dk_x \, dk_y \, dk_z}{\phi_{\mathrm{II}}(k_x, k_y, k_z^2)}. \tag{128.5}$$

This integral is easily seen to converge, so that the fluctuations remain finite. Thus a body having such a density function could in theory exist, but it is not known whether such bodies do in fact exist in Nature.

§129. Symmetry with respect to orientation of molecules

The condition ϱ = constant is necessary but certainly not sufficient for a body to be isotropic. This is clear from the following example. Let us imagine a body consisting of elongated molecules, all positions in space of a molecule as a whole (i.e. of its centre of mass) being equally probable, but the axes of the molecules being predominantly oriented in one direction. Such a body is obviously anisotropic, despite the fact that ϱ = constant for each atom present in the molecule.

The property whose symmetry is here under consideration may be formulated in terms of a mutual correlation between the positions of the different atoms. Let $\varrho_{12}\, dV_2$ be the probability of finding an atom 2 in the volume element dV_2 for a given position of atom 1 (atoms of different types usually being involved); ϱ_{12} is a function of the radius vector $\mathbf{r}_{12}$ between the two atoms, and the symmetry properties of this function determine the symmetry of the body (in which ϱ = constant).

The fact that the density function ϱ is constant signifies that a relative displacement of parts of the body (without change of volume) does not affect the equilibrium state of the body, i.e. does not change its thermodynamic quantities. This is precisely the characteristic property of liquids (and gases). We must therefore regard bodies with ϱ = constant and an anisotropic function $\varrho_{12}(\mathbf{r}_{12})$ as *liquid crystals*, that is, anisotropic fluids.

When the length of the vector $\mathbf{r}_{12}$ varies without change in its direction, the functions ϱ_{12} do not, of course, display any periodicity, though they may undergo oscillations. Thus these functions do not possess translational symmetry, and their symmetry groups can consist only of certain rotations and reflections, i.e. are what are called *point groups*.†

† See *Quantum Mechanics*, §93.

Regarding liquid crystals as bodies with an anisotropic correlation ϱ_{12}, we can therefore say that their possible types of symmetry are classified in accordance with the point groups, and the order of the axes of symmetry in these groups is arbitrary. In particular, liquid crystals are possible with an axis of complete axial symmetry (the groups $\boldsymbol{C}_\infty$, $\boldsymbol{C}_{\infty h}$, $\boldsymbol{C}_{\infty v}$, $\boldsymbol{D}_\infty$, $\boldsymbol{D}_{\infty h}$)†; it is customary to suppose that all known liquid crystals are of these types, but it must be borne in mind that optical observations do not enable us to distinguish an axis of complete axial symmetry from one of order $n > 2$.

Finally, it may be mentioned that in ordinary isotropic liquids also there are two different types of symmetry. If the liquid consists of a substance which does not have stereoisomers, it is completely symmetrical not only under a rotation through any angle about any axis but also under a reflection in any plane, i.e. its symmetry group is the complete group of rotations about a point, together with a centre of symmetry (group $\boldsymbol{K}_h$). If the substance has two stereoisomeric forms, however, and the liquid contains different numbers of molecules of the two isomers, it will not possess a centre of symmetry and therefore will not allow reflections in planes. Its symmetry group is just the complete group of rotations about a point (group $\boldsymbol{K}$).

§130. Symmetry elements of a crystal lattice

Proceeding to study the symmetry of a crystal lattice, we must first of all ascertain which elements can contribute to this symmetry.

The symmetry of a crystal lattice is based on its spatial periodicity, the property of being unchanged by a parallel displacement or *translation* through certain distances in certain directions‡; translational symmetry will be further discussed in §131.

As well as translational symmetry, the lattice may also be symmetrical under certain rotations and reflections; the corresponding symmetry elements (*axes of symmetry*, *planes of symmetry*, and *rotary-reflection axes*) are the same as those which can occur in symmetrical bodies of finite size.||

In addition, however, crystal lattices can also possess symmetry elements consisting of combinations of parallel translations with rotations and reflections. Let us first consider combinations of translations with the axes of symmetry. The combination of an axis of symmetry with a translation in a direction perpendicular to the axis does not give a new type of symmetry element. It is easy to see that a rotation through a certain angle followed by a translation perpendicular to the axis is equivalent to a rotation through the same

† It will be recalled that, of these, only $\boldsymbol{C}_{\infty v}$ and $\boldsymbol{D}_{\infty h}$ can appear as the symmetry groups of a single molecule; see *Quantum Mechanics*, §98.

‡ Here the crystal lattice must be regarded as infinite, ignoring the faces of the crystal.

|| See *Quantum Mechanics*, §91.

angle about an axis parallel to the first. The combination of a rotation about an axis and a translation along that axis leads to a new type of symmetry element, a *screw axis*. The lattice has a screw axis of order n if it is unchanged by rotation through an angle $2\pi/n$ about the axis, accompanied by translation through a certain distance d along the axis.

After n rotations, with accompanying translations, about a screw axis of order n, the lattice is simply shifted along the axis by a distance nd. Thus, when there is a screw axis, the lattice must certainly also have a simple periodicity along this axis with a period not exceeding nd. This means that screw axes of order n can be correlated only with translations through distances $d = pa/n$ ($p = 1, 2, \ldots, n-1$), where a is the smallest period of the lattice in the direction of the axis. For example, a screw axis of order 2 can be of only one type, the translation being through half a period; screw axes of order 3 can be correlated with translations by $\frac{1}{3}$ or $\frac{2}{3}$ period, and so on.

Similarly, we can combine translations with planes of symmetry. Reflection in a plane together with translation in a direction perpendicular to the plane does not give a new type of symmetry element, since such a transformation is easily seen to be equivalent to a reflection in a plane parallel to the first. The combination of a reflection with a translation along a direction lying in the reflection plane leads to a new type of symmetry element, a *glide-reflection plane* or *glide plane*. The lattice has a glide-reflection plane if it is unchanged by a reflection in this plane, accompanied by a translation through a certain distance d in a certain direction lying in this plane.

A twofold reflection in a glide-reflection plane amounts to a translation through a distance $2d$. It is therefore clear that a lattice can have only glide-reflection planes such that the translation distance $d = \frac{1}{2}a$, where a is the smallest period of the lattice in the direction of the translation.

The combination of rotary-reflection axes with translations does not lead to new types of symmetry element, since in this case any translation can be resolved into two parts, one perpendicular to the axis and the other parallel to it and therefore perpendicular to the reflection plane. Thus a rotary-reflection transformation followed by a translation is always equivalent to another rotary-reflection transformation about an axis parallel to the first.

§131. The Bravais lattice

The translational periods of a lattice can be represented by vectors **a** whose directions are those of the respective translations and whose magnitudes are equal to the distances concerned. The lattice has an infinity of different lattice vectors. These vectors are not all independent, however: one can always choose in a crystal lattice three basic lattice vectors (corresponding to the three dimensions of space) which do not lie in one plane, and then any other lattice vector can be represented as a sum of three vectors each an integral multiple of one of the basic vectors. If the basic vectors are denoted by $\mathbf{a}_1$, $\mathbf{a}_2$, $\mathbf{a}_3$, an arbitrary lattice vector $\mathbf{a}$ will be of the form

$$\mathbf{a} = n_1\mathbf{a}_1+n_2\mathbf{a}_2+n_3\mathbf{a}_3, \tag{131.1}$$

where n_1, n_2, n_3 are any positive or negative integers or zero.

The choice of the basic lattice vectors is not unique. On the contrary, these vectors may be chosen in an infinity of ways. Let $\mathbf{a}_1$, $\mathbf{a}_2$, $\mathbf{a}_3$ be basic lattice vectors, and let us replace them by other vectors $\mathbf{a}_1'$, $\mathbf{a}_2'$, $\mathbf{a}_3'$, defined by the formulae

$$\mathbf{a}_i' = \sum_k \alpha_{ik}\mathbf{a}_k \qquad (i, k = 1, 2, 3), \tag{131.2}$$

where the α_{ik} are some integers. If the new periods $\mathbf{a}_i'$ are also basic lattice vectors, then, in particular, the vectors $\mathbf{a}_i$ must be expressible in terms of the $\mathbf{a}_i'$ as linear functions with integral coefficients; then any other lattice vector can also be expressed in terms of the $\mathbf{a}_i'$. In other words, if we express the $\mathbf{a}_i$ in terms of the $\mathbf{a}_i'$ in accordance with (131.2), the resulting formulae must be of the type

$$\mathbf{a}_i = \sum_k \beta_{ik}\mathbf{a}_k',$$

with the β_{ik} again integral. The determinant $|\beta_{ik}|$ is the reciprocal of the determinant $|\alpha_{ik}|$, and since both are integers it follows that the equation

$$|\alpha_{ik}| = \pm 1 \tag{131.3}$$

is a necessary and sufficient condition for the $\mathbf{a}_i'$ to be basic lattice vectors.

Let us choose a lattice point and mark off from it three basic lattice vectors. The parallelepiped formed by the three vectors is called a *unit cell* of the lattice. The whole lattice can then be regarded as a regular assembly of such parallelepipeds. All the unit cells, are of course, identical in their properties; they have the same shape and size, and each contains the same number of atoms of each kind identically arranged.

It is evident that identical atoms will be found at every vertex of every unit cell. All these vertices, therefore are, equivalent lattice points, and each can be brought to the position of any other by translation through a lattice vector. A set of all such equivalent points which can be brought into coincidence by a translation forms what is called a ***Bravais lattice*** of the crystal. This clearly does not include every point of the crystal lattice; indeed, in general it does not even include all equivalent points, since the lattice may contain equivalent points which can be made to coincide only by transformations involving rotations or reflections.

The Bravais lattice can be constructed by selecting any crystal lattice point and performing all possible translations. By taking initially some other point not in the first Bravais lattice we should obtain another Bravais lattice displaced relative to the first. It is therefore clear that the crystal lattice in general consists of several interpenetrating Bravais lattices, each corresponding to atoms of a particular type and position. All these lattices, regarded as sets of points (i.e. purely geometrically), are completely identical.

Let us return now to the unit cells. Because the choice of the basic lattice vectors is arbitrary, that of the unit cell is also not unique. The unit cell can be constructed from any basic vectors. The resulting cells are, of course, of varying shapes, but their volumes are all equal. This is most simply seen as follows. It is clear from the above discussion that each unit cell contains one point belonging to each of the Bravais lattices that can be constructed in the crystal concerned. Consequently, the number of unit cells in a given volume is always equal to the number of atoms of a particular type and position, i.e. is independent of the choice of cell. The volume of each cell is therefore the same, and equal to the total volume divided by the number of cells.

§132. Crystal systems

Let us now consider the possible types of symmetry of the Bravais lattices.

First, we shall prove a general theorem concerning the symmetry of crystal lattices with respect to rotations. Let us see which axes of symmetry the lattice can have. Let A (Fig. 56) be a point of a Bravais lattice, lying on an axis of symmetry perpendicular to the plane of the diagram. If B is another point separated from A by one of the possible translations, a similar axis of symmetry must pass through B.

Let us now perform a rotation through an angle $\phi = 2\pi/n$ about the axis through A, where n is the order of the axis. Then the point B and the axis through it will move to B'. Similarly, a rotation about B carries A into A'. From their construction, the points A' and B' belong to the same Bravais lattice as A and B, and so can be made to coincide by a translation. The distance $A'B'$ must therefore also be a translational period of the lattice. If a is the shortest period in the direction concerned, the distance $A'B'$ must therefore be equal to pa with p integral. It is seen from the figure that this gives

$$a+2a \sin(\phi-\tfrac{1}{2}\pi) = a-2a\cos\phi$$
$$= pa,$$

or $\cos\phi = \frac{1}{2}(1-p)$.

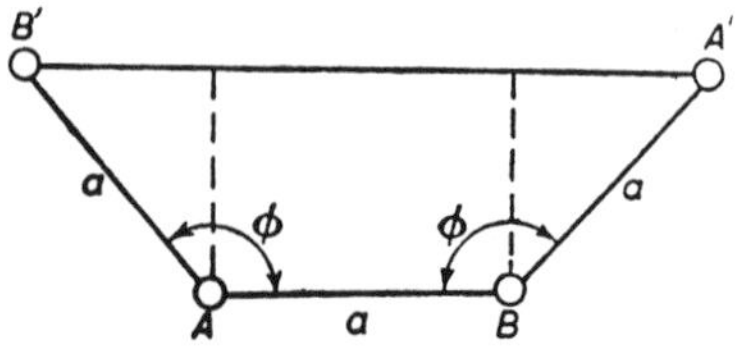

FIG. 56

Since $|\cos\phi| \leqslant 1$, p can be 3, 2, 1 or 0. These values correspond to $\phi = 2\pi/n$ with $n = 2$, 3, 4 or 6. Thus the crystal lattice can have axes of symmetry only of order 2, 3, 4 and 6.

Let us now examine the possible types of symmetry of the Bravais lattice under rotations and reflections. These types of symmetry are called *crystal systems*, and each corresponds to a certain set of axes and planes of symmetry, i.e. is a point group.

It is easy to see that every point of a Bravais lattice is a centre of symmetry thereof. For to each atom in a Bravais lattice there corresponds another atom collinear with that atom and with the lattice point considered, and such that the two atoms are equidistant from this lattice point. If the centre of symmetry is the only symmetry element of the Bravais lattice (apart from translations), we have

1. The *triclinic system*. This system, the least symmetrical of all, corresponds to the point group $\boldsymbol{C}_i$. The points of a triclinic Bravais lattice lie at the vertices of equal parallelepipeds with edges of arbitrary lengths and arbitrary angles between edges. Such a parallelepiped is shown in Fig. 57.

The Bravais lattices are customarily denoted by special symbols; that of the triclinic system is denoted by Γ_t.

2. The *monoclinic system* is next in degree of symmetry. Its symmetry elements are a second-order axis and a plane of symmetry perpendicular to this axis, forming the point group $\boldsymbol{C}_{2h}$. This is the symmetry of a right parallelepiped with a base of any shape. The Bravais lattice for this system can be constructed in two ways. In one, called the simple monoclinic Bravais lattice (Γ_m), the lattice points are at the vertices of right parallelepipeds with the *ac* face an arbitrary parallelogram (Fig. 57). In the other, the base-centred lattice (Γ_m^b), the lattice points are not only at the vertices but also at the centres of opposite rectangular faces of the parallelepipeds.

3. The *orthorhombic sytem* corresponds to the point group $\boldsymbol{D}_{2h}$. This is the symmetry of a rectangular parallelepiped with edges of any length. The system has four types of Bravais lattice. In the simple orthorhombic lattice (Γ_o), the lattice points are at the vertices of rectangular parallelepipeds. In the base-centred lattice (Γ_o^b), there are in addition lattice points at the centre of two opposite faces of each parallelepiped. In the body-centred lattice (Γ_o^v), the points are at the vertices and centres of the parallelepipeds; finally, in the face-centred lattice (Γ_o^f), the points are at the vertices and at the centre of each face.

4. The *tetragonal system* represents the point group $\boldsymbol{D}_{4h}$; this is the symmetry of a right square prism. The Bravais lattice for this system can be constructed in two ways, giving the simple and body-centred tetragonal Bravais lattices (Γ_q and Γ_q^v), whose points lie respectively at the vertices and at the vertices and centres of right square prisms.

5. The *rhombohedral system* corresponds to the point group $\boldsymbol{D}_{3d}$; this is the symmetry of a rhombohedron (a solid formed from a cube by stretching or compressing it along a spatial diagonal). In the only Bravais lattice possible in this system (Γ_{rh}) the lattice points are at the vertices of rhombohedra.

6. The *hexagonal system* corresponds to the point group $\boldsymbol{D}_{6h}$; this is the symmetry of a regular hexagonal prism. The Bravais lattice for this system (Γ_h) can be constructed in only one way; its lattice points are at the vertices

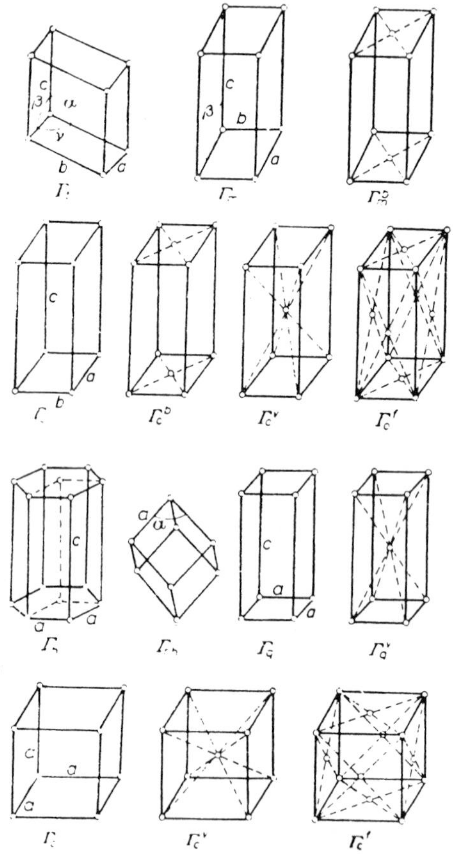

FIG. 57

of regular hexagonal prisms and at the centres of their hexagonal bases. It is useful to mention the following difference between the rhombohedral and hexagonal Bravais lattices. In both, the lattice points lie in planes perpendicular to the axis of order 3 or 6, and form a network of equilateral triangles; but in the hexagonal lattice the points are directly superimposed in successive such planes (in the direction of the C_6 axis); these planes are shown in plan in Fig. 58. In the rhombohedral lattice, on the other hand, the points in each plane lie above the centres of the triangles formed by the points in the previous plane, as shown by the circles and crosses in Fig. 58.

7. The *cubic system* corresponds to the point group $\boldsymbol{O}_h$; this is the symmetry of a cube. This system has three types of Bravais lattice: the simple cubic (Γ_c), the body-centred cubic (Γ_c^v) and the face-centred cubic (Γ_c^f).

In the sequence of systems: triclinic, monoclinic, orthorhombic, tetragonal, cubic, each has higher symmetry than those which precede it, i.e. each contains all the symmetry elements which appear in the preceding ones. The rhombohedral system is similarly of higher symmetry than the monoclinic, while at

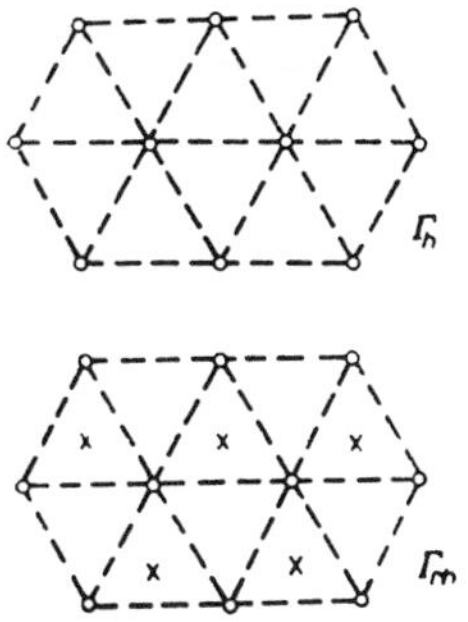

FIG. 58

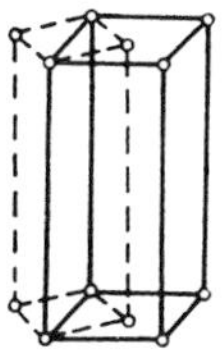

FIG. 59

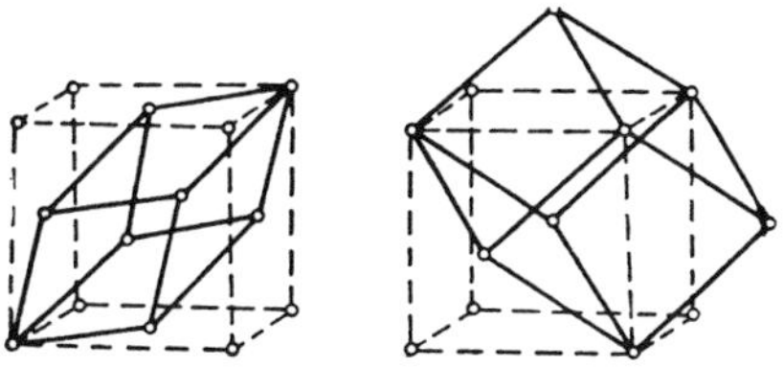

FIG. 60

the same time it is of lower symmetry than the cubic and hexagonal systems: its symmetry elements are present in both of the latter, which are the two systems of highest symmetry.

We may also mention the following fact. It might appear at first sight that further types of Bravais lattice beyond the fourteen listed above are possible. For instance, if we add to the simple tetragonal lattice a point at the centre of each opposite square base of the prisms, the lattice would again be of tetragonal symmetry. However, it is easy to see that this would not give a new Bravais lattice. For, on joining the points of such a lattice in the manner indicated in Fig. 59 by the broken lines, we see that the new lattice is again a simple tetragonal one. The same is easily found to be true in all similar cases.

The Bravais lattice parallelepipeds shown in Fig. 57 themselves have all the symmetry elements of the system to which they belong. However, it must be remembered that, for all the Bravais lattices except the simple ones, these parallelepipeds are not unit cells: the lattice vectors from which they are constructed are not basic ones. As the basic lattice vectors in the face-centred Bravais lattices we can take the vectors from any vertex of the parallelepiped to the centres of the faces, in the body-centred lattices from a vertex to the centres of the parallelepipeds, and so on. Fig. 60 shows the unit cells for the cubic lattices Γ_c^f and Γ_c^v; these cells are rhombohedra and do not themselves possess all the symmetry elements of the cubic system.

In order to define completely the triclinic Bravais lattice, it is necessary to specify six quantities: the lengths of the edges of its parallelepipeds and the angles between the edges. In the monoclinic system four quantities are sufficient, since two of the angles between the edges are always right angles. Similarly, we easily find that the Bravais lattices of the various systems are defined by the following numbers of quantities (lengths of edges of parallelepipeds or angles between edges): triclinic 6, monoclinic 4, orthorhombic 3, tetragonal 2, rhombohedral 2, hexagonal 2, cubic 1.

§133. **Crystal classes**

In many effects which may be called macroscopic, a crystal behaves as a homogeneous and continuous body. The macroscopic properties of the crystal depend only on the direction considered in it. For example, the properties of the passage of light through a crystal depend only on the direction of the light ray; the thermal expansion of a crystal is in general different in different directions; finally, the elastic deformations of a crystal under various external forces also depend on direction.

On the other hand, the symmetry of crystals brings about an equivalence of various directions in them. All macroscopic properties of a crystal will be exactly the same in such directions. We can therefore say that the macroscopic properties of the crystal are determined by the symmetry of directions in it. For instance, if the crystal has a centre of symmetry, every direction in it will be equivalent to the opposite direction.

Translational symmetry of the lattice does not lead to equivalence of directions, since parallel displacements do not affect directions. For the same reason, the difference between screw axes and simple axes of symmetry, and between simple planes of symmetry and glide-reflection planes, does not affect the symmetry of directions.

Thus the symmetry of directions, and therefore that of the macroscopic properties of the crystal, are determined by its axes and planes of symmetry, with screw axes and glide planes regarded as ordinary axes and planes. Such sets of symmetry elements are called *crystal classes*.

As we already know, an actual crystal may be regarded as a set of several interpenetrating identical Bravais lattices. Because of this superposition of the Bravais lattices, the symmetry of an actual crystal is in general different from that of the corresponding Bravais lattice.

In particular, the set of symmetry elements forming the class of a given crystal is in general different from its system. It is evident that the addition of further points to a Bravais lattice can only eliminate some of its axes or planes of symmetry, not introduce new ones. Thus the crystal class contains fewer (or at most the same number of) symmetry elements than the corresponding system, i.e. the set of axes and planes of symmetry of the Bravais lattice of the crystal in question.

From this we can derive a method of finding all the classes belonging to a given system. To do so, we must find all the point groups which contain some or all of the symmetry elements of the system. It may happen, however, that a point group thus obtained comprises symmetry elements present in more than one system. For example, we have seen in §132 that all Bravais lattices have a centre of symmetry. The point group C_i is therefore present in all systems. Nevertheless, the distribution of crystal classes among systems is usually physically unique: each class must be assigned to the system of lowest symmetry among those which contain it. For example, the class C_i must be assigned to the triclinic system, which has no symmetry element except a centre of inversion. With this method of assigning the classes, a crystal having a certain Bravais lattice will never be placed in a class which could be constructed from a Bravais lattice of a system of lower symmetry—with one exception (see below).

The necessity of satisfying this condition is physically evident: it is physically most improbable that the atoms in a crystal which belong to its Bravais lattice should be arranged more symmetrically than is required by the symmetry of the crystal. Moreover, even if such a configuration were to occur by chance, any external perturbation, even a weak one (heating, for example), would be sufficient to destroy this configuration, since it is not imposed by the symmetry of the crystal. For instance, if a cubic Bravais lattice were to occur in a crystal belonging to a class for which the tetragonal system was sufficient, even a slight interaction would be capable of lengthening or shortening one of the edges of the cubic cell, converting it into a right square prism.

From this example we see the importance of the fact that the Bravais lattice of a system of higher symmetry can be converted to that of a system of lower symmetry by means of an arbitrarily small deformation. There is one exceptional case, however, where such a transformation is not possible: a hexagonal Bravais lattice can not be converted by any infinitesimal deformation into the lattice of the rhombohedral system, which is of lower symmetry. For we see from Fig. 58 that, to transform the hexagonal into the rhombohedral lattice, it is necessary to move the vertices in alternate layers by a finite amount from the vertices to the centres of the triangles. In consequence, all the classes of the rhombohedral system can be obtained with either a hexagonal or a rhombohedral Bravais lattice.†

Thus, to find all the crystal classes, we must first look for the point groups of the triclinic system, which has the lowest symmetry, and then go on in turn

† Crystals of rhombohedral classes with a hexagonal Bravais lattice are usually assigned to the rhombohedral system.

to systems of higher symmetry, omitting those of their point groups (i.e. classes) which have already been assigned to systems of lower symmetry. It is found that there are altogether 32 classes; a list of these arranged according to systems is as follows.

System	Classes				
Triclinic	$\boldsymbol{C}_1$,	$\boldsymbol{C}_i$			
Monoclinic	$\boldsymbol{C}_s$,	$\boldsymbol{C}_2$,	$\boldsymbol{C}_{2h}$		
Orthorhombic	$\boldsymbol{C}_{2v}$,	$\boldsymbol{D}_2$,	$\boldsymbol{D}_{2h}$		
Tetragonal	$\boldsymbol{S}_4$,	$\boldsymbol{D}_{2d}$,	$\boldsymbol{C}_4$,	$\boldsymbol{C}_{4h}$,	
	$\boldsymbol{C}_{4v}$,	$\boldsymbol{D}_4$,	$\boldsymbol{D}_{4h}$		
Rhombohedral	$\boldsymbol{C}_3$,	$\boldsymbol{S}_6$,	$\boldsymbol{C}_{3v}$,	$\boldsymbol{D}_3$,	$\boldsymbol{D}_{3d}$
Hexagonal	$\boldsymbol{C}_{3h}$,	$\boldsymbol{D}_{3h}$,	$\boldsymbol{C}_6$,	$\boldsymbol{C}_{6h}$,	
	$\boldsymbol{C}_{6v}$,	$\boldsymbol{D}_6$,	$\boldsymbol{D}_{6h}$		
Cubic	$\boldsymbol{T}$,	$\boldsymbol{T}_h$,	$\boldsymbol{T}_d$,	$\boldsymbol{O}$,	$\boldsymbol{O}_h$

In each of these sets of classes the last is the one of highest symmetry, and contains all the symmetry elements of the corresponding system. The classes whose symmetry is equal to that of the system are called *holohedral* classes. Those whose number of different symmetry transformations (rotations and reflections, including the identical transformation), is less than for a holohedral class by a factor of two or four are called *hemihedral* and *tetartohedral* classes respectively. For example, in the cubic system the class $\boldsymbol{O}_h$ is holohedral, $\boldsymbol{O}$, $\boldsymbol{T}_h$ and $\boldsymbol{T}_d$ are hemihedral, and $\boldsymbol{T}$ is tetartohedral.

§134. Space groups

Having studied the symmetry of the Bravais lattices and the symmetry of directions in the crystal, we can, finally, go on to consider the complete actual symmetry of crystal lattices. This symmetry may be termed microscopic, in contradistinction to the macroscopic symmetry of crystals discussed in §133. The microscopic symmetry determines those properties of a crystal which depend on the arrangement of the atoms in its lattice (e.g. the scattering of X-rays by the crystal).

The set of (actual) symmetry elements of the crystal lattice is called its *space group*. The lattice always has a certain translational symmetry, and may also have simple, rotary-reflection and screw axes of symmetry and simple and glide-reflection planes of symmetry. The translational symmetry of the lattice is entirely determined by its Bravais lattice, since by the definition of the latter the crystal lattice can have no translational periods except those of its Bravais lattice. Hence, to determine the space group of a crystal, it is sufficient to find the Bravais lattice and to enumerate the symmetry elements which involve rotations and reflections, including of course the relative position of these axes and planes of symmetry. It must also be remembered that the translational symmetry of the crystal lattice means that, if the lattice possesses an axis or plane of symmetry, there exists an infinity of parallel axes or planes which are carried into one another by displacements through the lattice vectors. Finally, in addition to these axes (or planes) of symmetry separated by lattice vectors, the simultaneous presence of translational symmetry and the axes (or planes) of symmetry results in the existence of other axes (or planes) which can not be made to coincide with the former by a translation through any lattice vector. For example, the presence of a plane of symmetry involves not only planes parallel to it at distances equal to the lattice vector but also planes of symmetry which bisect each lattice vector: it is easily seen that reflection in any plane followed by translation through a distance d in a direction perpendicular to the plane is equivalent to reflection in a plane parallel to the first and at a distance $\frac{1}{2}d$ from it.

The possible space groups can be divided among the crystal classes, each space group being assigned to the class where the set of axes and planes of symmetry is the same as in the space group when no distinction is made in the latter between simple and screw axes and between simple and glide planes. Altogether 230 different space groups are possible; they were first found by E. S. FEDOROV (1895). The space groups are distributed among classes as shown in Table 1.

Table 1

Class	Number of groups	Class	Number of groups
$\boldsymbol{C}_1$	1	$\boldsymbol{S}_6$	2
$\boldsymbol{C}_i$	1	$\boldsymbol{C}_{3v}$	6
$\boldsymbol{C}_s$	4	$\boldsymbol{D}_3$	7
$\boldsymbol{C}_2$	3	$\boldsymbol{D}_{3d}$	6
$\boldsymbol{C}_{2h}$	6	$\boldsymbol{C}_{3h}$	1
$\boldsymbol{C}_{2v}$	22	$\boldsymbol{C}_6$	6
$\boldsymbol{D}_2$	9	$\boldsymbol{C}_{6h}$	2
$\boldsymbol{D}_{2h}$	28	$\boldsymbol{D}_{3h}$	4
$\boldsymbol{S}_4$	2	$\boldsymbol{C}_{6v}$	4
$\boldsymbol{C}_4$	6	$\boldsymbol{D}_6$	6
$\boldsymbol{C}_{4h}$	6	$\boldsymbol{D}_{6h}$	4
$\boldsymbol{D}_{2d}$	12	$\boldsymbol{T}$	5
$\boldsymbol{C}_{4v}$	12	$\boldsymbol{T}_h$	7
$\boldsymbol{D}_4$	10	$\boldsymbol{T}_d$	6
$\boldsymbol{D}_{4h}$	20	$\boldsymbol{O}$	8
$\boldsymbol{C}_3$	4	$\boldsymbol{O}_h$	10

We shall not pause here to enumerate the symmetry elements of all the space groups, which would be a very lengthy process. They may be found in manuals of crystallography.†

† A full account of the space groups is given, for example, by G. YU. LYUBARSKIĬ, *The Application of Group Theory in Physics*, Pergamon, Oxford 1960, and in the *International*

Here the following point may be noted. Among the space groups there are some which differ only in the direction of rotation about their screw axes. There are in all 11 such pairs of space groups.

§135. The reciprocal lattice

All physical quantities which describe the properties of a crystal lattice have the same periodicity as the lattice itself. Such quantities are, for example, the electromagnetic field created in the lattice by the atoms forming it, the charge density due to the electrons in these atoms, the probability of finding an atom at a particular point in the lattice, and so on.

Let U be any such quantity. U is a function of the co-ordinates x, y, z of the point in the crystal or, as we shall write it, of the radius vector $\mathbf{r}$ of the point. The function $U(\mathbf{r})$ must be periodic, with the same periods as those of the lattice itself. This means that we must have

$$U(\mathbf{r}+n_1\mathbf{a}_1+n_2\mathbf{a}_2+n_3\mathbf{a}_3) = U(\mathbf{r}) \tag{135.1}$$

for any integral n_1, n_2, n_3 ($\mathbf{a}_1$, $\mathbf{a}_2$, $\mathbf{a}_3$ being the basic vectors of the lattice).

Let us expand the periodic function $U(\mathbf{r})$ as a triple Fourier series, which may be written

$$U = \sum_{\mathbf{b}} U_{\mathbf{b}} e^{2\pi i\mathbf{b}\cdot\mathbf{r}}, \tag{135.2}$$

where the summation is over all possible values of the vector $\mathbf{b}$. These are determined from the requirement that the function U, when put in the form of the series (135.2), satisfies the periodicity condition (135.1). This means that the exponential factors must be left unchanged when $\mathbf{r}$ is replaced by $\mathbf{r}+\mathbf{a}$, $\mathbf{a}$ being any lattice vector. For this to be so it is necessary that the scalar product $\mathbf{a}\cdot\mathbf{b}$ should always be integral. Taking $\mathbf{a}$ successively as the basic vectors $\mathbf{a}_1$, $\mathbf{a}_2$, $\mathbf{a}_3$, we must therefore have $\mathbf{a}_1\cdot\mathbf{b} = p_1$, $\mathbf{a}_2\cdot\mathbf{b} = p_2$, $\mathbf{a}_3\cdot\mathbf{b} = p_3$, where p_1, p_2, p_3 are positive or negative integers or zero. The solution of these three equations has the form

$$\mathbf{b} = p_1\mathbf{b}_1+p_2\mathbf{b}_2+p_3\mathbf{b}_3, \tag{135.3}$$

where the vectors $\mathbf{b}_i$ are given in terms of the $\mathbf{a}_i$ by

$$\mathbf{b}_1 = \mathbf{a}_2\times\mathbf{a}_3/v, \quad \mathbf{b}_2 = \mathbf{a}_3\times\mathbf{a}_1/v, \quad \mathbf{b}_3 = \mathbf{a}_1\times\mathbf{a}_2/v, \quad v = \mathbf{a}_1\cdot\mathbf{a}_2\times\mathbf{a}_3. \tag{135.4}$$

We have thus determined the possible values of the vector $\mathbf{b}$. The summation in (135.2) is taken over all integral values of p_1, p_2, p_3.

Geometrically, the product $v = \mathbf{a}_1\cdot\mathbf{a}_2\times\mathbf{a}_3$ represents the volume of the parallelepiped formed by the vectors $\mathbf{a}_1$, $\mathbf{a}_2$, $\mathbf{a}_3$, i.e. the volume of the unit cell;

Tables for the Determination of Crystal Structures, Bell, London 1935. The latter also lists the equivalent points for each space group.

the products $\mathbf{a}_1 \times \mathbf{a}_2$ etc. represent the areas of the three faces of this cell. The vectors $\mathbf{b}_i$ therefore have the dimensions of reciprocal length, and in magnitude are equal to the reciprocal altitudes of the parallelepiped formed by the vectors $\mathbf{a}_1, \mathbf{a}_2, \mathbf{a}_3$.

From (135.4) it is seen that $\mathbf{b}_i$ and $\mathbf{a}_i$ are related by

$$\begin{aligned} \mathbf{a}_i \cdot \mathbf{b}_k &= 0 \quad \text{if} \quad i \neq k, \\ &= 1 \quad \text{if} \quad i = k. \end{aligned} \tag{135.5}$$

Hence the vector $\mathbf{b}_1$ is perpendicular to $\mathbf{a}_2$ and $\mathbf{a}_3$, and similarly for $\mathbf{b}_2$ and $\mathbf{b}_3$.

Having defined the vectors $\mathbf{b}_i$, we can formally construct a lattice with $\mathbf{b}_1, \mathbf{b}_2, \mathbf{b}_3$ as basic vectors. This is called the *reciprocal lattice*, and the vectors $\mathbf{b}_1, \mathbf{b}_2, \mathbf{b}_3$ are called the (basic) vectors of the reciprocal lattice.

It is evident that the reciprocal lattice cell corresponding to a triclinic Bravais lattice will also be an arbitrary parallelepiped. Similarly, the reciprocal lattices of the simple Bravais lattices of the other systems are also simple lattices of the same system; for example, the reciprocal lattice of a simple cubic Bravais lattice also has a simple cubic cell. It is also easy to see by a straightforward construction that the reciprocal lattices of the face-centred Bravais lattices (orthorhombic, tetragonal and cubic) are body-centred lattices of the corresponding systems, and conversely that the body-centred Bravais lattices have face-centred reciprocal lattices. Finally, base-centred lattices have reciprocal lattices which are also base-centred.

Let us calculate the "volume" of the unit cell of the reciprocal lattice. This is $v' = \mathbf{b}_1 \cdot \mathbf{b}_2 \times \mathbf{b}_3$. Substitution of the expressions (135.4) gives

$$\begin{aligned} v' &= \frac{1}{v^3}\, \mathbf{a}_2 \times \mathbf{a}_3 \cdot (\mathbf{a}_3 \times \mathbf{a}_1) \times (\mathbf{a}_1 \times \mathbf{a}_2) \\ &= \frac{1}{v^3}\, (\mathbf{a}_2 \times \mathbf{a}_3 \cdot \mathbf{a}_1)(\mathbf{a}_3 \times \mathbf{a}_1 \cdot \mathbf{a}_2) \\ &= 1/v. \end{aligned} \tag{135.6}$$

Thus the volume of the unit cell of the reciprocal lattice is the reciprocal of that of the original lattice.

An equation of the form $\mathbf{b} \cdot \mathbf{r} = $ constant, where $\mathbf{b}$ is a given vector, represents a plane perpendicular to the vector $\mathbf{b}$ and at a distance from the origin equal to the constant divided by b. Let us take the origin at any of the Bravais lattice points, and let $\mathbf{b} = p_1\mathbf{b}_1 + p_2\mathbf{b}_2 + p_3\mathbf{b}_3$ be any vector of the reciprocal lattice (p_1, p_2, p_3 being integers). Also writing $\mathbf{r}$ in the form $\mathbf{r} = n_1\mathbf{a}_1 + n_2\mathbf{a}_2 + n_3\mathbf{a}_3$, we obtain the equation of a plane:

$$\mathbf{b} \cdot \mathbf{r} = n_1 p_1 + n_2 p_2 + n_3 p_3 = m, \tag{135.7}$$

where m is a given constant. If this equation represents a plane containing an infinity of Bravais lattice points (called a *crystal plane*), it must be satisfied by a set of integers n_1, n_2, n_3. For this to be so, the constant m must clearly be an integer also. For given p_1, p_2, p_3, when the constant m takes various integral values, equation (135.7) successively defines an infinity of crystal planes which are all parallel. A particular family of parallel crystal planes thus defined corresponds to each reciprocal lattice vector.

The numbers p_1, p_2, p_3 in (135.7) can always be taken as mutually prime, i.e. as having no common divisor except unity. If there were such a divisor, both sides of the equation could be divided by it, leaving an equation of the same form. The numbers p_1, p_2, p_3 are called the *Miller indices* of the family of crystal planes in question and are written as $(p_1p_2p_3)$.

The plane (135.7) intersects the co-ordinate axes (taken along the basic lattice vectors $\mathbf{a}_1$, $\mathbf{a}_2$, $\mathbf{a}_3$) at the points ma_1/p_1, ma_2/p_2, ma_3/p_3. The ratio of the intercepts (measured in units of a_1, a_2, a_3 respectively) is $1/p_1 : 1/p_2 : 1/p_3$, i.e. they are in inverse proportion to the Miller indices. For instance, the Miller indices of planes parallel to the co-ordinate planes (i.e. having intercepts in the ratio $\infty : \infty : 1$) are (100), (010), (001) for the three co-ordinate planes respectively. Planes parallel to the diagonal plane of the basic parallelepiped of the lattice have indices (111), and so on.

It is easy to find the distance between two successive planes of the same family. The distance of the plane (135.7) from the origin is m/b, where b is the "length" of the reciprocal lattice vector concerned. The distance of the next plane from the origin is $(m+1)/b$, and the distance d between these two planes is $(m+1)/b - m/b$, or

$$d = 1/b. \tag{135.8}$$

It is the reciprocal of the length of the vector $\mathbf{b}$.

§136. Irreducible representations of space groups

The physical applications of the theory of symmetry generally involve using the mathematical formalism of what are called *representations* of groups.† In particular, such applications will be encountered in the next chapter. Since the symmetry of crystals will be involved, it is necessary to discuss first the question of the classification and method of constructing the irreducible representations of the space groups.

Each space group contains a sub-group of translations comprising an infinity of all possible parallel displacements which leave the crystal lattice unchanged; this sub-group is the mathematical expression of the Bravais lattice of the crystal. The complete space group is obtained from this sub-group by adding n elements involving rotations and reflections, where n is

† The reader is assumed familiar with group theory to the extent given, for example, in *Quantum Mechanics*, Chapter XII.

the number of symmetry transformations of the corresponding crystal class; we shall call these the "rotational" elements. Every element of the space group may be represented as the product of one of the elements of translational symmetry and one of the "rotational" elements.

If the space group does not contain essential screw axes and glide planes (see below), the n rotational elements can be taken simply as the n symmetry transformations (rotations and reflections) of the crystal class; in this case these same elements also form a sub-group of the space group. In the contrary case, the rotational elements are rotations and reflections combined with a simultaneous translation through a certain fraction of one of the basic vectors of the lattice. In such space groups, the rotational elements of symmetry are "interlinked" with translations and do not themselves form a sub-group; for example, a repeated reflection in a glide plane is not an identical transformation but a translation through one of the basic vectors of the lattice.

Any irreducible representation of the space group can be given by a set of functions of the form†

$$\phi_{\mathbf{k}\alpha} = u_{\mathbf{k}\alpha} e^{i\mathbf{k}\cdot\mathbf{r}}, \tag{136.1}$$

where the $\mathbf{k}$ are constant vectors, the $u_{\mathbf{k}\alpha}$ are periodic functions with the same periods as those of the lattice, and the suffix $\alpha = 1, 2, \ldots$ labels functions with the same $\mathbf{k}$.

As a result of a parallel displacement, i.e. a transformation of the form $\mathbf{r} \to \mathbf{r}+\mathbf{a}$ (where $\mathbf{a}$ is any vector of the lattice), the functions (136.1) are multiplied by constants $e^{i\mathbf{k}\cdot\mathbf{a}}$. In other words, the matrices of translations are diagonal in the representation given by the functions (136.1). It is evident that two vectors $\mathbf{k}$ which differ by $2\pi\mathbf{b}$ (where $\mathbf{b}$ is any vector of the reciprocal lattice) will give the same law of transformation of the functions $\phi_{\mathbf{k}\alpha}$ by translations: since $\mathbf{a}\cdot\mathbf{b}$ is an integer, $e^{2\pi i\mathbf{a}\cdot\mathbf{b}} = 1$. Such vectors $\mathbf{k}$ will be said to be *equivalent*. If we imagine the vectors $\mathbf{k}/2\pi$ drawn from a vertex of a reciprocal lattice cell to various points, the non-equivalent vectors will correspond to the points in one unit cell.

By the application of a rotational element of symmetry, the function $\phi_{\mathbf{k}\alpha}$ is transformed into a linear combination of the functions $\phi_{\mathbf{k}'\alpha}$ with various values of α and a vector $\mathbf{k}'$ which is obtained from $\mathbf{k}$ by means of the rotation or reflection in question, performed in the reciprocal lattice.‡ The set of all (non-equivalent) vectors $\mathbf{k}$ which can be obtained from one another by the application of all n rotational elements of the group is called the *star* of the vector $\mathbf{k}$. In the general case of arbitrary $\mathbf{k}$ the star contains n vectors. The functions $\phi_{\mathbf{k}\alpha}$ which form the basis of an irreducible representation must always include functions having all the different vectors of the star of $\mathbf{k}$: it is

† The arguments below are due to F. Seitz (1936).

‡ In transforming the vector $\mathbf{k}$ in the reciprocal lattice, all axes and planes of symmetry must, of course, be treated as simple ones.

clear that, since functions with non-equivalent **k** are multiplied by different factors under translations, no choice of linear combinations of them can bring about a decrease in the number of functions which are transformed into combinations of one another.

For certain values of **k** the number of vectors in its star may be less than n, since it may happen that some of the rotational elements of symmetry leave **k** unchanged or transform it into an equivalent vector. For example, if the vector **k** is along an axis of symmetry, it is unchanged by rotations about this axis; a vector of the form $\mathbf{k} = \pi\mathbf{b}_i$, where $\mathbf{b}_i$ is one of the basic vectors of the reciprocal lattice, is transformed by inversion into the equivalent vector $-\pi\mathbf{b}_i = \pi\mathbf{b}_i - 2\pi\mathbf{b}_i$.

The set of rotational elements of symmetry (regarded as all being simple) which appear in a given space group and which do not alter the vector **k** (or which transform it into an equivalent vector) will be called the *proper symmetry group* of the vector **k**, or simply the group of **k**; it is one of the ordinary point symmetry groups.

Let us first consider the simple case where the space group includes no screw axes or glide planes. The base functions of an irreducible representation of such a group can be written as products

$$\phi_{\mathbf{k}\alpha} = u_\alpha \psi_{\mathbf{k}}, \tag{136.2}$$

where the u_α are periodic functions and the $\psi_{\mathbf{k}}$ are linear combinations of the expressions $e^{i\mathbf{k}\cdot\mathbf{r}}$ (with equivalent **k**) invariant with respect to the transformations in the proper symmetry group of the vector **k**; in (136.2) this vector takes all the values in its star. In translations the periodic functions u_α are unchanged, but the functions $\psi_{\mathbf{k}}$, and therefore the $\phi_{\mathbf{k}\alpha}$, are multiplied by $e^{i\mathbf{k}\cdot\mathbf{a}}$. In rotations and reflections belonging to the group of **k**, the functions $\psi_{\mathbf{k}}$ are unchanged but the functions u_α are transformed into combinations of one another. Thus the functions u_α give one of the irreducible representations of the point group of **k**, these being called in this connection *small representations*. Finally, rotational elements which are not in the group of **k** transform sets of functions (136.2) with non-equivalent **k** into combinations of one another. The dimension of the representation of the space group thus constructed is equal to the number of vectors in the star of **k** multiplied by the dimension of the small representation.

Thus the problem of finding all irreducible representations of space groups (having no screw axes or glide planes) reduces entirely to the classification of the vectors **k** with respect to their proper symmetry and the known problem of discovering the irreducible representations of finite point groups.

Let us now consider space groups which have screw axes or glide planes. The presence of such elements of symmetry is still unimportant if the vector **k** is such that it remains unchanged (i.e. is not transformed into an equivalent vector) under all the transformations in its group.† In this case the corresponding representations are again given by functions of the form (136.2), in which the u_α form the basis of a representation of the point group of the vector **k**. The only difference from the previous case will be that under rotational transformations the functions $\psi_{\mathbf{k}} = e^{i\mathbf{k}\cdot\mathbf{r}}$ in (136.2) will not remain unchanged, but will be multiplied by $e^{i\mathbf{k}\cdot\boldsymbol{\tau}}$, where $\boldsymbol{\tau}$ is the part of the lattice period through which the translation occurs which pertains to a screw axis or glide plane.

For all other values of **k**, functions of the form (136.2) become inapplicable. In a rotational transformation with a simultaneous translation $\boldsymbol{\tau}$, functions $e^{i\mathbf{k}\cdot\mathbf{r}}$ with equivalent but different values of **k** are multiplied by different factors (since $\mathbf{b}\cdot\boldsymbol{\tau}$ is not integral), and therefore their linear combinations $\psi_{\mathbf{k}}$ will not be transformed into combinations of one another.

In such cases it is no longer possible to consider the rotational elements and the translations separately, but of the infinity of translations it is sufficient to consider a finite number only. We shall call the *extended group* of the vector **k** the group consisting of the corresponding rotational transformations (together with the relevant translations through fractions of a period $\boldsymbol{\tau}$) and those translations for which $\mathbf{k}\cdot\mathbf{a}/2\pi$ is not integral; the translations for which $\mathbf{k}\cdot\mathbf{a}/2\pi$ is integral are regarded as identical transformations. The functions $\phi_{\mathbf{k}\alpha}$ which give irreducible representations of the finite group thus formed, together with the corresponding functions $\phi_{\mathbf{k}'\alpha}$ for other vectors in the star of **k**, give irreducible representations of the space group. The dimension of these representations is equal to the dimension of the representation of the extended group of the vector **k**, multiplied by the number of vectors in the star.

This procedure will now be demonstrated for a specific example. In order to characterise explicitly the elements of the space group, it is convenient to denote them by symbols $(P\,|\,\mathbf{t})$, where P is any rotation or reflection, and $\mathbf{t}$ the vector of a simultaneous translation; the effect of this element on the radius vector $\mathbf{r}$ of any point is shown by $(P\,|\,\mathbf{t})\mathbf{r} = P\mathbf{r}+\mathbf{t}$. The multiplication of elements follows the obvious rule $(P'\,|\,\mathbf{t}')(P\,|\,\mathbf{t}) = (P'P\,|\,P'\mathbf{t}+\mathbf{t}')$. In particular, the element inverse to $(P\,|\,\mathbf{t})$ is $(P\,|\,\mathbf{t})^{-1} = (P^{-1}|-P^{-1}\mathbf{t})$.

Let us consider the space group (D_{2h}^2) which corresponds to the simple orthorhombic Bravais lattice and contains the following rotational elements:

$$\begin{array}{cccc} (1\,|\,0), & (C_2^x\,|\,0), & (C_2^y\,|\,0), & (C_2^z\,|\,0), \\ (I\,|\,\boldsymbol{\tau}), & (\sigma_x\,|\,\boldsymbol{\tau}), & (\sigma_y\,|\,\boldsymbol{\tau}), & (\sigma_z\,|\,\boldsymbol{\tau}), \end{array}$$

the x, y and z axes being taken along the three basic vectors of the lattice; $\boldsymbol{\tau} = \frac{1}{2}(\mathbf{a}_1+\mathbf{a}_2+\mathbf{a}_3)$; the axes of symmetry C_2 are simple axes but the planes σ perpendicular to them are glide planes.

† This always includes, in particular, the vector $\mathbf{k} = 0$ and a vector in a general position in which the unit element of its group is the identical transformation.

Let us take, for example, the vector $\mathbf{k}/2\pi = (\frac{1}{2}, 0, 0)$; the three numbers in parentheses give the components of the vector along the x, y and z axes, measured in units of the edge lengths ($b_i = 1/a_i$) of the reciprocal lattice cell. Its proper symmetry includes all the axes and planes of the point group $\boldsymbol{D}_{2h}$, and so this vector is its own star. The only translation (other than multiples of itself) with $\mathbf{k}\cdot\mathbf{a}/2\pi$ not integral is the translation $(1|\mathbf{a}_1)$. Thus we obtain a group of 16 elements in 10 classes as shown in the upper line in Table 2; for brevity the rotational elements are denoted simply by C_2, σ, I, and the translation $(1|\mathbf{a}_1)$ by a_1. The fact that the elements $C_2{}^y$ and $a_1C_2{}^y$, for example, are conjugate (i.e. belong to the same class) may be seen as follows.† We have

$$\begin{aligned}(I|\boldsymbol{\tau})^{-1}(C_2{}^y|0)(I|\boldsymbol{\tau}) &= (I|-\boldsymbol{\tau})(C_2{}^y|0)(I|\boldsymbol{\tau})\\ &= (I|-\boldsymbol{\tau})(C_2{}^yI|C_2{}^y\boldsymbol{\tau})\\ &= (C_2{}^y|-\boldsymbol{\tau}+C_2{}^y\boldsymbol{\tau}).\end{aligned}$$

But

$$\begin{aligned}C_2{}^y\boldsymbol{\tau} &= \tfrac{1}{2}(-\mathbf{a}_1+\mathbf{a}_2-\mathbf{a}_3),\\ -\boldsymbol{\tau}+C_2{}^y\boldsymbol{\tau} &= -\mathbf{a}_1-\mathbf{a}_3\\ &= \mathbf{a}_1-(2\mathbf{a}_1+\mathbf{a}_3),\end{aligned}$$

and, since translations through $\mathbf{a}_3$ and $2\mathbf{a}_1$ must be regarded as identical transformations, we have

$$(I|\boldsymbol{\tau})^{-1}(C_2{}^y|0)(I|\boldsymbol{\tau}) = (C_2{}^y|\mathbf{a}_1).$$

Table 2

	1	a_1	$C_2{}^x$	$a_1C_2{}^x$	$C_2{}^y$ $a_1C_2{}^y$	$C_2{}^z$ $a_1C_2{}^z$	I a_1I	σ_x $a_1\sigma_x$	σ_y $a_1\sigma_y$	σ_z $a_1\sigma_z$
Γ_1	2	-2	2	-2	0	0	0	0	0	0
Γ_2	2	-2	-2	2	0	0	0	0	0	0

From the numbers of elements and classes in the group we find that it has 8 one-dimensional and 2 two-dimensional irreducible representations ($8\cdot1^2+2\cdot2^2 = 16$). All the one-dimensional representations are obtained from representations of the point group $\boldsymbol{D}_{2h}$, the translation a_1 being assigned the character $\chi(a_1) = 1$. These representations, however, occur here as "spurious" representations and must be rejected. They do not solve the problem in question: their base functions are invariant under all translations, whereas the functions $e^{i\mathbf{k}\cdot\mathbf{r}}$ with given $\mathbf{k}$ are certainly not invariant with respect to the translation a_1. Thus there remain only two irreducible

† Two elements A and B are said to be *conjugate* if $A = C^{-1}BC$, where C is another element of the group.

representations, whose characters are shown in Table 2.† The base functions of these representations can be taken as

$$\Gamma_1 : \cos \pi x, \ \sin \pi x,$$
$$\Gamma_2 : \cos \pi x \cos 2\pi y, \ \sin \pi x \cos 2\pi y.$$

Let us also consider the representations corresponding to the star of two vectors $(\frac{1}{2}, 0, \varkappa)$, $(\frac{1}{2}, 0, -\varkappa)$ with proper symmetry $\boldsymbol{C}_{2v}$ (the axis C_2 being along the z-axis); $\varkappa$ is an arbitrary number between 0 and 1 (other than $\frac{1}{2}$). The extended group of $\mathbf{k}$ contains 8 elements in 5 classes (Table 3). (The dependence of the base functions of the representations of this group on the co-ordinate z reduces to a common factor $e^{i\varkappa z}$ or $e^{-i\varkappa z}$, which is invariant under all transformations of the group; it is therefore unnecessary to extend the group by translations along the z-axis.) There are 4 one-dimensional and 1 two-dimensional representation of this group.‡ The one-dimensional representations must be rejected for the same reasons as previously, leaving only one representation, whose characters are shown in Table 3. Its base functions can be taken as

$$e^{\pm i\varkappa z} \cos \pi x, \qquad e^{\pm i\varkappa z} \sin \pi x,$$

with the plus sign in the exponent for the vector $(\frac{1}{2}, 0, \varkappa)$ and the minus sign for the vector $(\frac{1}{2}, 0, -\varkappa)$; the complete irreducible representation of the whole space group is four-dimensional, and is given by all four of these functions together.

Table 3

1	a_1	C_2^z $a_1C_2^z$	σ_x $a_1\sigma_x$	σ_y $a_1\sigma_y$
2	-2	0	0	0

† This group is isomorphous with the "double" point group $\boldsymbol{D}_{2h}'$; see *Quantum Mechanics*, §99.

‡ It is isomorphous with the "double" point group $\boldsymbol{D}_2'$.

CHAPTER XIV

PHASE TRANSITIONS OF THE SECOND KIND

§137. Phase transitions of the second kind

IT HAS already been mentioned in §83 that the transition between phases of different symmetry (crystal and liquid; different crystal modifications) cannot occur in a continuous manner such as is possible for a liquid and a gas. In every state the body has either one symmetry or the other, and therefore we can always assign it to one of the two phases.

The transition between different crystal modifications is usually effected by means of a phase transition in which there is a sudden rearrangement of the

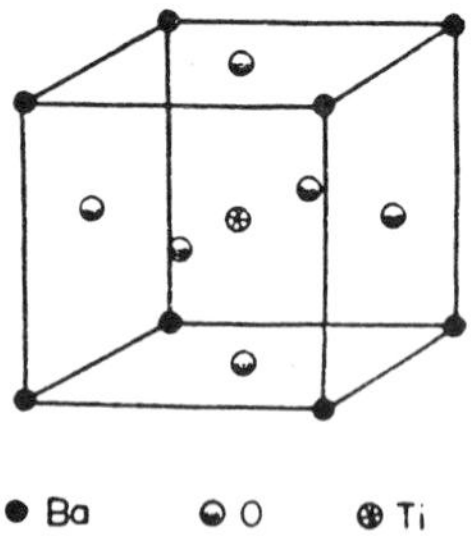

FIG. 61

crystal lattice and the state of the body changes discontinuously. As well as such discontinuous transitions, however, another type of transition involving a change of symmetry is also possible.

To elucidate the nature of these transitions, let us consider a specific example. At high temperatures, $BaTiO_3$ has a cubic lattice whose unit cell is as shown in Fig. 61 (the barium atoms are at the vertices, the oxygen atoms at the centres of the faces, and the titanium atoms at the centres of the cells). As the temperature decreases below a certain value, the titanium and oxygen atoms begin to move relative to the barium atoms parallel to an edge of the cube. It is clear that, as soon as this movement begins, the symmetry of the lattice is affected, and it becomes tetragonal instead of cubic.

This example is typical in that there is no discontinuous change in state of the body. The configuration of atoms in the crystal† changes continuously.

† To simplify the discussion, we shall conventionally speak of the configuration of the

However, an arbitrarily small displacement of the atoms from their original symmetrical positions is sufficient to change the symmetry of the lattice. The resulting transition from one crystal modification to another is called a *phase transition of the second kind*, in contrast to ordinary phase transitions, which in this case are said to be of the first kind.†

Thus a phase transition of the second kind is continuous in the sense that the state of the body changes continuously. It should be emphasised, however, that the symmetry, of course, changes discontinuously at the transition point, and at any instant we can say to which of the two phases the body belongs. But whereas at a phase transition point of the first kind bodies in two different states are in equilibrium, the states of the two phases are the same at a transition point of the second kind.

As well as cases where the change in symmetry of the body occurs by a displacement of the atoms (as in the example given above), the change in symmetry in a phase transition of the second kind may result from a change in the ordering of the crystal. It has already been mentioned in §61 that the concept of ordering arises if the number of lattice points that can be occupied by atoms of a given kind exceeds the number of such atoms. We shall use the word "own" for the places occupied by atoms of the kind in question in a completely ordered crystal, in contrast to the "other" places which are taken by some of the atoms when the crystal becomes disordered. In many cases, which will be of interest in connection with transitions of the second kind, it is found that the "own" and "other" lattice sites are geometrically identical and differ only in that they have different probabilities of containing atoms of the kind in question.‡ If now these probabilities become equal (they will not be unity, of course), all such sites become equivalent, and therefore new symmetry elements appear, i.e. the symmetry of the lattice is increased. Such a crystal will be said to be *disordered.*

The foregoing may be illustrated by an example. The completely ordered alloy CuZn has a cubic lattice with the zinc atoms at the vertices, say, and the copper atoms at the centres of the cubic cells (Fig. 62a; a simple cubic Bravais lattice). When the alloy is heated and becomes disordered, copper and zinc atoms change places, i.e. non-zero probabilities of finding atoms of either kind exist at every lattice site. Until the probabilities of finding copper (or zinc) atoms at the vertices and at the centres of the cells become equal

atoms or its symmetry as if the atoms were at rest. In reality we should speak of the probability distribution for various configurations of the atoms in space, and of the symmetry of this distribution.

† Phase transition points of the second kind are also called *Curie points* or *λ points.*

‡ We may note that in this case it can always be assumed that the probability of finding an atom at one of its "own" sites is greater than at one of the "other" sites simply because, if it were not, we could transpose the nomenclature of the sites.

(that is, while the crystal is ordered but not completely ordered), these sites remain non-equivalent, and the symmetry of the lattice is unchanged. But when the probabilities become equal, all sites become equivalent, and the symmetry of the crystal is raised: a new lattice vector appears, from a vertex to the centre of a cell, and the crystal acquires a body-centred cubic Bravais lattice (Fig. 62b).

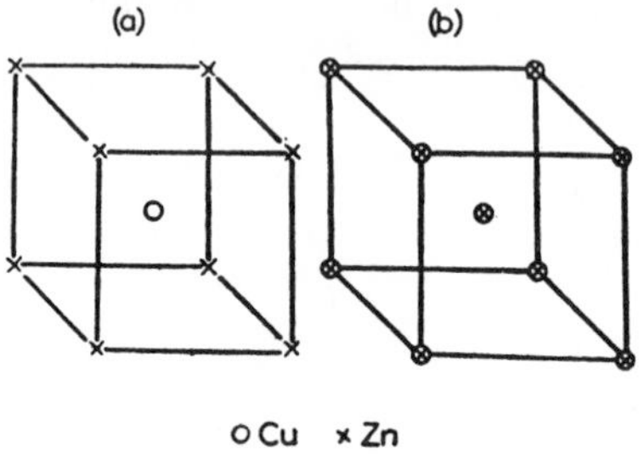

FIG. 62

For each state of ordering we can define a quantitative characteristic, the *degree of ordering* η, such that it is zero in a disordered phase, and takes various positive or negative non-zero values in crystals with various degrees of ordering. For instance, in the above example of the alloy CuZn, this quantity may be defined as

$$\eta = (w_{Cu}-w_{Zn})/(w_{Cu}+w_{Zn}),$$

where w_{Cu} and w_{Zn} are the probabilities of finding a copper atom and a zinc atom respectively at any given lattice site.

It must again be emphasised that the symmetry of the crystal is changed (namely, increased) only when η becomes exactly zero; any non-zero degree of ordering, however small, brings about the same symmetry as that of a completely ordered crystal.

If, as the temperature is increased, the degree of ordering becomes zero discontinuously from some finite value, the change from an ordered to a disordered crystal will be a phase transition of the first kind, but if the degree of ordering becomes zero continuously, we have a phase transition of the second kind.†

† Cases are in principle possible where the occurrence of ordering does not cause a change in the symmetry of the crystal. A phase transition of the second kind is then impossible: even if the transition from the ordered to the disordered state of the crystal were to occur continuously, there would still be no discontinuity of specific heat (see below). In such cases a phase transition of the first kind is, of course, possible.

The statement occurs in the literature that there is a relation between phase transitions of the second kind and the appearance of rotating molecules (or radicals) in the crystal. This view is incorrect, since at a transition point of the second kind the state of the body must change continuously, and so there can be no sharp change in the nature of the motion. If a phase transition which involves rotations of molecules in the crystal is considered, the

So far we have discussed only transitions between different crystal modifications, but phase transitions of the second kind need not necessarily involve a change in symmetry of the configuration of atoms in the lattice. A transition of the second kind can also bring about a transformation between two phases differing in some other property of symmetry, as for example at the Curie points of ferromagnetic substances (the points at which they become paramagnetic). In this case there is a change in symmetry of the arrangement of the elementary magnetic moments in the body, or more precisely a disappearance of the currents **j** in it; see the first footnote to §128. Other phase transitions of the second kind are the transition of a metal to the superconducting state (in the absence of a magnetic field) and that of liquid helium to the superfluid state. In both these cases the state of the body changes continuously, but it acquires a qualitatively new property at the transition point.

Since the states of the two phases are the same at a transition point of the second kind, it is clear that the symmetry of the body at the transition point itself must contain all the symmetry elements of both phases. It will be shown below that the symmetry at the transition point itself is the same as the symmetry everywhere on one side of that point, i.e. the symmetry of one of the phases. Thus the change in symmetry of the body in a phase transition of the second kind has the following very important general property: the symmetry of one phase is higher than that of the other.† It should be emphasised that in a phase transition of the first kind the change in symmetry of the body is subject to no restriction, and the symmetries of the two phases may be unrelated.

In the great majority of the known instances of phase transitions of the second kind, the more symmetrical phase corresponds to higher temperatures and the less symmetrical one to lower temperatures. In particular, a transition of the second kind from an ordered to a disordered state always occurs with increasing temperature. This is not a law of thermodynamics, however, and exceptions are therefore possible.‡

The absence of any discontinuous change of state at a phase transition point of the second kind has the result that the thermodynamic functions of

difference between the two phases must be that in the more symmetrical phase the probabilities of different orientations of the molecules are equal, while in the less symmetrical one they are different.

† It will be recalled that the term "higher symmetry" refers to a symmetry which includes all the symmetry elements (rotations, reflections and translational periods) of the lower symmetry, together with additional elements.

The condition mentioned is necessary but not sufficient for a phase transition of the second kind to be possible; we shall see later that the possible changes of symmetry in such a transition are subject to further restrictions.

‡ One exception, for example, is the "lower Curie point" of Rochelle salt, below which the crystal is orthorhombic, but above which it is monoclinic.

the state of the body (its entropy, energy, volume, etc.) vary continuously as the transition point is passed. Hence, in particular, a phase transition of the second kind, unlike one of the first kind, is not accompanied by evolution or absorption of heat. We shall see below, however, that the derivatives of these thermodynamic quantities (i.e. the specific heat of the body, the thermal expansion coefficient, the compressibility, etc.) are discontinuous at a transition point of the second kind.

We must expect that mathematically a phase transition point of the second kind is a singularity of the thermodynamic quantities, and in particular of the thermodynamic potential Φ; the nature of this singularity is not yet known. In order to see this, let us first recall that (see §83) a phase transition point of the first kind is not a singularity; it is a point at which the thermodynamic potentials $\Phi_1(P, T)$ and $\Phi_2(P, T)$ of the two phases are equal, and each of the functions Φ_1 and Φ_2 on either side of the transition point corresponds to an equilibrium (though possibly metastable) state of the body. In a phase transition of the second kind, however, the thermodynamic potential of each phase, if formally regarded on the far side of the transition point, corresponds to no equilibrium state, i.e. to no minimum of Φ; we shall see in §138 that the thermodynamic potential of the more symmetrical phase would indeed correspond to a maximum of Φ beyond the transition point.

This last result implies that superheating and supercooling effects are impossible in phase transitions of the second kind (whereas they can occur in ordinary phase transitions). In this case neither phase can exist beyond the transition point (here we ignore, of course, the time needed to establish the equilibrium distribution of atoms, which in solid crystals may be considerable).

PROBLEM

Let c be the concentration of atoms of one component of a binary solid solution, and c_0 the concentration of these atoms' "own" sites. If $c \neq c_0$ the crystal cannot be completely ordered. Assuming the difference $c-c_0$ small and the crystal almost completely ordered, determine the concentration λ of atoms at "other" sites, expressing it in terms of the value λ_0 which it would have at $c = c_0$ for given P and T (C. Wagner and W. Schottky 1930).

SOLUTION. Considering throughout only the atoms of one component, we use the concentration λ of atoms at "other" sites and the concentration λ' of their "own" sites not occupied by these atoms; concentrations are defined with respect to the total number of all atoms in the crystal. Clearly

$$c-c_0 = \lambda-\lambda'. \tag{1}$$

We shall regard the crystal as a "solution" of atoms at "other" sites and of "own" sites not occupied by atoms, the "solvent" being represented by atoms at their "own" sites. The transition of atoms from "other" to their "own" sites can then be regarded as a "chemical reaction" between the "solutes" (with small concentrations λ and λ') to form the "solvent" (with concentration $\cong 1$). Applying to this "reaction" the law of mass action, we obtain $\lambda\lambda' = K$, where K depends only on P and T. For $c = c_0$ we must have $\lambda = \lambda' \equiv \lambda_0$; hence $K = \lambda_0^2$, and so

$$\lambda\lambda' = \lambda_0^2. \tag{2}$$

From (1) and (2) we find the required concentrations:

$$\lambda = \tfrac{1}{2}[(c-c_0)+\sqrt{\{(c-c_0)^2+4\lambda_0^2\}}],$$
$$\lambda' = \tfrac{1}{2}[-(c-c_0)+\sqrt{\{(c-c_0)^2+4\lambda_0^2\}}].$$

§138. The discontinuity of specific heat†

To give a mathematical description of a phase transition of the second kind, we define a quantity η which is to represent the extent to which the configuration of the atoms in the less symmetrical phase differs from that in the more symmetrical phase; in the latter phase $\eta = 0$, and in the less symmetrical phase η has positive or negative values. For example, in transitions which involve a change in the ordering of the crystal, η may be taken as the degree of ordering; in transitions where there is a movement of the atoms (as in $BaTiO_3$‡), η may be taken as the amount of displacement, and so on.

For brevity we shall arbitrarily call the more symmetrical phase simply the symmetrical one, and the less symmetrical phase the unsymmetrical one.

Considering the thermodynamic quantities of the crystal for given deviations from the symmetrical state (i.e. for given η), we can represent the thermodynamic potential Φ as a function of P, T and η. Here it must of course be remembered that in the function $\Phi(P, T, \eta)$ the variable η is in one sense not on the same footing as the variables P and T: whereas the pressure and temperature can be specified arbitrarily, the value of η which actually occurs must itself be determined from the condition of thermal equilibrium, i.e. the condition that Φ is a minimum (for given P and T).

The continuity of the change of state in a phase transition of the second kind is expressed mathematically by the fact that the quantity η takes arbitrarily small values near the transition point. Considering the neighbourhood of this point, we expand $\Phi(P, T, \eta)$ in powers of η:

$$\Phi(P, T, \eta) = \Phi_0+\alpha\eta+A\eta^2+B\eta^3+C\eta^4+\ldots, \tag{138.1}$$

where the coefficients α, A, B, C, ... are functions of P and T.

It must be emphasised, however, that the possibility of such an expansion is by no means obvious *a priori*. Moreover, since, as already mentioned, a transition point of the second kind must be a singularity of the thermodynamic potential, there is every reason to suppose that such an expansion can not be continued to terms of arbitrarily high order, and that the expansion coefficients can have singularities as functions of P and T. A complete elucidation of the nature of the singularity of the thermodynamic potential at the

† The theory given in this and the following sections is due to L. D. Landau (1937).

‡ To avoid misunderstanding it may be noted that in the particular case of $BaTiO_3$ the displacement of the atoms has a small but finite discontinuity at the transition point, and so the transition is in fact of the first kind.

transition point offers great difficulties and has not yet been achieved. Here we shall give a theory based on the assumption that the presence of the singularity does not affect the terms of the expansion that are used. Until an exhaustive theory is developed it is difficult to say which of the results thus obtained may undergo modification, and to what extent.

It can be shown (see §139) that, if the states with $\eta = 0$ and $\eta \neq 0$ are of different symmetry (as we assume), the first-order term in the expansion (138.1) is identically zero: $\alpha \equiv 0$. The coefficient $A(P, T)$ in the second-order term is easily seen to vanish at the transition point, since in the symmetrical phase the value $\eta = 0$ must correspond to a minimum of Φ, and for this to be so it is evident that $A > 0$ is necessary, while on the other side of the transition point, in the unsymmetrical phase, non-zero values of η must correspond to the stable state (i.e. to the minimum of Φ), and this is possible

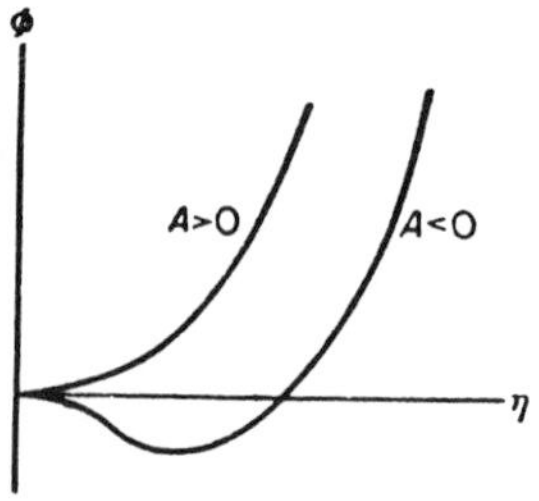

FIG. 63

only if $A < 0$; Fig. 63 shows the form of the function $\Phi(\eta)$ for $A < 0$ and $A > 0$. Since A is positive on one side of the transition point and negative on the other, it must vanish at the transition point itself:

$$A_c(P, T) = 0, \tag{138.2}$$

where the suffix c refers to the transition point.

But if the transition point itself is a stable state, i.e. if Φ as a function of η is a minimum at $\eta = 0$, it is necessary that the third-order term should be zero and the fourth-order term positive there:

$$B_c(P, T) = 0, \qquad C_c(P, T) > 0. \tag{138.3}$$

The coefficient C, being positive at the transition point, is of course also positive in the neighbourhood of that point.

Two cases can occur. In one, the third-order term is identically zero owing to the symmetry of the crystal: $B(P, T) \equiv 0$. Then there remains at the transition point only the one condition $A(P, T) = 0$, which determines P as a function of T or *vice versa*. Thus in the PT-plane there is a line of phase transition points of the second kind.†

If, however, B is not identically zero, the transition points are determined by the two equations $A(P, T) = 0$, $B(P, T) = 0$. In this case, therefore, the continuous phase transitions occur only at isolated points.

The most interesting case is, of course, that where there is a line of continuous-transition points. In what follows we shall take the discussion of phase transitions of the second kind to refer only to this case, which will now be considered.‡ That is, we shall suppose that $B(P, T) \equiv 0$ and the expansion of the thermodynamic potential has the form

$$\Phi(P, T, \eta) = \Phi_0(P, T)+A(P, T)\eta^2+C(P, T)\eta^4+\dots. \tag{138.4}$$

Here $C > 0$, while $A > 0$ in the symmetrical phase and $A < 0$ in the unsymmetrical phase; the transition points are determined by the equation $A(P, T) = 0$.

If we consider a transition at a given value of the pressure, then near the transition point (the temperature of which is denoted by T_c) we can write

$$A(T) = a(T-T_c), \tag{138.5}$$

where $a = [\partial A/\partial T]_{T=T_c}$ is a constant. The coefficient $C(T)$ can be put simply equal to a constant $C(T_c)$.

The dependence of η on the temperature near the transition point, in the unsymmetrical phase, is determined from the condition for Φ to be a minimum as a function of η. Equating the derivative $\partial\Phi/\partial\eta$ to zero, we obtain $\eta(A+2C\eta^2) = 0$, and hence

$$\eta^2 = -A/2C = a(T_c-T)/2C; \tag{138.6}$$

the solution $\eta = 0$ corresponds to the symmetrical phase.||

Next, let us determine the entropy of the body near the transition point. Neglecting higher powers of η, we have from (138.4)

$$S = -\partial\Phi/\partial T = S_0-(\partial A/\partial T)\eta^2,$$

where $S_0 = -\partial\Phi_0/\partial T$; the term containing the temperature derivative of η is zero, because $\partial\Phi/\partial\eta = 0$. In the symmetrical phase $\eta = 0$ and $S = S_0$; in

† The condition that there is no term in η^3 in the expansion (138.1) is in fact necessary but not sufficient for phase transitions of the second kind to be possible; see the sixth footnote to §139.

‡ It can be shown (L. Landau, *Zhurnal éksperimental'noĭ i teoreticheskoĭ fiziki* **7**, 627, 1937; translation in *Collected Papers of L. D. Landau*, p. 209, Pergamon, Oxford 1965) that a phase transition of the second kind between a liquid and a solid (crystal) is always impossible, since there is a third-order term in the expansion of the thermodynamic potential.

|| It should be noted that for $A < 0$ the value $\eta = 0$ would correspond to a maximum of Φ.

the unsymmetrical phase $\eta^2 = -A/2C$ and

$$S = S_0 + \frac{A}{2C}\frac{\partial A}{\partial T} = S_0 + \frac{a^2}{2C}(T-T_c). \tag{138.7}$$

At the transition point itself this expression becomes S_0, and the entropy is therefore continuous, as it should be.

Finally, let us determine the specific heats $C_p = T(\partial S/\partial T)_P$ of the two phases at the transition point. For the unsymmetrical phase we have, differentiating (138.7),

$$C_p = C_{p0} + a^2T_c/2C. \tag{138.8}$$

For the symmetrical phase $S = S_0$, and therefore $C_p = C_{p0}$. Thus we conclude that the specific heat is discontinuous at a phase transition point of the second kind. Since $C > 0$, $C_p > C_{p0}$ at the transition point, i.e. the specific heat is greater in the unsymmetrical phase than in the symmetrical one.

Other quantities besides C_p are discontinuous: C_v, the thermal expansion coefficient, the compressibility, etc. There is no difficulty in deriving relations between the discontinuities of all these quantities. First of all we note that the volume and the entropy are continuous at the transition point, i.e. their discontinuities ΔV and ΔS are zero:

$$\Delta V = 0, \qquad \Delta S = 0.$$

We differentiate these equations with respect to temperature along the curve of transition points, i.e. assuming the pressure to be the function of temperature given by this curve. The result is

$$\begin{aligned} \Delta(\partial V/\partial T)_P + (\mathrm{d}P/\mathrm{d}T)\Delta(\partial V/\partial P)_T &= 0, \\ \Delta C_p/T - (\mathrm{d}P/\mathrm{d}T)\Delta(\partial V/\partial T)_P &= 0, \end{aligned} \tag{138.9}$$

since $(\partial S/\partial P)_T = -(\partial V/\partial T)_P$. These two equations relate the discontinuities of the specific heat C_p, the thermal expansion coefficient and the compressibility at a phase transition point of the second kind (W. H. KEESOM, and P. EHRENFEST, 1933).

Differentiating along the curve of transition points the equations $\Delta S = 0$ and $\Delta P = 0$ (the pressure is, of course, unchanged in the transition), but with temperature and volume as independent variables, we find

$$\begin{aligned} \Delta(\partial P/\partial T)_V + (\mathrm{d}V/\mathrm{d}T)\Delta(\partial P/\partial V)_T &= 0, \\ \Delta C_v/T + (\mathrm{d}V/\mathrm{d}T)\Delta(\partial P/\partial T)_V &= 0. \end{aligned} \tag{138.10}$$

From (138.9) and (138.10) we can express the discontinuities of C_p, C_v, $(\partial P/\partial T)_V$ and $(\partial V/\partial T)_P$ in terms of that of $(\partial V/\partial P)_T$:

$$\left.\begin{aligned} \Delta(\partial V/\partial T)_P &= -(\mathrm{d}P/\mathrm{d}T)\Delta(\partial V/\partial P)_T, \\ \Delta C_p &= -T(\mathrm{d}P/\mathrm{d}T)^2\Delta(\partial V/\partial P)_T, \\ \Delta(\partial P/\partial T)_V &= -(\mathrm{d}V/\mathrm{d}T)\Delta[1/(\partial V/\partial P)_T], \\ \Delta C_v &= T(\mathrm{d}V/\mathrm{d}T)^2\Delta[1/(\partial V/\partial P)_T]. \end{aligned}\right\} \tag{138.11}$$

We may note that, according to these formulae, the discontinuities of the specific heat C_p and the compressibility $-(\partial V/\partial P)_T$ have the same sign. Hence it follows, from the previous statement about the discontinuity of the specific heat, that the compressibility decreases discontinuously on going from the unsymmetrical to the symmetrical phase.

The foregoing thermodynamic theory (with the reservation made at the beginning of this section) answers the problem of the nature of the changes in the thermodynamic quantities in a continuous transition between phases of different symmetry. We see that the first derivatives of quantities such as entropy and volume must be discontinuous in a transition of this type.†

PROBLEM

Find the relation between the discontinuities of specific heat and heat of solution in a transition of the second kind in a solution (I. M. Lifshitz 1950).

Solution. The heat of solution per molecule of solute is given by $q = \partial W/\partial n - w_0'$, where W is the heat function of the solution and w_0' the heat function per particle of the pure solute. Since w_0' is not affected by the phase transition in solution, we have for the discontinuity of q

$$\Delta q = \Delta(\partial W/\partial n) = \Delta \frac{\partial}{\partial n}\left(\Phi - T\frac{\partial \Phi}{\partial T}\right) = -T\Delta(\partial^2\Phi/\partial n\,\partial T),$$

where we have used the fact that the chemical potential $\mu' = \partial\Phi/\partial n$ is continuous at the transition. On the other hand, differentiation of the equation $\Delta(\partial\Phi/\partial T) = 0$ (continuity of entropy) along the curve of the transition temperature as a function of the concentration c at constant pressure gives

$$\frac{\mathrm{d}T_c}{\mathrm{d}c}\Delta\frac{\partial^2\Phi}{\partial T^2} + N\Delta\frac{\partial^2\Phi}{\partial N\partial T} = 0.$$

Hence we have the required relation

$$N\,\Delta q = (\mathrm{d}T_0/\mathrm{d}c)\,\Delta C_p.$$

We may note that in the derivation of this relation no assumption has been made concerning the concentration of the solution.

§139. Change in symmetry in a phase transition of the second kind

In the theory given in §138 we have considered a phase transition of the second kind with some definite change in symmetry of the body, assuming *a priori* that such a transition is possible. Such a theory, however, does not say whether a given change of symmetry can in fact occur by a transition of the second kind. The theory developed in the present section is designed to answer this question; it starts from a different statement of the problem, whereby a certain symmetry of the body at the transition point itself is specified, and we ask what symmetry is possible on either side of this point.

† This renders pointless a discussion of transitions involving discontinuities only of higher-order derivatives.

For definiteness, we shall speak of phase transitions involving a change in structure of the crystal lattice, i.e. a change in the symmetry of the configuration of atoms in it. Let $\varrho(x, y, z)$ be the "density function" (defined in §128), which gives the probability distribution of various positions of the atoms in the crystal. The symmetry of the crystal lattice is the set or group of all transformations of the co-ordinates under which the function $\varrho(x, y, z)$ is invariant. Here we mean, of course, the complete symmetry of the lattice, including rotations, reflections and also the infinite (discrete) set of all possible parallel displacements (translations); that is, we are concerned with one of the 230 space groups.

Let G_0 be the symmetry group of the crystal at the transition point itself. As we know from group theory, an arbitrary function $\varrho(x, y, z)$ can be represented as a linear combination of several functions $\phi_1, \phi_2, \ldots$ having the property of being transformed into combinations of one another by all the transformations in the group concerned. In general the number of these functions is equal to the number of elements in the group, but when the function ϱ itself has a certain symmetry the functions ϕ_i may be fewer in number.

Bearing this in mind, we write the density function $\varrho(x, y, z)$ of the crystal as the sum

$$\varrho = \sum_i c_i \phi_i,$$

where the functions ϕ_i are transformed into combinations of one another by all transformations in the group G_0. The matrices of these transformations form a representation of the group G_0. The choice of the functions ϕ_i is not unique; they can obviously be replaced by any linear combinations of themselves. The functions ϕ_i can always be so chosen as to form a number of independent sets containing the minimum number of functions, the functions in each set being transformed only into combinations of one another by all transformations in the group G_0. The matrices of the transformations of the functions in each of these sets form irreducible representations of the group G_0, and the functions themselves are the basis of these representations. Thus we can write

$$\varrho = \sum_n \sum_i c_i^{(n)} \phi_i^{(n)}, \tag{139.1}$$

n being the number of the irreducible representation and i the number of the function in its basis. In what follows we shall assume the functions $\phi_i^{(n)}$ to be normalised in some definite manner.

The functions $\phi_i^{(n)}$ always include one which is invariant under all the transformations in the group G_0 and gives what is called the unit representation of the group. Thus this function (which we denote by ϱ_0) has the symmetry of $\boldsymbol{G}_0$. Denoting the remaining part of ϱ by $\delta\varrho$, we can write

$$\varrho = \varrho_0 + \delta\varrho, \qquad \delta\varrho = \sum_n{}' \sum_i c_i^{(n)}\phi_i^{(n)}, \tag{139.2}$$

where now the unit representation is excluded from the summation (as indicated by the prime to the summation sign). The function $\delta\varrho$ has a lower symmetry than that of $\boldsymbol{G}_0$, since $\delta\varrho$ may also remain invariant under some transformations in this group but certainly does not do so under all. We may note that the symmetry $\boldsymbol{G}$ of the function ϱ (which clearly is the symmetry of $\delta\varrho$) has, strictly speaking, been assumed from the start to be lower than that of $\boldsymbol{G}_0$, since otherwise the sum (139.1) would include only one term, the function ϱ itself, which gives the unit representation.

Some of the irreducible representations of the space group may be complex (i.e. the transformations of the group transform the base functions into linear combinations of one another with complex coefficients). Each such representation is accompanied by its complex conjugate representation (given by the complex conjugate functions). Since the physical density $\delta\varrho$ must be real and must remain real under all transformations, it is clear that two complex conjugate irreducible representations must be physically regarded as one representation of twice the dimension (number of base functions). The density $\delta\varrho$ must be a real linear combination of all these complex conjugate functions. Throughout the following discussion we shall assume that this is so and that the functions $\phi_i^{(n)}$ are taken to be real.†

The thermodynamic potential Φ of a crystal whose density function ϱ is given by (139.2) is a function of temperature, pressure and the coefficients $c_i^{(n)}$ (and depends, of course, on the specific form of the functions $\phi_i^{(n)}$ themselves). The actual values of the $c_i^{(n)}$ as functions of P and T are determined thermodynamically from the conditions of equilibrium, i.e. the conditions for Φ to be a minimum. This determines also the symmetry $\boldsymbol{G}$ of the crystal, since it is clear that the symmetry of the function (139.2), with functions $\phi_i^{(n)}$ whose laws of transformation are known, is determined by the values of the coefficients in the linear combination of the $\phi_i^{(n)}$.

If the crystal is to have the symmetry $\boldsymbol{G}_0$ at the transition point itself, it is necessary that all the $c_i^{(n)}$ should be zero there, i.e. $\delta\varrho = 0$, $\varrho = \varrho_0$. Since the change in state of the crystal in a phase transition of the second kind is continuous, $\delta\varrho$ must tend continuously to zero at the transition point, not discontinuously, i.e. the coefficients $c_i^{(n)}$ must tend to zero through arbitrarily

† The method of constructing irreducible representations of the space groups has been discussed in §136. The remark just made shows that to obtain the "physically irreducible" (real) representations we must include in the star of **k** the vector $-\mathbf{k}$ as well as each **k**. In other words, in order to obtain the whole required star of **k** we must apply to some initial **k** all the elements of the crystal class, together with a centre of symmetry if this is not already present.

small values near the transition point. Accordingly, we can expand the potential $\Phi(P, T, c_i^{(n)})$ in powers of the $c_i^{(n)}$ near the transition point.

First of all let us note that, since the functions $\phi_i^{(n)}$ (belonging to the basis of each irreducible representation) are transformed into combinations of one another by the transformations in the group G_0, these transformations can be regarded as transforming (in the same manner) the coefficients $c_i^{(n)}$ instead of the functions $\phi_i^{(n)}$. Next, since the thermodynamic potential of the body must obviously be independent of the choice of co-ordinates, it must be invariant under any transformation of the co-ordinate system, and in particular under the transformations of the group G_0. Thus the expansion of Φ in powers of the $c_i^{(n)}$ can contain in each term only an invariant combination of the $c_i^{(n)}$ that is of the appropriate power.

No linear invariant can be formed from quantities which are transformed according to a (non-unit) irreducible representation of a group.† Only one second-order invariant exists for each representation: a positive-definite form in the $c_i^{(n)}$, which can always be reduced to a sum of squares.

Thus the leading terms in the expansion of Φ are of the form

$$\Phi = \Phi_0 + \sum_n{}' A^{(n)} \sum_i [c_i^{(n)}]^2, \tag{139.3}$$

where the $A^{(n)}$ are functions of P and T.

At the transition point itself, the crystal must have the symmetry G_0, i.e. the equilibrium values of the $c_i^{(n)}$ must be zero. It is evident that Φ can have a minimum when every $c_i^{(n)} = 0$ only if all the $A^{(n)}$ are non-negative.

If all the $A^{(n)}$ were positive at the transition point, they would also be positive near that point, so that the $c_i^{(n)}$ would remain zero and there would be no change of symmetry. For some $c_i^{(n)}$ to be non-zero, i.e. for the symmetry of the body to change, one of the coefficients $A^{(n)}$ must change sign, and this coefficient must therefore vanish at the transition point.‡ (Two

† For otherwise that representation would contain the unit representation, i.e. would be reducible.

‡ Strictly speaking, this condition should be more accurately stated as follows. The coefficients $A^{(n)}$ depend, of course, on the particular form of the functions $\phi_i^{(n)}$, being quadratic functionals of these which depend on P and T as parameters. On one side of the transition point, all these functionals $A^{(n)}\{\phi_i^{(n)}; P, T\}$ are positive-definite. The transition point is defined as that at which (as P or T varies gradually) one of the $A^{(n)}$ can vanish:

$$A^{(n)}\{\phi_i^{(n)}; P, T\} \geqslant 0.$$

This vanishing corresponds to a definite set of functions $\phi_i^{(n)}$, which may in principle be determined by solving the appropriate variational problem. These will also be the functions $\phi_i^{(n)}$ which determine the change $\delta\varrho$ at the transition point. Substituting these functions in $A^{(n)}\}\phi_i^{(n)}; P, T\}$, we obtain just the function $A^{(n)}(P, T)$ which satisfies the condition of vanishing at the transition point. The functions $\phi_i^{(n)}$ may then be regarded as given, as will be assumed below; the allowance for the variation of the $\phi_i^{(n)}$ with P and T would lead to correction terms of higher order than those of interest here.

coefficients $A^{(n)}$ can vanish simultaneously only at an isolated point in the *PT*-plane, which is the intersection of more than one line of transitions of the second kind.)

Thus on one side of the transition point all the $A^{(n)} > 0$, and on the other side one of the coefficients $A^{(n)}$ is negative. Accordingly, all the $c_i^{(n)}$ are always zero on one side of the transition point, and on the other side non-zero $c_i^{(n)}$ appear. We conclude, therefore, that on one side of the transition point the crystal has the higher symmetry $\boldsymbol{G}_0$, which is retained at the transition point itself, while on the other side of the transition point the symmetry is lower, and so the group $\boldsymbol{G}$ is a sub-group of the group $\boldsymbol{G}_0$.

The change in sign of one of the $A^{(n)}$ causes the appearance of non-zero $c_i^{(n)}$ belonging to the nth representation. Thus the crystal with symmetry $\boldsymbol{G}_0$ becomes one with density $\varrho = \varrho_0 + \delta\varrho$, where

$$\delta\varrho = \sum_i c_i^{(n)} \phi_i^{(n)} \tag{139.4}$$

is a linear combination of the base functions of any one of the irreducible representations of the group $\boldsymbol{G}_0$ (other than the unit representation). Accordingly we shall henceforward omit the index n which gives the number of the representation, meaning always the one which corresponds to the transition considered.

We shall use the notation

$$\eta^2 = \sum_i c_i^2, \qquad c_i = \eta\gamma_i, \tag{139.5}$$

(so that $\sum \gamma_i^2 = 1$) and write the expansion of Φ as

$$\Phi = \Phi_0(P, T) + \eta^2 A(P, T) + \eta^3 \sum_\alpha B_\alpha(P, T) f_\alpha^{(3)}(\gamma_i) + \\ + \eta^4 \sum_\alpha C_\alpha(P, T) f_\alpha^{(4)}(\gamma_i) + \ldots, \tag{139.6}$$

where $f_\alpha^{(3)}, f_\alpha^{(4)}, \ldots$ are invariants of the third, fourth etc. orders formed from the quantities γ_i; in the sums over α there are as many terms as there are independent invariants of the appropriate order which can be formed from the γ_i. In this expansion of the thermodynamic potential, the coefficient A must vanish at the transition point. In order that the transition point itself should be a stable state (i.e. in order that Φ should have a minimum at that point when $c_i = 0$), the third-order terms must vanish and the fourth-order terms must be positive-definite. As has been mentioned in §138, a line of phase transitions of the second kind (in the *PT*-plane) can exist only if the third-order terms in the expansion of Φ vanish identically. This condition may now be formulated as requiring that it should be impossible to construct from the

c_i third-order invariants which are transformed according to the corresponding irreducible representation of the group G_0.†

Assuming this condition to be satisfied, we write the expansion as far as the fourth-order terms in the form

$$\Phi = \Phi_0 + A(P, T)\eta^2 + \eta^4 \sum_\alpha C_\alpha(P, T) f_\alpha^{(4)}(\gamma_i). \tag{139.7}$$

Since the second-order term does not involve the γ_i, the latter are determined simply from the condition for a minimum of the fourth-order terms, i.e. of the coefficient of η^4 in (139.7).‡ Denoting the minimum value of this coefficient simply by $C(P, T)$ (which must be positive, as shown above), we return to the expansion of Φ in the form (138.4), η being determined from the condition that Φ is a minimum regarded as a function of η alone, as in §138. The values of the γ_i thus found determine the symmetry of the function

$$\delta\varrho = \eta \sum_i \gamma_i \phi_i, \tag{139.8}$$

i.e. the symmetry G of the crystal which is formed in the transition of the second kind from a crystal of symmetry G_0.||

The conditions derived above, however, are not yet sufficient to ensure the possibility of a phase transition of the second kind. A further essential condition is obtained if we consider a fact (hitherto deliberately ignored) relating to the classification properties of representations of space groups.+ We have seen in §136 that these representations can be classified not only by a discrete parameter (such as the number of the small representation) but also by the parameter **k**, which takes a continuous series of values. The coefficients $A^{(n)}$ in the expansion (139.3) must therefore depend not only on the discrete number n but also on the continuous variable **k**.

Let a phase transition correspond to the vanishing (as a function of P and T) of the coefficient $A^{(n)}(\mathbf{k})$ with a given number n and a given $\mathbf{k} = \mathbf{k}_0$. In

† In the language of the theory of representations, this signifies that the symmetric cube $[\Gamma^3]$ of the representation Γ in question must not contain the unit representation.

‡ It may happen that there is only one fourth-order invariant, $(\sum c_i^2)^2 = \eta^4$. In this case, the fourth-order term is independent of the γ_i, and higher-order terms must be used to determine the γ_i.

|| In §138 we have considered a transition with a given change of symmetry. Using the concepts defined here, we can say that the quantities γ_i were assumed to have given values (so that the function $\delta\varrho$ had a given symmetry). With the problem stated in these terms, the absence of the third-order term (in the expansion (138.4)) could not be a sufficient condition for the existence of a line of transition points of the second kind, since it does not exclude the possibility that there are third-order terms in the general expansion with respect to several c_i (if the irreducible representation is not one-dimensional). For example, if there are three c_i and the product $\gamma_1\gamma_2\gamma_3$ is invariant, the expansion of Φ contains a third-order term, whereas this term will vanish if the function $\delta\varrho$ has a certain symmetry which requires that one or two γ_i should be zero.

+ The results and examples given below are due to E. M. LIFSHITZ (1941). Further examples will be found in *Zhurnal éksperimental'noĭ i teoreticheskoĭ fiziki* **11**, 255, 269, 1941 (*Journal of Physics* **6**, 61, 251, 1942).

order that the transition should actually occur, it is necessary that $A^{(n)}$ as a function of $\mathbf{k}$ should have a minimum for $\mathbf{k} = \mathbf{k}_0$ (and therefore for all vectors of the star of $\mathbf{k}_0$), i.e. the expansion of $A^{(n)}(\mathbf{k})$ in powers of $\mathbf{k}-\mathbf{k}_0$ about $\mathbf{k}_0$ should contain no linear terms. Otherwise, some coefficients $A^{(n)}(\mathbf{k})$ necessarily vanish before $A^{(n)}(\mathbf{k}_0)$ and a transition of the type in question cannot occur. A convenient formulation of this condition can be obtained on the basis of the following arguments.

The value of $\mathbf{k}_0$ determines the translational symmetry of the functions ϕ_i, and therefore that of the function $\delta\varrho$ (139.8), i.e. it determines the periodicity of the lattice of the new phase. This structure must be stable in comparison with those which correspond to values of $\mathbf{k}$ close to $\mathbf{k}_0$. But a structure with $\mathbf{k} = \mathbf{k}_0+\boldsymbol{\varkappa}$ (where $\boldsymbol{\varkappa}$ is small) differs from that with $\mathbf{k} = \mathbf{k}_0$ by a spatial "modulation" in the periodicity of the latter, that is, by the appearance of non-uniformity over distances ($\sim 1/\varkappa$) which are large compared with the periods (cell dimensions) of the lattice. Such non-uniformity can be macroscopically described by regarding the coefficients c_i as slowly varying functions of the co-ordinates (whereas the functions ϕ_i oscillate over interatomic distances). Thus we obtain the requirement that the state of the crystal should be stable with respect to loss of macroscopic homogeneity.

When the quantities c_i are not constant in space, the thermodynamic potential per unit volume of the crystal will depend not only on the c_i but also on their derivatives with respect to the co-ordinates (in the first approximation, on the first derivatives). Accordingly Φ (for unit volume) must be expanded in powers of the c_i and of their gradients ∇c_i near the transition point. If the thermodynamic potential (of the whole crystal) is to be a minimum for constant c_i, it is necessary that the first-order terms in the gradients in this expansion should vanish identically. (The terms quadratic in the derivatives must be positive-definite, but this imposes no restriction on the c_i, since such a quadratic form exists for c_i which are transformed by any of the irreducible representations.)

Among the terms linear in the derivatives, the only ones that can be of interest are those simply proportional to $\partial c_i/\partial x, \ldots$, and those containing the products $c_i\, \partial c_k/\partial x, \ldots$. The higher-order terms are clearly of no importance. The thermodynamic potential of the whole crystal, i.e. the integral $\int \Phi\, dV$ over the whole volume, is to be a minimum. The integration of all the total derivatives in Φ gives a constant which does not affect the determination of the minimum of the integral. We can therefore omit all terms in Φ which are simply proportional to derivatives of the c_i. Among the terms containing products $c_i \partial c_k/\partial x, \ldots$ we can omit all symmetrical combinations $c_k \partial c_i/\partial x + c_i \partial c_k/\partial x = \partial(c_i c_k)/\partial x, \ldots$, leaving the antisymmetrical parts

$$c_k \frac{\partial c_i}{\partial x} - c_i \frac{\partial c_k}{\partial x}, \ \ldots . \tag{139.9}$$

The expansion of Φ can contain only invariant linear combinations of the quantities (139.9). Hence the condition for a phase transition to be possible is that such invariants do not appear.

The components of the gradients ∇c_i are transformed as the products of the components of a vector and the quantities c_i. The differences (139.9) are therefore transformed as the products of the components of a vector and the antisymmetrised products of the quantities c_i. Consequently the requirement that no linear scalar can be formed from the quantities (139.9) is equivalent to the requirement that no combinations which transform as the components of a vector can be formed from the antisymmetrised products

$$\chi_{ik} = \phi_i \phi_k' - \phi_k \phi_i'; \tag{139.10}$$

here the ϕ_i and ϕ_i' are the same base functions of the relevant irreducible representation, which we regard as taken at two different points x, y, z and x', y', z' in order that the difference shall not be identically zero.† Labelling the base functions by the two suffixes $\mathbf{k}\alpha$ (as in §136), we write the difference (139.10) in the form

$$\chi_{\mathbf{k}\alpha,\mathbf{k}'\beta} = \phi_{\mathbf{k}\alpha}\phi_{\mathbf{k}'\beta}' - \phi_{\mathbf{k}\alpha}'\phi_{\mathbf{k}'\beta}, \tag{139.11}$$

where $\mathbf{k}$, $\mathbf{k}'$,... are vectors of the same star.

Let the vector $\mathbf{k}$ occupy the most general position and have no proper symmetry. The star of $\mathbf{k}$ contains n vectors according to the number of rotational elements in the group (or $2n$ if the space group itself does not include inversion), each $\mathbf{k}$ being accompanied by the different vector $-\mathbf{k}$. The corresponding irreducible representation is given by the same number of functions $\phi_\mathbf{k}$ (one for each $\mathbf{k}$, and so we omit the suffix α). The quantities

$$\chi_{\mathbf{k},-\mathbf{k}} = \phi_\mathbf{k}\phi_{-\mathbf{k}}' - \phi_\mathbf{k}'\phi_{-\mathbf{k}} \tag{139.12}$$

are invariant under translations. Under the rotational elements, these n (or $2n$) quantities are transformed into combinations of one another, giving a representation of the corresponding point group (crystal class) with dimension equal to the order of the group. But this representation (called a *regular* representation) contains all the irreducible representations of the group, including those by which the components of a vector are transformed.

Similar considerations show that it is possible to form a vector from the quantities $\chi_{\mathbf{k}\alpha,-\mathbf{k}\beta}$ in cases where the group of the vector $\mathbf{k}$ contains one axis and planes of symmetry passing through that axis.

This discussion becomes inapplicable, however, if the group of the vector $\mathbf{k}$ contains axes which intersect one another or intersect planes of symmetry, or

† In the language of the theory of representations, we can say that the antisymmetric square $\{\Gamma^2\}$ of the representation Γ in question must not contain the irreducible representations by which the components of a vector are transformed.

contains inversion; such groups will be said to have a central point. In such cases the question of constructing a vector from the quantities (139.11) requires separate treatment in each particular case. In particular, such a vector certainly can not be constructed if the group of **k** contains inversion, so that **k** and $-$**k** are equivalent, and only one function $\phi_{\mathbf{k}}$ corresponds to each **k** in the star; in this case there are no $\chi_{\mathbf{kk}'}$ invariant under translations (as the components of a vector must necessarily be).

Thus the requirement formulated above greatly restricts the possible changes of symmetry in a phase transition of the second kind. Of the infinity of different irreducible representations of the group G_0, we need consider only a comparatively small number for which the group of the vector **k** has a central point.

A proper symmetry of this kind can, of course, occur only for vectors $\mathbf{k}/2\pi$ which occupy certain exceptional positions in the reciprocal lattice, their components being equal to certain simple fractions ($\frac{1}{2}$, $\frac{1}{3}$, $\frac{1}{4}$) of the basic vectors of that lattice. This means that the change in the translational symmetry of the crystal (i.e. in its Bravais lattice) in a phase transition of the second kind must consist in an increase by a small factor in some of the basic lattice vectors. Investigation shows that in the majority of cases the only possible change in the Bravais lattice is a doubling of the lattice vectors. In addition, in body-centred orthorhombic, tetragonal and cubic and face-centred cubic lattices some lattice vectors can be quadrupled, and in a hexagonal lattice tripled. The volume of the unit cell can be increased by a factor of 2, 4 or 8, in a face-centred cubic lattice also by 16 or 32, and in a hexagonal lattice by 3 or 6.†

Transitions are, of course, also possible without change of Bravais lattice (corresponding to irreducible representations with $\mathbf{k} = 0$). The change in symmetry then consists in a decrease in the number of rotational elements, i.e. a change in the crystal class.‡

We may note the following general theorem. A phase transition of the second kind can occur for any change in structure which halves the number of symmetry transformations; such a change may occur either by a doubling of the unit cell for a given crystal class or by a halving of the number of rotations

† It has been pointed out by I. E. Dzyaloshinskiĭ that a very unusual situation can arise for transitions which involve a change in magnetic structure. Here there may be physical reasons why the coefficient, in the expansion of Φ, of the invariant (if any exists) formed from the quantities (139.9) should be anomalously small. This means that a phase transition is possible but does not lead to a state described by a vector $\mathbf{k}_0$ (satisfying the conditions imposed above); instead, it leads to a state with vector $\mathbf{k}_0+\boldsymbol{\varkappa}$, where the small quantity $\boldsymbol{\varkappa}$ is determined by the relative magnitude of the terms in the expansion of Φ which are of the first and second order in the derivatives. The resulting structure does not differ from that corresponding to the vector $\mathbf{k}_0$ on the small scale (of the order of atomic distances); that is, it corresponds to change in the lattice vectors by a small factor. On this structure is superposed a "hyperstructure" consisting of "beats" in the basic structure with a period ($\sim 1/\varkappa$) much greater than interatomic distances.

‡ Such cases can actually occur; see V. L. Indenbom, *Soviet Physics Crystallography* **5**, 106, 1960.

and reflections for a given unit cell. The proof is based on the fact that, if the group $\boldsymbol{G}_0$ has a sub-group $\boldsymbol{G}$ of half the order, then the irreducible representations of $\boldsymbol{G}_0$ always include a one-dimensional representation given by a function which is invariant under all transformations of the sub-group $\boldsymbol{G}$ and changes sign under all the remaining transformations of the group $\boldsymbol{G}_0$. It is clear that in this case there are no odd-order invariants, and no quantities of the type (139.11) can be formed from one function.

The following theorem also appears to be valid. Phase transitions of the second kind cannot occur for changes in structure which reduce to one-third the number of symmetry transformations, owing to the presence of third-order terms in the expansion of Φ.

Finally, to illustrate the practical applications of the general theory given above, let us consider the occurrence of ordering in alloys which, in the disordered state, have a body-centred cubic lattice with atoms at the vertices and centres of cubic cells, as in Fig. 62b (§137).† The problem is to determine the possible types of ordering (called *superlattices* in crystallography) which can appear in such a lattice in a phase transition of the second kind.

For a body-centred cubic lattice, the reciprocal lattice is face-centred cubic. If the edge of the body-centred cubic lattice cell is taken as the unit of length, the edge length of the cubic cell in the reciprocal lattice is $\frac{1}{2}$, and in this lattice the following vectors $\mathbf{k}/2\pi$ have intrinsic symmetry groups with a central point:

$$\left.\begin{array}{lllll}
\text{(a)} & (000) & & & \boldsymbol{O}_h \\
\text{(b)} & (\frac{1}{2}\,\frac{1}{2}\,\frac{1}{2}) & & & \boldsymbol{O}_h \\
\text{(c)} & (\frac{1}{4}\,\frac{1}{4}\,\frac{1}{4}), & (\frac{\bar{1}}{4}\,\frac{\bar{1}}{4}\,\frac{\bar{1}}{4}) & & \boldsymbol{T}_d \\
\text{(d)} & (0\,\frac{1}{4}\,\frac{1}{4}), & (\frac{1}{4}\,0\,\frac{1}{4}), & (\frac{1}{4}\,\frac{1}{4}\,0), & \\
 & (0\,\frac{1}{4}\,\frac{\bar{1}}{4}), & (\frac{\bar{1}}{4}\,0\,\frac{1}{4}), & (\frac{1}{4}\,\frac{\bar{1}}{4}\,0) & \boldsymbol{D}_{2h}
\end{array}\right\} \qquad (139.13)$$

These symbols show the components of the vectors $\mathbf{k}/2\pi$ along the edges of the cubic cell (x, y, z axes) as fractions of the edge lengths; a bar over a quantity indicates a negative value. In order to obtain the vectors $\mathbf{k}$ in the units specified above, these numbers must be multiplied by $2\cdot 2\pi = 4\pi$. In (139.13) only non-equivalent vectors are shown, i.e. the vectors of each star.

The subsequent discussion is greatly simplified by the fact that not all small representations need be considered in solving the problem in question. The reason is that we are concerned only with the possible changes of symmetry that can occur by the formation of a superlattice, that is, by an ordered arrangement of atoms at existing lattice sites without relative displacement. In this case the unit cell of the disordered lattice contains only one atom. Hence the

† Such a lattice belongs to the space group $\boldsymbol{O}_h^9$; it has no screw axes or glide planes.

appearance of the superlattice can only mean that the lattice points in different cells become non-equivalent. The change $\delta\varrho$ in the density distribution function must therefore be invariant under all rotational transformations of the group of **k** (without simultaneous translation). Thus only the unit small representation is admissible, and accordingly u_α may be replaced by unity in the base functions (136.2).

Let us now consider in turn the stars listed in (139.13).

(a) The function with $\mathbf{k} = 0$ has complete translational invariance, i.e. in this case the unit cell is unchanged, and since each cell contains only one atom no change of symmetry can occur.

(b) The function $e^{2\pi i(x+y+z)}$ corresponds to this **k**. The linear combination (of this function and the functions obtained from it by all rotations and reflections) which has the symmetry $\boldsymbol{O}_h$ of the group of **k** is

$$\phi = \cos 2\pi x \cos 2\pi y \cos 2\pi z. \tag{139.14}$$

The symmetry of the phase formed is that of the density function $\varrho = \varrho_0 + \delta\varrho$, $\delta\varrho = \eta\phi$.† The function ϕ is invariant under all transformations of the class $\boldsymbol{O}_h$ and under translations along any edge of the cubic cell, but not under a translation through half the space diagonal, $(\frac{1}{2}\frac{1}{2}\frac{1}{2})$. Hence the ordered phase has a simple cubic Bravais lattice with two non-equivalent points in the unit cell, (000) and $(\frac{1}{2}\frac{1}{2}\frac{1}{2})$; these will be occupied by different atoms. The alloys which can be completely ordered in this way have the composition AB (e.g. the alloy CuZn mentioned in §137).

(c) The functions corresponding to these **k** which have the symmetry $\boldsymbol{T}_d$ are

$$\begin{aligned} \phi_1 &= \cos \pi x \cos \pi y \cos \pi z, \\ \phi_2 &= \sin \pi x \sin \pi y \sin \pi z. \end{aligned} \tag{139.15}$$

From these we can form two fourth-order invariants: $(\phi_1^2+\phi_2^2)^2$ and $(\phi_1^4+\phi_2^4)$. The expansion of Φ (139.7) therefore has the form

$$\Phi = \Phi_0 + A\eta^2 + C_1\eta^4 + C_2\eta^4(\gamma_1^2+\gamma_2^2). \tag{139.16}$$

Here two cases must be distinguished. Let $C_2 < 0$; then Φ as a function of γ_1 and γ_2, with the added condition $\gamma_1^2+\gamma_2^2 = 1$, has a minimum for $\gamma_1 = 1$, $\gamma_2 = 0$.‡ The function $\delta\varrho = \eta\phi_1$ has the symmetry of the class $\boldsymbol{O}_h$ with a face-centred Bravais lattice, whose cubic cell has a volume 8 times that of the original cubic lattice cell. The unit cell contains 4 atoms; the cubic cell, 16 atoms. By placing like atoms at equivalent lattice sites we find that this superlattice

† This does not mean, of course, that the change $\delta\varrho$ in an actual crystal is given by the function (139.14). Only the symmetry of the expression (139.14) is important.

‡ Or for $\gamma_1 = 0$, $\gamma_2 = 1$. But the function $\delta\varrho = \eta\phi_2$ has the same symmetry as $\eta\phi_1$, differing from it only in that the origin is shifted by one lattice vector.

corresponds to a ternary alloy of composition ABC_2 with atoms in the following positions:

4A (000), $(0\,\frac{1}{2}\,\frac{1}{2})$ & cyclic,

4B $(\frac{1}{2}\,\frac{1}{2}\,\frac{1}{2})$, $(0\,0\,\frac{1}{2})$ & cyclic,

8C $(\frac{1}{4}\,\frac{1}{4}\,\frac{1}{4})$, $(\frac{3}{4}\,\frac{3}{4}\,\frac{3}{4})$, $(\frac{1}{4}\,\frac{3}{4}\,\frac{3}{4})$ & cyclic, $(\frac{1}{4}\,\frac{1}{4}\,\frac{3}{4})$ & cyclic;

here the co-ordinates of the atoms are given in units of the edge length of the new cubic lattice cell, which is twice that of the original cell (see Fig. 64a); "& cyclic" denotes cyclic interchange. If the B and C atoms are identical we obtain an ordered lattice of composition AB_3.

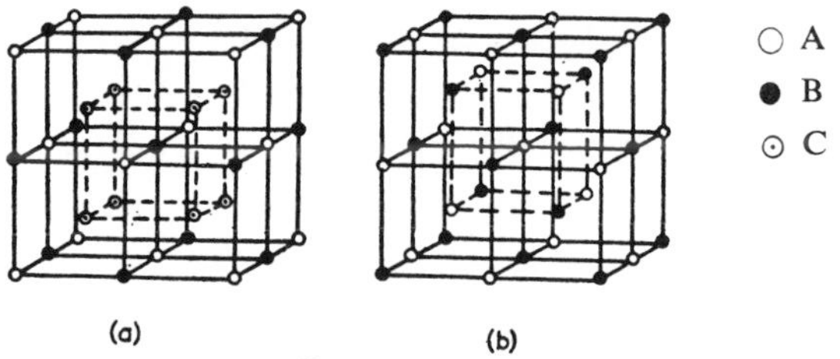

FIG. 64

Now let $C_2 > 0$. Then Φ has a minimum at $\gamma_1^2 = \gamma_2^2 = \frac{1}{2}$, so that $\delta\varrho = \eta(\phi_1+\phi_2)/\sqrt{2}$ (or $\eta(\phi_1-\phi_2)/\sqrt{2}$, which leads to the same result). This function has the symmetry of the class $\boldsymbol{O}_h$ with the same face-centred Bravais lattice as in the preceding case but only two sets of equivalent points, which can be occupied by atoms of two kinds A and B:

8A (000), $(\frac{1}{4}\,\frac{1}{4}\,\frac{1}{4})$, $(\frac{1}{4}\,\frac{3}{4}\,\frac{3}{4})$ & cyclic, $(0\,\frac{1}{2}\,\frac{1}{2})$ & cyclic

8B $(\frac{1}{2}\,\frac{1}{2}\,\frac{1}{2})$, $(\frac{3}{4}\,\frac{3}{4}\,\frac{3}{4})$, $(\frac{1}{4}\,\frac{1}{4}\,\frac{3}{4})$ & cyclic, $(0\,0\,\frac{1}{2})$ & cyclic

(see Fig. 64b).†

(d) The following functions with the required symmetry $\boldsymbol{D}_{2h}$ correspond to these vectors **k**:

$$\phi_1 = \cos\pi(y-z), \quad \phi_3 = \cos\pi(x-y), \quad \phi_5 = \cos\pi(x-z),$$
$$\phi_2 = \cos\pi(y+z), \quad \phi_4 = \cos\pi(x+y), \quad \phi_6 = \cos\pi(x+z).$$

From these we can form one third-order invariant and four fourth-order invariants, and so the expansion (139.6) becomes

$$\begin{aligned}\Phi = \Phi_0 &+ A\eta^2 + B\eta^3(\gamma_1\gamma_3\gamma_5+\gamma_2\gamma_3\gamma_6+\gamma_1\gamma_4\gamma_6+\gamma_2\gamma_4\gamma_5)+\\ &+C_1\eta^4+C_2\eta^4(\gamma_1^4+\gamma_2^4+\gamma_3^4+\gamma_4^4+\gamma_5^4+\gamma_6^4)+\\ &+C_3\eta^4(\gamma_1^2\gamma_2^2+\gamma_3^2\gamma_4^2+\gamma_5^2\gamma_6^2)+\\ &+C_4\eta^4(\gamma_1\gamma_2\gamma_3\gamma_4+\gamma_3\gamma_4\gamma_5\gamma_6+\gamma_1\gamma_2\gamma_5\gamma_6).\end{aligned}$$

† The structures in Figs. 64a and b belong to space groups $\boldsymbol{O}_h^5$ and $\boldsymbol{O}_h^7$. The former is the structure of the *Heusler alloys*.

Because cubic terms are present, a phase transition of the second kind is impossible in this case. To examine whether isolated points of continuous transition can exist and the properties of such points (see §140) it would be necessary to investigate the behaviour of the function Φ near its minimum; we shall not pause to do so here.

The above example shows what rigid limitations are imposed by the thermodynamic theory on the possibility of phase transitions of the second kind; for example, in this case they can exist only when superlattices of three types are formed.

The following fact may also be pointed out. In case (c), when $C_2 < 0$, the actual change in the density function, $\delta\varrho = \eta\phi_1$, corresponds to only one of the two parameters γ_1, γ_2 which appear in the thermodynamic potential (139.16). This illustrates an important feature of the foregoing theory: in considering a particular change in the lattice in a phase transition of the second kind, it may be necessary to take account of other, "virtually possible", changes.

§140. Isolated and critical points of continuous transition

The curve of phase transitions of the second kind in the *PT*-plane separates phases of different symmetry, and cannot, of course, simply terminate at some point, but it may pass into a curve of phase transitions of the first kind. A point at which this happens may be called a *critical point* of a transition of the second kind; it is in some ways analogous to an ordinary critical point.

The nature of the temperature dependence of the thermodynamic quantities near the critical point can, in principle, be investigated by the method given in §138, but this problem is even more subject to the remark made in §138 that the results thus obtained are of uncertain validity. We shall therefore not give the corresponding investigation in detail here, but simply describe briefly the results.

In the expansion (138.4) the critical point is given by the vanishing of the two coefficients $A(P, T)$ and $C(P, T)$; if $A = 0$ but $C > 0$ we have a transition of the second kind, and so the curve of such transitions terminates only where C changes sign. Thus, to examine the neighbourhood of the critical point, we must discuss the expansion of the thermodynamic potential as far as the sixth-order terms. It can be shown that the curve of phase transitions of the second kind passes smoothly into the curve of transitions of the first kind, i.e. the derivative $\mathrm{d}T/\mathrm{d}P$ does not have a discontinuity on the curve, although the second derivative $\mathrm{d}^2T/\mathrm{d}P^2$ does. At the critical point, the specific heat C_p of the less symmetrical phase becomes infinite inversely as the square root of the distance from the critical point.

Finally, it remains to consider the case where the third-order terms in the expansion of the thermodynamic potential do not vanish identically. In this case the condition for the existence of a point of continuous phase transition requires that the coefficients $B_\alpha(P, T)$ of the third-order invariants in the expansion (139.6) should vanish, as well as $A(P, T)$. It is evident that this is possible only if there is not more than one third-order invariant, since otherwise we should obtain more than two equations for the two unknowns P and T.† When there is only one third-order invariant, the two equations $A(P, T) = 0$, $B(P, T) = 0$ determine pairs of values of P and T, i.e. there are isolated points of continuous phase transition.

Since these points are isolated, they must lie in a certain way at the intersection of curves (in the PT-plane) of phase transitions of the first kind. Since such isolated points of continuous transition have not yet been observed experimentally, we shall not pause to give a detailed discussion here, but simply mention the results.‡

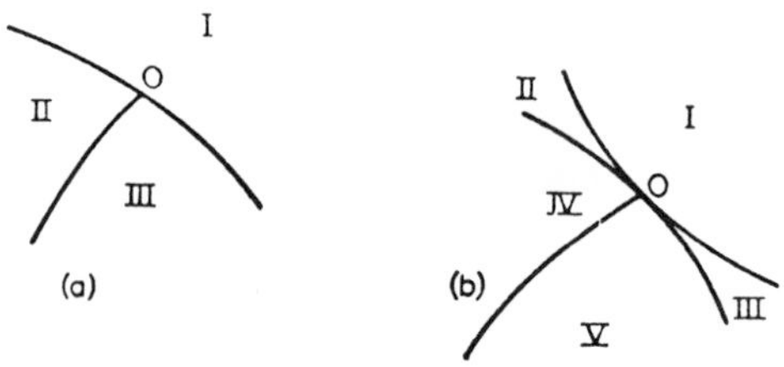

FIG. 65

The simplest type is that shown in Fig. 65a. Phase I has the higher symmetry, and phases II and III the same lower symmetry, these two phases differing only in the sign of η. At the point of continuous transition (O in Fig. 65) all three phases become identical.

In more complex cases two or more curves of phase transition of the first kind (e.g. two in Fig. 65b) touch at the point of continuous transition. Phase I has the highest symmetry, phases II and III a lower symmetry, phases IV and V another lower symmetry, these pairs of phases differing only in the sign of η.‖

† It is apparently true (though we have not obtained a general proof) that there can never be more than one third-order invariant for the representations of the space groups.

‡ See L. LANDAU, *Zhurnal éksperimental'noĭ i teoreticheskoĭ fiziki* 7, 19, 1937; translation in *Collected Papers of L. D. Landau*, p. 193, Pergamon, Oxford 1965.

‖ There is reason to suppose that even isolated points of continuous phase transition are impossible for transitions between a liquid and a solid crystal.

§141. Phase transitions of the second kind in a two-dimensional lattice

Considerable theoretical interest attaches to the examination of phase transitions of the second kind in two-dimensional systems also.† Although no general analysis of this problem has been made, it is likely that the general nature of the singularity of the thermodynamic quantities at the transition point will appear from the solution of the transition problem in a simple particular model of a two-dimensional lattice; such a problem was first solved by L. ONSAGER (1944).‡ This topic is the more interesting in that no similar problem has yet been solved for the three-dimensional case.

The model considered is a plane square lattice having N points, at each of which is a "dipole" with its axis perpendicular to the lattice plane. The dipole can have two opposite orientations, so that the total number of possible configurations of the dipoles in the lattice is 2^N. To describe the various configurations we proceed as follows. To each lattice point (with integral co-ordinates k, l) we assign a variable σ_{kl} which takes two values ± 1, corresponding to the two possible orientations of the dipole. If we take into account only the interaction between adjoining dipoles, the energy of the configuration may be written

$$E(\sigma) = -J \sum_{k,l=1}^{L} (\sigma_{kl}\sigma_{k,l+1} + \sigma_{kl}\sigma_{k+1,l}), \tag{141.1}$$

where L is the number of points in a lattice line, the lattice being regarded as a large square, and $N = L^2$.|| The parameter $J\,(> 0)$ determines the energy of interaction of a pair of adjoining dipoles, which is $-J$ and $+J$ for like and unlike orientations of the two dipoles respectively. Then the configuration with the least energy is the "completely polarised" (ordered) configuration, in which all the dipoles are oriented in the same direction. This configuration is reached at absolute zero; as the temperature increases, the degree of ordering decreases, becoming zero at the transition point, when the two orientations of each dipole become equally probable.

The determination of the thermodynamic quantities requires the calculation of the partition function

$$Z = \sum_{(\sigma)} e^{-E(\sigma)/T} = \sum_{(\sigma)} \exp\{\theta \sum_{k,l} (\sigma_{kl}\sigma_{k,l+1} + \sigma_{kl}\sigma_{k+1,l})\}, \tag{141.2}$$

† In addition to its purely theoretical interest, this problem is closely related to that of the behaviour of crystals with a markedly stratified structure and of adsorbed films (cf. §147).

‡ The original method used by ONSAGER was extremely complex. Later, various authors simplified the solution. The method described below (which in part makes use of certain ideas in the method of KAC and WARD (1952)) is due to N. V. VDOVICHENKO.

|| The number L is, of course, assumed macroscopically large, and edge effects (due to the special properties of points near the edges of the lattice) will be neglected throughout the following discussion.

taken over all the 2^N possible configurations ($\theta = J/T$). The equation

$$\exp(\theta\sigma_{kl}\sigma_{k'l'}) = \cosh\theta + \sigma_{kl}\sigma_{k'l'}\sinh\theta = \cosh\theta(1+\sigma_{kl}\sigma_{k'l'}\tanh\theta)$$

is easily verified by expanding both sides in powers of θ and using the fact that all the $\sigma_{kl}^2 = 1$. The expression (141.2) can therefore be written

$$Z = (1-x^2)^{-N}S, \tag{141.3}$$

where

$$S = \sum_{(\sigma)} \prod_{k,l=1}^{L} (1+x\sigma_{kl}\sigma_{k,l+1})(1+x\sigma_{kl}\sigma_{k+1,l}) \tag{141.4}$$

and $x = \tanh\theta$.

The summand in (141.4) is a polynomial in the variables x and σ_{kl}. Since each point (k, l) has four neighbours, each σ_{kl} can appear in the polynomial in powers from zero to four. After summation over all the $\sigma_{kl} = \pm 1$ the terms containing odd powers of σ_{kl} vanish, and so a non-zero contribution comes only from terms containing σ_{kl} in powers 0, 2 or 4. Since $\sigma_{kl}^0 = \sigma_{kl}^2 = \sigma_{kl}^4 = 1$, each term of the polynomial which contains all the variables σ_{kl} in even powers gives a contribution to the sum which is proportional to the total number of configurations, 2^N.

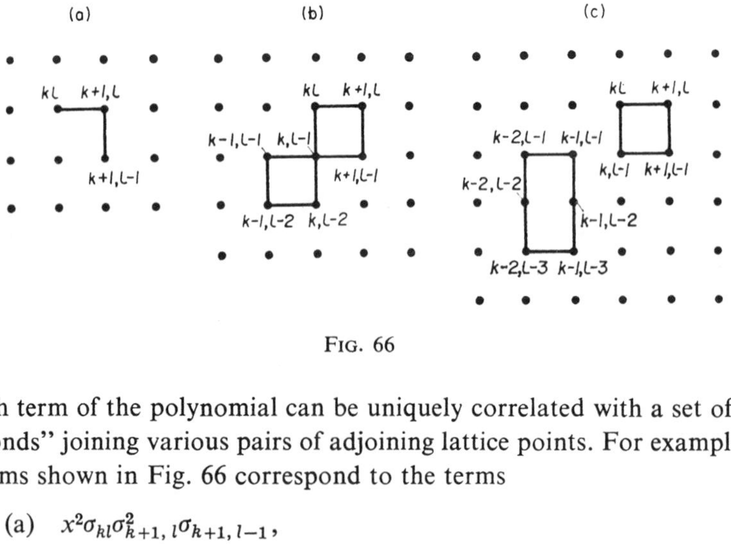

FIG. 66

Each term of the polynomial can be uniquely correlated with a set of lines or "bonds" joining various pairs of adjoining lattice points. For example, the diagrams shown in Fig. 66 correspond to the terms

(a) $x^2\sigma_{kl}\sigma_{k+1,l}^2\sigma_{k+1,l-1}$,

(b) $x^8\sigma_{kl}^2\sigma_{k+1,l}^2\sigma_{k+1,l-1}^2\sigma_{k,l-1}^4\sigma_{k,l-2}^2\sigma_{k-1,l-1}^2\sigma_{k-1,l-2}^2$,

(c) $x^{10}\sigma_{kl}^2\sigma_{k+1,l}^2\sigma_{k+1,l-1}^2\sigma_{k,l-1}^2\sigma_{k-2,l-1}^2\sigma_{k-1,l-1}^2 \times$

$\times\sigma_{k-1,l-2}^2\sigma_{k-1,l-3}^2\sigma_{k-2,l-3}^2\sigma_{k-2,l-2}^2$.

Each line in the diagram is assigned a factor x and each end of each line a factor σ_{kl}.

The fact that a non-zero contribution to the partition function comes only from terms in the polynomial which contain all the σ_{kl} in even powers signifies geometrically that either 2 or 4 bonds must end at each point in the diagram. Hence the summation is taken only over closed diagrams, which may be self-intersecting (as at the point k, $l-1$ in Fig. 66b).

Thus the sum S may be expressed in the form

$$S = 2^N \sum_r x^r g_r, \tag{141.5}$$

where g_r is the number of closed diagrams formed from an (even) number r of bonds, each multiple diagram (e.g. Fig. 66c) being counted as one.

The subsequent calculation is in two stages: (1) the sum over diagrams of this type is converted into one over all possible closed loops, (2) the resulting sum is calculated by reducing it to the problem of the "random walk" of a point in the lattice.

We shall regard each diagram as consisting of one or more closed loops. For non-self-intersecting diagrams this is obvious; for example, the diagram in Fig. 66c consists of two loops For self-intersecting diagrams, however, the resolution into loops is not unique: a given diagram may consist of different numbers of loops for different ways of construction. This is illustrated by Fig. 67, which shows three ways of representing the diagram in Fig. 66b as one or two non-self-intersecting loops or as one self-intersecting loop. Any intersection may similarly be traversed in three ways on more complicated diagrams.

It is easy to see that the sum (141.5) can be extended to all possible sets of loops if, in computing the number of diagrams g_r, each diagram is taken with the sign $(-1)^n$, where n is the total number of self-intersections in the loops of a given set, since when this is done all the extra terms in the sum

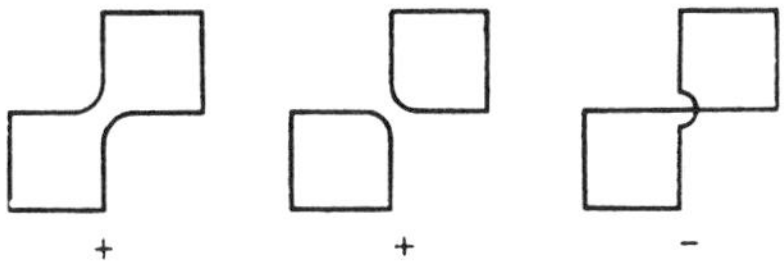

FIG. 67

necessarily cancel. For example, the three diagrams in Fig. 67 have signs $+, +, -$ respectively, so that two of them cancel, leaving a single contribution to the sum, as they should. The new sum will also include diagrams with "repeated bonds", of which the simplest example is shown in Fig. 68a. These diagrams are not permissible, since some points have an odd number of bonds meeting at them, namely three, but in fact they cancel from the sum, as they should: when the loops corresponding to such a diagram are constructed, each bond in common can be traversed in two ways, without intersection (as in Fig. 68b) and with self-intersection (Fig. 68c); the resulting sets of loops appear in the sum with opposite signs, and so cancel. We can also avoid the need to take into account explicitly the number of intersections by using the geometrical result that the total angle of rotation of the tangent in going round a closed plane loop is $2\pi(l+1)$, where l is a (positive or negative) integer whose parity is the same as that of the number ν of self-intersections of the loop. Hence, if we assign a factor $e^{\frac{1}{2}i\phi}$ to each point of the loop (with the angle of rotation there $\phi = 0, \pm\frac{1}{2}\pi$), the product of these factors after going round the whole loop will be $(-1)^{\nu+1}$, and for a set of s loops the resultant factor is $(-1)^{n+s}$, where $n = \sum\nu$.

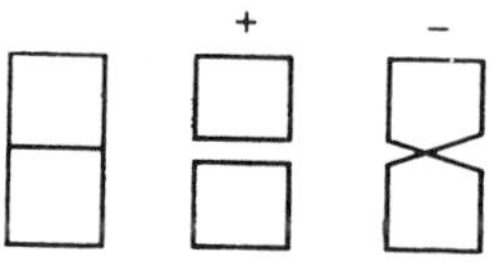

FIG. 68

Thus the number of intersections need not be considered if each point on the loop is taken with a factor $e^{\frac{1}{2}i\phi}$ and a further factor $(-1)^s$ is taken for the whole diagram (set of loops) in order to cancel the same factor in $(-1)^{n+s}$.

Let f_r denote the sum over all single loops of length r (i.e. consisting of r bonds), each loop having a factor $e^{\frac{1}{2}i\phi}$ at each point on it. Then the sum over all pairs of loops with total number of bonds r is

$$\frac{1}{2!}\sum_{r_1+r_2=r} f_{r_1} f_{r_2};$$

the factor 1/2! takes into account the fact that the same pair of loops is obtained when the suffixes r_1 and r_2 are interchanged, and similarly for groups of three or more loops. Thus the sum S becomes

$$S = \sum_{s=0}^{\infty}(-1)^s \frac{1}{s!} \sum_{r_1, r_2, \ldots=1}^{\infty} x^{r_1+\ldots+r_s} f_{r_1}\ldots f_{r_s}.$$

Since S includes sets of loops with every total length $r_1+r_2+\ldots$, the numbers $r_1, r_2, \ldots$ in the inner sum take independently all values from 1 to ∞.† Hence

$$\sum_{r_1,\ldots,r_s} x^{r_1+\ldots+r_s} f_{r_1}\ldots f_{r_s} = \Big(\sum_{r=1}^{\infty} x^r f_r\Big)^s$$

and S becomes

$$S = \exp\Big(-\sum_{r=1}^{\infty} x^r f_r\Big). \tag{141.6}$$

This completes the first stage of the calculation.

† Loops with more than N points make no contribution to the sum, since they must necessarily contain repeated bonds.

It is now convenient to assign to each lattice point the four possible directions from it and to number them by a quantity $\nu = 1, 2, 3, 4$, say as follows:

$$\begin{array}{ccc} & 2 & \\ & \uparrow & \\ 3 \leftarrow & \cdot & \rightarrow 1 \\ & \downarrow & \\ & 4 & \end{array}$$

We define as an auxiliary quantity $W_r(k, l, \nu)$ the sum over all possible paths of length r from some given point k_0, l_0, ν_0 to a point k, l, ν (each bond having as usual the factor $e^{\frac{1}{2}i\phi}$, where ϕ is the change of direction to the next bond); the final step to the point k, l, ν must not be from the point to which the arrow ν is directed.† With this definition, $W_r(k_0, l_0, \nu_0)$ is the sum over all loops leaving the point k_0, l_0 in the direction ν_0 and returning to that point. It is evident that

$$f_r = \frac{1}{2r} \sum_{k_0,\, l_0,\, \nu_0} W_r(k_0, l_0, \nu_0): \tag{141.7}$$

both sides contain the sum over all single loops, but $\sum W_r$ contains each loop $2r$ times, since it can be traversed in two opposite directions and can be assigned to each of r starting points on it.

From the definition of $W_r(k, l, \nu)$ we have the recurrence relations

$$\left.\begin{aligned} W_{r+1}(k, l, 1) &= W_r(k-1, l, 1) + e^{-\frac{1}{4}i\pi} W_r(k, l-1, 2) + \\ &\qquad + 0 + e^{\frac{1}{4}i\pi} W_r(k, l+1, 4), \\ W_{r+1}(k, l, 2) &= e^{\frac{1}{4}i\pi} W_r(k-1, l, 1) + W_r(k, l-1, 2) + \\ &\qquad + e^{-\frac{1}{4}i\pi} W_r(k+1, l, 3) + 0, \\ W_{r+1}(k, l, 3) &= 0 + e^{\frac{1}{4}i\pi} W_r(k, l-1, 2) + \\ &\qquad + W_r(k+1, l, 3) + e^{-\frac{1}{4}i\pi} W_r(k, l+1, 4), \\ W_{r+1}(k, l, 4) &= e^{-\frac{1}{4}i\pi} W_r(k-1, l, 1) + 0 + \\ &\qquad + e^{\frac{1}{4}i\pi} W_r(k+1, l, 3) + W_r(k, l+1, 4). \end{aligned}\right\} \tag{141.8}$$

The method of constructing these relations is evident: for example, the point $k, l, 1$ can be reached by taking the last $(r+1)$th step from the left, from below or from above, but not from the right; the coefficients of W_r arise from the factors $e^{\frac{1}{2}i\phi}$.

Let Λ denote the matrix of the coefficients in equations (141.8) (with all k, l), written in the form

$$W_{r+1}(k, l, \nu) = \sum_{k', l', \nu'} \Lambda(kl\nu|k'l'\nu') W_r(k', l', \nu').$$

† In fact $W_r(k, l, \nu)$ depends, of course, only on the differences $k - k_0$, $l - l_0$.

The method of constructing these equations enables us to associate with this matrix an intuitive picture of a point moving step by step through the lattice with a "transition probability" per step from one point to another which is equal to the corresponding element of the matrix Λ; its elements are in fact zero except when either k or l changes by ± 1 and the other remains constant, i.e. the point traverses only one bond per step. It is evident that the "probability" of traversing a length r will be given by the matrix Λ^r. In particular, the diagonal elements of this matrix give the "probability" that the point will return to its original position after traversing a loop of length r, i.e. they are equal to $W_r(k_0, l_0, \nu_0)$. Hence

$$\operatorname{tr} \Lambda^r = \sum_{k_0, l_0, \nu_0} W_r(k_0, l_0, \nu_0).$$

Comparison with (141.7) shows that

$$f_r = \frac{1}{2r} \operatorname{tr} \Lambda^r = \frac{1}{2r} \sum_i \lambda_i^r,$$

where the λ_i are the eigenvalues of the matrix Λ. Substituting this expression in (141.6) and interchanging the order of summation over i and r, we obtain

$$\begin{aligned} S &= \exp \{-\tfrac{1}{2} \sum_i \sum_{r=1}^{\infty} \frac{1}{r} x^r \lambda_i^r\} \\ &= \exp \{\tfrac{1}{2} \sum_i \log(1 - x\lambda_i)\} \\ &= \prod_i \sqrt{(1 - x\lambda_i)}. \end{aligned} \tag{141.9}$$

The matrix Λ is easily diagonalised with respect to the suffixes k and l by using a Fourier transformation:

$$W_r(p, q, \nu) = \sum_{k, l=0}^{L} e^{-2\pi i(pk+ql)/L} W_r(k, l, \nu). \tag{141.10}$$

Taking Fourier components on both sides of equations (141.8), we find that each equation contains only $W_r(p, q, \nu)$ with the same p, q, so that the matrix Λ is diagonal with respect to p and q. For given p, q its elements are

$$\Lambda(pq\nu \mid pq\nu') = \begin{bmatrix} \varepsilon^{-p} & \alpha^{-1}\varepsilon^{-q} & 0 & \alpha\varepsilon^{q} \\ \alpha\varepsilon^{-p} & \varepsilon^{-q} & \alpha^{-1}\varepsilon^{p} & 0 \\ 0 & \alpha\varepsilon^{-q} & \varepsilon^{p} & \alpha^{-1}\varepsilon^{q} \\ \alpha^{-1}\varepsilon^{-p} & 0 & \alpha\varepsilon^{p} & \varepsilon^{q} \end{bmatrix},$$

where $\alpha = e^{\frac{1}{4}i\pi}$, $\varepsilon = e^{2\pi i/L}$.

For given p, q a simple calculation of the determinant shows that

$$\prod_{i=1}^{4} (1-x\lambda_i) = \det (\delta_{\nu\nu'}-x\Lambda_{\nu\nu'})$$

$$= (1+x^2)^2-2x(1-x^2)\left(\cos\frac{2\pi p}{L}+\cos\frac{2\pi q}{L}\right).$$

Hence, from (141.3) and (141.9), we finally obtain the partition function

$$Z = 2^N(1-x^2)^{-N}\prod_{p,q=0}^{L}\left[(1+x^2)^2-\right.$$

$$\left.-2x(1-x^2)\left(\cos\frac{2\pi p}{L}+\cos\frac{2\pi q}{L}\right)\right]^{1/2}. \qquad (141.11)$$

The thermodynamic potential is†

$$\Phi = -T\log Z$$

$$= -NT\log 2+NT\log(1-x^2)-$$

$$-\tfrac{1}{2}T\sum_{p,q=0}^{L}\log\left[(1+x^2)^2-2x(1-x^2)\left(\cos\frac{2\pi p}{L}+\cos\frac{2\pi q}{L}\right)\right]$$

or, changing from summation to integration,

$$\Phi = -NT\log 2+NT\log(1-x^2)-$$

$$-\frac{NT}{2(2\pi)^2}\int_0^{2\pi}\int_0^{2\pi}\log[(1+x^2)^2-2x(1-x^2)(\cos\omega_1+\cos\omega_2)]d\omega_1 d\omega_2 \qquad (141.12)$$

(remembering that $x = \tanh(J/T)$).

Let us now examine this expression. The function $\Phi(T)$ has a singularity at the value of x for which the argument of the logarithm in the integrand can vanish. As a function of ω_1 and ω_2, this argument is a minimum for $\cos\omega_1 = \cos\omega_2 = 1$, when it equals $(1+x^2)^2-4x(1-x^2) = (x^2+2x-1)^2$. This expression has a minimum value of zero for only one (positive) value of x, $x_c = \sqrt{2}-1$; the corresponding temperature T_c ($\tanh(J/T_c) = x_c$) is the phase transition point.

The expansion of $\Phi(T)$ in powers of $t = T-T_c$ near the transition point includes a singular term as well as the regular part. Here we are interested only in the singular term, the regular part being simply replaced by its value at $t = 0$. To find the form of the singular term, we expand the argument of the logarithm in (141.12) in powers of ω_1, ω_2 and t about the minimum; the

† In the model under discussion the temperature affects only the ordering of dipole orientations, not the distances between dipoles (the "thermal expansion coefficient" of the lattice is zero). It is then immaterial whether we consider the free energy or the thermodynamic potential.

integral then becomes

$$\int_0^{2\pi}\int_0^{2\pi} \log\,[c_1t^2+c_2(\omega_1^2+\omega_2^2)]\mathrm{d}\omega_1\mathrm{d}\omega_2,$$

where c_1 and c_2 are constants. Carrying out the integration, we find that the thermodynamic potential near the transition point has the form

$$\Phi \cong a-\tfrac{1}{2}b(T-T_c)^2 \log|T-T_c|, \tag{141.13}$$

where a and b are further constants (with $b > 0$). The potential itself is continuous at the transition point, but the specific heat becomes infinite in accordance with the formula

$$C \cong b \log|T-T_c|, \tag{141.14}$$

which is symmetrical about the transition point.

In an actual two-dimensional structure we must expect singularities of the same type in the thermodynamic quantities, the coefficients a, b and the transition temperature T_c being functions of the "pressure". A singularity of the type (141.14) will also occur in the compressibility and in the thermal expansion coefficient of the lattice.

The degree of ordering η of the lattice is represented in this model by the mean dipole moment at a lattice point (the "spontaneous polarisation" of the lattice), which is non-zero below the transition point and zero above it. The temperature dependence of this quantity can also be determined.† Without pausing to give the derivation, we shall simply state the final result for the manner in which the degree of ordering tends to zero as the transition point is approached:

$$\eta = \text{constant} \times (T_c-T)^{1/8}. \tag{141.15}$$

† This problem also was first solved by L. Onsager (1947). The simplest method of solution is given by N. V. Vdovichenko, *Soviet Physics JETP* **21**, 350, 1965.

CHAPTER XV

SURFACES

§142. Surface tension

HITHERTO we have entirely neglected effects resulting from the presence of surfaces of separation between different bodies.† Since, as the size of a body (i.e. the number of particles in it) increases, surface effects increase much more slowly than volume effects, the neglect of surface effects in the study of volume properties of bodies is entirely justified. There are, however, a number of phenomena which depend in fact on the properties of surfaces of separation.

The thermodynamic properties of such an interface are entirely described by one quantity, a function of the state of the bodies, defined as follows. We denote by $\mathfrak{s}$ the area of the interface, and consider a process whereby this area undergoes a reversible change by an infinitesimal amount $\mathrm{d}\mathfrak{s}$. The work done in such a process is obviously proportional to $\mathrm{d}\mathfrak{s}$, and so can be written as

$$\mathrm{d}R = \alpha \mathrm{d}\mathfrak{s}. \tag{142.1}$$

The quantity α thus defined is a fundamental characteristic of the interface, and is called the *surface-tension coefficient*.

Formula (142.1) is exactly analogous to the formula $\mathrm{d}R = -P\,\mathrm{d}V$ for the work done in a reversible change in the volume of a body. We may say that α plays the same part in relation to the surface as the pressure does in relation to the volume. In particular, we can easily show that the force on unit length of the perimeter of any part of the interface is equal in magnitude to α and is directed tangentially to the surface and along the inward normal to the perimeter.

Here we have assumed that α is positive. The fact that it must indeed always be positive is shown immediately by the following argument. If $\alpha < 0$, the contour bounding the surface would be subject to forces along the outward normal, i.e. tending to "stretch" the surface; the interface between two phases would therefore tend to increase without limit, and the phases would mix and cease to exist separately. If $\alpha > 0$, on the other hand, the interface tends to become as small as possible (for a given volume of the two phases). Hence,

† In reality, of course, phases in contact are separated by a thin transition layer, but the structure of this is of no interest here, and we may regard it as a geometrical surface.

for example, if one isotropic phase is surrounded by another, it will take the form of a sphere (the effect of an external field, e.g. gravity, being neglected, of course).

Let us now consider in more detail the surface tension at the interface between two isotropic phases, liquid and vapour, of the same substance. If an interface between two phases in equilibrium is concerned, it must be remembered that their pressure and temperature are in a definite functional relation given by the equation of the phase equilibrium curve, and α is then essentially a function of only one independent variable, not of two.

At a critical point, the liquid and gas phases become identical. The interface between them ceases to exist, and α must become zero. The law governing this vanishing of α is not yet known.

We can apply qualitatively the law of corresponding states (§85) to the surface tension between the liquid and its vapour. According to this law we should expect that the dimensionless ratio of α to a quantity of dimensions erg/cm² formed from the critical temperature and critical pressure would be a universal function of the reduced temperature T/T_c:†

$$\alpha/(T_c P_c^2)^{1/3} = f(T/T_c). \tag{142.2}$$

When surface effects are neglected, the differential of the energy of a system of two phases (of the same substance), for a given total volume V of the system, is $\mathrm{d}E = T\,\mathrm{d}S + \mu\,\mathrm{d}N$; in equilibrium, the temperatures T and chemical potentials μ of the two phases are equal, and this equation can therefore be written for the whole system. When the presence of surface effects is taken into account, the right-hand side of the equation must clearly include also the expression (142.1):

$$\mathrm{d}E = T\,\mathrm{d}S + \mu\,\mathrm{d}N + \alpha\,\mathrm{d}\mathfrak{s}. \tag{142.3}$$

It is, however, more convenient to take as the fundamental thermodynamic quantity not the energy but the potential Ω, the thermodynamic potential in terms of the independent variables T and μ (and the volume V). The convenience of Ω in this case arises because T and μ are quantities which have equal values in the two phases, whereas the pressures are not in general equal when surface effects are taken into account; see §144. For the differential $\mathrm{d}\Omega$, again with V = constant, we have

$$\mathrm{d}\Omega = -S\,\mathrm{d}T - N\,\mathrm{d}\mu + \alpha\,\mathrm{d}\mathfrak{s}. \tag{142.4}$$

The thermodynamic quantities (such as E, Ω and S) for the system under consideration can be written as the sum of a "volume" part and a "surface" part. This division, however, is not unique, since the number of particles in

† At temperatures considerably below the critical temperature this ratio is approximately equal to 4.

each phase is indeterminate to the extent of the number of particles in the transition layer between the phases; the same is true of the volumes of the phases. This indeterminacy is of just the same order of magnitude as the surface effects with which we are concerned. The division can be made unique if the following reasonable condition is imposed: the volumes V_1 and V_2 of the two phases are defined so that, in addition to the equation $V_1+V_2 = V$, where V is the total volume of the system, the equation $n_1V_1+n_2V_2 = N$ is satisfied, where N is the total number of particles in the system, and $n_1 = n_1(\mu, T)$ and $n_2 = n_2(\mu, T)$ are the numbers of particles per unit volume in each phase (the phases being regarded as unbounded).

These two equations determine the choice of the volumes V_1 and V_2 (and the numbers of particles $N_1 = n_1V_1$, $N_2 = n_2V_2$), and hence also the volume parts of all other thermodynamic quantities. We shall denote volume parts by the suffix v, and surface parts by the suffix s; by definition, $N_s = 0$.

From (142.4) we have, for constant T and μ (and therefore constant α), $d\Omega = \alpha\, d\mathfrak{s}$; it is therefore evident that $\Omega_s = \alpha\mathfrak{s}$, and so

$$\Omega = \Omega_v+\alpha\mathfrak{s}. \tag{142.5}$$

Since the entropy $S = -(\partial\Omega/\partial T)_{\mu,\mathfrak{s}}$, the surface part of it is†

$$S_s = -\partial\Omega_s/\partial T = -\mathfrak{s}\, d\alpha/dT. \tag{142.6}$$

Next, let us find the surface free energy; since $F = \Omega+N\mu$ and $N_s = 0$,

$$F_s = \alpha\mathfrak{s}. \tag{142.7}$$

The surface energy is

$$E_s = F_s+TS_s = (\alpha-T\, d\alpha/dT)\mathfrak{s}. \tag{142.8}$$

The quantity of heat absorbed in a reversible isothermal change of surface area from $\mathfrak{s}_1$ to $\mathfrak{s}_2$ is

$$Q = T(S_{s2}-S_{s1}) = -T\,(d\alpha/dT)(\mathfrak{s}_2-\mathfrak{s}_1). \tag{142.9}$$

The sum of the heat Q and the work $R = \alpha(\mathfrak{s}_2-\mathfrak{s}_1)$ in this process is equal to the change in energy $E_{s2}-E_{s1}$, as it should be.

PROBLEM

Find the limiting law of temperature dependence of the surface tension of liquid helium at low temperatures (K. R. Atkins 1953).

Solution. We calculate the surface part $F_s = \mathfrak{s}\alpha$ of the free energy by means of formula (61.1), in which the frequencies ω_α now relate to oscillations of the liquid surface. In the

† The coefficient α is a function of only one independent variable; for such a function the partial derivatives with respect to μ and T have no meaning. But, by putting $N_s = -(\partial\Omega_s/\partial\mu)_T = 0$, we have formally assumed that $(\partial\alpha/\partial\mu)_T = 0$; in this case we clearly have $d\alpha/dT = (\partial\alpha/\partial T)_\mu$, and this has been used in (142.6).

two-dimensional case the change from summation to integration requires a factor $\mathfrak{s}(2\pi k\,dk)/(2\pi)^2$. Integrating by parts, we find

$$F_s = \mathfrak{s}\alpha_0+\mathfrak{s}(T/2\pi)\int \log(1-e^{-\hbar\omega/T})k\,dk = \mathfrak{s}\alpha_0-\mathfrak{s}(\hbar/4\pi)\int k^2\,d\omega/(e^{\hbar\omega/T}-1),$$

where α_0 is the surface tension at $T = 0$. At sufficiently low temperatures, only the oscillations at low frequencies are important, i.e. those with small wave numbers (long wavelengths). Such oscillations are hydrodynamic capillary waves, for which $\omega^2 = (\alpha/\varrho)k^3 \cong (\alpha_0/\varrho)k^3$, where ϱ is the density of the liquid.† Hence

$$\alpha = \alpha_0-\frac{\hbar}{4\pi}\left(\frac{\varrho}{\alpha_0}\right)^{2/3}\int_0^\infty \frac{\omega^{4/3}\,d\omega}{e^{\hbar\omega/T}-1};$$

since the integral converges rapidly, the upper limit may be taken as infinity. The integration (carried out as in the second footnote to §57) gives

$$\alpha = \alpha_0 - \frac{T^{7/3}}{4\pi\hbar^{4/3}}\left(\frac{\varrho}{\alpha}\right)^{2/3}\Gamma(7/3)\zeta(7/3) = \alpha_0-0.13T^{7/3}\varrho^{2/3}/\hbar^{4/3}\alpha^{2/3}.$$

§143. Surface tension of crystals

The surface tension of an anisotropic body, a crystal‡, is different at different faces; it may be said to depend on the direction of the face, i.e. on its Miller indices. The form of this dependence is somewhat unusual. Firstly, the difference in the values of α for two crystal planes with arbitrarily close directions is itself arbitrarily small, i.e. the surface tension can be represented as a continuous function of the direction of the face. It can nevertheless be shown that this function nowhere has a definite derivative. For example, if we consider a set of crystal planes intersecting along a common line, and denote by ϕ the angle of rotation around this line, which defines the direction of the plane, we find that the function $\alpha = \alpha(\phi)$ has two different derivatives for every value of ϕ, one for increasing and the other for decreasing values of the argument.||

Let us suppose that the surface tension is a known function of the direction of the faces. The question arises how this function can be used to determine the equilibrium form of the crystal. (It must be emphasised that the crystal shape observed under ordinary conditions is determined by the conditions of growth of the crystal and is not the equilibrium shape.) The equilibrium form is determined by the condition for the potential Ω to be a minimum

† See *Fluid Mechanics*, §61. The derivation given here applies only to liquid He^4 and temperatures so low that the whole mass of the liquid may be regarded as superfluid. In a Fermi liquid (liquid He^3), capillary waves of this kind do not exist, because the viscosity increases indefinitely as $T \to 0$.

‡ That is, the surface tension at an interface between the crystal and a gas or liquid.

|| This is discussed in more detail by L. D. Landau, *Sbornik v chest' 70-letiya A. F. Ioffe*, p. 44, Moscow 1950; translation in *Collected Papers of L. D. Landau*, p. 540, Pergamon, Oxford 1965.

(for given T, μ and volume V of the crystal) or, what is the same thing, by the condition for its surface part to be a minimum. The latter is

$$\Omega_s = \oint \alpha \, d\mathfrak{s},$$

the integral being taken over the whole surface of the crystal; for an isotropic body $\alpha =$ constant, $\Omega_s = \alpha\mathfrak{s}$, and the equilibrium form is determined simply by the condition for the total area $\mathfrak{s}$ to be a minimum, i.e. it is a sphere.

Let $z = z(x, y)$ be the equation of the surface of the crystal, and let $p = \partial z/\partial x$, $q = \partial z/\partial y$ denote the derivatives which determine the direction of the surface at each point; α can be expressed as a function of these, $\alpha = \alpha(p, q)$. The equilibrium form is given by the condition

$$\int \alpha(p, q) \sqrt{(1+p^2+q^2)} \, dx \, dy = \text{minimum} \tag{143.1}$$

with the added condition of constant volume

$$\int z \, dx \, dy = \text{constant}. \tag{143.2}$$

This variational problem leads to the differential equation

$$\frac{\partial}{\partial x}\frac{\partial f}{\partial p} + \frac{\partial}{\partial y}\frac{\partial f}{\partial q} = 2\lambda, \tag{143.3}$$

where

$$f(p, q) = \alpha(p, q) \sqrt{(1+p^2+q^2)} \tag{143.4}$$

and λ is a constant.

Next, we have by definition $dz = p\,dx + q\,dy$; with the auxiliary function

$$\zeta = px + qy - z, \tag{143.5}$$

we find $d\zeta = x\,dp + y\,dq$ or

$$x = \partial\zeta/\partial p, \qquad y = \partial\zeta/\partial q, \tag{143.6}$$

ζ being here regarded as a function of p and q. Writing the derivatives with respect to x and y in (143.3) as Jacobians, multiplying both sides by $\partial(x, y)/\partial(p, q)$ and using (143.6), we obtain the equation

$$\frac{\partial(\partial f/\partial p, \partial\zeta/\partial q)}{\partial(p, q)} + \frac{\partial(\partial\zeta/\partial p, \partial f/\partial q)}{\partial(p, q)} = 2\lambda \frac{\partial(\partial\zeta/\partial p, \partial\zeta/\partial q)}{\partial(p, q)}.$$

This equation has an integral $f = \lambda\zeta = \lambda(px+qy-z)$, or

$$z = \frac{1}{\lambda}\left(p\frac{\partial f}{\partial p} + q\frac{\partial f}{\partial q} - f\right). \tag{143.7}$$

This is just the equation of the envelope of the family of planes

$$px + qy - z = \alpha(p, q) \sqrt{(1+p^2+q^2)}/\lambda, \tag{143.8}$$

where p and q act as parameters.

This result can be expressed in terms of the following geometrical construction. On each radius vector from the origin we mark off a segment of length proportional to $\alpha(p, q)$, where p and q correspond to the direction of that radius vector.† A plane is drawn through the end of each segment at right angles to the radius vector; then the envelope of these planes gives the equilibrium form of the crystal (G. V. VUL'F).

It can be shown‡ that the unusual behaviour of the function α mentioned at the beginning of this section may have the result that the equilibrium form of the crystal determined by this procedure will include a number of plane areas corresponding to crystal planes with small values of the Miller indices. The size of the plane areas rapidly decreases as the Miller indices increase. In practice this means that the equilibrium shape will consist of a small number of plane areas which are joined by rounded regions instead of intersecting at sharp angles.

§144. Surface pressure

The condition for the pressures of two phases in contact to be equal has been derived (in §12) from the equality of the forces exerted on the interface by the two phases; as elsewhere, surface effects were neglected. It is clear, however, that if the interface is not plane a displacement of it will in general change its area and therefore the surface energy. In other words, the existence of a curved interface between the phases leads to additional forces, as a result of which the pressures of the two phases will not be equal. The difference between them is called the *surface pressure*.

Thus the conditions of equilibrium now require only that the temperature and the chemical potential should be constant throughout the system. For given values of these quantities, and of the total volume of the system, the thermodynamic potential Ω must be a minimum (with respect to a displacement of the interface).

Let us consider two isotropic phases (two liquids, or a liquid and a vapour). Having in view only the thermodynamic aspects of the problem, we shall assume that one of the phases (phase 1) is a sphere surrounded by the other phase.‖ Then the pressure is constant within each phase, and the total thermodynamic potential Ω of the system is

$$\Omega = -P_1V_1 - P_2V_2 + \alpha\mathfrak{s} \tag{144.1}$$

the first two terms forming the volume part of the potential; the suffixes 1 and 2 refer to the two phases.

† The direction cosines of the radius vector are proportional to p, q, -1.

‡ See the paper quoted in the second footnote to this section.

‖ The general case of an arbitrary shape of the interface is discussed in *Fluid Mechanics*, §60.

The pressures of two phases in equilibrium satisfy the equations $\mu_1(P, T) = \mu_2(P_2, T) \equiv \mu$, where μ is the common value of the two chemical potentials. Hence, for constant μ and T, we must regard P_1 and P_2 as constant also, and likewise the surface-tension coefficient α. Since V_1+V_2 is constant, we find as the condition for Ω to be a minimum

$$d\Omega = -(P_1-P_2)\, dV_1+\alpha\, d\mathfrak{s} = 0.$$

Finally, substituting $V_1 = 4\pi r^3/3$, $\mathfrak{s} = 4\pi r^2$ (where r is the radius of the sphere), we obtain the required formula:

$$P_1-P_2 = 2\alpha/r. \tag{144.2}$$

For a plane interface ($r \to \infty$) the two pressures are equal, as we should expect.

Formula (144.2) determines only the difference between the pressures in the two phases. We shall now calculate each of them separately.

The pressures P_1 and P_2 satisfy the equation $\mu_1(P_1, T) = \mu_2(P_2, T)$. The common pressure P_0 in the two phases when the interface is plane is determined at the same temperature by the relation $\mu_1(P_0, T) = \mu_2(P_0, T)$. Subtraction of these two equations gives

$$\mu_1(P_1, T)-\mu_1(P_0, T) = \mu_2(P_2, T)-\mu_2(P_0, T). \tag{144.3}$$

Assuming that the differences $\delta P_1 = P_1-P_0$, $\delta P_2 = P_2-P_0$ are relatively small and expanding the two sides of equation (144.3) in terms of δP_1 and δP_2, we find

$$v_1\delta P_1 = v_2\delta P_2, \tag{144.4}$$

where v_1 and v_2 are the molecular volumes (see (24.12)). Combining this with formula (144.2) written in the form $\delta P_1-\delta P_2 = 2\alpha/r$, we find the required δP_1 and δP_2 as

$$\delta P_1 = \frac{2\alpha}{r}\frac{v_2}{v_2-v_1}, \qquad \delta P_2 = \frac{2\alpha}{r}\frac{v_1}{v_2-v_1}. \tag{144.5}$$

For a drop of liquid in a vapour, $v_1 \ll v_2$; regarding the vapour as an ideal gas, we have $v_2 = T/P_2 \cong T/P_0$, and so

$$\delta P_l = 2\alpha/r, \qquad \delta P_g = 2v_l\alpha P_0/rT, \tag{144.6}$$

where for clarity the suffixes l and g are used in place of 1 and 2. Thus we see that the vapour pressure over the drop is greater than the saturated vapour pressure over a plane liquid surface, and increases with decreasing radius of the drop.

When the drop is sufficiently small and $\delta P_g/P_0$ is no longer small, formulae (144.6) become invalid, since the large variation of the vapour volume with pressure means that the expansion used to derive (144.4) from (144.3) is no longer permissible. For a liquid, whose compressibility is small, the effect of a change of pressure is slight, and the left-hand side of (144.3) can again be replaced by $v_l \delta P_l$. On the right-hand side we substitute the chemical potential of the vapour in the form $\mu = T \log P_g + \chi(T)$, obtaining

$$\delta P_l = P_l - P_0 = (T/v_l) \log (P_g/P_0).$$

Since in this case $\delta P_l \gg \delta P_g$, the difference $P_l - P_0$ can be replaced by $P_l - P_g$; using formula (144.2) for the surface pressure, we then have finally

$$\log (P_g/P_0) = 2\alpha v_l/rT. \tag{144.7}$$

For a bubble of vapour in a liquid we similarly obtain the same formulae (144.6), (144.7) but with the opposite signs.

§145. Surface tension of solutions

Let us now consider an interface between a liquid solution and a gas phase (a gas and a solution of it in a liquid, a liquid solution and its vapour, etc.).

As in §142, we divide all thermodynamic quantities for the system under consideration into volume and surface parts, the manner of division being determined by the conditions $V = V_1 + V_2$, $N = N_1 + N_2$ for the volume and number of solvent particles. That is, the total volume V of the system is divided between the two phases in such a way that, on multiplying V_1 and V_2 by the corresponding numbers of solvent particles per unit volume, and adding we obtain just the total number N of solvent particles in the system. Thus by definition the surface part $N_s = 0$.

As well as other quantities, the number of solute particles will also be written as a sum of two parts, $n = n_v + n_s$. We may say that n_v is the quantity of solute which would be contained in the volumes V_1 and V_2 if it were distributed in each with a constant concentration equal to the volume concentration of the corresponding solution. The number n_v thus defined may be either greater or less than the actual total number n of solute particles. If $n_s = n - n_v > 0$, this means that the solute accumulates at a higher concentration in the surface layer (called *positive adsorption*). If $n_s < 0$, the concentration in the surface layer is less than in the volume (*negative adsorption*).

The surface-tension coefficient of the solution is a function of two independent variables, not one. Since the derivative of the potential Ω with respect to the chemical potential is minus the corresponding number of particles, n_s can be found by differentiating $\Omega_s = \alpha \mathfrak{s}$ with respect to the chemical potential μ' of the solute:†

$$n_s = -\partial \Omega_s / \partial \mu' = -\mathfrak{s}(\partial \alpha / \partial \mu')_T. \tag{145.1}$$

† The coefficient α is now a function of two independent variables, e.g. μ' and T; the derivative $\partial \Omega_s / \partial \mu'$ must be taken at constant T and chemical potential μ of the solvent. The condition $N_s = -(\partial \Omega_s / \partial \mu)_{\mu', T} = 0$ used here implies that we formally take $(\partial \alpha / \partial \mu)_{\mu', T} = 0$ and therefore we can write equation (145.1) (cf. the third footnote to §142).

Let us assume that the pressure of the gas phase is so small that its effect on the properties of the liquid phase may be neglected. Then the derivative of α in formula (145.1), which must be taken along the phase equilibrium curve at the temperature concerned, can be replaced by the derivative at constant (viz.zero) pressure (and constant T). Regarding α as a function of the temperature and the concentration c of the solution, we can rewrite formula (145.1) as

$$n_s = -\mathfrak{s}(\partial\alpha/\partial c)_T(\partial c/\partial\mu')_{T,P}. \qquad (145.2)$$

According to the thermodynamic inequality (98.7), the derivative $(\partial\mu'/\partial c)_{T,P}$ is always positive. Hence it follows from (145.2) that n_s and $(\partial\alpha/\partial c)_T$ have opposite signs. This means that, if the solute raises the surface tension (α increases with increasing concentration of the solution), it is negatively adsorbed. Substances which lower the surface tension are positively adsorbed.

If the solution is a weak one, the chemical potentiai of the solute is of the form $\mu' = T\log c+\psi(P, T)$; substituting this in (145.2), we find

$$n_s = -\mathfrak{s}(c/T)(\partial\alpha/\partial c)_T. \qquad (145.3)$$

A similar formula,

$$n_s = -\mathfrak{s}(P/T)(\partial\alpha/\partial P)_T, \qquad (145.4)$$

is obtained for the adsorption of a gas (at pressure P) by a liquid surface.

If not only the solution but also the adsorption from it is weak, α can be expanded in powers of c; we have approximately $\alpha = \alpha_0+\alpha_1 c$, where α_0 is the surface tension at the interface between two phases of pure solvent. Then we have from (145.3) $\alpha_1 = -n_s T/\mathfrak{s}c$, and hence

$$\alpha-\alpha_0 = -n_s T/\mathfrak{s}. \qquad (145.5)$$

The resemblance between this formula and VAN 'T HOFF's formula for the osmotic pressure should be noticed (the volume being here replaced by the surface area).

§146. Surface tension of solutions of strong electrolytes

The change in surface tension of a liquid when a strong electrolyte is dissolved in it can be calculated in a general form for weak solutions (L. ONSAGER and N. N. T. SAMARAS 1934).

Let $w_a(x)$ denote the additional energy of an ion (of the ath kind) because of the free surface at a distance x from the ion ($w_a(x)$ tending to zero as $x \to \infty$). The ion concentration near the surface differs from its value c_a within the solution by a factor $e^{-w_a/T} \cong 1-w_a/T$. The contribution of the surface to the total number of these ions in the liquid is therefore

$$n_{as} = -\frac{\mathfrak{z} c_a}{vT}\int\limits_0^\infty w_a\,\mathrm{d}x, \tag{146.1}$$

where v is the molecular volume of the solvent.

To calculate the surface tension, we begin from the relation

$$\mathfrak{z}\,\mathrm{d}\alpha = -\sum_a n_{as}\,\mathrm{d}\mu'_a, \tag{146.2}$$

where the summation is over all the kinds of ion in the solution. For weak solutions ($\mu'_a = T\log c_a + \psi_a$),

$$\mathfrak{z}\,\mathrm{d}\alpha = -T\sum_a \frac{n_{as}}{c_a}\,\mathrm{d}c_a. \tag{146.3}$$

Substitution of (146.1) gives,

$$\mathrm{d}\alpha = \frac{1}{v}\sum_a \mathrm{d}c_a\int\limits_0^\infty w_a\,\mathrm{d}x. \tag{146.4}$$

It will be seen from the subsequent results that the main contribution to the integral comes from distances x which are large compared with the distances between molecules but small compared with the Debye–Hückel length $1/\varkappa$.

The energy w_a consists of two parts:

$$w_a = \frac{\varepsilon-1}{\varepsilon(\varepsilon+1)}\,\frac{e^2 z_a^2}{4x} + e z_a \phi(x). \tag{146.5}$$

The first term arises from the "image force" on a charge ez_a in a medium with dielectric constant ε at a distance x from its surface. Since $x \ll 1/\varkappa$, the screening effect of the ion cloud round the charge does not alter this energy. In the second term, $\phi(x)$ denotes the change (owing to the presence of the surface) in the field potential due to all the other ions in the solution. This term is unimportant here, however, since it disappears on substituting (146.5) in (146.4) because of the electrical neutrality of the solution ($\sum c_a z_a = 0$, and therefore $\sum z_a\,\mathrm{d}c_a = 0$).

Thus, carrying out the integration in (146.4), we find

$$\mathrm{d}\alpha = \frac{(\varepsilon-1)e^2}{4\varepsilon(\varepsilon+1)v}\sum_a \log\frac{1}{\varkappa a_a}\,\mathrm{d}(z_a^2 c_a).$$

The logarithmic divergence of the integral at both limits confirms the statement made above concerning the range of integration; we have naturally taken as the upper limit the screening length $1/\varkappa$, and as the lower limit a quantity a_a of the order of atomic dimensions, but different for the different kinds of ion. Since $\varkappa^2$ is proportional to the sum $\sum z_a^2 c_a$, we see that the expression obtained is a total differential and so can be integrated directly, giving

$$\alpha - \alpha_0 = \frac{(\varepsilon-1)e^2}{8\varepsilon(\varepsilon+1)v} \sum_a c_a z_a^2 \log \frac{\lambda_a z_a^2}{\sum_b c_b z_b^2}, \tag{146.6}$$

where α_0 is the surface tension of the pure solvent and the λ_a are dimensionless constants.

This formula gives the solution of the problem. It should be noticed that the dissolution of a strong electrolyte increases the surface tension of the liquid.

§147. Adsorption

Adsorption is the restricted sense includes cases where the solute is almost entirely concentrated at the surface of a solid or liquid *adsorbent*†, and hardly any of it enters the volume of the adsorbent. The "adsorbed film" thus formed can be described by the *surface concentration* γ, defined as the number of particles of the *adsorbate* (adsorbed substance) per unit surface area. At low pressures of the gas from which adsorption occurs, the concentration γ must be proportional to the pressure‡; at high pressures, however, γ increases less rapidly and tends to a limiting value corresponding to the formation of a *monomolecular layer* with the adsorbate molecules closely packed together.

Let μ' be the chemical potential of the adsorbate. By the same method as was used in §98 for ordinary solutions we can derive for adsorption the thermodynamic inequality

$$(\partial\mu'/\partial\gamma)_T > 0, \tag{147.1}$$

which is entirely analogous to (98.7). From (145.1) we have

$$\gamma = -(\partial\alpha/\partial\mu')_T = -(\partial\alpha/\partial\gamma)_T(\partial\gamma/\partial\mu')_T, \tag{147.2}$$

and (147.1) therefore implies that

$$(\partial\alpha/\partial\gamma)_T < 0, \tag{147.3}$$

i.e. the surface tension decreases as the surface concentration increases.

The minimum work which must be done to form the adsorbed film is equal to the corresponding change in the thermodynamic potential Ω:

$$R_{\min} = \mathfrak{s}(\alpha-\alpha_0), \tag{147.4}$$

† For definiteness we shall consider adsorption from a gas phase.

‡ This rule is, however, not obeyed in practice for adsorption on a solid surface, since such a surface is never sufficiently homogeneous.

where α_0 is the surface tension on the surface before adsorption. Hence, using (92.4), we find the heat of adsorption

$$Q = -\mathfrak{s}T^2 \left(\frac{\partial}{\partial T} \frac{\alpha - \alpha_0}{T}\right)_P. \tag{147.5}$$

The adsorbed film may be regarded as a kind of "two-dimensional" thermodynamic system, which may be either isotropic or anisotropic, despite the isotropy of the two volume phases.† The question arises of the possible types of symmetry of the film.

The analogue of ordinary solid crystals would be a "solid crystalline" film with the atoms regularly arranged at the points of a two-dimensional (plane) lattice. This arrangement could be described by a two-dimensional "density function" $\varrho(x, y)$ (cf. §128). However, an investigation similar to that given in §128 for the three-dimensional case shows that such a lattice can not exist, since it would be "smoothed out" by thermal fluctuations (and so the only possibility is ϱ = constant): the mean square of the fluctuation displacement is given by an integral of the same form (128.2) as for the three-dimensional crystal lattice:

$$\overline{u^2} \sim T \iint \frac{dk_x\, dk_y}{\phi_{\text{II}}(k_x, k_y)},$$

but in the two-dimensional case this integral is logarithmically divergent for small values of the wave number.

To avoid misunderstanding, however, the following comment is necessary. The investigation just mentioned shows only that the fluctuation deformations become infinite as the dimensions (area) of the system increase without limit‡, whereas for a three-dimensional crystal lattice the characteristic property is that these deformations remain finite even for an infinite system. In practice, however, the size of the film for which the fluctuations still remain small may be quite large. In such cases a film of finite size might exhibit the properties of a solid crystal and we could approximately describe it as a two-dimensional lattice.

For a two-dimensional film regarded as an infinite structure we can strictly speak only of the symmetry of the correlation between the positions of different molecules when the position of one of them is specified; in this sense the anisotropic film is a two-dimensional analogue of three-dimensional liquid crystals (see §129). Accordingly, the types of symmetry of films must be

† Here we are considering adsorption on a liquid surface; adsorption on a solid surface is of no interest from this point of view, since, as mentioned above, such a surface is almost always inhomogeneous.

It may be noted that anisotropy of the interface between two isotropic phases (liquid and vapour) of the same pure substance is also possible in principle.

‡ Thus allowing us to consider arbitrarily small wave numbers.

classified according to point groups (combinations of planes and axes of symmetry). The rotations about axes and reflections in planes must, of course, leave the plane of the film unchanged and furthermore must leave the relative position of the two phases on either side of the film unchanged (this means that a plane of symmetry coinciding with the plane of the film is not possible). Thus the film can have only an axis of symmetry perpendicular to its plane and planes of symmetry passing through this axis, i.e. the possible types of symmetry of the film are restricted to the point groups $\boldsymbol{C}_n$ and $\boldsymbol{C}_{nv}$.

As in three-dimensional bodies, so also in two-dimensional films there can exist different phases between which transitions of either the first or the second kind can occur. Transitions of the second kind are possible only between phases of different symmetry, but transitions of the first kind can occur between any phases, of either different or the same symmetry, including transitions between two isotropic phases of the gas-liquid type. The equilibrium conditions for the two phases of the film require that their surface tensions as well as their temperatures and chemical potentials should be equal. The condition concerning the surface tensions corresponds to the condition of equal pressures for volume phases and simply expresses the equality of the forces exerted by the two phases on each other.

§148. Wetting

Let us consider adsorption on the surface of a solid from a vapour at a pressure close to the saturation value. The equilibrium concentration γ is determined by the condition that the chemical potential of the adsorbate μ' is equal to that of the vapour μ_g. Various cases can occur according to the dependence of μ' on γ.

Let us suppose that the quantity of adsorbate gradually increases and the adsorbed layer becomes a liquid film of macroscopic thickness. The "surface concentration" γ then becomes a conventionally defined quantity proportional to the film thickness d: $\gamma = \varrho d/m$, where m is the mass of a molecule and ϱ the density of the liquid. As the film thickness increases, the chemical potential of the substance forming it tends to μ_l, the chemical potential of the liquid in bulk. We shall measure the value of μ' (for given P and T) from this limiting value, i.e. write $\mu'+\mu_l$ in place of μ'; thus, by definition, $\mu' \to 0$ as $\gamma \to \infty$.

The chemical potential of the vapour can be written as $\mu_g = \mu_l(T)+ T\log(p/p_0)$, where $p_0(T)$ is the saturated vapour pressure; here we have used the fact that the saturated vapour is, by definition, in equilibrium with the liquid, i.e. we must have $\mu_g = \mu_l$ when $p = p_0$.† The surface concentration

† The liquid itself is regarded as incompressible, i.e. we neglect the dependence of its chemical potential on the pressure.

is determined by the condition $\mu' + \mu_l = \mu_g$, or

$$\mu'(\gamma) = T \log (p/p_0). \tag{148.1}$$

If this equation is satisfied by several values of γ, the one which corresponds to a stable state is that for which the potential Ω_s is a minimum. Taking the value per unit area of the surface, we obtain a quantity which may be called (in the general case of any film thickness) the "effective surface-tension coefficient" α at the solid-vapour boundary, and which takes into account the presence of the layer between them. Integrating the relation (147.2), we can write

$$\alpha(\gamma) = \int_{\gamma}^{\infty} \gamma \frac{d\mu'}{d\gamma} d\gamma + \alpha_{sl} + \alpha_{lg}. \tag{148.2}$$

The constant is so chosen that as $\gamma \to \infty$ the function $\alpha(\gamma)$ becomes the sum of the surface tensions at the "bulk"-phase (solid-liquid and liquid-gas) interfaces.

It may also be recalled that a necessary condition for the thermodynamic stability of a state is the inequality (147.1), which is valid for any γ.

Let us now consider some typical cases which may occur, depending on the nature of the function $\mu'(\gamma)$. In the diagrams given below, the continuous curve shows the form of this function in the region of macroscopically thick films of liquid, while the broken curve is that for adsorbed films of "molecular" thickness. It is, of course, not strictly possible to represent the function in these two regions in one diagram to the same scale, and to this extent the diagrams are a convention.

In the first case shown (Fig. 69a) the function $\mu'(\gamma)$ decreases monotonically with increasing γ (i.e. with increasing film thickness) in the range of macroscopic thicknesses. For molecular dimensions the function $\mu'(\gamma)$ always tends to $-\infty$ as $T \log \gamma$ when $\gamma \to 0$, this law corresponding to a "weak solution" of the adsorbate on the surface. The equilibrium concentration is determined, according to (148.1), by the point of intersection of the curve with a horizontal line $\mu' = \text{constant} \leqslant 0$. In this case, this can occur only at molecular concentrations, i.e. ordinary molecular adsorption must occur, as discussed in §147.

If $\mu'(\gamma)$ increases monotonically but is everywhere negative (Fig. 69b), then in equilibrium a liquid film of macroscopic thickness is formed on the surface of the adsorbent. In particular, when the pressure $p = p_0$ (saturated vapour), the film formed must be so thick that the properties of the substance in it do not differ from those of the liquid in bulk, i.e. the saturated vapour must be in contact with its own liquid phase. In that case we say that the liquid *completely wets* the solid surface in question.

More complicated cases are also theoretically possible. For example, if the function $\mu'(\gamma)$ passes through zero and has a maximum (Fig. 69c) we have a case of wetting but with formation of a film stable only at thicknesses below a certain limit. The film of finite thickness corresponding to the point *A* is in

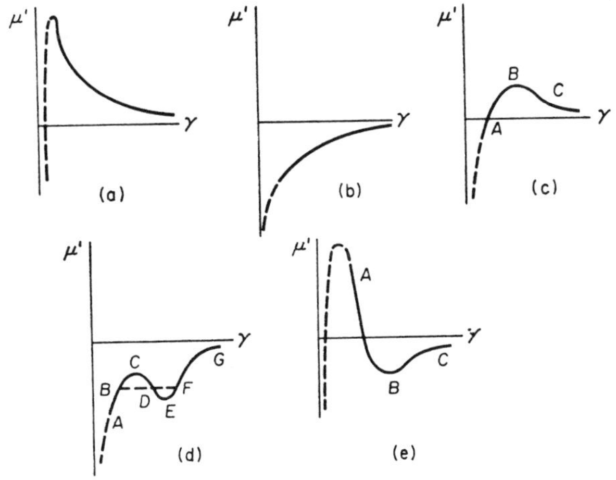

FIG. 69

equilibrium with the saturated vapour. This state is separated from the other stable state (equilibrium of the solid wall with the bulk liquid) by a metastable region *AB* and a region of complete instability *BC*.

The type of curve shown in Fig. 69d corresponds to a film which is unstable over a certain range of thickness. The line *BF* which cuts off equal areas *BCD* and *DEF* joins points *B* and *F* which have equal values of α (and equal μ'), as is easily seen from (148.2). The branches *AB* and *FG* correspond to stable films; the range *CE* is completely unstable, while *BC* and *EF* are metastable.

The two boundaries of the instability range (the points *B* and *F*) correspond in this case to macroscopic film thicknesses. Instability in the range from a certain macroscopic thickness to molecular thicknesses would correspond to a curve of the type shown in Fig. 69e, but such a curve would more likely lead to non-wetting, since the limit of stability would correspond to a point on *BC* where a horizontal line cuts off equal areas below the upper part and above the lower part of the curve. But this is usually impossible, since the latter area, which is related to the van der Waals forces (see below), is small compared with the former, which is related to the considerably greater forces at molecular distances. This means that the surface tension everywhere on *BC* is greater than that which would correspond to molecular adsorption on a solid surface, and the film will therefore be metastable.

The chemical potential of the liquid film (measured from μ_l) represents the difference between the energy of the substance in the film and that in the bulk liquid. It is therefore clear that μ' is determined by the interaction forces between atoms at distances large compared with atomic dimensions and $\sim d$ (van der Waals forces). The potential $\mu'(d)$ can be calculated in a general form, the result being expressed in terms of electromagnetic properties of the solid wall and the liquid, namely their permittivities.†

§149. The angle of contact

Let us consider three bodies in contact, solid, liquid and gas (or one solid and two liquid), distinguishing them by suffixes 1, 2, 3 respectively and denoting the surface-tension coefficients at the interfaces by α_{12}, α_{13}, α_{23} (Fig. 70).

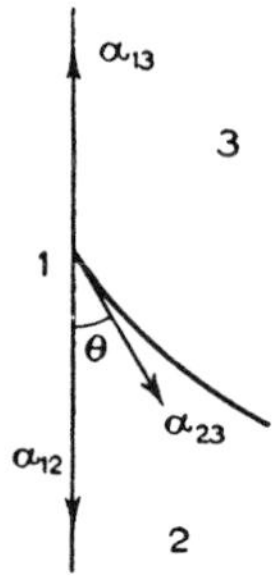

FIG. 70

Three surface-tension forces act on the line where all three bodies meet, each force being in the interface between the corresponding pair of bodies. We denote by θ the angle between the surface of the liquid and the plane surface of the solid, called the *angle of contact*. The value of this angle is determined by the condition of mechanical equilibrium: the resultant of the three surface-tension forces must have no component along the surface of the solid. Thus $\alpha_{13} = \alpha_{12}+\alpha_{23}\cos\theta$, whence

$$\cos\theta = (\alpha_{13}-\alpha_{12})/\alpha_{23}. \tag{149.1}$$

If $\alpha_{13} > \alpha_{12}$, i.e. the surface tension between the gas and the solid is greater than that between the solid and the liquid, then $\cos\theta > 0$ and the angle of contact is acute, as in Fig. 70. If $\alpha_{13} < \alpha_{12}$, however, the angle of contact is obtuse.

From the expression (149.1) we see that in any actual case of stable contact the inequality

$$|\alpha_{13}-\alpha_{12}| \leqslant \alpha_{23} \tag{149.2}$$

must necessarily hold, since otherwise the condition of equilibrium would lead

† See I. E. DZYALOSHINSKIĬ, E. M. LIFSHITZ and L. P. PITAEVSKIĬ, *Soviet Physics Uspekhi* **4**, 153, 1961; *Advances of Physics* **10**, 165, 1961.

to an imaginary value of the angle θ, which has no meaning. On the other hand, if α_{12}, α_{13}, α_{23} are regarded as the values of the corresponding coefficients for each pair of bodies by themselves, without the third one, then it may well happen that the condition (149.2) is not satisfied. Actually, however, it must be remembered that when three different substances are in contact there may in general be an adsorbed film of each substance on the interface between the other two, and this lowers the surface tension. The resulting coefficients α will certainly satisfy the inequality (149.2), and such adsorption will necessarily occur if the inequality would not be satisfied without it.

If the liquid completely wets the solid surface, then a macroscopically thick liquid film, not an adsorbed film, is formed on the surface. The gas will therefore be everywhere in contact with the same liquid substance, and the surface tension between the solid and the gas is not involved at all. The condition of mechanical equilibrium gives simply $\cos\theta = 1$, i.e. the angle of contact is zero.

Similar arguments are valid for contact between three bodies of which none is solid: a liquid drop (3 in Fig. 71) on the surface of another liquid (1)

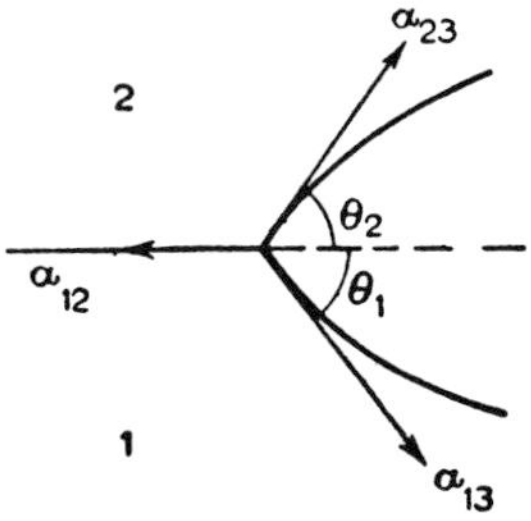

FIG. 71

adjoining a gas (2). In this case the angles of contact θ_1 and θ_2 are determined by the vanishing of the resultant of the three surface-tension forces, i.e. of their vector sum:

$$\mathbf{a}_{12}+\mathbf{a}_{13}+\mathbf{a}_{23} = 0. \tag{149.3}$$

Here it is evident that none of the quantities α_{12}, α_{13}, α_{23} can be greater than the sum or less than the difference of the other two.

§150. Nucleation in phase transitions

If a substance is in a metastable state, it will sooner or later enter another state which is stable. For example, supercooled vapour in time condenses to a liquid; a superheated liquid is converted into vapour. This change occurs in the following manner. Owing to fluctuations, small quantities of a new phase are formed in an originally homogeneous phase; for example, droplets of liquid form in a vapour. If the vapour is the stable phase, these droplets are always unstable and eventually disappear. If the vapour is supercooled, however, then when the droplets formed in it are sufficiently large they are stable and in time begin to grow and form a kind of centre of condensation for the vapour. The droplets must be sufficiently large in order to compensate the unfavourable energy change when a liquid-vapour interface is formed.†

Thus there is a certain minimum or "critical" size necessary for a *nucleus*, as it is called, of a new phase formed in a metastable phase, in order for it to become a centre for formation of the new phase. Since one phase or the other is stable for sizes less than and greater than the critical, the "critical nucleus" is in unstable equilibrium with the metastable phase. In what follows we shall discuss the probability that such nuclei occur.‡ Because of the rapid decrease in the probability of fluctuations of increasing size, the beginning of the phase transition is determined by the probability that nuclei of just this minimum necessary size occur.

Let us consider the formation of nuclei in isotropic phases: the formation of liquid droplets in supercooled vapour, or of vapour bubbles in superheated liquid. A nucleus may be regarded as spherical, since, owing to its very small size, the effect of gravity on its shape is entirely negligible. For a nucleus in equilibrium with the surrounding medium we have, from (144.2), $P'-P = 2\alpha/r$, and the radius of the nucleus is therefore

$$r_{cr} = 2\alpha/(P'-P); \qquad (150.1)$$

the primed and unprimed letters everywhere refer to the nucleus and to the main (metastable) phase respectively.

According to the general formula (114.1), the probability w of a fluctuation producing a nucleus is proportional to $\exp(-R_{min}/T)$, where R_{min} is the minimum work needed to form the nucleus. Since the temperature and chemical potential of the nucleus have the same values as in the surrounding medium (the main phase), this work is given by the change in the potential Ω in the process. Before the formation of the nucleus, the volume of the metastable phase was $V+V'$ and its potential $\Omega = -P(V+V')$; after the formation of the nucleus of volume V', the potential Ω of the whole system is $-PV-P'V'+\alpha\mathfrak{s}$. We therefore have

$$R_{min} = -(P'-P)V'+\alpha\mathfrak{s}. \qquad (150.2)$$

For a spherical nucleus $V' = 4\pi r^3/3$ and $\mathfrak{s} = 4\pi r^2$, and replacing r by (150.1) we find

$$R_{min} = 16\pi\alpha^3/3(P'-P)^2. \qquad (150.3)$$

† It should be borne in mind that this mechanism of formation of a new phase can actually occur only in a sufficiently pure substance. In practice, the centres of formation of the new phase are usually various kinds of "impurity" (dust particles, ions, etc.).

‡ The calculation of the probability that nuclei of any size occur is given in Problem 2, and illustrates the relationships described.

As in §144, we denote by P_0 the pressure of both phases (at a given temperature T) when the interface between them is plane; in other words, P_0 is the pressure for which the given value of T is the ordinary phase transition point, from which the superheating or supercooling is measured. If the metastable phase is only slightly superheated or supercooled, the differences $\delta P = P - P_0$, $\delta P' = P' - P_0$ are relatively small and satisfy the equation (144.4):

$$v'\delta P' = v\delta P, \tag{150.4}$$

where v' and v are the molecular volumes of the nucleus and the metastable phase. Replacing $P' - P$ by $\delta P' - \delta P$ in (150.3) and expressing $\delta P'$ in terms of δP from (150.4), we find for the probability of formation of a nucleus in a slightly superheated or supercooled phase:

$$w \sim \exp\left\{-\frac{16\pi\alpha^3 v'^2}{3T(v-v')^2(\delta P)^2}\right\}. \tag{150.5}$$

In the formation of vapour bubbles in a superheated liquid we can neglect v in this formula in comparison with v':

$$w \sim \exp\left\{-\frac{16\pi\alpha^3}{3T(\delta P)^2}\right\}. \tag{150.6}$$

In the formation of liquid droplets in a supercooled vapour we can neglect v' in (150.5) in comparison with v, and substitute $v = T/P \cong T/P_0$. This gives

$$w \sim \exp\left\{-\frac{16\pi\alpha^3 v'^2 P_0^2}{3T^3(\delta P)^2}\right\}. \tag{150.7}$$

The degree of metastability can be defined by the difference $\delta T = T - T_0$ between the temperature T of the metastable phase (with which the nucleus is in equilibrium) and the temperature T_0 of equilibrium of the two phases when the interface is plane, instead of by δP. According to the Clapeyron–Clausius formula, δT and δP are related by

$$\delta P = \frac{q}{T_0(v-v')}\delta T,$$

where q is the molecular heat of the transition from the metastable phase to the nucleus phase. Substituting δP in (150.5), we obtain for the probability of formation of a nucleus

$$w \sim \exp\left\{-\frac{16\pi\alpha^3 v'^2 T_0}{3q^2(\delta T)^2}\right\}. \tag{150.8}$$

If saturated vapour is in contact with a solid surface (the wall of a vessel) which is completely wetted by the liquid, condensation of the vapour will occur without nucleation, directly on this surface. The formation of a liquid film on the solid surface in this case does not require work to be done to form the interface, and so the existence of a metastable phase (supercooling of the vapour) is impossible.

For the same reason, superheating of a solid with a free surface is in general impossible. This is because usually liquids completely wet the surface of a solid phase of the same substance; consequently, the formation of a liquid layer on the surface of a melting body does not require work to be done to form a new surface.

The formation of nuclei within a crystal on melting can, however, occur if the necessary conditions of heating are maintained: the body must be heated internally and its surface kept at a temperature below the melting point. The probability of formation of nuclei then depends on elastic deformations accompanying the creation of liquid droplets within the solid.†

PROBLEMS

PROBLEM 1. Determine the probability of formation of a nucleus of a liquid on a solid surface for a given (non-zero) value of the angle of contact θ.

SOLUTION. The nucleus will have the shape of a segment of a sphere with base radius $r \sin\theta$, r being the radius of the sphere. Its volume is $V = \frac{1}{3}\pi r^3(1-\cos\theta)^2(2+\cos\theta)$, and the surface areas of the curved part and the base are respectively $2\pi r^2(1-\cos\theta)$ and $\pi r^2 \sin^2\theta$. Using the relation (149.1) for the angle of contact, we find that the change in Ω_s on formation of the nucleus is

$$\alpha \cdot 2\pi r^2(1-\cos\theta) - \alpha\cos\theta \cdot \pi r^2\sin^2\theta = \alpha\pi r^2(1-\cos\theta)^2(2+\cos\theta),$$

where α is the surface-tension coefficient between the liquid and the vapour. This change in Ω_s is the same as would occur in the formation, in the vapour, of a spherical nucleus of volume V and surface tension

$$\alpha_{\text{eff}} = \alpha\left(\frac{1-\cos\theta}{2}\right)^{2/3}(2+\cos\theta)^{1/3}.$$

Accordingly, the required formulae for the formation of nuclei are obtained from those derived in the text on replacing α by α_{eff}.

PROBLEM 2. Find the probability of formation of a nucleus of arbitrary dimensions.

SOLUTION. We regard the metastable phase as an external medium containing the nucleus, and calculate the work of formation of the nucleus from formula (20.2): $R_{\min} = \Delta(E - T_0S + P_0V)$ or, since in this case the process occurs at constant temperature equal to the temperature of the medium, $R_{\min} = \Delta(F + P_0V)$. To determine this quantity, it is sufficient to consider only the amount of substance which enters the other phase (since the state of the remaining substance in the metastable phase remains unchanged). Again denoting the quantities pertaining to the substance in the original and the new phase by unprimed and primed letters respectively, we have

$$R_{\min} = [F'(P') + PV' + \alpha\mathfrak{s}] - [F(P) + PV] = \Phi'(P') - \Phi(P) - (P'-P)V' + \alpha\mathfrak{s}; \qquad (1)$$

for a nucleus in unstable equilibrium with the metastable phase we should have $\Phi'(P') = \Phi(P)$ and thus return to (150.2).

† See I. M. LIFSHITZ and L. S. GULIDA, *Doklady Akademii nauk SSSR* **87**, 377, 523, 1952.

Assuming the degree of metastability to be small, we have $\Phi'(P') \cong \Phi'(P)+(P'-P)V'$, and (1) thus reduces to

$$R_{\min} = n[\mu'(P)-\mu(P)]+\alpha\mathfrak{s},$$

where $n = V'/v'$ is the number of particles in the nucleus. For a spherical nucleus,

$$R_{\min} = -\frac{4\pi r^3}{3v'}[\mu(P)-\mu'(P)]+4\pi r^2\alpha. \tag{2}$$

In the metastability range, $\mu(P) > \mu'(P)$ and hence the first term (the volume term) is negative. The expression (2) may be said to describe the potential barrier which has to be overcome for the formation of a stable nucleus. It has a maximum at

$$r = r_{cr} = 2\alpha v'/[\mu(P)-\mu'(P)],$$

corresponding to the critical radius of the nucleus. For $r < r_{cr}$ a decrease of r is energetically favourable and the nucleus is absorbed; for $r > r_{cr}$ an increase of r is favourable and the nucleus grows.†

§151. Fluctuations in the curvature of long molecules

In ordinary molecules, the strong interaction between atoms reduces the thermal motion within molecules to small oscillations of the atoms about their equilibrium positions, which have practically no effect on the shape of the molecule. Molecules consisting of very long chains of atoms (e.g. long polymer hydrocarbon chains) behave quite differently, however. The great length of the molecule, together with the relative weakness of the forces tending to preserve the equilibrium straight shape of the molecule, means that the fluctuation curvature of the molecule may become very large and even cause the molecule to "coil up". The great length of the molecule enables us to consider it as a kind of macroscopic linear system, and statistical methods may be used in order to calculate the mean values of quantities describing its curvature (S. E. Bresler and Ya. I. Frenkel' 1939).‡

We shall consider molecules having a uniform structure along their length (as is true of long polymer chains). Being concerned only with their shape, we can regard such molecules as uniform continuous threads. The shape of a thread is defined by specifying at each point in it the "curvature vector" $\boldsymbol{\varrho}$, which is along the principal normal to the curve and equal in magnitude to the reciprocal of the radius of curvature.

The curvature of the molecule is in general small at each point; since the molecule is of great length this does not, of course, exclude very large relative movements of distant points of it. For small values of the vector $\boldsymbol{\varrho}$, the free

† The calculation of $R_{\min}$ for $r = r_{cr}$ naturally leads to formula (150.5) if we note that under these conditions $\mu(P)-\mu'(P) \cong (v-v')\delta P$.

‡ In the theory given here, the molecule is regarded as an isolated system, its interaction with neighbouring molecules being neglected. In a solid or liquid substance this interaction may, of course, have a considerable effect on the shape of the molecules. Although the applicability of the results to actual substances is therefore very limited, their derivation is of considerable methodological interest.

energy per unit length of the curved molecule can be expanded in powers of the components of this vector. Since the free energy is a minimum in the equilibrium position (the straight shape, with $\varrho = 0$ at every point) linear terms do not appear in the expansion, and we have

$$F = F_0 + \tfrac{1}{2} \sum_{i,k} a_{ik} \varrho_i \varrho_k, \tag{151.1}$$

where the values of the coefficients a_{ik} represent the properties of the straight molecule (its "resistance to curvature") and are constant along its length, since the molecule is assumed homogeneous.

The vector ϱ is in the plane normal to the line of the molecule at the point considered, and has two independent components in that plane. Accordingly, the set of constants a_{ik} forms a symmetrical tensor of dimension two and rank two in this plane We refer this to its principal axes, and denote its principal values by a_1, a_2; the thread which represents the molecule need not be axially symmetrical in its properties, and so a_1 and a_2 need not be equal. The expression (151.1) then becomes

$$F = F_0 + \tfrac{1}{2}(a_1 \varrho_1^2 + a_2 \varrho_2^2),$$

where ϱ_1 and ϱ_2 are the components of ϱ in the direction of the corresponding principal axes.

Finally, integrating over the whole length of the molecule, we find the total change in its free energy due to a slight curvature:

$$\Delta F_t = \tfrac{1}{2} \int (a_1 \varrho_1^2 + a_2 \varrho_2^2)\, \mathrm{d}l, \tag{151.2}$$

where l is a co-ordinate along the thread. It is clear that a_1 and a_2 are necessarily positive.

Let $\mathbf{t}_a$ and $\mathbf{t}_b$ be unit vectors along the tangents at two points a and b on the thread separated by a section of length l, and let $\theta = \theta(l)$ denote the angle between these tangents, i.e. $\mathbf{t}_a \cdot \mathbf{t}_b = \cos\theta$.

Let us first consider a curvature so slight that the angle θ is small even for distant points. We draw two planes through the vector $\mathbf{t}_a$ and the two principal axes of the tensor a_{ik} in the normal plane (at the point a). For small θ, the square of this angle may be written

$$\theta^2 = \theta_1^2 + \theta_2^2, \tag{151.3}$$

where θ_1 and θ_2 are the angles of rotation of the vector $\mathbf{t}_b$ relative to $\mathbf{t}_a$ in these two planes. The components of the curvature vector are related to the functions $\theta_1(l)$ and $\theta_2(l)$ by $\varrho_1 = \mathrm{d}\theta_1(l)/\mathrm{d}l$, $\varrho_2 = \mathrm{d}\theta_2(l)/\mathrm{d}l$, and the change in the free energy due to the curvature of the molecule may be written

$$\Delta F_t = \tfrac{1}{2} \int \left[a_1 \left(\frac{\mathrm{d}\theta_1}{\mathrm{d}l}\right)^2 + a_2 \left(\frac{\mathrm{d}\theta_2}{\mathrm{d}l}\right)^2 \right] \mathrm{d}l. \tag{151.4}$$

In calculating the probability of a fluctuation with given values of $\theta_1(l) = \theta_1$ and $\theta_2(l) = \theta_2$ for a particular l, we must consider the most complete equilibrium possible for given θ_1 and θ_2 (see the first footnote to §113). That is, we must determine the minimum possible value of the free energy for given θ_1 and θ_2. An integral of the form

$$\int_0^l \left(\frac{d\theta_1}{dl}\right)^2 dl$$

for given values of the function $\theta_1(l)$ at both limits ($\theta_1(0) = 0$, $\theta_1(l) = \theta_1$) has a minimum value if $\theta_1(l)$ varies linearly. Then

$$\Delta F_t = \frac{a_1\theta_1^2}{2l} + \frac{a_2\theta_2^2}{2l};$$

the fluctuation probability $w \sim e^{-\Delta F_t/T}$ (see (119.1)), and so we obtain for the mean squares of the two angles

$$\overline{\theta_1^2} = lT/a_1, \qquad \overline{\theta_2^2} = lT/a_2.$$

The mean square of the angle $\theta(l)$ under consideration is

$$\overline{\theta^2} = lT\left(\frac{1}{a_1} + \frac{1}{a_2}\right). \tag{151.5}$$

In this approximation it is, as we should expect, proportional to the length of the section of the molecule between the two points.

Curvature with large values of the angles $\theta(l)$ may now be treated as follows. The angles between the tangents $\mathbf{t}_a$, $\mathbf{t}_b$, $\mathbf{t}_c$ at three points a, b, c on the thread are related by the trigonometrical formula

$$\cos\theta_{ac} = \cos\theta_{ab}\cos\theta_{bc} - \sin\theta_{ab}\sin\theta_{bc}\cos\phi,$$

where ϕ is the angle between the planes ($\mathbf{t}_a$, $\mathbf{t}_b$) and ($\mathbf{t}_b$, $\mathbf{t}_c$). Averaging this formula and bearing in mind that, in the approximation considered, the fluctuations of curvature of the sections ab and bc of the molecule (for a given direction of the tangent $\mathbf{t}_b$ at the middle point) are statistically independent, we obtain

$$\begin{aligned}\overline{\cos\theta_{ac}} &= \overline{\cos\theta_{ab}\cos\theta_{bc}}\\ &= \overline{\cos\theta_{ab}}\;\overline{\cos\theta_{bc}};\end{aligned}$$

the term containing $\cos\phi$ gives zero on averaging.

This relation shows that the mean value $\overline{\cos\theta(l)}$ must be a multiplicative function of the length l of the section of the molecule between two given points. But for small $\theta(l)$ we must have, according to (151.5),

$$\overline{\cos\theta(l)} \cong 1 - \tfrac{1}{2}\overline{\theta^2} = 1 - lT/a,$$

with the notation $2/a = 1/a_1 + 1/a_2$. The function which satisfies both these conditions is

$$\overline{\cos\theta(l)} = e^{-lT/a}, \tag{151.6}$$

and this is the required formula. For large distances l, the mean value $\overline{\cos\theta} \cong 0$, in accordance with the statistical independence of the directions of sufficiently distant parts of the molecule.

By means of formula (151.6) it is easy to determine the mean square of the distance R (measured in a straight line) between the two ends of the molecule. If $\mathbf{t}(l)$ is a unit vector along the tangent at an arbitrary point in the molecule, the radius vector between its ends is

$$\mathbf{R} = \int_0^L \mathbf{t}(l)\,\mathrm{d}l,$$

where L is the total length of the molecule. Writing the square of this integral as a double integral and averaging, we obtain

$$\overline{R^2} = \int_0^L\int_0^L \overline{\mathbf{t}(l_1)\cdot\mathbf{t}(l_2)}\,\mathrm{d}l_1\,\mathrm{d}l_2 = \int_0^L\int_0^L e^{-T|l_1-l_2|/a}\,\mathrm{d}l_1\,\mathrm{d}l_2.$$

The calculation of the integral gives the final formula

$$\overline{R^2} = 2\left(\frac{a}{T}\right)^2\left(\frac{LT}{a} - 1 + e^{-LT/a}\right). \tag{151.7}$$

For low temperatures ($LT \ll a$) this becomes

$$\overline{R^2} = L^2(1 - LT/3a); \tag{151.8}$$

as $T \to 0$ the mean square $\overline{R^2}$ tends to the square of the total length of the molecule, as it should. If $LT \gg a$ (high temperatures or great lengths L),

$$\overline{R^2} = 2La/T. \tag{151.9}$$

Then $\overline{R^2}$ is proportional to the length of the molecule, and as L increases the ratio $\overline{R^2}/L^2$ tends to zero.

§152. The impossibility of the existence of phases in one-dimensional systems

A problem of theoretical interest is that of the possibility of existence of more than one phase in one-dimensional (linear) systems, i.e. those in which the particles lie along a line. The following argument shows that in fact thermodynamic equilibrium between two homogeneous phases in contact at a single point and having arbitrarily large extent in length is not possible.

To prove this, let us imagine a linear system consisting of an alternation of sections formed by two different phases. Let Φ_0 be the thermodynamic potential of this system, without allowance for the existence of points of contact between different phases, i.e. the thermodynamic potential of the total amounts of the two phases without regard to their division into sections. To take into account the effect of the points of contact, we note that the system may be formally regarded as a "solution" of these points in the two phases. If the "solution" is weak, the thermodynamic potential Φ of the system will be

$$\Phi = \Phi_0 + nT \log (n/eL) + n\psi,$$

where n is the number of points of contact in a length L. Hence

$$\partial\Phi/\partial n = T \log (n/L) + \psi.$$

When the "concentration" n/L is sufficiently small (i.e. for a small number of sections of different phases), $\log (n/L)$ is large and negative, and therefore $\partial\Phi/\partial n < 0$. Thus Φ decreases with increasing n, and since Φ must tend to a minimum n will increase (until the derivative $\partial\Phi/\partial n$ becomes positive). That is, the two phases will tend to intermingle in shorter and shorter sections, and therefore can not exist as separate phases.

INDEX

Absolute temperature 34
Acoustic waves 183
Adiabatic processes 37–40
Adsorption 462, 465–7
Amorphous solids 170
Angle of contact 470–1
Anisotropic bodies 401
Averaging, statistical 4
Azeotropic mixture 307 n.

Barometric formula 109, 282
Black-body radiation 162–9
Boltzmann distribution 107–9
Boltzmann's constant 34
Boltzmann's formula 109
Boltzmann's H theorem 114
Boltzmann's law 165
Bose
  gas 146 ff., 234–40
  statistics 145 ff., 186, 187–91
Bose–Einstein condensation 160 n., 251
Boyle point 219 n.
Bravais lattice 407 ff.

Canonical distribution 77
Carnot cycle 56
Characteristic functions 47
Characteristic temperature 177
Chemical constant 119
Chemical equilibrium 316
  constant 318
Chemical potential 68, 273
Chemical reactions 316 ff.
Clapeyron–Clausius formula 261
Classes, crystal 412–14
Collisions, molecular 110–12
Concentration
  point of equal 301
  of solution 273, 276
Condensation
  Bose–Einstein 160 n., 251
  retrograde 308
Correlation functions 228–30, 360, 370, 381
Corresponding states 179, 271

Critical line 299
Critical point 263 ff., 299, 301–5, 362–5, 445–6
Critical pressure 263
Critical temperature 263
Cross-sections 111
Crystal
  classes 412–14
  planes 417
  space groups 414–16
  systems 408–12
Crystals 170
  liquid 404
  ordering of 170–1, 425–7
  surface tension of 458–60
  symmetry of 401 ff.
Curie point 425 n.

Debye
  function 178
  interpolation formula 178
  temperature 177
Degenerate
  frequency 142
  gas 151–62, 234–56
Degrees
  of freedom, thermodynamic 275
  of temperature 34–35
Density
  function 401–4
  matrix 16
  matter at high 327 ff.
Diatomic gases 129–40
Dielectrics, electronic spectra of 210–11
Disordered crystal 425
Dispersion relation 183
Dissipation 378–81
Distribution function 3, 76 ff.
  Boltzmann 107–9
  canonical 77
  Gaussian 344 ff.
  Gibbs 77; *see* Gibbs distribution
  Maxwellian 80
  microcanonical 12, 22, 76
  Planck's 163

Distribution law 281
Dulong and Petit's law 175

Efficiency of Carnot cycle 57
Elastic waves 183
Electrolytes, strong 287
Electron–positron pair production 324–6
Endothermic reaction 320
Energy 11–14
  free 46; *see* Free energy
  gravitational 336–8
Ensemble, statistical 9 n.
Enthalpy 45 n.
Entropy 24–32
  law of increase of 28
  of mixing 291
Equation of state 49
  at high density 327–9
  of ideal gas 116
Equilibrium
  curves 301
  incomplete 13
  of bodies of large mass 330–6
  of "neutron" sphere 338–42
  partial 13
  statistical 5
  surface 300
  thermal 5
  thermodynamic 5
Equipartition, law of 124
Equivalent lattice points 402
Equivalent vectors 419
Ergodic hypothesis 12 n.
Eutectic point 309
Exchange effect 106, 228
Excitations, elementary 188
Excitons 210
Exothermic reaction 320
Expansion, thermal 180–1
Extended group 421

Fermi
  gas 145 ff.
  sphere 197
  statistics 144 ff., 196–211
  surface 200 ff.
Fluctuations 343 ff.
  in curvature of molecules 475–8
  relative 8
  root-mean-square (r.m.s.) 7
  thermodynamic 372
Free energy 46, 47 n.
  in Gibbs distribution 86–89
  of ideal Boltzmann gas 114–15
Freedom, thermodynamic degrees of 275
Friction 378–81
Fugacity 220

Gas
  completely ionised 225–8
  constant 116 n.
  ideal 106 ff., 289–91, 354–5, 365–70
  non-ideal 215 ff.
  see also Bose gas; Degenerate gas; Fermi gas
Gaussian distribution 344 ff.
Gibbs distribution 77
  free energy in 86–89
  for rotating bodies 99–101
  for variable number of particles 101–3
Gibbs free energy 47 n.
Gibbs' phase rule 275
Glide plane 406
Glide-reflection plane 406
Gravitational collapse 341
Gravitational energy 336–8
Gravitational radius 341
Group, symmetry 401; *see* Symmetry groups

Heat
  bath 26
  content 45 n.
  function 45
  quantity of 43
  of reaction 320
  of solution 283
  specific 45, 429–33
  of transition 259
Helium 171, 189–96, 457–8
Helmholtz free energy 47 n.
Henry's law 282
Hydrogen molecules 134–5
Hyperfine splitting 129

Ideal gases 106 ff., 289–91, 354–5, 365–70
Ideal mixtures 293
Incomplete equilibrium 13
Integral
  of the motion 11
  over states 88
Ionisation equilibrium 323–4
Irreversible processes 32
Isobaric process 120
Isochoric process 120
Isolation, thermal 37

Isotopes, mixtures of 291–4
Isotropic bodies 401

Joule–Thomson process 54, 219, 225

Kinetic coefficients 375–8
Kinetics 6
Kirchhoff's law 166
Kramers and Kronig's formulae 390

λ-point 195, 425 n.
Latent heat of transition 259
Le Chatelier's principle 64
Lever rule 258, 306
Liouville's theorem 10
Liquid crystals 404

Macroscopic motion 35
Macroscopic states 13
Magnetic structure 401 n.
Mass
  action, law of 318
  effective 200
  equilibrium of bodies of large 330–6
Matrix
  density 16
  statistical 17
Maxwellian distribution 80
Mechanical invariants 11
Metals, electronic spectra of 203–10
Metastable states 43, 63
Microcanonical distribution 12, 22, 76
Miller indices 418
Mixed states 18
Mixtures 289 ff.; *see also* Solutions
Monatomic gases 125–9

Nernst's theorem 66, 127
"Neutron sphere", equilibrium of 338–42
Non-ideal gases 215 ff.
Normal co-ordinates 84
Normal modes 142
Nucleation 471–5
Nucleus 472

Occupation numbers 106
Optical vibrations 184
Ordering of crystals 170–1, 425–7
Oscillator 84
  probability distribution for 83–86
Osmotic pressure 277

Pair production 324–6
Partial equilibrium 13
Partition function 87
Pauli's principle 144
Perturbation theory, thermodynamic 90–93
Phase
  diagram 306
  equilibrium 257 ff.
  of matter 257
  point 2
  rule 275
  space 2
  trajectory 2
  transitions of the second kind 424 ff.
Phonons 185
Photon gas 162 ff.
Planck's distribution 163
Planck's formula 163
Plasmas 225–8
Point groups 404 ff.
Poisson adiabatic 119
Poisson's formula 357
Polaron 211
Polyatomic gases 140–3
Polytropic process 120
Potential
  chemical 68, 273
  thermodynamic 47
Pressure 41
  critical 263
  negative 42
  osmotic 277
  surface 460–2
Proper symmetry group 420
Pure states 18

Quantum
  liquids 188–203
  statistics 14 ff., 66, 171, 185–211, 230–56, 343–4
Quasi-closed subsystems 6
Quasi-momentum 185, 205
Quasi-particles 185, 188

Raoult's law 281
Rayleigh–Jeans formula 163
Reactions, chemical 316 ff.
Reciprocal lattice 417
Reduced quantities 271
Regular representations 440
Relativistic thermodynamic relations 73–75, 125, 157–9, 324–6, 328–9
Relaxation time 6

Representations of space groups 418–23, 434–42
small 420
Retrograde condensation 308
Reversible processes 32
Root-mean-square (r.m.s.) fluctuation 7
Rotary-reflection axis 405
Rotating bodies 71–73, 99–101, 118
Rotons 190

Screw axis 406
Second quantisation method 365
Solids 170 ff.
amorphous 170
crystalline 170
at high temperatures 174–6
at low temperatures 170–4
thermal expansion of 180–1
Solute 275
Solutions 273 ff., 358–9
surface tension of 462–5
weak 275–87
Solvent 273
Sound propagation in solids 172–3, 181–7
Space groups 414–16
extended 421
representations of 418–23, 434–42
Specific heat 45, 429–33
Star of a vector 419
Statistical averaging 4
Statistical distribution function 3, 22
Statistical ensemble 9 n.
Statistical equilibrium 5
Statistical independence 6
Statistical laws 1
Statistical matrix 17
Statistical physics 1 ff.
Statistical weight 24, 125
Statistics 1 ff.
Subsystems 2, 101
quasi-closed 6
Sum over states 87
Superfluidity 191–6, 250
Superlattice 442
Surface
pressure 460–2
tension 455 ff.
Surfaces 455 ff.
Susceptibility, generalised 385–91

Symmetry
axis of 405–6
of crystals 401 ff., 433–45
groups 401
point 404 ff.
proper 420
space 414–16, 418–23, 434–42
plane of 405–6
transformations 401
Systems, crystal 408–12

Temperature 34–37
critical 263
Debye 177
negative 211–14
thermodynamic scale of 53
Thermal equilibrium 5
Thermal expansion 180–1
Thermal isolation 37
Thermodynamic degrees of freedom 275
Thermodynamic equilibrium 5
Thermodynamic fluctuations 372
Thermodynamic inequalities 62, 296–9
Thermodynamic perturbation theory 90–93
Thermodynamic potentials 47
Thermodynamic quantities 33 ff.
Thermodynamic relations 33
Thermodynamics, second law of 28
Thermostat 26
Three-phase line 300, 303
Triple points 259, 300
line of 300, 303

Unit cell 407
Utilisation coefficient 57

Van der Waals' equation 224
reduced 271
Van 't Hoff's formula 278
Virial coefficients 220, 230–4
Virial theorem 89–90

Wetting 467–70
Wien's formula 164
Work 43
maximum 55–60
Working medium 56

CDE7987654321